SECOND EDITION

Arduino Cookbook

Michael Margolis

O'REILLY®
Beijing • Cambridge • Farnham • Köln • Sebastopol • Tokyo

Arduino Cookbook, Second Edition
by Michael Margolis

Printed in the United States of America.

Published by O'Reilly Media, Inc., 1005 Gravenstein Highway North, Sebastopol, CA 95472.

O'Reilly books may be purchased for educational, business, or sales promotional use. Online editions are also available for most titles (*http://my.safaribooksonline.com*). For more information, contact our corporate/institutional sales department: (800) 998-9938 or *corporate@oreilly.com*.

Editors: Shawn Wallace and Brian Jepson
Production Editor: Teresa Elsey
Proofreader: Kiel Van Horn
Indexer: Lucie Haskins
Cover Designer: Karen Montgomery
Interior Designer: David Futato
Illustrator: Robert Romano

March 2011: First Edition.
December 2011: Second Edition.

Revision History for the Second Edition:
2011-12-09 First release
See *http://oreilly.com/catalog/errata.csp?isbn=9781449313876* for release details.

ISBN: 978-1-449-31387-6

[LSI]

1323455802

Table of Contents

Preface

This book was written by Michael Margolis with Nick Weldin to help you explore the amazing things you can do with Arduino.

Arduino is a family of microcontrollers (tiny computers) and a software creation environment that makes it easy for you to create programs (called *sketches*) that can interact with the physical world. Things you make with Arduino can sense and respond to touch, sound, position, heat, and light. This type of technology, often referred to as *physical computing*, is used in all kinds of things from the iPhone to automobile electronics systems. Arduino makes it possible for anyone with an interest—even people with no programming or electronics experience—to use this rich and complex technology.

Who This Book Is For

Unlike in most technical cookbooks, experience with software and hardware is not assumed. This book is aimed at readers interested in using computer technology to interact with the environment. It is for people who want to quickly find the solution to hardware and software problems. The recipes provide the information you need to accomplish a broad range of tasks. It also has details to help you customize solutions to meet your specific needs. There is insufficient space in a book limited to 700 pages to cover general theoretical background, so links to external references are provided throughout the book. See "What Was Left Out" on page xiv for some general references for those with no programming or electronics experience.

If you have no programming experience—perhaps you have a great idea for an interactive project but don't have the skills to develop it—this book will help you learn what you need to know to write code that works, using examples that cover over 200 common tasks.

If you have some programming experience but are new to Arduino, the book will help you become productive quickly by demonstrating how to implement specific Arduino capabilities for your project.

People already using Arduino should find the content helpful for quickly learning new techniques, which are explained using practical examples. This will help you to embark on more complex projects by showing how to solve problems and use capabilities that may be new to you.

Experienced C/C++ programmers will find examples of how to use the low-level AVR resources (interrupts, timers, I2C, Ethernet, etc.) to build applications using the Arduino environment.

How This Book Is Organized

The book contains information that covers the broad range of the Arduino's capabilities, from basic concepts and common tasks to advanced technology. Each technique is explained in a recipe that shows you how to implement a specific capability. You do not need to read the content in sequence. Where a recipe uses a technique covered in another recipe, the content in the other recipe is referenced rather than repeating details in multiple places.

Chapter 1, *Getting Started*, introduces the Arduino environment and provides help on getting the Arduino development environment and hardware installed and working.

The next couple of chapters introduce Arduino software development. Chapter 2, *Making the Sketch Do Your Bidding*, covers essential software concepts and tasks, and Chapter 3, *Using Mathematical Operators*, shows how to make use of the most common mathematical functions.

Chapter 4, *Serial Communications*, describes how to get Arduino to connect and communicate with your computer and other devices. Serial is the most common method for Arduino input and output, and this capability is used in many of the recipes throughout the book.

Chapter 5, *Simple Digital and Analog Input*, introduces a range of basic techniques for reading digital and analog signals. Chapter 6, *Getting Input from Sensors*, builds on this with recipes that explain how to use devices that enable Arduino to sense touch, sound, position, heat, and light.

Chapter 7, *Visual Output*, covers controlling light. Recipes cover switching on one or many LEDs and controlling brightness and color. This chapter explains how you can drive bar graphs and numeric LED displays, as well as create patterns and animations with LED arrays. In addition, the chapter provides a general introduction to digital and analog output for those who are new to this.

Chapter 8, *Physical Output*, explains how you can make things move by controlling motors with Arduino. A wide range of motor types is covered: solenoids, servo motors, DC motors, and stepper motors.

Chapter 9, *Audio Output*, shows how to generate sound with Arduino through an output device such as a speaker. It covers playing simple tones and melodies and playing WAV files and MIDI.

Chapter 10, *Remotely Controlling External Devices*, describes techniques that can be used to interact with almost any device that uses some form of remote controller, including TV, audio equipment, cameras, garage doors, appliances, and toys. It builds on techniques used in previous chapters for connecting Arduino to devices and modules.

Chapter 11, *Using Displays*, covers interfacing text and graphical LCD displays. The chapter shows how you can connect these devices to display text, scroll or highlight words, and create special symbols and characters.

Chapter 12, *Using Time and Dates*, covers built-in Arduino time-related functions and introduces many additional techniques for handling time delays, time measurement, and real-world times and dates.

Chapter 13, *Communicating Using I2C and SPI*, covers the Inter-Integrated Circuit (I2C) and Serial Peripheral Interface (SPI) standards. These standards provide simple ways for digital information to be transferred between sensors and Arduino. This chapter shows how to use I2C and SPI to connect to common devices. It also shows how to connect two or more Arduino boards, using I2C for multiboard applications.

Chapter 14, *Wireless Communication*, covers wireless communication with XBee and other wireless modules. This chapter provides examples ranging from simple wireless serial port replacements to mesh networks connecting multiple boards to multiple sensors.

Chapter 15, *Ethernet and Networking*, describes the many ways you can use Arduino with the Internet. It has examples that demonstrate how to build and use web clients and servers and shows how to use the most common Internet communication protocols with Arduino.

Arduino software libraries are a standard way of adding functionality to the Arduino environment. Chapter 16, *Using, Modifying, and Creating Libraries*, explains how to use and modify software libraries. It also provides guidance on how to create your own libraries.

Chapter 17, *Advanced Coding and Memory Handling*, covers advanced programming techniques, and the topics here are more technical than the other recipes in this book because they cover things that are usually concealed by the friendly Arduino wrapper. The techniques in this chapter can be used to make a sketch more efficient—they can help improve performance and reduce the code size of your sketches.

Chapter 18, *Using the Controller Chip Hardware*, shows how to access and use hardware functions that are not fully exposed through the documented Arduino language. It covers low-level usage of the hardware input/output registers, timers, and interrupts.

Appendix A, *Electronic Components*, provides an overview of the components used throughout the book.

Appendix B, *Using Schematic Diagrams and Data Sheets*, explains how to use schematic diagrams and data sheets.

Appendix C, *Building and Connecting the Circuit*, provides a brief introduction to using a breadboard, connecting and using external power supplies and batteries, and using capacitors for decoupling.

Appendix D, *Tips on Troubleshooting Software Problems*, provides tips on fixing compile and runtime problems.

Appendix E, *Tips on Troubleshooting Hardware Problems*, covers problems with electronic circuits.

Appendix F, *Digital and Analog Pins*, provides tables indicating functionality provided by the pins on standard Arduino boards.

Appendix G, *ASCII and Extended Character Sets*, provides tables showing ASCII characters.

Appendix H, *Migrating to Arduino 1.0*, explains how to modify code written for previous releases to run correctly with Arduino 1.0.

What Was Left Out

There isn't room in this book to cover electronics theory and practice, although guidance is provided for building the circuits used in the recipes. For more detail, readers may want to refer to material that is widely available on the Internet or to books such as the following:

- *Make: Electronics* by Charles Platt (O'Reilly; search for it on oreilly.com)
- *Getting Started in Electronics* by Forrest M. Mims III (Master Publishing)
- *Physical Computing* by Dan O'Sullivan and Tom Igoe (Cengage)
- *Practical Electronics for Inventors* by Paul Scherz (McGraw-Hill)

This cookbook explains how to write code to accomplish specific tasks, but it is not an introduction to programming. Relevant programming concepts are briefly explained, but there is insufficient room to cover the details. If you want to learn more about programming, you may want to refer to the Internet or to one of the following books:

- *Practical C Programming* by Steve Oualline (O'Reilly; search for it on oreilly.com)
- *A Book on C* by Al Kelley and Ira Pohl (Addison-Wesley)

My favorite, although not really a beginner's book, is the book I used to learn C programming:

- *The C Programming Language* by Brian W. Kernighan and Dennis M. Ritchie (Prentice Hall)

Code Style (About the Code)

The code used throughout this book has been tailored to clearly illustrate the topic covered in each recipe. As a consequence, some common coding shortcuts have been avoided, particularly in the early chapters. Experienced C programmers often use rich but terse expressions that are efficient but can be a little difficult for beginners to read. For example, the early chapters increment variables using explicit expressions that are easy for nonprogrammers to read:

```
result = result + 1; // increment the count
```

Rather than the following, commonly used by experienced programmers, that does the same thing:

```
result++;  // increment using the post increment operator
```

Feel free to substitute your preferred style. Beginners should be reassured that there is no benefit in performance or code size in using the terse form.

Some programming expressions are so common that they are used in their terse form. For example, the loop expressions are written as follows:

```
for(int i=0; i < 4; i++)
```

This is equivalent to the following:

```
int i;
for(i=0; i < 4; i = i+1)
```

See Chapter 2 for more details on these and other expressions used throughout the book.

Good programming practice involves ensuring that values used are valid (garbage in equals garbage out) by checking them before using them in calculations. However, to keep the code focused on the recipe topic, very little error-checking code has been included.

Arduino Platform Release Notes

This edition has been updated for Arduino 1.0. All of the code has been tested with the latest Arduino 1.0 release candidate at the time of going to press (RC2). The download code for this edition will be updated online if necessary to support the final 1.0 release, so check the book's website (*http://shop.oreilly.com/product/0636920022244.do*) to get the latest code. The download contains a file named *changelog.txt* that will indicate code that has changed from the published edition.

Although many of the sketches will run on earlier Arduino releases, you need to change the extension from *.ino* to *.pde* to load the sketch into a pre-1.0 IDE. If you have not migrated to Arduino 1.0 and have good reason to stick with an earlier release, you can use the example code from the first edition of this book (available at *http://shop.oreilly.com/product/9780596802486.do*), which has been tested with releases from 0018 to 0022. Note that many recipes in the second edition have been enhanced, so we encourage you to upgrade to Arduino 1.0. If you need help migrating older code, see Appendix H.

There's also a link to errata on that site. Errata give readers a way to let us know about typos, errors, and other problems with the book. Errata will be visible on the page immediately, and we'll confirm them after checking them out. O'Reilly can also fix errata in future printings of the book and on Safari, making for a better reader experience pretty quickly.

If you have problems making examples work, check the *changelog.txt* file in the latest code download to see if the sketch has been updated. If that doesn't fix the problem, see Appendix D, which covers troubleshooting software problems. The Arduino forum is a good place to post a question if you need more help: *http://www.arduino.cc*.

If you like—or don't like—this book, by all means, please let people know. Amazon reviews are one popular way to share your happiness or other comments. You can also leave reviews at the O'Reilly site for the book.

Conventions Used in This Book

The following font conventions are used in this book:

Italic
: Indicates pathnames, filenames, and program names; Internet addresses, such as domain names and URLs; and new items where they are defined

`Constant width`
: Indicates command lines and options that should be typed verbatim; names and keywords in programs, including method names, variable names, and class names; and HTML element tags

`Constant width bold`
: Indicates emphasis in program code lines

`Constant width italic`
: Indicates text that should be replaced with user-supplied values

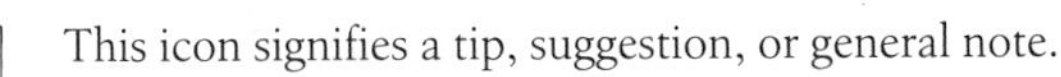

This icon signifies a tip, suggestion, or general note.

This icon indicates a warning or caution.

Using Code Examples

This book is here to help you make things with Arduino. In general, you may use the code in this book in your programs and documentation. You do not need to contact us for permission unless you're reproducing a significant portion of the code. For example, writing a program that uses several chunks of code from this book does not require permission. Selling or distributing a CD-ROM of examples from this book *does* require permission. Answering a question by citing this book and quoting example code does not require permission. Incorporating a significant amount of example code from this book into your product's documentation *does* require permission.

We appreciate, but do not require, attribution. An attribution usually includes the title, author, publisher, and ISBN. For example: "*Arduino Cookbook, Second Edition,* by Michael Margolis with Nick Weldin (O'Reilly). Copyright 2012 Michael Margolis, Nicholas Weldin, 978-1-4493-1387-6."

If you feel your use of code examples falls outside fair use or the permission given here, feel free to contact us at *permissions@oreilly.com*.

Safari® Books Online

Safari Books Online

Safari Books Online is an on-demand digital library that lets you easily search over 7,500 technology and creative reference books and videos to find the answers you need quickly.

With a subscription, you can read any page and watch any video from our library online. Read books on your cell phone and mobile devices. Access new titles before they are available for print, and get exclusive access to manuscripts in development and post feedback for the authors. Copy and paste code samples, organize your favorites, download chapters, bookmark key sections, create notes, print out pages, and benefit from tons of other time-saving features.

O'Reilly Media has uploaded this book to the Safari Books Online service. To have full digital access to this book and others on similar topics from O'Reilly and other publishers, sign up for free at *http://my.safaribooksonline.com*.

How to Contact Us

We have tested and verified the information in this book to the best of our ability, but you may find that features have changed (or even that we have made a few mistakes!). Please let us know about any errors you find, as well as your suggestions for future editions, by writing to:

O'Reilly Media, Inc.
1005 Gravenstein Highway North
Sebastopol, CA 95472
800-998-9938 (in the United States or Canada)
707-829-0515 (international/local)
707-829-0104 (fax)

We have a web page for this book, where we list errata, examples, and any additional information. You can access this page at:

http://shop.oreilly.com/product/0636920022244.do

To comment or ask technical questions about this book, send email to:

bookquestions@oreilly.com

For more information about our books, courses, conferences, and news, see our website at *http://www.oreilly.com*.

Find us on Facebook: *http://facebook.com/oreilly*

Follow us on Twitter: *http://twitter.com/oreillymedia*

Watch us on YouTube: *http://www.youtube.com/oreillymedia*

Acknowledgments

Nick Weldin's contribution was invaluable for the completion of this book. It was 90 percent written when Nick came on board—and without his skill and enthusiasm, it would still be 90 percent written. His hands-on experience running Arduino workshops for all levels of users enabled us to make the advice in this book practical for our broad range of readers. Thank you, Nick, for your knowledge and genial, collaborative nature.

Simon St. Laurent was the editor at O'Reilly who first expressed interest in this book. And in the end, he is the man who pulled it together. His support and encouragement kept us inspired as we sifted our way through the volumes of material necessary to do the subject justice.

Brian Jepson helped me get started with the writing of this book. His vast knowledge of things Arduino and his concern and expertise for communicating about technology in plain English set a high standard. He was an ideal guiding hand for shaping the book and making technology readily accessible for readers. We also have Brian to thank for the XBee content in Chapter 14.

Brian Jepson and Shawn Wallace were technical editors for this second edition and provided excellent advice for improving the accuracy and clarity of the content.

Audrey Doyle worked tirelessly to stamp out typos and grammatical errors in the initial manuscript and untangle some of the more convoluted expressions.

Philip Lindsay collaborated on content for Chapter 15 in the first edition. Adrian McEwen, the lead developer for many of the Ethernet enhancements in Release 1.0, provided valuable advice to ensure this Chapter reflected all the changes in that release.

Mikal Hart wrote recipes covering GPS and software serial. Mikal was the natural choice for this—not only because he wrote the libraries, but also because he is a fluent communicator, an Arduino enthusiast, and a pleasure to collaborate with.

Arduino is possible because of the creativity of the core Arduino development team: Massimo Banzi, David Cuartielles, Tom Igoe, Gianluca Martino, and David Mellis. On behalf of all Arduino users, I wish to express our appreciation for their efforts in making this fascinating technology simple and their generosity in making it free.

Special thanks to Alexandra Deschamps-Sonsino, whose Tinker London workshops provided important understanding of the needs of users. Thanks also to Peter Knight, who has provided all kinds of clever Arduino solutions as well as the basis of a number of recipes in this book.

On behalf of everyone who has downloaded user-contributed Arduino libraries, I would like to thank the authors who have generously shared their knowledge.

The availability of a wide range of hardware is a large part of what makes Arduino exciting—thanks to the suppliers for stocking and supporting a broad range of great devices. The following were helpful in providing hardware used in the book: SparkFun, Maker Shed, Gravitech, and NKC Electronics. Other suppliers that have been helpful include Modern Device, Liquidware, Adafruit, MakerBot Industries, Mindkits, Oomlout, and SK Pang.

Nick would like to thank everyone who was involved with Tinker London, particularly Alexandra, Peter, Brock Craft, Daniel Soltis and all the people who assisted on workshops over the years.

Nick's final thanks go to his family, Jeanie, Emily, and Finn, who agreed to let him do this over their summer holiday, and of course, much longer after that than they originally thought, and to his parents, Frank and Eva, for bringing him up to take things apart.

Last but not least, I express thanks to the following people:

Joshua Noble for introducing me to O'Reilly. His book, *Programming Interactivity*, is highly recommended for those interested in broadening their knowledge in interactive computing.

Robert Lacy-Thompson for offering advice early on with the first edition.

Mark Margolis for his support and help as a sounding board in the book's conception and development.

I thank my parents for helping me to see that the creative arts and technology were not distinctive entities and that, when combined, they can lead to extraordinary results.

And finally, this book would not have been started or finished without the support of my wife, Barbara Faden. My grateful appreciation to her for keeping me motivated and for her careful reading and contributions to the manuscript.

Notes on the Second Edition

The second edition of this book has followed relatively quickly from the first, prompted by the release of Arduino 1.0. The stated purpose of 1.0 is to introduce significant change that will smooth the way for future enhancements but break some code written for older software. These have necessitated changes to code in many of the chapters of this book. Most changed are Chapter 15, *Ethernet and Networking*, and Chapter 13, *Communicating Using I2C and SPI*, but all of the recipes in this edition have been migrated to 1.0, with many being updated to use features new in this release. If you are using a release prior to Arduino 1.0, then you can download code from the first edition of this book. See "Arduino Platform Release Notes" on page xv for download details.

Appendix H, *Migrating to Arduino 1.0*, has been added to describe the changes introduced by Arduino Release 1.0. This describes how to update older code to use with Arduino 1.0.

Recipes for devices that are no longer widely available have been updated to use current replacements and some new sensors and wireless devices have been added.

Errata posted on the O'Reilly site has been corrected, thanks to readers taking the time to notify us of these.

We think you will like the improvements made in Arduino 1.0 as well as the enhancements made to this edition of the *Arduino Cookbook*. The first edition was well received; the constructive criticism being divided between people that wanted more technical content and those that preferred less. In a book that we limited to only 700 or so pages (to keep it affordable and portable), that seems to indicate that the right balance has been achieved.

CHAPTER 1

Getting Started

1.0 Introduction

The Arduino environment has been designed to be easy to use for beginners who have no software or electronics experience. With Arduino, you can build objects that can respond to and/or control light, sound, touch, and movement. Arduino has been used to create an amazing variety of things, including musical instruments, robots, light sculptures, games, interactive furniture, and even interactive clothing.

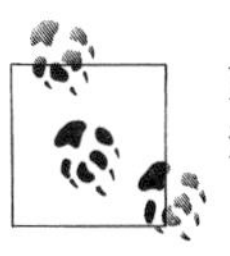

If you're not a beginner, please feel free to skip ahead to recipes that interest you.

Arduino is used in many educational programs around the world, particularly by designers and artists who want to easily create prototypes but do not need a deep understanding of the technical details behind their creations. Because it is designed to be used by nontechnical people, the software includes plenty of example code to demonstrate how to use the Arduino board's various facilities.

Though it is easy to use, Arduino's underlying hardware works at the same level of sophistication that engineers employ to build embedded devices. People already working with microcontrollers are also attracted to Arduino because of its agile development capabilities and its facility for quick implementation of ideas.

Arduino is best known for its hardware, but you also need software to program that hardware. Both the hardware and the software are called "Arduino." The combination enables you to create projects that sense and control the physical world. The software is free, open source, and cross-platform. The boards are inexpensive to buy, or you can build your own (the hardware designs are also open source). In addition, there is an active and supportive Arduino community that is accessible worldwide through the Arduino forums and the wiki (known as the Arduino Playground). The forums and the wiki offer project development examples and solutions to problems that can provide inspiration and assistance as you pursue your own projects.

The recipes in this chapter will get you started by explaining how to set up the development environment and how to compile and run an example sketch.

Source code containing computer instructions for controlling Arduino functionality is usually referred to as a *sketch* in the Arduino community. The word *sketch* will be used throughout this book to refer to Arduino program code.

The Blink sketch, which comes with Arduino, is used as an example for recipes in this chapter, though the last recipe in the chapter goes further by adding sound and collecting input through some additional hardware, not just blinking the light built into the board. Chapter 2 covers how to structure a sketch for Arduino and provides an introduction to programming.

If you already know your way around Arduino basics, feel free to jump forward to later chapters. If you're a first-time Arduino user, patience in these early recipes will pay off with smoother results later.

Arduino Software

Software programs, called *sketches*, are created on a computer using the Arduino integrated development environment (IDE). The IDE enables you to write and edit code and convert this code into instructions that Arduino hardware understands. The IDE also transfers those instructions to the Arduino board (a process called *uploading*).

Arduino Hardware

The Arduino board is where the code you write is executed. The board can only control and respond to electricity, so specific components are attached to it to enable it to interact with the real world. These components can be sensors, which convert some aspect of the physical world to electricity so that the board can sense it, or actuators, which get electricity from the board and convert it into something that changes the world. Examples of sensors include switches, accelerometers, and ultrasound distance sensors. Actuators are things like lights and LEDs, speakers, motors, and displays.

There are a variety of official boards that you can use with Arduino software and a wide range of Arduino-compatible boards produced by members of the community.

The most popular boards contain a USB connector that is used to provide power and connectivity for uploading your software onto the board. Figure 1-1 shows a basic board that most people start with, the Arduino Uno.

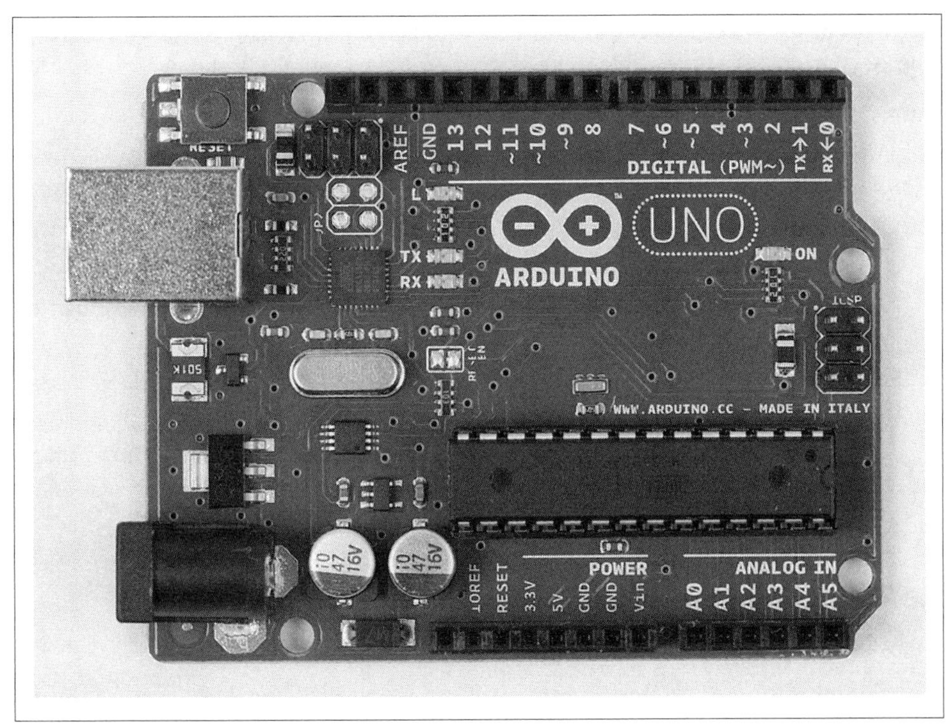

Figure 1-1. Basic board: the Arduino Uno. Photograph courtesy todo.to.it.

The Arduino Uno has a second microcontroller onboard to handle all USB communication; the small surface-mount chip (the ATmega8U2) is located near the USB socket on the board. This can be programmed separately to enable the board to appear as different USB devices (see Recipe 18.14 for an example). The Arduino Leonardo board replaces the ATmega8U2 and the ATmega328 controllers with a single ATmega32u4 chip that implements the USB protocol in software. The Arduino-compatible Teensy and Teensy+ boards from PJRC (*http://www.pjrc.com/teensy/*) are also capable of emulating USB devices. Older boards, and most of the Arduino-compatible boards, use a chip from the FTDI company that provides a hardware USB solution for connection to the serial port of your computer.

You can get boards as small as a postage stamp, such as the Arduino Mini and Pro Mini; larger boards that have more connection options and more powerful processors, such as the Arduino Mega; and boards tailored for specific applications, such as the LilyPad for wearable applications, the Fio for wireless projects, and the Arduino Pro for embedded applications (standalone projects that are often battery-operated).

Recent additions to the range include the Arduino ADK, which has a USB host socket on it and is compatible with the Android Open Accessory Development Kit, the officially approved method of attaching hardware to Android devices. The Leonardo board uses a controller chip (the ATmega32u4) that is able to present itself as various HID devices. The Ethernet board includes Ethernet connectivity, and has a Power Over Ethernet option, so it is possible to use a single cable to connect and power the board.

Other Arduino-compatible boards are also available, including the following:

- Arduino Nano, a tiny board with USB capability, from Gravitech (*http://store.gravitech.us/arna30wiatn.html*)
- Bare Bones Board, a low-cost board available with or without USB capability, from Modern Device (*http://www.moderndevice.com/products/bbb-kit*)
- Boarduino, a low-cost breadboard-compatible board, from Adafruit Industries (*http://www.adafruit.com/*)
- Seeeduino, a flexible variation of the standard USB board, from Seeed Studio Bazaar (*http://www.seeedstudio.com/*)
- Teensy and Teensy++, tiny but extremely versatile boards, from PJRC (*http://www.pjrc.com/teensy/*)

A list of Arduino-compatible boards is available at *http://www.freeduino.org/*.

See Also

An overview of Arduino boards: *http://www.arduino.cc/en/Main/Hardware*.

Online guides for getting started with Arduino are available at *http://arduino.cc/en/Guide/Windows* for Windows, *http://arduino.cc/en/Guide/MacOSX* for Mac OS X, and *http://www.arduino.cc/playground/Learning/Linux* for Linux.

A list of over a hundred boards that can be used with the Arduino development environment can be found at: *http://jmsarduino.blogspot.com/2009/03/comprehensive-arduino-compatible.html*

1.1 Installing the Integrated Development Environment (IDE)

Problem

You want to install the Arduino development environment on your computer.

Solution

The Arduino software for Windows, Mac, and Linux can be downloaded from *http://arduino.cc/en/Main/Software*.

The Windows download is a ZIP file. Unzip the file to any convenient directory—*Program Files/Arduino* is a sensible place.

A free utility for unzipping files, called 7-Zip, can be downloaded from *http://www.7-zip.org/*.

Unzipping the file will create a folder named *Arduino-00<nn>* (where *<nn>* is the version number of the Arduino release you downloaded). The directory contains the executable file (named *Arduino.exe*), along with various other files and folders. Double-click the *Arduino.exe* file and the splash screen should appear (see Figure 1-2), followed by the main program window (see Figure 1-3). Be patient, as it can take some time for the software to load.

Figure 1-2. Arduino splash screen (Version 1.0 in Windows 7)

The Arduino download for the Mac is a disk image (*.dmg*); double-click the file when the download is complete. The image will mount (it will appear like a memory stick

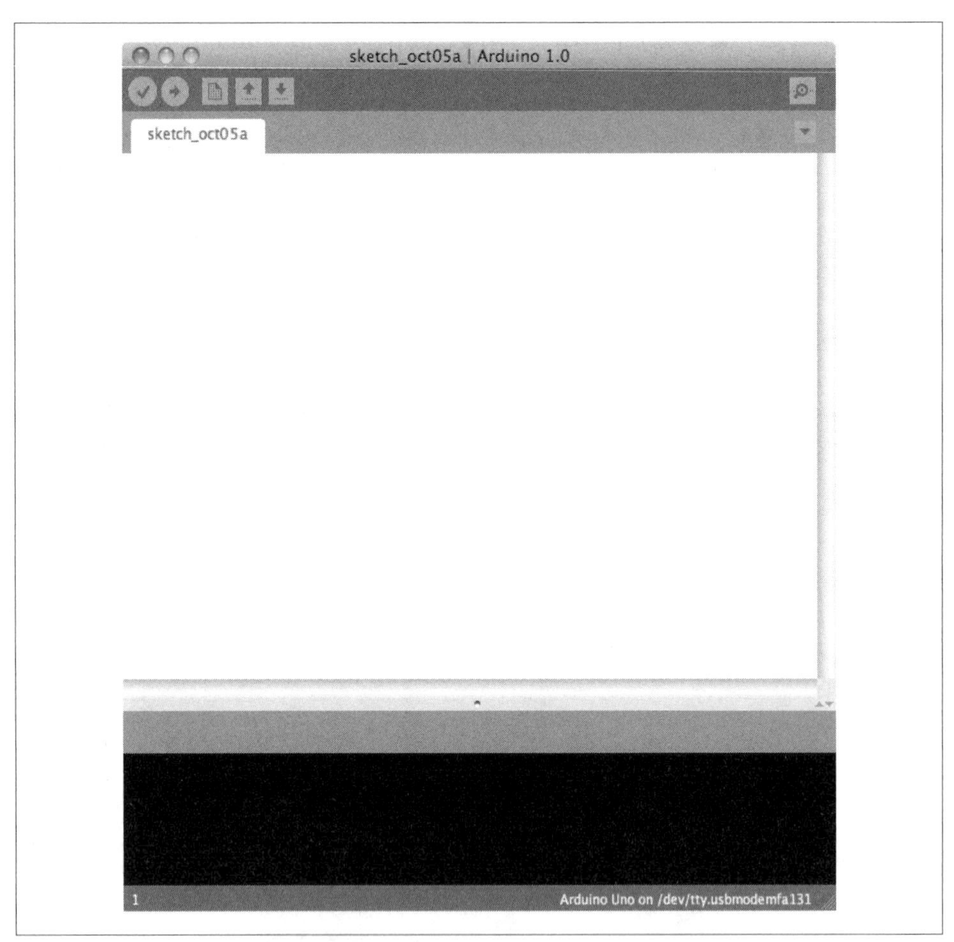

Figure 1-3. IDE main window (Arduino 1.0 on a Mac)

on the desktop). Inside the disk image is the Arduino application. Copy this to somewhere convenient—the *Applications* folder is a sensible place. Double-click the application once you have copied it over (it is not a good idea to run it from the disk image). The splash screen will appear, followed by the main program window.

Linux installation varies depending on the Linux distribution you are using. See the Arduino wiki for information (*http://www.arduino.cc/playground/Learning/Linux*).

To enable the Arduino development environment to communicate with the board, you need to install drivers.

On Windows, use the USB cable to connect your PC and the Arduino board and wait for the Found New Hardware Wizard to appear. If you are using an Uno board, let the wizard attempt to find and install drivers. It will fail to do this (don't worry, this is the expected behavior). To fix it you now need to go to Start Menu→Control Panel→System and Security. Click on System, and then open Device Manager. In the listing that is displayed find the entry in COM and LPT named `Arduino UNO (COM nn)`. `nn` will be the number Windows has assigned to the port created for the board. You will see a warning logo next to this because the appropriate drivers have not yet been assigned. Right click on the entry and select Update Driver Software. Choose the "Browse my computer for driver software" option, and navigate to the *Drivers* folder inside the Arduino folder you just unzipped. Select the `ArduinoUNO.inf` file and windows should then complete the installation process.

If you are using an earlier board (any board that uses FTDI drivers) with Windows Vista or Windows 7 and are online, you can let the wizard search for drivers and they will install automatically. On Windows XP (or if you don't have Internet access), you should specify the location of the drivers. Use the file selector to navigate to the *FTDI USB Drivers* directory, located in the directory where you unzipped the Arduino files. When this driver has installed, the Found New Hardware Wizard will appear again, saying a new serial port has been found. Follow the same process as before.

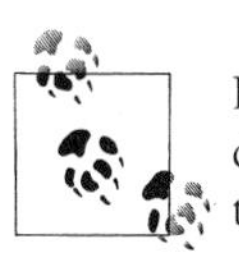

It is important that you go through the sequence of steps to install the drivers two times, or the software will not be able to communicate with the board.

On the Mac, the latest Arduino boards, such as the Uno, can be used without additional drivers. When you first plug the board in a notification will pop up saying a new network port has been found, you can dismiss this. If you are using earlier boards (boards that need FTDI drivers), you will need to install driver software. There is a package named *FTDIUSBSerialDriver*, with a range of numbers after it, inside the disk image. Double-click this and the installer will take you through the process. You will need to know an administrator password to complete the process.

On Linux, most distributions have the driver already installed, but follow the Linux link given in this chapter's introduction for specific information for your distribution.

Discussion

If the software fails to start, check the troubleshooting section of the Arduino website, *http://arduino.cc/en/Guide/Troubleshooting*, for help solving installation problems.

See Also

Online guides for getting started with Arduino are available at *http://arduino.cc/en/Guide/Windows* for Windows, *http://arduino.cc/en/Guide/MacOSX* for Mac OS X, and *http://www.arduino.cc/playground/Learning/Linux* for Linux.

1.2 Setting Up the Arduino Board

Problem

You want to power up a new board and verify that it is working.

Solution

Plug the board in to a USB port on your computer and check that the green LED power indicator on the board illuminates. Standard Arduino boards (Uno, Duemilanove, and Mega) have a green LED power indicator located near the reset switch.

An orange LED near the center of the board (labeled "Pin 13 LED" in Figure 1-4) should flash on and off when the board is powered up (boards come from the factory preloaded with software to flash the LED as a simple check that the board is working).

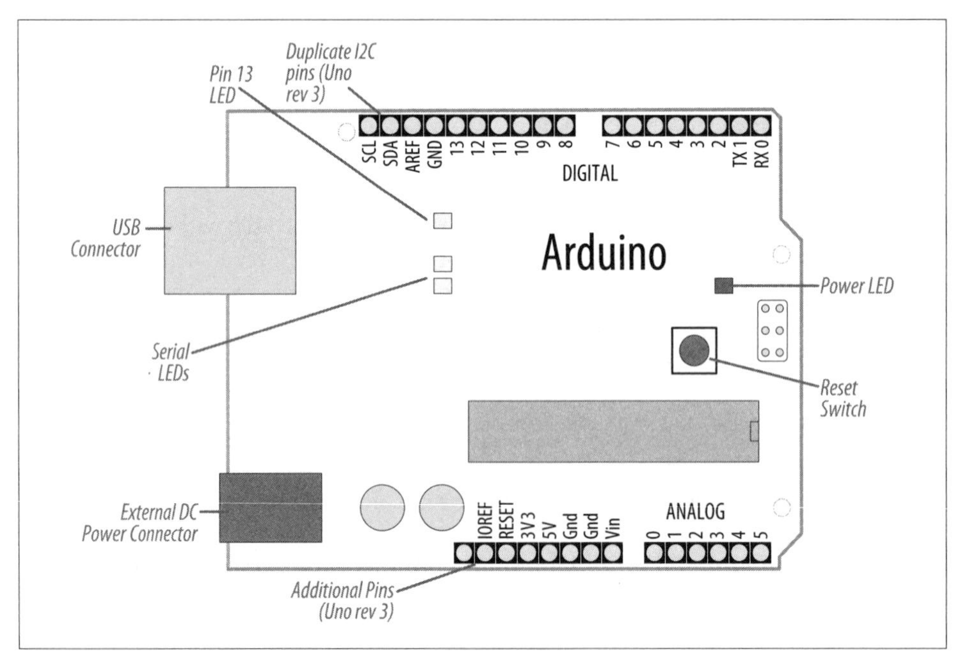

Figure 1-4. Basic Arduino board (Duemilanove and Uno)

New boards such as Leonardo have the LEDs located near the USB connector; see Figure 1-5. Recent boards have duplicate pins for use with I2C (marked `SCL` and `SDA`). These boards also have a pin marked `IOREF` that can be used to determine the operating voltage of the chip.

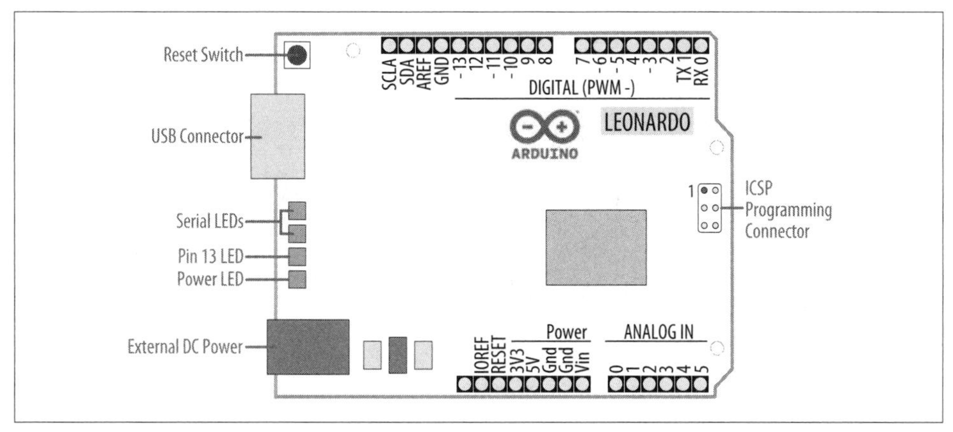

Figure 1-5. Leonardo Board

The latest boards have three additional connections in the new standard for connector layout on the board. This does not affect the use of older shields (they will all continue to work with the new boards, just as they did with earlier boards). The new connections provide a pin (IOREF) for shields to detect the analog reference voltage (so that analog input values can be calibrated to the supply voltage), SCL and SDA pins to enable a consistent connection for I2C devices (the location of the I2C pins has differed on previous boards due to different chip configurations). Shields designed for the new layout should work on any board that uses the new pin locations. An additional pin (next to the IOREF pin) is not being used at the moment, but enables new functionality to be implemented in the future without needing to change the pin layout again.

Discussion

If the power LED does not illuminate when the board is connected to your computer, the board is probably not receiving power.

The flashing LED (connected to digital output pin 13) is being controlled by code running on the board (new boards are preloaded with the Blink example sketch). If the pin 13 LED is flashing, the sketch is running correctly, which means the chip on the board is working. If the green power LED is on but the pin 13 LED is not flashing, it could be that the factory code is not on the chip; follow the instructions in Recipe 1.3 to load the Blink sketch onto the board to verify that the board is working. If you are not using a standard board, it may not have a built-in LED on pin 13, so check the documentation for details of your board. The Leonardo board fades the LED up and down (it looks like the LED is "breathing") to show that the board is working.

See Also

Online guides for getting started with Arduino are available at *http://arduino.cc/en/Guide/Windows* for Windows, *http://arduino.cc/en/Guide/MacOSX* for Mac OS X, and *http://www.arduino.cc/playground/Learning/Linux* for Linux.

A troubleshooting guide can be found at *http://arduino.cc/en/Guide/Troubleshooting*.

1.3 Using the Integrated Development Environment (IDE) to Prepare an Arduino Sketch

Problem

You want to get a sketch and prepare it for uploading to the board.

Solution

Use the Arduino IDE to create, open, and modify sketches that define what the board will do. You can use buttons along the top of the IDE to perform these actions (shown in Figure 1-6), or you can use the menus or keyboard shortcuts (shown in Figure 1-7).

The Sketch Editor area is where you view and edit code for a sketch. It supports common text-editing keys such as Ctrl-F (⌘+F on a Mac) for find, Ctrl-Z (⌘+Z on a Mac) for undo, Ctrl-C (⌘+C on a Mac) to copy highlighted text, and Ctrl-V (⌘+V on a Mac) to paste highlighted text.

Figure 1-7 shows how to load the Blink sketch (the sketch that comes preloaded on a new Arduino board).

After you've started the IDE, go to the File→Examples menu and select 1. Basics→Blink, as shown in Figure 1-7. The code for blinking the built-in LED will be displayed in the Sketch Editor window (refer to Figure 1-6).

Before the code can be sent to the board, it needs to be converted into instructions that can be read and executed by the Arduino controller chip; this is called *compiling*. To do this, click the compile button (the top-left button with a tick inside), or select Sketch→Verify/Compile (Ctrl-R; ⌘+R on a Mac).

You should see a message that reads "Compiling sketch..." and a progress bar in the message area below the text-editing window. After a second or two, a message that reads "Done Compiling" will appear. The black console area will contain the following additional message:

```
Binary sketch size: 1026 bytes (of a 32256 byte maximum)
```

The exact message may differ depending on your board and Arduino version; it is telling you the size of the sketch and the maximum size that your board can accept.

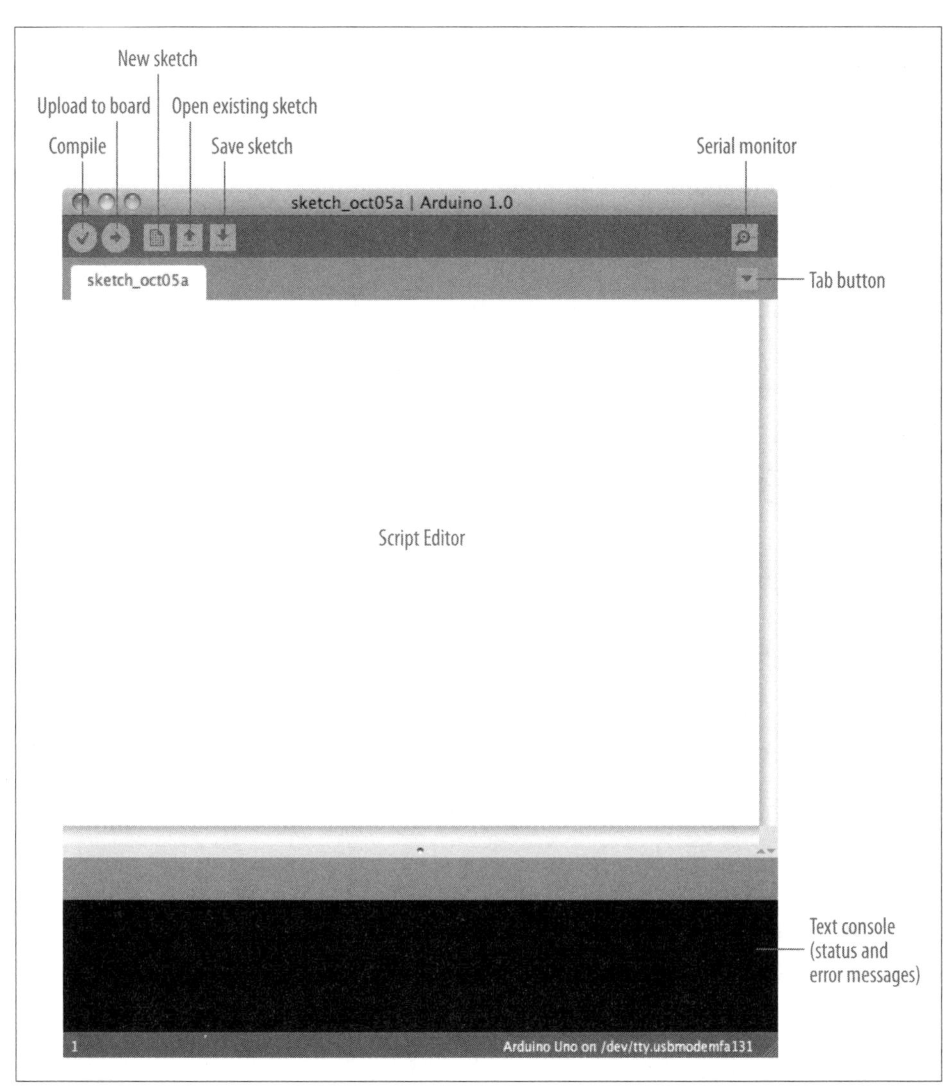

Figure 1-6. Arduino IDE

Discussion

Source code for Arduino is called a *sketch*. The process that takes a sketch and converts it into a form that will work on the board is called *compilation*. The IDE uses a number of command-line tools behind the scenes to compile a sketch. For more information on this, see Recipe 17.1.

The final message telling you the size of the sketch indicates how much program space is needed to store the controller instructions on the board. If the size of the compiled

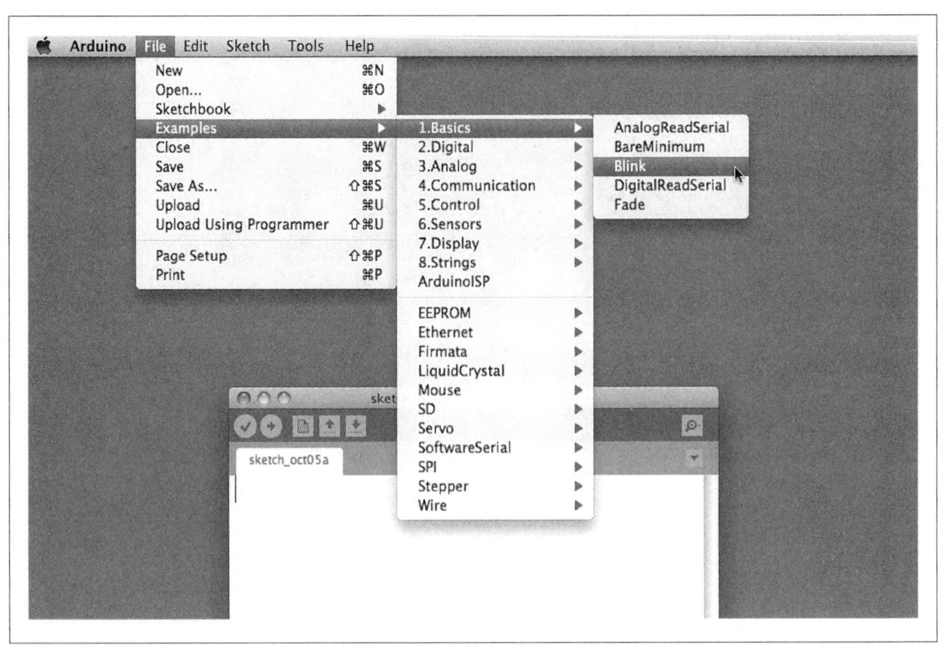

Figure 1-7. IDE menu (selecting the Blink example sketch)

sketch is greater than the available memory on the board, the following error message is displayed:

```
Sketch too big; see http://www.arduino.cc/en/Guide/Troubleshooting#size
  for tips on reducing it.
```

If this happens, you need to make your sketch smaller to be able to put it on the board, or get a board with higher capacity.

If there are errors in the code, the compiler will print one or more error messages in the console window. These messages can help identify the error—see Appendix D on software errors for troubleshooting tips.

To prevent accidental overwriting of the examples, the Arduino IDE does not allow you to save changes to the provided example sketches. You must rename them using the Save As menu option. You can save sketches you write yourself with the Save button (see Recipe 1.5).

As you develop and modify a sketch, you should also consider using the File→Save As menu option and using a different name or version number regularly so that as you implement each bit, you can go back to an older version if you need to.

Code uploaded onto the board cannot be downloaded back onto your computer. Make sure you save your sketch code on your computer. You cannot save changes back to the example files; you need to use Save As and give the changed file another name.

See Also

Recipe 1.5 shows an example sketch. Appendix D has tips on troubleshooting software problems.

1.4 Uploading and Running the Blink Sketch

Problem

You want to transfer your compiled sketch to the Arduino board and see it working.

Solution

Connect your Arduino board to your computer using the USB cable. Load the Blink sketch into the IDE as described in Recipe 1.3.

Next, select Tools→Board from the drop-down menu and select the name of the board you have connected (if it is the standard Uno board, it is probably the first entry in the board list).

Now select Tools→Serial Port. You will get a drop-down list of available serial ports on your computer. Each machine will have a different combination of serial ports, depending on what other devices you have used with your computer.

On Windows, they will be listed as numbered COM entries. If there is only one entry, select it. If there are multiple entries, your board will probably be the last entry.

On the Mac, your board will be listed twice if it is an Uno board:

```
/dev/tty.usbmodem-XXXXXXX
/dev/cu.usbmodem-XXXXXXX
```

If you have an older board, it will be listed as follows:

```
/dev/tty.usbserial-XXXXXXX
/dev/cu.usbserial-XXXXXXX
```

Each board will have different values for *XXXXXXX*. Select either entry.

Click on the upload button (in Figure 1-6, it's the second button from the left), or choose File→Upload to I/O board (Ctrl-U, ⌘+U on a Mac).

The software will compile the code, as in Recipe 1.3. After the software is compiled, it is uploaded to the board. If you look at your board, you will see the LED stop flashing, and two lights (labeled as Serial LEDs in Figure 1-4) just below the previously flashing

LED should flicker for a couple of seconds as the code uploads. The original light should then start flashing again as the code runs.

Discussion

For the IDE to send the compiled code to the board, the board needs to be plugged in to the computer, and you need to tell the IDE which board and serial port you are using.

When an upload starts, whatever sketch is running on the board is stopped (if you were running the Blink sketch, the LED will stop flashing). The new sketch is uploaded to the board, replacing the previous sketch. The new sketch will start running when the upload has successfully completed.

Older Arduino boards and some compatibles do not automatically interrupt the running sketch to initiate upload. In this case, you need to press the Reset button on the board just after the software reports that it is done compiling (when you see the message about the size of the sketch). It may take a few attempts to get the timing right between the end of the compilation and pressing the Reset button.

The IDE will display an error message if the upload is not successful. Problems are usually due to the wrong board or serial port being selected or the board not being plugged in. The currently selected board and serial port are displayed in the status bar at the bottom of the Arduino window

If you have trouble identifying the correct port on Windows, try unplugging the board and then selecting Tools→Serial Port to see which COM port is no longer on the display list. Another approach is to select the ports, one by one, until you see the lights on the board flicker to indicate that the code is uploading.

See Also

The Arduino troubleshooting page: *http://www.arduino.cc/en/Guide/Troubleshooting*.

1.5 Creating and Saving a Sketch

Problem

You want to create a sketch and save it to your computer.

Solution

To open an editor window ready for a new sketch, launch the IDE (see Recipe 1.3), go to the File menu, and select New. Paste the following code into the Sketch Editor window (it's similar to the Blink sketch, but the blinks last twice as long):

```
const int ledPin =  13;    // LED connected to digital pin 13

void setup()
{
  pinMode(ledPin, OUTPUT);
}

void loop()
{
  digitalWrite(ledPin, HIGH);   // set the LED on
  delay(2000);                  // wait for two seconds
  digitalWrite(ledPin, LOW);    // set the LED off
  delay(2000);                  // wait for two seconds
}
```

Compile the code by clicking the compile button (the top-left button with a triangle inside), or select Sketch→Verify/Compile (see Recipe 1.3).

Upload the code by clicking on the upload button, or choose File→Upload to I/O board (see Recipe 1.4). After uploading, the LED should blink, with each flash lasting two seconds.

You can save this sketch to your computer by clicking the Save button, or select File→Save.

You can save the sketch using a new name by selecting the Save As menu option. A dialog box will open where you can enter the filename.

Discussion

When you save a file in the IDE, a standard dialog box for the operating system will open. It suggests that you save the sketch to a folder called *Arduino* in your *My Documents* folder (or your *Documents* folder on a Mac). You can replace the default sketch name with a meaningful name that reflects the purpose of your sketch. Click Save to save the file.

The default name is the word *sketch* followed by the current date. Sequential letters starting from *a* are used to distinguish sketches created on the same day. Replacing the default name with something meaningful helps you to identify the purpose of a sketch when you come back to it later.

If you use characters that the IDE does not allow (e.g., the space character), the IDE will automatically replace these with valid characters.

Arduino sketches are saved as plain text files with the extension *.ino*. Older versions of the IDE used the *.pde* extension, also used by Processing. They are automatically saved in a folder with the same name as the sketch.

You can save your sketches to any folder on your computer, but if you use the default folder (the *Arduino* folder in your *Documents* folder) your sketches will automatically appear in the Sketchbook menu of the Arduino software and be easier to locate.

If you have edited one of the examples from the Arduino download, you will not be able to save the changed file using the same filename. This preserves the standard examples intact. If you want to save a modified example, you will need to select another location for the sketch.

After you have made changes, you will see a dialog box asking if you want to save the sketch when a sketch is closed.

The § symbol following the name of the sketch in the top bar of the IDE window indicates that the sketch code has changes that have not yet been saved on the computer. This symbol is removed when you save the sketch.

The Arduino software does not provide any kind of version control, so if you want to be able to revert to older versions of a sketch, you can use Save As regularly and give each revision of the sketch a slightly different name.

Frequent compiling as you modify or add code is a good way to check for errors as you write your code. It will be easier to find and fix any errors because they will usually be associated with what you have just written.

Once a sketch has been uploaded onto the board there is no way to download it back to your computer. Make sure you save any changes to your sketches that you want to keep.

If you try and save a sketch file that is not in a folder with the same name as the sketch, the IDE will inform you that this can't be opened as is and suggest you click OK to create the folder for the sketch with the same name.

Sketches must be located in a folder with the same name as the sketch. The IDE will create the folder automatically when you save a new sketch.

Sketches made with older versions of Arduino software have a different file extension (*.pde*). The IDE will open them, when you save the sketch it will create a file with the new extension (*.ino*). Code written for early versions of the IDE may not be able to compile in version 1.0. Most of the changes to get old code running are easy to do. See Appendix H for more details.

See Also

The code in this recipe and throughout this book use the `const int` expression to provide meaningful names (`ledPin`) for constants instead of numbers (`13`). See Recipe 17.5 for more on the use of constants.

1.6 Using Arduino

Problem

You want to get started with a project that is easy to build and fun to use.

Solution

This recipe provides a taste of some of the techniques that are covered in detail in later chapters.

The sketch is based on the LED blinking code from the previous recipe, but instead of using a fixed delay, the rate is determined by a light-sensitive sensor called a light dependent resistor or LDR (see Recipe 6.2). Wire the LDR as shown in Figure 1-8.

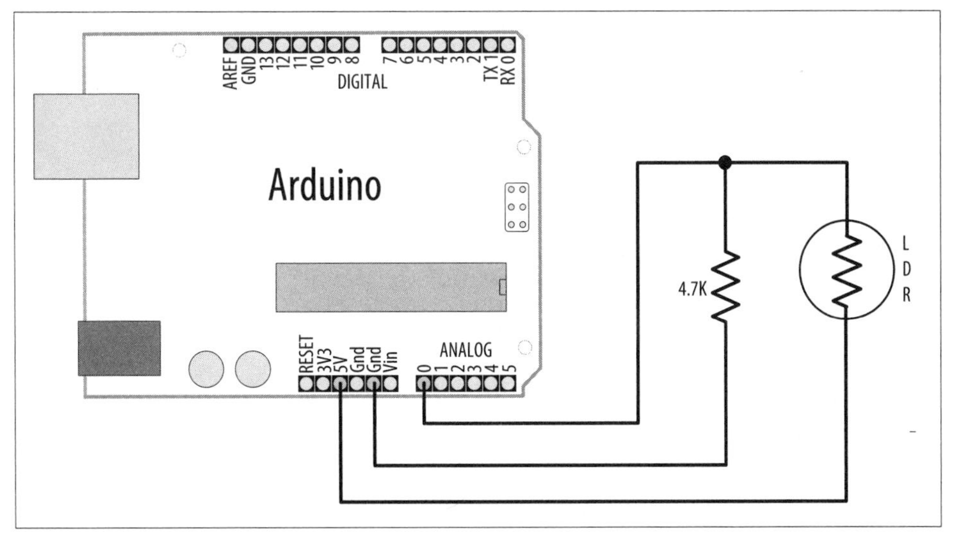

Figure 1-8. Arduino with light dependent resistor

If you are not familiar with building a circuit from a schematic, see Appendix B for step-by-step illustrations on how to make this circuit on a breadboard.

The following sketch reads the light level of an LDR connected to analog pin 0. The light level striking the LDR will change the blink rate of the internal LED connected to pin 13:

```
const int ledPin =  13;     // LED connected to digital pin 13
const int sensorPin = 0;    // connect sensor to analog input 0

void setup()
{
  pinMode(ledPin, OUTPUT); // enable output on the led pin
}

void loop()
{
  int rate = analogRead(sensorPin);    // read the analog input
  digitalWrite(ledPin, HIGH);   // set the LED on
  delay(rate);                  // wait duration dependent on light level
  digitalWrite(ledPin, LOW);    // set the LED off
  delay(rate);
}
```

Discussion

The value of the 4.7K resistor is not critical. Anything from 1K to 10K can be used. The light level on the LDR will change the voltage level on analog pin 0. The `analogRead` command (see Chapter 6) provides a value that ranges from around 200 when the LDR is dark to 800 or so when it is very bright. This value determines the duration of the LED on and off times, so the blink time increases with light intensity.

You can scale the blink rate by using the Arduino `map` function as follows:

```
const int ledPin =  13;     // LED connected to digital pin 13
const int sensorPin = 0;    // connect sensor to analog input 0

// the next two lines set the min and max delay between blinks
const int minDuration = 100;  // minimum wait between blinks
const int maxDuration = 1000;  // maximum wait between blinks


void setup()
{
  pinMode(ledPin, OUTPUT); // enable output on the led pin
}

void loop()
{
  int rate = analogRead(sensorPin);    // read the analog input
  // the next line scales the blink rate between the min and max values
  rate = map(rate, 200,800,minDuration, maxDuration); // convert to blink rate
  rate = constrain(rate, minDuration,maxDuration);    // constrain the value
```

```
  digitalWrite(ledPin, HIGH);   // set the LED on
  delay(rate);                  // wait duration dependent on light level
  digitalWrite(ledPin, LOW);    // set the LED off
  delay(rate);
}
```

Recipe 5.7 provides more details on using the map function to scale values. Recipe 3.5 has details on using the constrain function to ensure values do not exceed a given range.

If you want to view the value of the rate variable on your computer, you can print this to the Arduino Serial Monitor as shown in the revised loop code that follows. The sketch will display the blink rate in the Serial Monitor. You open the Serial Monitor window in the Arduino IDE by clicking on the icon on the right of the top bar (see Chapter 4 for more on using the Serial Monitor):

```
const int ledPin =  13;     // LED connected to digital pin 13
const int sensorPin = 0;    // connect sensor to analog input 0

// the next two lines set the min and max delay between blinks
const int minDuration = 100;  // minimum wait between blinks
const int maxDuration = 1000;  // maximum wait between blinks


void setup()
{
  pinMode(ledPin, OUTPUT);  // enable output on the led pin
  Serial.begin(9600);       // initialize Serial
}

void loop()
{
  int rate = analogRead(sensorPin);     // read the analog input
  // the next line scales the blink rate between the min and max values
  rate = map(rate, 200,800,minDuration, maxDuration); // convert to blink rate
  rate = constrain(rate, minDuration,maxDuration);    // constrain the value

  Serial.println(rate);         // print rate to serial monitor
  digitalWrite(ledPin, HIGH);   // set the LED on
  delay(rate);                  // wait duration dependent on light level
  digitalWrite(ledPin, LOW);    // set the LED off
  delay(rate);
}
```

You can use the LDR to control the pitch of a sound by connecting a small speaker to the pin, as shown in Figure 1-9.

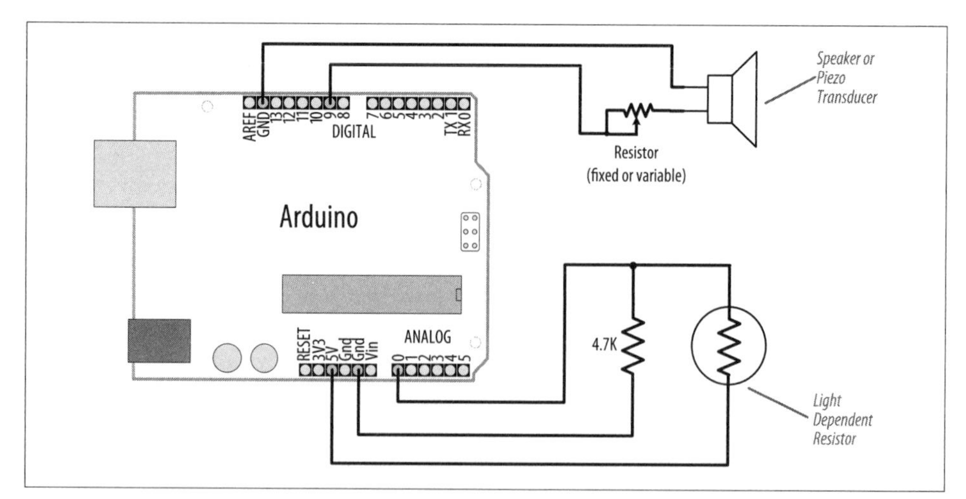

Figure 1-9. Connections for a speaker with the LDR circuit

You will need to increase the on/off rate on the pin to a frequency in the audio spectrum. This is achieved, as shown in the following code, by decreasing the min and max durations:

```
const int outputPin = 9;     // Speaker connected to digital pin 9
const int sensorPin = 0;     // connect sensor to analog input 0

const int minDuration = 1;   // 1ms on, 1ms off (500 Hz)
const int maxDuration = 10;  // 10ms on, 10ms off (50 hz)

void setup()
{
  pinMode(outputPin, OUTPUT);  // enable output on the led pin
}

void loop()
{
  int sensorReading = analogRead(sensorPin);     // read the analog input
  int rate = map(sensorReading, 200,800,minDuration, maxDuration);
  rate = constrain(rate, minDuration,maxDuration);     // constrain the value

  digitalWrite(outputPin, HIGH);    // set the LED on
  delay(rate);                      // wait duration dependent on light level
  digitalWrite(outputPin, LOW);     // set the LED off
  delay(rate);
}
```

See Also

See Recipe 3.5 for details on using the `constrain` function.

See Recipe 5.7 for a discussion on the `map` function.

If you are interested in creating sounds, see Chapter 9 for a full discussion on audio output with Arduino.

CHAPTER 2

Making the Sketch Do Your Bidding

2.0 Introduction

Though much of an Arduino project will involve integrating the Arduino board with supporting hardware, you need to be able to tell the board what to do with the rest of your project. This chapter introduces core elements of Arduino programming, shows nonprogrammers how to use common language constructs, and provides an overview of the language syntax for readers who are not familiar with C or C++, the language that Arduino uses.

Since making the examples interesting requires making Arduino do something, the recipes use physical capabilities of the board that are explained in detail in later chapters. If any of the code in this chapter is not clear, feel free to jump forward, particularly to Chapter 4 for more on serial output and Chapter 5 for more on using digital and analog pins. You don't need to understand all the code in the examples, though, to see how to perform the specific capabilities that are the focus of the recipes. Here are some of the more common functions used in the examples that are covered in the next few chapters:

`Serial.println(value);`
: Prints the value to the Arduino IDE's Serial Monitor so you can view Arduino's output on your computer; see Recipe 4.1.

`pinMode(pin, mode);`
: Configures a digital pin to read (input) or write (output) a digital value; see the introduction to Chapter 5.

`digitalRead(pin);`
: Reads a digital value (`HIGH` or `LOW`) on a pin set for input; see Recipe 5.1.

`digitalWrite(pin, value);`
: Writes the digital value (`HIGH` or `LOW`) to a pin set for output; see Recipe 5.1.

2.1 Structuring an Arduino Program

Problem

You are new to programming and want to understand the building blocks of an Arduino program.

Solution

Programs for Arduino are usually referred to as *sketches*; the first users were artists and designers and *sketch* highlights the quick and easy way to have an idea realized. The terms *sketch* and *program* are interchangeable. Sketches contain code—the instructions the board will carry out. Code that needs to run only once (such as to set up the board for your application) must be placed in the `setup` function. Code to be run continuously after the initial setup has finished goes into the `loop` function. Here is a typical sketch:

```
const int ledPin =  13;     // LED connected to digital pin 13

 // The setup() method runs once, when the sketch starts
 void setup()
 {
   pinMode(ledPin, OUTPUT);      // initialize the digital pin as an output
 }

 // the loop() method runs over and over again,
 void loop()
 {
   digitalWrite(ledPin, HIGH);   // turn the LED on
   delay(1000);                  // wait a second
   digitalWrite(ledPin, LOW);    // turn the LED off
   delay(1000);                  // wait a second
 }
```

When the Arduino IDE finishes uploading the code, and every time you power on the board after you've uploaded this code, it starts at the top of the sketch and carries out the instructions sequentially. It runs the code in `setup` once and then goes through the code in `loop`. When it gets to the end of `loop` (marked by the closing bracket, }) it goes back to the beginning of `loop`.

Discussion

This example continuously flashes an LED by writing `HIGH` and `LOW` outputs to a pin. See Chapter 5 to learn more about using Arduino pins. When the sketch begins, the code in `setup` sets the pin mode (so it's capable of lighting an LED). After the code in `setup` is completed, the code in `loop` is repeatedly called (to flash the LED) for as long as the Arduino board is powered on.

You don't need to know this to write Arduino sketches, but experienced C/C++ programmers may wonder where the expected `main()` entry point function has gone. It's there, but it's hidden under the covers by the Arduino build environment. The build process creates an intermediate file that includes the sketch code and the following additional statements:

```
int main(void)
{
    init();

    setup();

    for (;;)
        loop();

    return 0;
}
```

The first thing that happens is a call to an `init()` function that initializes the Arduino hardware. Next, your sketch's `setup()` function is called. Finally, your `loop()` function is called over and over. Because the `for` loop never terminates, the `return` statement is never executed.

See Also

Recipe 1.4 explains how to upload a sketch to the Arduino board.

Chapter 17 and *http://www.arduino.cc/en/Hacking/BuildProcess* provide more on the build process.

2.2 Using Simple Primitive Types (Variables)

Problem

Arduino has different types of variables to efficiently represent values. You want to know how to select and use these Arduino data types.

Solution

Although the `int` (short for *integer*, a 16-bit value in Arduino) data type is the most common choice for the numeric values encountered in Arduino applications, you can use Table 2-1 to determine the data type that fits the range of values your application expects.

Table 2-1. Arduino data types

Numeric types	Bytes	Range	Use
`int`	2	–32768 to 32767	Represents positive and negative integer values.
`unsigned int`	2	0 to 65535	Represents only positive values; otherwise, similar to `int`.
`long`	4	–2147483648 to 2147483647	Represents a very large range of positive and negative values.
`unsigned long`	4	4294967295	Represents a very large range of positive values.
`float`	4	3.4028235E+38 to –3.4028235E+38	Represents numbers with fractions; use to approximate real-world measurements.
`double`	4	Same as `float`	In Arduino, `double` is just another name for `float`.
`boolean`	1	`false` (0) or `true` (1)	Represents true and false values.
`char`	1	–128 to 127	Represents a single character. Can also represent a signed value between –128 and 127.
`byte`	1	0 to 255	Similar to `char`, but for unsigned values.

Other types	Use
`String`	Represents arrays of `char`s (characters) typically used to contain text.
`void`	Used only in function declarations where no value is returned.

Discussion

Except in situations where maximum performance or memory efficiency is required, variables declared using `int` will be suitable for numeric values if the values do not exceed the range (shown in the first row in Table 2-1) and if you don't need to work with fractional values. Most of the official Arduino example code declares numeric variables as `int`. But sometimes you do need to choose a type that specifically suits your application.

Sometimes you need negative numbers and sometimes you don't, so numeric types come in two varieties: `signed` and `unsigned`. `unsigned` values are always positive. Variables without the keyword `unsigned` in front are signed so that they can represent negative and positive values. One reason to use `unsigned` values is when the range of `signed` values will not fit the range of the variable (an unsigned variable has twice the capacity of a signed variable). Another reason programmers choose to use `unsigned` types is to clearly indicate to people reading the code that the value expected will never be a negative number.

`boolean` types have two possible values: `true` or `false`. They are commonly used for things like checking the state of a switch (if it's pressed or not). You can also use `HIGH` and `LOW` as equivalents to `true` and `false` where this makes more sense; `digitalWrite(pin, HIGH)` is a more expressive way to turn on an LED than `digitalWrite(pin, true)` or `digitalWrite(pin,1)`, although all of these are treated identically when the sketch actually runs, and you are likely to come across all of these forms in code posted on the Web.

See Also

The Arduino reference at *http://www.arduino.cc/en/Reference/HomePage* provides details on data types.

2.3 Using Floating-Point Numbers

Problem

Floating-point numbers are used for values expressed with decimal points (this is the way to represent fractional values). You want to calculate and compare these values in your sketch.

Solution

The following code shows how to declare floating-point variables, illustrates problems you can encounter when comparing floating-point values, and demonstrates how to overcome them:

```
/*
 * Floating-point example
 * This sketch initialized a float value to 1.1
 * It repeatedly reduces the value by 0.1 until the value is 0
 */


float  value = 1.1;

void setup()
{
  Serial.begin(9600);
}

void loop()
{
  value = value - 0.1;    // reduce value by 0.1 each time through the loop
  if( value == 0)
     Serial.println("The value is exactly zero");
  else if(almostEqual(value, 0))
  {
    Serial.print("The value ");
    Serial.print(value,7); // print to 7 decimal places
    Serial.println(" is almost equal to zero");
  }
  else
    Serial.println(value);

  delay(100);
```

```
}

// returns true if the difference between a and b is small
// set value of DELTA to the maximum difference considered to be equal
boolean almostEqual(float a, float b)
{
  const float DELTA = .00001; // max difference to be almost equal
  if (a == 0) return fabs(b) <= DELTA;
  if (b == 0) return fabs(a) <= DELTA;
  return fabs((a - b) / max(fabs(a), fabs(b))) <= DELTA ;
}
```

Discussion

Floating-point math is not exact, and values returned can have a small approximation error. The error occurs because floating-point values cover a huge range, so the internal representation of the value can only hold an approximation. Because of this, you need to test if the values are within a range of tolerance rather than exactly equal.

The Serial Monitor output from this sketch is as follows:

```
1.00
0.90
0.80
0.70
0.60
0.50
0.40
0.30
0.20
0.10
The value -0.0000001 is almost equal to zero
-0.10
-0.20
```

The output continues to produce negative numbers.

You may expect the code to print `"The value is exactly zero"` after `value` is `0.1` and then 0.1 is subtracted from this. But `value` never equals exactly zero; it gets very close, but that is not good enough to pass the test: `if (value == 0)`. This is because the only memory-efficient way that floating-point numbers can contain the huge range in values they can represent is by storing an approximation of the number.

The solution to this is to check if a variable is close to the desired value, as shown in this recipe's Solution.

The `almostEqual` function tests if the variable `value` is within 0.00001 of the desired target and returns true if so. The acceptable range is set with the constant `DELTA`, you can change this to smaller or larger values as required. The function named `fabs` (short for *floating-point absolute value*) returns the absolute value of a floating-point variable and this is used to test the difference between the given parameters.

Floating point approximates numbers because it only uses 32 bits to hold all values within a huge range. Eight bits are used for the decimal multiplier (the exponent), and that leaves 24 bits for the sign and value—only enough for seven significant decimal digits.

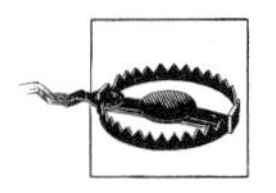

Although `float` and `double` are exactly the same on Arduino, `doubles` do have a higher precision on many other platforms. If you are importing code that uses `float` and `double` from another platform, check that there is sufficient precision for your application.

See Also

The Arduino reference for `float`: *http://www.arduino.cc/en/Reference/Float*.

2.4 Working with Groups of Values

Problem

You want to create and use a group of values (called *arrays*). The arrays may be a simple list or they could have two or more dimensions. You want to know how to determine the size of the array and how to access the elements in the array.

Solution

This sketch creates two arrays: an array of integers for pins connected to switches and an array of pins connected to LEDs, as shown in Figure 2-1:

```
/*
 array sketch
 an array of switches controls an array of LEDs
 see Chapter 5 for more on using switches
 see Chapter 7 for information on LEDs
 */

int inputPins[] = {2,3,4,5};  // create an array of pins for switch inputs

int ledPins[] = {10,11,12,13};  // create array of output pins for LEDs

void setup()
{
  for(int index = 0; index < 4; index++)
  {
    pinMode(ledPins[index], OUTPUT);          // declare LED as output
    pinMode(inputPins[index], INPUT);         // declare pushbutton as input

    digitalWrite(inputPins[index],HIGH);      // enable pull-up resistors
    // (see Recipe 5.2)
  }
}
```

```
void loop(){
  for(int index = 0; index < 4; index++)
  {
    int val = digitalRead(inputPins[index]);  // read input value
    if (val == LOW)                           // check if the switch is pressed
    {
      digitalWrite(ledPins[index], HIGH); // turn LED on if switch is pressed
    }
    else
    {
      digitalWrite(ledPins[index], LOW);  // turn LED off
    }
  }
}
```

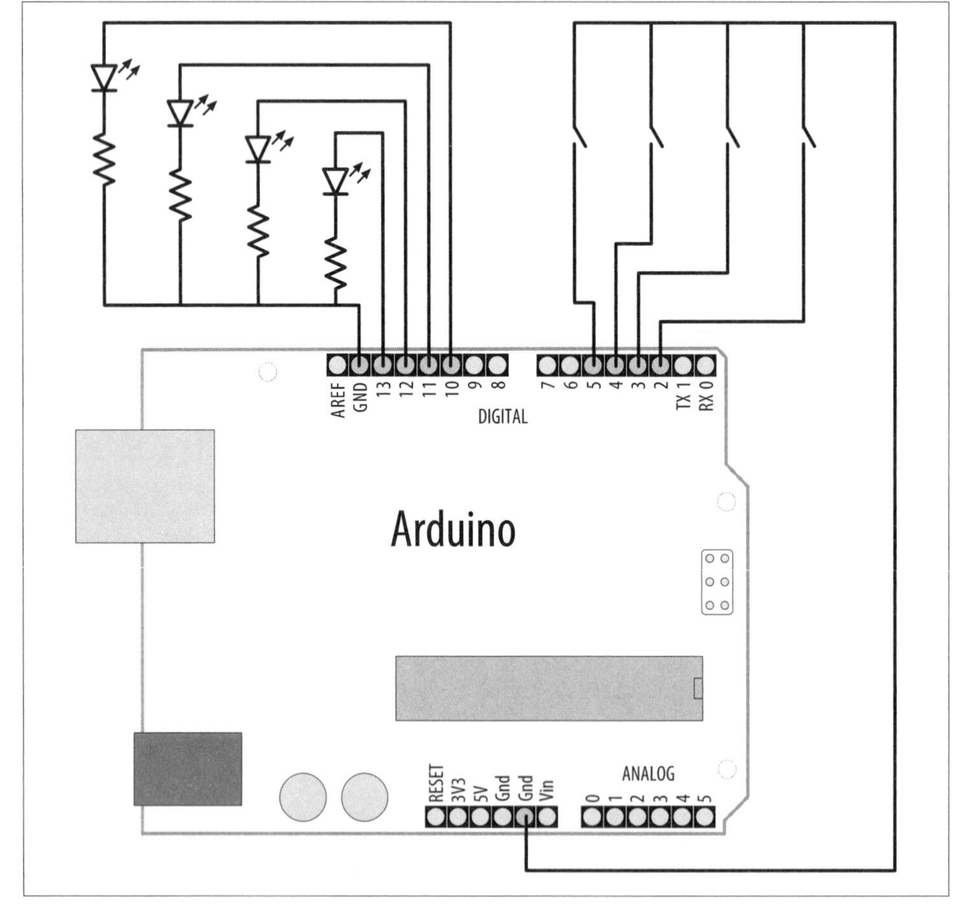

Figure 2-1. Connections for LEDs and switches

Discussion

Arrays are collections of consecutive variables of the same type. Each variable in the collection is called an *element*. The number of elements is called the *size* of the array.

The Solution demonstrates a common use of arrays in Arduino code: storing a collection of pins. Here the pins connect to switches and LEDs (a topic covered in more detail in Chapter 5). The important parts of this example are the declaration of the array and access to the array elements.

The following line of code declares (creates) an array of integers with four elements and initializes each element. The first element is set equal to 2, the second to 3, and so on:

```
int inputPins[] = {2,3,4,5};
```

If you don't initialize values when you declare an array (for example, when the values will only be available when the sketch is running), you must change each element individually. You can declare the array as follows:

```
int inputPins[4];
```

This declares an array of four elements with the initial value of each element set to zero. The number within the square brackets ([]) is the size, and this sets the number of elements. This array has a size of four and can hold, at most, four integer values. The size can be omitted if array declaration contains initializers (as shown in the first example) because the compiler figures out how big to make the array by counting the number of initializers.

The first element of the array is `element[0]`:

```
int firstElement = inputPins[0];  // this is the first element

inputPins[0] = 2; // set the value of this element equal to 2
```

The last element is one less than the size, so in the preceding example, with a size of four, the last element is element 3:

```
int lastElement = inputPins[3]; // this is the last element
```

It may seem odd that an array with a size of four has the last element accessed using `array[3]`, but because the first element is `array[0]`, the four elements are:

```
inputPins[0],inputPins[1],inputPins[2],inputPins[3]
```

In the previous sketch, the four elements are accessed using a `for` loop:

```
for(int index = 0; index < 4; index++)
{
    //get the pin number by accessing each element in the pin arrays
    pinMode(ledPins[index], OUTPUT);         // declare LED as output
    pinMode(inputPins[index], INPUT);        // declare pushbutton as input
}
```

This loop will step through the variable `index` with values starting at 0 and ending at 3. It is a common mistake to accidentally access an element that is beyond the actual size of the array. This is a bug that can have many different symptoms and care must be taken to avoid it. One way to keep your loops under control is to set the size of an array by using a constant as follows:

```
const int PIN_COUNT = 4;  // define a constant for the number of elements
int inputPins[PIN_COUNT] = {2,3,4,5};

  for(int index = 0; index < PIN_COUNT; index++)
    pinMode(inputPins[index], INPUT);
```

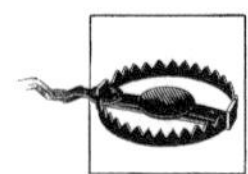

The compiler will not report an error if you accidentally try to store or read beyond the size of the array. You must be careful that you only access elements that are within the bounds you have set. Using a constant to set the size of an array and in code referring to its elements helps your code stay within the bounds of the array.

Another use of arrays is to hold a string of text characters. In Arduino code, these are called *character strings* (*strings* for short). A character string consists of one or more characters, followed by the null character (the value 0) to indicate the end of the string.

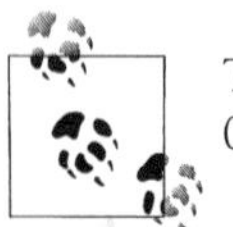

The null at the end of a character string is not the same as the character 0. The null has an ASCII value of 0, whereas 0 has an ASCII value of 48.

Methods to use strings are covered in Recipes 2.5 and 2.6.

See Also

Recipe 5.2; Recipe 7.1

2.5 Using Arduino String Functionality

Problem

You want to manipulate text. You need to copy it, add bits together, and determine the number of characters.

Solution

The previous recipe mentioned how arrays of characters can be used to store text: these character arrays are usually called strings. Arduino has a capability called `String` that adds rich functionality for storing and manipulating text.

The word *String* with an uppercase *S* refers to the Arduino text capability provided by the Arduino String library. The word *string* with a lowercase *s* refers to the group of characters rather than the Arduino `String` functionality.

This recipe demonstrates how to use Arduino `Strings`.

The `String` capability was introduced in version 0019 alpha (older than 1.0) of Arduino. If you are using an older version, you can use the Text-String library; see the link at the end of this recipe.

Load the following sketch onto your board, and open the Serial Monitor to view the results:

```
/*
  Basic_Strings sketch
 */

String text1 = "This string";
String text2 = " has more text";
String text3;  // to be assigned within the sketch

void setup()
{
  Serial.begin(9600);

  Serial.print( text1);
  Serial.print(" is ");
  Serial.print(text1.length());
  Serial.println(" characters long.");

  Serial.print("text2 is ");
  Serial.print(text2.length());
  Serial.println(" characters long.");

  text1.concat(text2);
  Serial.println("text1 now contains: ");
  Serial.println(text1);
}

void loop()
{
}
```

Discussion

This sketch creates three variables of type `String`, called `text1`, `text2`, and `text3`. Variables of type `String` have built-in capabilities for manipulating text. The statement `text1.length()` returns (provides the value of) the length (number of characters) in the string `text1`.

`text1.concat(text2)` combines the contents of strings; in this case, it appends the contents of `text2` to the end of `text1` (`concat` is short for *concatenate*).

The Serial Monitor will display the following:

```
This string is 11 characters long.
text2 is 14 characters long.
text1 now contains:
This string has more text
```

Another way to combine strings is to use the string addition operator. Add these two lines to the end of the `setup` code:

```
text3 = text1 + " and more";
Serial.println(text3);
```

The new code will result in the Serial Monitor adding the following line to the end of the display:

```
This is a string with more text and more
```

You can use the `indexOf` and `lastIndexOf` functions to find an instance of a particular character in a string.

Because the `String` class is a recent addition to Arduino, you will come across a lot of code that uses arrays of characters rather than the `String` type. See Recipe 2.6 for more on using arrays of characters without the help of the Arduino `String` functionality.

If you see a line such as the following:

```
char oldString[] = "this is a character array";
```

the code is using C-style character arrays (see Recipe 2.6). If the declaration looks like this:

```
String newString = "this is a string object";
```

the code uses Arduino `Strings`. To convert a C-style character array to an Arduino `String`, just assign the contents of the array to the `String` object:

```
char oldString[] = "I want this character array in a String object";
String newString = oldString;
```

To use any of the functions listed in Table 2-2, you need to invoke them upon an existing string object, as in this example:

```
int len = myString.length();
```

Table 2-2. Brief overview of Arduino String functions

`charAt(n)`	Returns the *n*th character of the `String`
`compareTo(S2)`	Compares the `String` to the given `String` S2
`concat(S2)`	Returns a new `String` that is the combination of the `String` and S2
`endsWith(S2)`	Returns true if the `String` ends with the characters of S2
`equals(S2)`	Returns true if the `String` is an exact match for S2 (case-sensitive)
`equalsIgnoreCase(S2)`	Same as `equals` but is not case-sensitive
`getBytes(buffer,len)`	Copies `len`(gth) characters into the supplied byte buffer
`indexOf(S)`	Returns the index of the supplied `String` (or character) or –1 if not found
`lastIndexOf(S)`	Same as `indexOf` but starts from the end of the `String`
`length()`	Returns the number of characters in the `String`
`replace(A,B)`	Replaces all instances of `String` (or character) A with B
`setCharAt(index,c)`	Stores the character c in the `String` at the given index
`startsWith(S2)`	Returns true if the `String` starts with the characters of S2
`substring(index)`	Returns a `String` with the characters starting from index to the end of the `String`
`substring(index,to)`	Same as above, but the substring ends at the character location before the 'to' position
`toCharArray(buffer,len)`	Copies up to `len` characters of the `String` to the supplied buffer
`toInt()`	Returns the integer value of the numeric digits in the `String`
`toLowerCase()`	Returns a `String` with all characters converted to lowercase
`toUpperCase()`	Returns a `String` with all characters converted to uppercase
`trim()`	Returns a `String` with all leading and trailing whitespace removed

See the Arduino reference pages for more about the usage and variants for these functions.

Choosing between Arduino Strings and C character arrays

Arduino's built-in `String` datatype is easier to use than C character arrays, but this is achieved through complex code in the String library, which makes more demands on your Arduino, and is, by nature, more prone to problems.

The `String` datatype is so flexible because it makes use of *dynamic memory allocation*. That is, when you create or modify a `String`, Arduino requests a new region of memory from the C library, and when you're done using a `String`, Arduino needs to release that memory. This usually works smoothly, but in practice there are many cracks through which memory can leak. Bugs in the String library can result in some or all of the memory not being returned properly. When this happens, the memory available to Arduino will slowly decrease (until you reboot the Arduino). And even if there were no memory leaks, it's complicated to write code to check if a `String` request failed due to insufficient memory (the `String` functions mimic those in Processing, but unlike that platform, Arduino does not have runtime error exception handling). Running out of dynamic memory is a bug that can be very difficult to track down because the sketch can run without problems for days or weeks before it starts misbehaving through insufficient memory.

If you use C character arrays, you are in control of memory usage: you're allocating a fixed (static) amount of memory at compile time so you don't get memory leaks. Your Arduino sketch will have the same amount of memory available to it all the time it's running. And if you do try to allocate more memory than available, finding the cause is easier because there are tools that tell you how much static memory you have allocated (see the reference to `avr-objdump` in Recipe 17.1).

However, with C character arrays, it's easier for you to have another problem: C will not prevent you from modifying memory beyond the bounds of the array. So if you allocate an array as `myString[4]`, and assign `myString[4] = 'A'` (remember, `myString[3]` is the end of the array), nothing will stop you from doing this. But who knows what piece of memory `myString[4]` refers to? And who knows whether assigning `'A'` to that memory location will cause you a problem? Most likely, it will cause your sketch to misbehave.

So, Arduino's built-in String library, by virtue of using dynamic memory, runs the risk of eating up your available memory. C's character arrays require care on your part to ensure that you do not exceed the bounds of the arrays you use. So use Arduino's built-in String library if you need rich text handling capability and you won't be creating and modifying `String`s over and over again. If you need to create and modify them in a loop that is constantly repeating, you're better off allocating a large C character array and writing your code carefully so you don't write past the bounds of that array.

Another instance where you may prefer C character arrays over Arduino `Strings` is in large sketches that need most of the available RAM or flash. The Arduino `String ToInt` example code uses almost 2 KB more flash than equivalent code using a C character array and `atoi` to convert to an `int`. The Arduino `String` version also requires a little more RAM to store allocation information in addition to the actual string.

If you do suspect that the String library, or any other library that makes use of dynamically allocated memory, might be leaking memory, you can determine how much memory is free at any given time; see Recipe 17.2. Check the amount of RAM when your sketch starts, and monitor it to see whether it's decreasing over time. If you suspect a problem with the String library, search the list of open bugs (*http://code.google.com/p/arduino/issues/list*) for "String."

See Also

The Arduino distribution provides `String` example sketches (File→Examples→Strings).

The `String` reference page can be found at *http://arduino.cc/en/Reference/StringObject*.

Tutorials for the new String library are available at *http://arduino.cc/en/Tutorial/HomePage*, and a tutorial for the original String library (only needed if you are using a version of Arduino older than 0019 alpha) is available at *http://www.arduino.cc/en/Tutorial/TextString*.

2.6 Using C Character Strings

Problem

You want to understand how to use raw character strings: you want to know how to create a string, find its length, and compare, copy, or append strings. The core C language does not support the Arduino-style `String` capability, so you want to understand code from other platforms written to operate with primitive character arrays.

Solution

Arrays of characters are sometimes called *character strings* (or simply *strings* for short). Recipe 2.4 describes Arduino arrays in general. This recipe describes functions that operate on character strings.

You declare strings like this:

```
char stringA[8]; // declare a string of up to 7 chars plus terminating null
char stringB[8]  = "Arduino"; // as above and init(ialize) the string to "Arduino"
char stringC[16] = "Arduino"; // as above, but string has room to grow
char stringD[ ]  = "Arduino"; // the compiler inits the string and calculates size
```

Use `strlen` (short for *string length*) to determine the number of characters before the terminating null:

```
int length = strlen(string);  // return the number of characters in the string
```

`length` will be 0 for `stringA` and 7 for the other strings shown in the preceding code. The null that indicates the end of the string is not counted by `strlen`.

Use `strcpy` (short for *string copy*) to copy one string to another:

```
strcpy(destination, source);   // copy string source to destination
```

Use `strncpy` to limit the number of characters to copy (useful to prevent writing more characters than the destination string can hold). You can see this used in Recipe 2.7:

```
// copy up to 6 characters from source to destination
strncpy(destination, source, 6);
```

Use `strcat` (short for *string concatenate*) to append one string to the end of another:

```
// append source string to the end of the destination string
strcat(destination, source);
```

Always make sure there is enough room in the destination when copying or concatenating strings. Don't forget to allow room for the terminating null.

Use `strcmp` (short for *string compare*) to compare two strings. You can see this used in Recipe 2.7:

```
if(strcmp(str, "Arduino") == 0)
   // do something if the variable str is equal to "Arduino"
```

Discussion

Text is represented in the Arduino environment using an array of characters called strings. A string consists of a number of characters followed by a null (the value 0). The null is not displayed, but it is needed to indicate the end of the string to the software.

See Also

See one of the many online C/C++ reference pages, such as *http://www.cplusplus.com/reference/clibrary/cstring/* and *http://www.cppreference.com/wiki/string/c/start*.

2.7 Splitting Comma-Separated Text into Groups

Problem

You have a string that contains two or more pieces of data separated by commas (or any other separator). You want to split the string so that you can use each individual part.

Solution

This sketch prints the text found between each comma:

```
/*
 * SplitSplit sketch
 * split a comma-separated string
 */

String  text = "Peter,Paul,Mary";  // an example string
String  message = text; // holds text not yet split
int     commaPosition;  // the position of the next comma in the string

void setup()
{
  Serial.begin(9600);

  Serial.println(message);          // show the source string
  do
  {
```

```
        commaPosition = message.indexOf(',');
        if(commaPosition != -1)
        {
            Serial.println( message.substring(0,commaPosition));
            message = message.substring(commaPosition+1, message.length());
        }
        else
        {  // here after the last comma is found
           if(message.length() > 0)
             Serial.println(message);  // if there is text after the last comma,
                                       // print it
        }
     }
     while(commaPosition >=0);
}

void loop()
{
}
```

The Serial Monitor will display the following:

```
Peter,Paul,Mary
Peter
Paul
Mary
```

Discussion

This sketch uses `String` functions to extract text from between commas. The following code:

```
commaPosition = message.indexOf(',');
```

sets the variable `commaPosition` to the position of the first comma in the `String` named `message` (it will be set to –1 if no comma is found). If there is a comma, the `substring` function is used to print the text from the beginning of the string up to, but excluding, the comma. The text that was printed, and its trailing comma, are removed from `message` in this line:

```
message = message.substring(commaPosition+1, message.length());
```

`substring` returns a string starting from `commaPosition+1` (the position just after the first comma) up to the length of the message. This results in that message containing only the text following the first comma. This is repeated until no more commas are found (`commaPosition` will be equal to –1).

If you are an experienced programmer, you can also use the low-level functions that are part of the standard C library. The following sketch has similar functionality to the preceding one using Arduino strings:

```
/*
 * SplitSplit sketch
 * split a comma-separated string
 */
```

```
const int  MAX_STRING_LEN = 20; // set this to the largest string
                                // you will process

char stringList[] = "Peter,Paul,Mary"; // an example string

char stringBuffer[MAX_STRING_LEN+1]; // a static buffer for computation and output

void setup()
{
  Serial.begin(9600);
}

void loop()
{
  char *str;
  char *p;
  strncpy(stringBuffer, stringList, MAX_STRING_LEN); // copy source string
  Serial.println(stringBuffer);                       // show the source string

  for( str = strtok_r(stringBuffer, ",", &p);         // split using comma
       str;                                           // loop while str is not null
       str = strtok_r(NULL, ",", &p)                  // get subsequent tokens
     )
  {
    Serial.println(str);
  }
    delay(5000);
}
```

The core functionality comes from the function named `strtok_r` (the name of the version of `strtok` that comes with the Arduino compiler). The first time you call `strtok_r`, you pass it the string you want to *tokenize* (separate into individual values). But `strtok_r` overwrites the characters in this string each time it finds a new token, so it's best to pass a copy of the string as shown in this example. Each call that follows uses a `NULL` to tell the function that it should move on to the next token. In this example, each token is printed to the serial port.

If your tokens consist only of numbers, see Recipe 4.5. This shows how to extract numeric values separated by commas in a stream of serial characters.

See Also

See *http://www.nongnu.org/avr-libc/user-manual/group__avr__string.html* for more on C string functions such as `strtok_r` and `strcmp`.

Recipe 2.5; online references to the C/C++ functions `strtok_r` and `strcmp`.

2.8 Converting a Number to a String

Problem

You need to convert a number to a string, perhaps to show the number on an LCD or other display.

Solution

The `String` variable will convert numbers to strings of characters automatically. You can use literal values, or the contents of a variable. For example, the following code will work:

```
String myNumber = 1234;
```

As will this:

```
int value = 127;
String myReadout = "The reading was ";
myReadout.concat(value);
```

Or this:

```
int value = 127;
String myReadout = "The reading was ";
myReadout += value;
```

Discussion

If you are converting a number to display as text on an LCD or serial device, the simplest solution is to use the conversion capability built in to the LCD and Serial libraries (see Recipe 4.2). But perhaps you are using a device that does not have this built-in support (see Chapter 13) or you want to manipulate the number as a string in your sketch.

The Arduino `String` class automatically converts numerical values when they are assigned to a `String` variable. You can combine (concatenate) numeric values at the end of a string using the `concat` function or the string + operator.

The + operator is used with number types as well as strings, but it behaves differently with each.

The following code results in `number` having a value of `13`:

```
int number = 12;
number += 1;
```

With a `String`, as shown here:

```
String textNumber = "12";
textNumber += 1;
```

textNumber is the text string "121".

Prior to the introduction of the String class, it was common to find Arduino code using the itoa or ltoa function. The names come from "integer to ASCII" (itoa) and "long to ASCII" (ltoa). The String version described earlier is easier to use, but the following can be used if you prefer working with C character arrays as described in Recipe 2.6.

itoa or ltoa take three parameters: the value to convert, the buffer that will hold the output string, and the number base (10 for a decimal number, 16 for hex, and 2 for binary).

The following sketch illustrates how to convert numeric values using ltoa:

```
/*
 * NumberToString
 * Creates a string from a given number
 */

void setup()
{
  Serial.begin(9600);
}

char buffer[12];  // long data type has 11 characters (including the
                  // minus sign) and a terminating null
void loop()
{
  long value = 12345;
  ltoa(value, buffer, 10);
  Serial.print( value);
  Serial.print(" has  ");
  Serial.print(strlen(buffer));
  Serial.println(" digits");
  value = 123456789;
  ltoa(value, buffer, 10);
  Serial.print( value);
  Serial.print(" has  ");
  Serial.print(strlen(buffer));
  Serial.println(" digits");
  delay(1000);
}
```

Your buffer must be large enough to hold the maximum number of characters in the string. For 16-bit base 10 (decimal) integers, that is seven characters (five digits, a possible minus sign, and a terminating 0 that always signifies the end of a string); 32-bit long integers need 12 character buffers (10 digits, the minus sign, and the terminating 0). No warning is given if you exceed the buffer size; this is a bug that can cause all kinds of strange symptoms, because the overflow will corrupt some other part of memory that may be used by your program. The easiest way to handle this is to always use a 12-character buffer and always use ltoa because this will work on both 16-bit and 32-bit values.

2.9 Converting a String to a Number

Problem

You need to convert a string to a number. Perhaps you have received a value as a string over a communication link and you need to use this as an integer or floating-point value.

Solution

There are a number of ways to solve this. If the string is received as serial data, it can be converted on the fly as each character is received. See Recipe 4.3 for an example of how to do this using the serial port.

Another approach to converting text strings representing numbers is to use the C language conversion function called `atoi` (for `int` variables) or `atol` (for `long` variables).

This code fragment terminates the incoming digits on any character that is not a digit (or if the buffer is full). For this to work, though, you'll need to enable the newline option in the Serial Monitor or type some other terminating character:

```
/*
 * StringToNumber
 * Creates a number from a string
 */

const int ledPin = 13; // pin the LED is connected to

int   blinkDelay;     // blink rate determined by this variable
char strValue[6];     // must be big enough to hold all the digits and the
                      // 0 that terminates the string
int index = 0;        // the index into the array storing the received digits

void setup()
{
 Serial.begin(9600);
 pinMode(ledPin,OUTPUT); // enable LED pin as output
}

void loop()
{
  if( Serial.available())
  {
    char ch = Serial.read();
    if(index < 5 && isDigit(ch) ){
      strValue[index++] = ch; // add the ASCII character to the string;
    }
    else
    {
      // here when buffer full or on the first non digit
      strValue[index] = 0;        // terminate the string with a 0
      blinkDelay = atoi(strValue);  // use atoi to convert the string to an int
      index = 0;
    }
```

```
  }
  blink();
}

void blink()
{
   digitalWrite(ledPin, HIGH);
   delay(blinkDelay/2);  // wait for half the blink period
   digitalWrite(ledPin, LOW);
   delay(blinkDelay/2);  // wait for the other half
}
```

Discussion

The obscurely named `atoi` (for ASCII to `int`) and `atol` (for ASCII to `long`) functions convert a string into integers or long integers. To use them, you have to receive and store the entire string in a character array before you can call the conversion function. The code creates a character array named `strValue` that can hold up to five digits (it's declared as `char strValue[6]` because there must be room for the terminating null). It fills this array with digits from `Serial.read` until it gets the first character that is not a valid digit. The array is terminated with a null and the `atoi` function is called to convert the character array into the variable `blinkRate`.

A function called `blink` is called that uses the value stored in `blinkDelay`.

As mentioned in the warning in Recipe 2.4, you must be careful not to exceed the bound of the array. If you are not sure how to do that, see the Discussion section of that recipe.

Arduino release 22 added the `toInt` method to convert a String to an integer:

```
String aNumber = "1234";
int value = aNumber.toInt();
```

Arduino 1.0 added the `parseInt` method that can be used to get integer values from Serial and Ethernet (or any object that derives from the `Stream` class). The following fragment will convert sequences of numeric digits into numbers. It is similar to the solution fragment but does not need a buffer (and does not limit the number of digits to 5):

```
int   blinkDelay;     // blink rate determined by this variablevoid loop()
{
  if( Serial.available())
  {
    blinkRate = Serial.parseInt();
  }
  blink();
  }
```

Stream-parsing methods such as `parseInt` use a timeout to return control to your sketch if data does not arrive within the desired interval. The default timeout is one second but this can be changed by calling the `setTimeout` method:

```
Serial.setTimeout(1000 * 60); // wait up to one minute
```

`parseInt` (and all other stream methods) will return whatever value was obtained prior to the timeout if no delimiter was received. The return value will consist of whatever values were collected; if no digits were received, the return will be zero. Arduino 1.0 does not have a way to determine if a parse method has timed out, but this capability is planned for a future release.

See Also

Documentation for `atoi` can be found at: *http://www.nongnu.org/avr-libc/user-manual/group__avr__stdlib.html*.

There are many online C/C++ reference pages covering these low-level functions, such as *http://www.cplusplus.com/reference/clibrary/cstdlib/atoi/* or *http://www.cppreference.com/wiki/string/c/atoi*.

See Recipe 4.3 and Recipe 4.5 for more about using `parseInt` with Serial.

2.10 Structuring Your Code into Functional Blocks

Problem

You want to know how to add functions to a sketch, and the correct amount of functionality to go into your functions. You also want to understand how to plan the overall structure of the sketch.

Solution

Functions are used to organize the actions performed by your sketch into functional blocks. Functions package functionality into well-defined *inputs* (information given to a function) and *outputs* (information provided by a function) that make it easier to structure, maintain, and reuse your code. You are already familiar with the two functions that are in every Arduino sketch: `setup` and `loop`. You create a function by declaring its *return type* (the information it provides), its name, and any optional parameters (values) that the function will receive when it is called.

The terms *functions* and *methods* are used to refer to well-defined blocks of code that can be called as a single entity by other parts of a program. The C language refers to these as functions. Object-oriented languages such as C++ that expose functionality through classes tend to use the term method. Arduino uses a mix of styles (the example sketches tend to use C-like style, libraries tend to be written to expose C++ class methods). In this book, the term function is usually used unless the code is exposed through a class. Don't worry; if that distinction is not clear to you, treat both terms as the same.

Here is a simple function that just blinks an LED. It has no parameters and doesn't return anything (the `void` preceding the function indicates that nothing will be returned):

```
// blink an LED once
void blink1()
{
   digitalWrite(13,HIGH);    // turn the LED on
   delay(500);               // wait 500 milliseconds
   digitalWrite(13,LOW);     // turn the LED off
   delay(500);               // wait 500 milliseconds
}
```

The following version has a parameter (the integer named `count`) that determines how many times the LED will flash:

```
// blink an LED the number of times given in the count parameter
void blink2(int count)
{
  while(count > 0 )  // repeat until count is no longer greater than zero
  {
    digitalWrite(13,HIGH);
    delay(500);
    digitalWrite(13,LOW);
    delay(500);
    count = count -1; // decrement count
  }
}
```

Experienced programmers will note that both functions could be blink because the compiler will differentiate them by the type of values used for the parameter. This behavior is called *function overloading*. The Arduino `print` discussed in Recipe 4.2 is a common example. Another example of overloading is in the discussion of Recipe 4.6.

That version checks to see if the value of `count` is `0`. If not, it blinks the LED and then reduces the value of `count` by one. This will be repeated until `count` is no longer greater than `0`.

A *parameter* is sometimes referred to as an *argument* in some documentation. For practical purposes, you can treat these terms as meaning the same thing.

Here is an example sketch that takes a parameter and returns a value. The parameter determines the length of the LED on and off times (in milliseconds). The function continues to flash the LED until a button is pressed, and the number of times the LED flashed is returned from the function:

```
/*
  blink3 sketch
  Demonstrates calling a function with a parameter and returning a value.
  Uses the same wiring as the pull-up sketch from
  Recipe 5.2

  The LED flashes when the program starts and stops when a switch connected
  to digital pin 2 is pressed.
  The program prints the number of times that the LED flashes.
 */

const int ledPin = 13;          // output pin for the LED
const int inputPin = 2;         // input pin for the switch

void setup() {
  pinMode(ledPin, OUTPUT);
  pinMode(inputPin, INPUT);
  digitalWrite(inputPin,HIGH); // use internal pull-up resistor (Recipe 5.2)
  Serial.begin(9600);
}

void loop(){
  Serial.println("Press and hold the switch to stop blinking");
  int count = blink3(250); // blink the LED 250ms on and 250ms off
  Serial.print("The number of times the switch blinked was ");
  Serial.println(count);
}

// blink an LED using the given delay period
// return the number of times the LED flashed
int blink3(int period)
{
  int result = 0;
  int switchVal = HIGH; //with pull-ups, this will be high when switch is up

  while(switchVal == HIGH)  // repeat this loop until switch is pressed
                            // (it will go low when pressed)
  {
    digitalWrite(13,HIGH);
    delay(period);
    digitalWrite(13,LOW);
    delay(period);
    result = result + 1; // increment the count
```

```
    switchVal = digitalRead(inputPin);  // read input value
  }
  // here when switchVal is no longer HIGH because the switch is pressed
  return result;  // this value will be returned
}
```

Discussion

The code in this recipe's Solution illustrates the three forms of function call that you will come across. `blink1` has no parameter and no return value. Its form is:

```
void blink1()
{
   // implementation code goes here...
}
```

`blink2` takes a single parameter but does not return a value:

```
void blink2(int count)
{
   // implementation code goes here...
}
```

`blink3` has a single parameter and returns a value:

```
int blink3(int period)
{
   // implementation code goes here...
}
```

The data type that precedes the function name indicates the return type (or no return type if `void`). When *declaring the function* (writing out the code that defines the function and its action), you do not put a semicolon following the parenthesis at the end. When you *use* (call) the function, you do need a semicolon at the end of the line that calls the function.

Most of the functions you come across will be some variation on these forms. For example, here is a function that takes a parameter and returns a value:

```
int sensorPercent(int pin)
{
int percent;

  int val = analogRead(pin);  // read the sensor (ranges from 0 to 1023)
  percent = map(val,0,1023,0,100); // percent will range from 0 to 100.
  return percent;
}
```

The function name is `sensorPercent`. It is given an analog pin number to read and returns the value as a percent (see Recipe 5.7 for more on `analogRead` and `map`). The `int` in front of the declaration tells the compiler (and reminds the programmer) that the function will return an integer. When creating functions, choose the return type appropriate to the action the function performs. This function returns an integer value from 0 to 100, so a return type of `int` is appropriate.

It is recommended that you give your functions meaningful names, and it is a common practice to combine words by capitalizing the first letter of each word, except for the first word. Use whatever style you prefer, but it helps others who read your code if you keep your naming style consistent.

`sensorPercent` has a parameter called `pin` (when the function is called, `pin` is given the value that is passed to the function).

The body of the function (the code within the brackets) performs the action you want—here it reads a value from an analog input pin and maps it to a percentage. In the preceding example, the percentage is temporarily held in a variable called `percent`. The following statement causes the value held in the temporary variable `percent` to be returned to the calling application:

```
return percent;
```

The same functionality can be achieved without using the `percent` temporary variable:

```
int sensorPercent(int pin)
{
  int val = analogRead(pin);  // read the sensor (ranges from 0 to 1023)
  return map(val,0,1023,0,100); // percent will ranges from 0 to 100.
}
```

Here is how the function can be called:

```
// print the percent value of 6 analog pins
for(int sensorPin = 0; sensorPin < 6; sensorPin++)
{
  Serial.print("Percent of sensor on pin ");
  Serial.print(sensorPin);
  Serial.print(" is ");
  int val = sensorPercent(sensorPin);
  Serial.print(val);
}
```

See Also

The Arduino function reference page: *http://www.arduino.cc/en/Reference/FunctionDeclaration*

2.11 Returning More Than One Value from a Function

Problem

You want to return two or more values from a function. Recipe 2.10 provided examples for the most common form of a function, one that returns just one value or none at all. But sometimes you need to modify or return more than one value.

Solution

There are various ways to solve this. The easiest to understand is to have the function change some global variables and not actually return anything from the function:

```
/*
  swap sketch
  demonstrates changing two values using global variables
 */

int x; // x and y are global variables
int y;

void setup() {
  Serial.begin(9600);
}

void loop(){
  x = random(10); // pick some random numbers
  y = random(10);

  Serial.print("The value of x and y before swapping are: ");
  Serial.print(x); Serial.print(","); Serial.println(y);
  swap();

  Serial.print("The value of x and y after swapping are: ");
  Serial.print(x); Serial.print(","); Serial.println(y);Serial.println();


  delay(1000);
}

// swap the two global values
void swap()
{
  int temp;
  temp = x;
  x = y;
  y = temp;
}
```

The swap function changes two values by using global variables. Global variables are easy to understand (global variables are values that are accessible everywhere and anything can change them), but they are avoided by experienced programmers because it's easy to inadvertently modify the value of a variable or to have a function stop working because you changed the name or type of a global variable elsewhere in the sketch.

A safer and more elegant solution is to pass references to the values you want to change and let the function use the references to modify the values. This is done as follows:

```
/*
  functionReferences sketch
  demonstrates returning more than one value by passing references
 */
```

```
void setup() {
  Serial.begin(9600);
}

void loop(){
  int x = random(10); // pick some random numbers
  int y = random(10);

  Serial.print("The value of x and y before swapping are: ");
  Serial.print(x); Serial.print(","); Serial.println(y);
  swap(x,y);

  Serial.print("The value of x and y after swapping are: ");
  Serial.print(x); Serial.print(","); Serial.println(y);Serial.println();

  delay(1000);
}

// swap the two given values
void swap(int &value1, int &value2)
{
  int temp;
  temp = value1;
  value1 = value2;
  value2 = temp;
}
```

Discussion

The `swap` function is similar to the functions with parameters described in Recipe 2.10, but the ampersand (&) symbol indicates that the parameters are *references*. This means changes in values within the function will also change the value of the variable that is given when the function is called. You can see how this works by first running the code in this recipe's Solution and verifying that the parameters are swapped. Then modify the code by removing the two ampersands in the function definition.

The changed line should look like this:

```
void swap(int value1, int value2)
```

Running the code shows that the values are not swapped—changes made within the function are local to the function and are lost when the function returns.

If you are using Arduino release 21 or earlier, you will need to create a function declaration to inform the compiler that your function is using references. The sketch for this recipe in the download for the first edition of this book shows how to create the function declaration:

```
// functions with references must be declared before use
// The declaration goes at the top, before your setup and loop code
```

```
// note the semicolon at the end of the declaration
  void swap(int &value1, int &value2);
```

A function declaration is a *prototype*—a specification of the name, the types of values that may be passed to the function, and the function's return type. The Arduino build process usually creates the declarations for you under the covers. But when you use nonstandard (for Arduino 21 and earlier, anyhow) syntax, the build process will not create the declaration and you need to add it to your code yourself, as done with the line just before `setup`.

A function definition is the function header and the function body. The function header is similar to the declaration except it does not have a semicolon at the end. The function body is the code within the brackets that is run to perform some action when the function is called.

2.12 Taking Actions Based on Conditions

Problem

You want to execute a block of code only if a particular condition is true. For example, you may want to light an LED if a switch is pressed or if an analog value is greater than some threshold.

Solution

The following code uses the wiring shown in Recipe 5.1:

```
/*
   Pushbutton sketch
   a switch connected to digital pin 2 lights the LED on pin 13
*/

const int ledPin = 13;            // choose the pin for the LED
const int inputPin = 2;           // choose the input pin (for a pushbutton)

void setup() {
  pinMode(ledPin, OUTPUT);         // declare LED pin as output
  pinMode(inputPin, INPUT);        // declare pushbutton pin as input
}

void loop(){
  int val = digitalRead(inputPin);  // read input value
  if (val == HIGH)                  // check if the input is HIGH
  {
    digitalWrite(ledPin, HIGH);     // turn LED on if switch is pressed
  }
}
```

Discussion

The `if` statement is used to test the value of `digitalRead`. An `if` statement must have a test within the parentheses that can only be true or false. In the example in this recipe's Solution, it's `val == HIGH`, and the code block following the `if` statement is only executed if the expression is true. A code block consists of all code within the brackets (or if you don't use brackets, the block is just the next executable statement terminated by a semicolon).

If you want to do one thing if a statement is true and another if it is false, use the `if...else` statement:

```
/*
   Pushbutton sketch
   a switch connected to pin 2 lights the LED on pin 13
*/

const int ledPin = 13;            // choose the pin for the LED
const int inputPin = 2;           // choose the input pin (for a pushbutton)

void setup() {
  pinMode(ledPin, OUTPUT);         // declare LED pin as output
  pinMode(inputPin, INPUT);        // declare pushbutton pin as input
}

void loop(){
  int val = digitalRead(inputPin);  // read input value
  if (val == HIGH)                  // check if the input is HIGH
  {
    // do this if val is HIGH
    digitalWrite(ledPin, HIGH);     // turn LED on if switch is pressed
  }
  else
  {
    // else do this if val is not HIGH
    digitalWrite(ledPin, LOW);      // turn LED off
  }
}
```

See Also

See the discussion on Boolean types in Recipe 2.2.

2.13 Repeating a Sequence of Statements

Problem

You want to repeat a block of statements while an expression is true.

Solution

A `while` loop repeats one or more instructions while an expression is true:

```
/*
 * Repeat
 * blinks while a condition is true
 */

const int ledPin    = 13; // digital pin the LED is connected to
const int sensorPin = 0; // analog input 0

void setup()
{
  Serial.begin(9600);
  pinMode(ledPin,OUTPUT); // enable LED pin as output
}

void loop()
{
  while(analogRead(sensorPin) > 100)
  {
    blink();     // call a function to turn an LED on and off
    Serial.print(".");
  }
  Serial.println(analogRead(sensorPin)); // this is not executed until after
                                         // the while loop finishes!!!
}

void blink()
{
   digitalWrite(ledPin, HIGH);
   delay(100);
   digitalWrite(ledPin, LOW);
   delay(100);
}
```

This code will execute the statements in the block within the brackets, {}, while the value from `analogRead` is greater than 100. This could be used to flash an LED as an alarm while some value exceeded a threshold. The LED is off when the sensor value is 100 or less; it flashes continuously when the value is greater than 100.

The {} symbols that define a block of code are given various names, including brackets, curly braces, and braces. This book refers to them as brackets.

Discussion

Brackets define the extent of the code block to be executed in a loop. If brackets are not used, only the first line of code will be repeated in the loop:

```
while(analogRead(sensorPin) > 100)
   blink();   // line immediately following the loop expression is executed
   Serial.print("."); // this is not executed until after the while loop finishes!!!
```

Loops without brackets can behave unexpectedly if you have more than one line of code.

The `do...while` loop is similar to the `while` loop, but the instructions in the code block are executed before the condition is checked. Use this form when you must have the code executed at least once, even if the expression is false:

```
do
{
   blink();  // call a function to turn an LED on and off
}
while (analogRead(sensorPin) > 100);
```

The preceding code will flash the LED at least once and will keep flashing as long as the value read from a sensor is greater than 100. If the value is not greater than 100, the LED will only flash once. This code could be used in a battery-charging circuit, if it were called once every 10 seconds or so: a single flash shows that the circuit is active, whereas continuous flashing indicates the battery is charged.

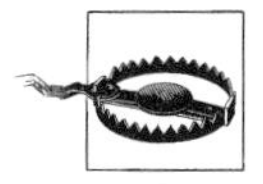

Only the code within a `while` or `do` loop will run until the conditions permit exit. If your sketch needs to break out of a loop in response to some other condition such as a timeout, sensor state, or other input, you can use `break`:

```
while(analogRead(sensorPin) > 100)
{
   blink();
   if(Serial.available())
     break; // any serial input breaks out of the while loop
}
```

See Also

Chapters 4 and 5

2.14 Repeating Statements with a Counter

Problem

You want to repeat one or more statements a certain number of times. The `for` loop is similar to the `while` loop, but you have more control over the starting and ending conditions.

Solution

This sketch counts from zero to three by printing the value of the variable `i` in a `for` loop:

```
/*
   ForLoop sketch
   demonstrates for loop
*/

void setup() {
  Serial.begin(9600);}

void loop(){
  Serial.println("for(int i=0; i < 4; i++)");
  for(int i=0; i < 4; i++)
  {
    Serial.println(i);
  }
}
```

The Serial Monitor output from this is as follows (it will be displayed over and over):

```
for(int i=0; i < 4; i++)
0
1
2
3
```

Discussion

A `for` loop consists of three parts: initialization, conditional test, and iteration (a statement that is executed at the end of every pass through the loop). Each part is separated by a semicolon. In the code in this recipe's Solution, `int i=0;` initializes the variable `i` to `0`; `i < 4;` tests the variable to see if it's less than 4; and `i++` increments `i`.

A `for` loop can use an existing variable, or it can create a variable for exclusive use inside the loop. This version uses the value of the variable `j` created earlier in the sketch:

```
int j;

Serial.println("for(j=0; j < 4; j++ )");
for(j=0; j < 4; j++ )
{
  Serial.println(j);
}
```

This is almost the same as the earlier example, but it does not have the `int` keyword in the initialization part because the variable `j` was already defined. The output of this version is similar to the output of the earlier version:

```
for(j=0; i < 4; i++)
0
1
2
3
```

You can leave out the initialization part completely if you want the loop to use the value of a variable defined earlier. This code starts the loop with `j` equal to 1:

```
int j = 1;

Serial.println("for(   ; j < 4; j++ )");
for(  ; j < 4; j++ )
{
  Serial.println(j);
}
```

The preceding code prints the following:

```
for(  ; j < 4; j++)
1
2
3
```

You control when the loop stops in the conditional test. The previous examples test whether the loop variable is less than 4 and will terminate when the condition is no longer true.

If your loop variable starts at 0 and you want it to repeat four times, your conditional statement should test for a value less than 4. The loop repeats while the condition is true and there are four values that are less than 4 with a loop starting at 0.

The following code tests if the value of the loop variable is less than or equal to 4. It will print the digits from 0 to 4:

```
Serial.println("for(int i=0; i <= 4; i++)");
for(int i=0; i <= 4; i++)
{
  Serial.println(i);
}
```

The third part of a `for` loop is the iterator statement that gets executed at the end of each pass through the loop. This can be any valid C/C++ statement. The following increases the value of `i` by two on each pass:

```
Serial.println("for(int i=0; i < 4; i+= 2)");
for(int i=0; i < 4; i+=2)
{
  Serial.println(i);
}
```

That expression only prints the values 0 and 2.

The *iterator* expression can be used to cause the loop to count from high to low, in this case from 3 to 0:

```
Serial.println("for(int i=3; i > = 0 ; i--)");
for(int i=3; i >= 0 ; i--)
{
```

```
    Serial.println(i);
  }
```

Like the other parts of a for loop, the iterator expression can be left blank (you must always have the two semicolons separating the three parts even if they are blank).

This version only increments i when an input pin is high. The for loop does not change the value of i; it is only changed by the if statement after Serial.print—you'll need to define inPin and set it to INPUT with pinMode():

```
Serial.println("for(int i=0; i < 4; )");
for(int i=0; i < 4; )
{
  Serial.println(i);
  if(digitalRead(inPin) == HIGH) {
    i++;  // only increment the value if the input is high
  }
}
```

See Also

Arduino reference for the for statement: *http://www.arduino.cc/en/Reference/For*

2.15 Breaking Out of Loops

Problem

You want to terminate a loop early based on some condition you are testing.

Solution

Use the following code:

```
while(analogRead(sensorPin) > 100)
{
   if(digitalRead(switchPin) == HIGH)
   {
      break;   //exit the loop if the switch is pressed
   }
   flashLED();              // call a function to turn an LED on and off
}
```

Discussion

This code is similar to the one using while loops, but it uses the break statement to exit the loop if a digital pin goes high. For example, if a switch is connected on the pin as shown in Recipe 5.1, the loop will exit and the LED will stop flashing even if the condition in the while loop is true.

See Also

Arduino reference for the `break` statement: *http://www.arduino.cc/en/Reference/Break*

2.16 Taking a Variety of Actions Based on a Single Variable

Problem

You need to do different things depending on some value. You could use multiple `if` and `else if` statements, but the code soon gets complex and difficult to understand or modify. Additionally, you may want to test for a range of values.

Solution

The `switch` statement provides for selection of a number of alternatives. It is functionally similar to multiple `if`/`else if` statements but is more concise:

```
/*
 * SwitchCase sketch
 * example showing switch statement by switching on chars from the serial port
 *
 * sending the character 1 blinks the LED once, sending 2 blinks twice
 * sending + turns the LED on, sending - turns it off
 * any other character prints a message to the Serial Monitor
 */
const int ledPin = 13; // the pin the LED is connected to

void setup()
{
  Serial.begin(9600); // Initialize serial port to send and
                      // receive at 9600 baud
  pinMode(ledPin, OUTPUT);
}

void loop()
{
  if ( Serial.available()) // Check to see if at least one
                           // character is available
  {
    char ch = Serial.read();
    switch(ch)
    {
    case '1':
      blink();
      break;
    case '2':
      blink();
      blink();
      break;
    case '+':
      digitalWrite(ledPin,HIGH);
      break;
```

```
      case '-':
        digitalWrite(ledPin,LOW);
        break;
      default :
        Serial.print(ch);
        Serial.println(" was received but not expected");
        break;
      }
    }
}

void blink()
{
  digitalWrite(ledPin,HIGH);
  delay(500);
  digitalWrite(ledPin,LOW);
  delay(500);
}
```

Discussion

The `switch` statement evaluates the variable `ch` received from the serial port and branches to the label that matches its value. The labels must be numeric constants (you can use strings in a `case` statement) and no two labels can have the same value. If you don't have a `break` statement following each expression, the execution will *fall through* into the statement:

```
case '1':
  blink();   // no break statement before the next label
case '2':
  blink();   // case '1' will continue here
  blink();
  break;     // break statement will exit the switch expression
```

If the `break` statement at the end of `case '1':` was removed (as shown in the preceding code), when `ch` is equal to the character `1` the `blink` function will be called three times. Accidentally forgetting the `break` is a common mistake. Intentionally leaving out the `break` is sometimes handy; it can be confusing to others reading your code, so it's a good practice to clearly indicate your intentions with comments in the code.

If your `switch` statement is misbehaving, check to ensure that you have not forgotten the `break` statements.

The `default:` label is used to catch values that don't match any of the `case` labels. If there is no `default` label, the `switch` expression will not do anything if there is no match.

See Also

Arduino reference for the `switch` and `case` statements: *http://www.arduino.cc/en/Reference/SwitchCase*

2.17 Comparing Character and Numeric Values

Problem

You want to determine the relationship between values.

Solution

Compare integer values using the relational operators shown in Table 2-3.

Table 2-3. Relational and equality operators

Operator	Test for	Example
`==`	Equal to	`2 == 3 // evaluates to false`
`!=`	Not equal to	`2 != 3 // evaluates to true`
`>`	Greater than	`2 > 3 // evaluates to false`
`<`	Less than	`2 < 3 // evaluates to true`
`>=`	Greater than or equal to	`2 >= 3 // evaluates to false`
`<=`	Less than or equal to	`2 <= 3 // evaluates to true`

The following sketch demonstrates the results of using the comparison operators:

```
/*
 * RelationalExpressions sketch
 * demonstrates comparing values
 */

int i = 1;  // some values to start with
int j = 2;

void setup() {
  Serial.begin(9600);
}

void loop(){
  Serial.print("i = ");
  Serial.print(i);
  Serial.print(" and j = ");
  Serial.println(j);

  if(i < j)
    Serial.println(" i is less than j");
  if(i <= j)
    Serial.println(" i is less than or equal to j");
```

```
    if(i != j)
      Serial.println(" i is not equal to j");
    if(i == j)
      Serial.println(" i is equal to j");
    if(i >= j)
      Serial.println(" i is greater than or equal to j");
    if(i > j)
      Serial.println(" i is greater than j");

    Serial.println();
    i = i + 1;
    if(i > j + 1)
      delay(10000);  // long delay after i is no longer close to j
  }
```

Here is the output:

```
i = 1 and j = 2
 i is less than j
 i is less than or equal to j
 i is not equal to j

i = 2 and j = 2
 i is less than or equal to j
 i is equal to j
 i is greater than or equal to j

i = 3 and j = 2
 i is not equal to j
 i is greater than or equal to j
 i is greater than j
```

Discussion

Note that the equality operator is the double equals sign, ==. One of the most common programming mistakes is to confuse this with the assignment operator, which uses a single equals sign.

The following expression will compare the value of `i` to 3. The programmer intended this:

```
if(i == 3)  // test if i equals 3
```

But he put this in the sketch:

```
if(i = 3)  // single equals sign used by mistake!!!!
```

This will always return `true`, because `i` will be set to `3`, so they will be equal when compared.

A tip to help avoid that trap when comparing variables to constants (fixed values) is to put the constant on the left side of the expression:

```
if(3 = i)  // single equals sign used by mistake!!!!
```

The compiler will tell you about this error because it knows that you can't assign a different value to a constant.

The error message is the somewhat unfriendly "value required as left operand of assignment." If you see this message, the compiler is telling you that you are trying to assign a value to something that cannot be changed.

See Also

Arduino reference for conditional and comparison operators: *http://www.arduino.cc/en/Reference/If*

2.18 Comparing Strings

Problem

You want to see if two character strings are identical.

Solution

There is a function to compare strings, called `strcmp` (short for *string compare*). Here is a fragment showing its use:

```
char string1[ ] = "left";
char string2[ ] = "right";

if(strcmp(string1, string2) == 0)
   Serial.print("strings are equal)
```

Discussion

`strcmp` returns the value `0` if the strings are equal and a value greater than zero if the first character that does not match has a greater value in the first string than in the second. It returns a value less than zero if the first nonmatching character in the first string is less than in the second. Usually you only want to know if they are equal, and although the test for zero may seem unintuitive at first, you'll soon get used to it.

Bear in mind that strings of unequal length will not be evaluated as equal even if the shorter string is contained in the longer one. So:

```
strcmp("left", "leftcenter") == 0)  // this will evaluate to false
```

You can compare strings up to a given number of characters by using the `strncmp` function. You give `strncmp` the maximum number of characters to compare and it will stop comparing after that many characters:

```
strncmp("left", "leftcenter", 4) == 0)  // this will evaluate to true
```

Unlike character strings, Arduino `Strings` can be directly compared as follows:

```
String stringOne = String("this");
if (stringOne == "this")
  Serial.println("this will be true");
if (stringOne == "that")
  Serial.println("this will be false");
```

A tutorial on Arduino `String` comparison is at *http://arduino.cc/en/Tutorial/StringComparisonOperators*.

See Also

More information on `strcmp` is available at *http://www.cplusplus.com/reference/clibrary/cstring/strcmp/*.

See Recipe 2.5 for an introduction to the Arduino `String`.

2.19 Performing Logical Comparisons

Problem

You want to evaluate the logical relationship between two or more expressions. For example, you want to take a different action depending on the conditions of an `if` statement.

Solution

Use the logical operators as outlined in Table 2-4.

Table 2-4. Logical operators

Symbol	Function	Comments
&&	Logical And	Evaluates as `true` if the conditions on both sides of the && operator are true
||	Logical Or	Evaluates as `true` if the condition on at least one side of the || operator is true
!	Not	Evaluates as `true` if the expression is false, and `false` if the expression is true

Discussion

Logical operators return true or false values based on the logical relationship. The examples that follow assume you have sensors wired to digital pins 2 and 3 as discussed in Chapter 5.

The logical And operator `&&` will return `true` if both its two operands are true, and `false` otherwise:

```
if( digitalRead(2) && digitalRead(3) )
  blink(); // blink of both pins are HIGH
```

The logical Or operator `||` will return `true` if either of its two operands are true, and `false` if both operands are false:

```
if( digitalRead(2) || digitalRead(3) )
  blink(); // blink of either pins is HIGH
```

The Not operator `!` has only one operand, whose value is inverted—it results in `false` if its operand is true and `true` if its operand is false:

```
if( !digitalRead(2)  )
  blink(); // blink of the pin is not HIGH
```

2.20 Performing Bitwise Operations

Problem

You want to set or clear certain bits in a value.

Solution

Use the bit operators as outlined in Table 2-5.

Table 2-5. Bit operators

Symbol	Function	Result	Example
&	Bitwise And	Sets bits in each place to 1 if both bits are 1; otherwise, bits are set to 0.	`3 & 1` equals `1` (`11 & 01` equals `01`)
|	Bitwise Or	Sets bits in each place to 1 if either bit is 1.	`3 | 1` equals `3` (`11 | 01` equals `11`)
^	Bitwise Exclusive Or	Sets bits in each place to 1 only if one of the two bits is 1.	`3 ^ 1` equals `2` (`11 ^ 01` equals `10`)
~	Bitwise Negation	Inverts the value of each bit. The result depends on the number of bits in the data type.	`~1` equals `254` (`~00000001` equals `11111110`)

Here is a sketch that demonstrates the example values shown in Table 2-5:

```
/*
 * bits sketch
 * demonstrates bitwise operators
 */

void setup() {
  Serial.begin(9600);
}

void loop(){
  Serial.print("3 & 1 equals "); // bitwise And 3 and 1
  Serial.print(3 & 1);         // print the result
  Serial.print(" decimal, or in binary: ");
  Serial.println(3 & 1 , BIN);   // print the binary representation of the result

  Serial.print("3 | 1 equals "); // bitwise Or 3 and 1
  Serial.print(3 | 1 );
  Serial.print(" decimal, or in binary: ");
  Serial.println(3 | 1 , BIN);   // print the binary representation of the result

  Serial.print("3 ^ 1 equals "); // bitwise exclusive or 3 and 1
  Serial.print(3 ^ 1);
  Serial.print(" decimal, or in binary: ");
  Serial.println(3 ^ 1 , BIN);   // print the binary representation of the result


  byte byteVal = 1;
  int intVal = 1;

  byteVal = ~byteVal;  // do the bitwise negate
  intVal = ~intVal;

  Serial.print("~byteVal (1) equals "); // bitwise negate an 8 bit value
  Serial.println(byteVal, BIN);  // print the binary representation of the result
  Serial.print("~intVal (1) equals "); // bitwise negate a 16 bit value
  Serial.println(intVal, BIN);   // print the binary representation of the result

  delay(10000);
}
```

This is what is displayed on the Serial Monitor:

```
3 & 1 equals 1 decimal, or in binary: 1
3 | 1 equals 3 decimal, or in binary: 11
3 ^ 1 equals 2 decimal, or in binary: 10
~byteVal (1) equals 11111110
~intVal (1) equals 11111111111111111111111111111110
```

Discussion

Bitwise operators are used to set or test bits. When you "And" or "Or" two values, the operator works on each individual bit. It is easier to see how this works by looking at the binary representation of the values.

Decimal 3 is binary 00000011, and decimal 1 is 00000001. Bitwise And operates on each bit. The rightmost bits are both 1, so the result of And-ing these is 1. Moving to the left, the next bits are 1 and 0; And-ing these results in 0. All the remaining bits are 0, so the bitwise result of these will be 0. In other words, for each bit position where there is a 1 in both places, the result will have a 1; otherwise, it will have a 0. So, `11 & 01` equals 1.

Tables 2-6, 2-7, and 2-8 should help to clarify the bitwise And, Or, and Exclusive Or values.

Table 2-6. Bitwise And

Bit 1	Bit 2	Bit 1 and Bit 2
0	0	0
0	1	0
1	0	0
1	1	1

Table 2-7. Bitwise Or

Bit 1	Bit 2	Bit 1 or Bit 2
0	0	0
0	1	1
1	0	1
1	1	1

Table 2-8. Bitwise Exclusive Or

Bit 1	Bit 2	Bit 1 ^ Bit 2
0	0	0
0	1	1
1	0	1
1	1	0

All the bitwise expressions operate on two values, except for the negation operator. This simply flips each bit, so 0 becomes 1 and 1 becomes 0. In the example, the `byte` (8-bit) value 00000001 becomes 11111110. The `int` value has 16 bits, so when each is flipped, the result is 15 ones followed by a single zero.

See Also

Arduino reference for the bitwise And, Or, and Exclusive Or operators: *http://www.arduino.cc/en/Reference/Bitwise*

2.21 Combining Operations and Assignment

Problem

You want to understand and use compound operators. It is not uncommon to see published code that uses expressions that do more than one thing in a single statement. You want to understand `a += b`, `a >>= b`, and `a &= b`.

Solution

Table 2-9 shows the compound assignment operators and their equivalent full expression.

Table 2-9. Compound operators

Operator	Example	Equivalent expression
`+=`	`value += 5;`	`value = value + 5; // add 5 to value`
`-=`	`value -= 4;`	`value = value - 4; // subtract 4 from value`
`*=`	`value *= 3;`	`value = value * 3; // multiply value by 3`
`/=`	`value /= 2;`	`value = value / 2; // divide value by 2`
`>>=`	`value >>= 2;`	`value = value >> 2; // shift value right two places`
`<<=`	`value <<= 2;`	`value = value << 2; // shift value left two places`
`&=`	`mask &= 2;`	`mask = mask & 2; // binary-and mask with 2`
`|=`	`mask |= 2;`	`mask = mask | 2; // binary-or mask with 2`

Discussion

These compound statements are no more efficient at runtime than the equivalent full expression, and if you are new to programming, using the full expression is clearer. Experienced coders often use the shorter form, so it is helpful to be able to recognize the expressions when you run across them.

See Also

See *http://www.arduino.cc/en/Reference/HomePage* for an index to the reference pages for compound operators.

CHAPTER 3

Using Mathematical Operators

3.0 Introduction

Almost every sketch uses mathematical operators to manipulate the value of variables. This chapter provides a brief overview of the most common mathematical operators. As the preceding chapter is, this summary is primarily for nonprogrammers or programmers who are not familiar with C or C++. For more details, see one of the C reference books mentioned in the Preface.

3.1 Adding, Subtracting, Multiplying, and Dividing

Problem

You want to perform simple math on values in your sketch. You want to control the order in which the operations are performed and you may need to handle different variable types.

Solution

Use the following code:

```
int myValue;
myValue = 1 + 2;  // addition
myValue = 3 - 2;  // subtraction
myValue = 3 * 2;  // multiplication
myValue = 3 / 2;  // division (the result is 1)
```

Discussion

Addition, subtraction, and multiplication for integers work much as you expect.

Make sure your result will not exceed the maximum size of the destination variable. See Recipe 2.2.

Integer division truncates the fractional remainder in the division example shown in this recipe's Solution; `myValue` will equal 1 after the division (see Recipe 2.3 if your application requires fractional results):

```
int value =   1 + 2 * 3 + 4;
```

Compound statements, such as the preceding statement, may appear ambiguous, but the *precedence* (order) of every operator is well defined. Multiplication and division have a higher precedence than addition and subtraction, so the result will be 11. It's advisable to use brackets in your code to make the desired calculation precedence clear. `int value = 1 + (2 * 3) + 4;` produces the same result but is easier to read.

Use parentheses if you need to alter the precedence, as in this example:

```
int value =   ((1 + 2) * 3) + 4;
```

The result will be 13. The expression in the inner parentheses is calculated first, so 1 gets added to 2, this then gets multiplied by 3, and finally is added to 4, yielding 13.

See Also

Recipe 2.2; Recipe 2.3

3.2 Incrementing and Decrementing Values

Problem

You want to increase or decrease the value of a variable.

Solution

Use the following code:

```
int myValue = 0;

myValue = myvalue + 1;  // this adds one to the variable myValue
myValue += 1;           // this does the same as the above

myValue = myvalue - 1;  // this subtracts one from the variable myValue
myValue -= 1;           // this does the same as the above

myValue = myvalue + 5;  // this adds five to the variable myValue
myValue += 5;           // this does the same as the above
```

Discussion

Increasing and decreasing the values of variables is one of the most common programming tasks, and the Arduino board has operators to make this easy. Increasing a value by one is called *incrementing*, and decreasing it by one is called *decrementing*. The longhand way to do this is as follows:

```
myValue = myvalue + 1;  // this adds one to the variable myValue
```

But you can also combine the increment and decrement operators with the assign operator, like this:

```
myValue += 1;           // this does the same as the above
```

See Also

Recipe 3.1

3.3 Finding the Remainder After Dividing Two Values

Problem

You want to find the remainder after you divide two values.

Solution

Use the % symbol (the modulus operator) to get the remainder:

```
int myValue0 =  20 % 10;  // get the modulus(remainder) of 20 divided by 10
int myValue1 =  21 % 10;  // get the modulus(remainder) of 21 divided by 10
```

myValue0 equals 0 (20 divided by 10 has a remainder of 0). myValue1 equals 1 (21 divided by 10 has a remainder of 1).

Discussion

The modulus operator is surprisingly useful, particularly when you want to see if a value is a multiple of a number. For example, the code in this recipe's Solution can be enhanced to detect when a value is a multiple of 10:

```
int myValue;
//... code here to set the value of myValue
if (myValue % 10 == 0)
{
  Serial.println("The value is a multiple of 10");
}
```

The preceding code takes the modulus of the myValue variable and compares the result to zero (see Recipe 2.17). If the result is zero, a message is printed saying the value is a multiple of 10.

Here is a similar example, but by using 2 with the modulus operator, the result can be used to check if a value is odd or even:

```
int myValue;
//... code here to set the value of myValue
if (myValue % 2 == 0)
{
  Serial.println("The value is even");
}
else
{
  Serial.println("The value is odd");
}
```

This example calculates the hour on a 24-hour clock for any given number of hours offset:

```
void printOffsetHour( int hourNow, int offsetHours)
{
   Serial.println((hourNow + offsetHours) % 24);
}

void printOffsetHour(int hourNow, int offsetHours)
{
    Serial.println((hourNow + offsetHours) % 24);
}
```

See Also

Arduino reference for `%` (the modulus operator): *http://www.arduino.cc/en/Reference/Modulo*

3.4 Determining the Absolute Value

Problem

You want to get the absolute value of a number.

Solution

`abs(x)` computes the absolute value of `x`. The following example takes the absolute value of the difference between readings on two analog input ports (see Chapter 5 for more on `analogRead()`):

```
int x = analogRead(0);
int y = analogRead(1);

if (abs(x-y) > 10)
{
  Serial.println("The analog values differ by more than 10");
}
```

Discussion

`abs(x-y);` returns the absolute value of the difference between `x` and `y`. It is used for integer (and `long` integer) values. To return the absolute value of floating-point values, see Recipe 2.3.

See Also

Arduino reference for `abs`: *http://www.arduino.cc/en/Reference/Abs*

3.5 Constraining a Number to a Range of Values

Problem

You want to ensure that a value is always within some lower and upper limit.

Solution

`constrain(x, min, max)` returns a value that is within the bounds of `min` and `max`:

```
myConstrainedValue  = constrain(myValue, 100, 200);
```

Discussion

`myConstrainedValue` is set to a value that will always be greater than or equal to 100 and less than or equal to 200. If `myValue` is less than 100, the result will be 100; if it is more than 200, it will be set to 200.

Table 3-1 shows some example output values using a `min` of 100 and a `max` of 200.

Table 3-1. Output from constrain with min = 100 and max = 200

`myValue` **(the input value)**	`constrain(myValue, 100, 200)`
99	100
100	100
150	150
200	200
201	200

See Also

Recipe 3.6

3.6 Finding the Minimum or Maximum of Some Values

Problem

You want to find the minimum or maximum of two or more values.

Solution

`min(x,y)` returns the smaller of two numbers. `max(x,y)` returns the larger of two numbers:

```
myValue = analogRead(0);
myMinValue = min(myValue, 200);  // myMinValue will be the smaller of
                                 // myVal or 200

myMaxValue = max(myValue, 100);  // myMaxValue will be the larger of
                                 // myVal or 100
```

Discussion

Table 3-2 shows some example output values using a `min` of 200. The table shows that the output is the same as the input (`myValue`) until the value becomes greater than 200.

Table 3-2. Output from min(myValue, 200)

`myValue` (the input value)	`min(myValue, 200)`
99	99
100	100
150	150
200	200
201	200

Table 3-3 shows the output using a `max` of 100. The table shows that the output is the same as the input (`myValue`) when the value is greater than or equal to 100.

Table 3-3. Output from max(myValue, 100)

`myValue` (the input value)	`max(myValue, 100)`
99	100
100	100
150	150
200	200
201	201

Use min when you want to limit the upper bound. That may be counterintuitive, but by returning the smaller of the input value and the minimum value, the output from min will never be higher than the minimum value (200 in the example).

Similarly, use max to limit the lower bound. The output from max will never be lower than the maximum value (100 in the example).

If you want to find the min or max value from more than two values, you can cascade the values as follows:

```
// myMinValue will be the smaller of the three analog readings:
int myMinValue = min(analogRead(0), min(analogRead(1), analogRead(2)) );
```

In this example, the minimum value is found for analog ports 1 and 2, and then the minimum of that and port 0. This can be extended for as many items as you need, but take care to position the parentheses correctly. The following example gets the maximum of four values:

```
int myMaxValue = max(analogRead(0), max(analogRead(1), max(analogRead(2),
                                                    analogRead(3))));
```

See Also

Recipe 3.5

3.7 Raising a Number to a Power

Problem

You want to raise a number to a power.

Solution

pow(x, y) returns the value of x raised to the power of y:

```
myValue =  pow(3,2);
```

This calculates 3^2, so myValue will equal 9.

Discussion

The pow function can operate on integer or floating-point values and it returns the result as a floating-point value:

```
Serial.print(pow(3,2)); // this prints 9.00
int z = pow(3,2);
Serial.println(z);      // this prints 9
```

The first output is 9.00 and the second is 9; they are not exactly the same because the first print displays the output as a floating-point number and the second treats the value as an integer before printing, and therefore displays without the decimal point.

If you use the pow function, you may want to read Recipe 2.3 to understand the difference between these and integer values.

Here is an example of raising a number to a fractional power:

```
float s = pow(2, 1.0 / 12); // the twelfth root of two
```

The twelfth root of two is the same as 2 to the power of 0.083333. The resultant value, s, is 1.05946 (this is the ratio of the frequency of two adjacent notes on a piano).

3.8 Taking the Square Root

Problem

You want to calculate the square root of a number.

Solution

The sqrt(x) function returns the square root of x:

```
Serial.print( sqrt(9) ); // this prints 3.00
```

Discussion

The sqrt function returns a floating-point number (see the pow function discussed in Recipe 3.7).

3.9 Rounding Floating-Point Numbers Up and Down

Problem

You want the next smallest or largest integer value of a floating-point number (floor or ceil).

Solution

floor(x) returns the largest integral value that is not greater than x. ceil(x) returns the smallest integral value that is not less than x.

Discussion

These functions are used for rounding floating-point numbers; use floor(x) to get the largest integer that is not greater than x. Use ceil to get the smallest integer that is greater than x.

Here is some example output using floor:

```
Serial.println( floor(1) );   // this prints  1.00
Serial.println( floor(1.1) ); // this prints  1.00
```

```
Serial.println( floor(0) );    // this prints  0.00
Serial.println( floor(.1) );   // this prints  0.00
Serial.println( floor(-1) );   // this prints -1.00
Serial.println( floor(-1.1) ); // this prints -2.00
```

Here is some example output using `ceil`:

```
Serial.println( ceil(1) );     // this prints  1.00
Serial.println( ceil(1.1) );   // this prints  2.00
Serial.println( ceil(0) );     // this prints  0.00
Serial.println( ceil(.1) );    // this prints  1.00
Serial.println( ceil(-1) );    // this prints -1.00
Serial.println( ceil(-1.1) );  // this prints -1.00
```

You can round to the nearest integer as follows:

```
if (floatValue  > 0.0)
  result = floor(floatValue + 0.5);
else
  result = ceil(num - 0.5);
```

You can truncate a floating-point number by *casting* (converting) to an `int`, but this does not round correctly. Negative numbers such as –1.9 should round down to –2, but when cast to an `int` they are rounded up to –1. The same problem exists with positive numbers: 1.9 should round up to 2 but will round down to 1. Use `floor` and `ceil` to get the correct results.

3.10 Using Trigonometric Functions

Problem

You want to get the sine, cosine, or tangent of an angle given in radians or degrees.

Solution

`sin(x)` returns the sine of angle `x`. `cos(x)` returns the cosine of angle `x`. `tan(x)` returns the tangent of angle `x`.

Discussion

Angles are specified in radians and the result is a floating-point number (see Recipe 2.3). The following example illustrates the trig functions:

```
float deg = 30;                // angle in degrees
float rad  = deg * PI / 180;   // convert to radians
Serial.println(rad);           // print the radians
Serial.println (sin(rad));     // print the sine
Serial.println (cos(rad));     // print the cosine
```

This converts the angle into radians and prints the sine and cosine. Here is the output with annotation added:

```
0.52   30 degrees is 0.5235988 radians, print only shows two decimal places
0.50   sine of 30 degrees is .5000000, displayed here to two decimal places
0.87   cosine is .8660254, which rounds up to 0.87
```

Although the sketch calculates these values using the full precision of floating-point numbers, the `Serial.print` routine shows the values of floating-point numbers to two decimal places.

The conversion from radians to degrees and back again is textbook trigonometry. `PI` is the familiar constant for π (3.14159265...). `PI` and `180` are both constants, and Arduino provides some precalculated constants you can use to perform degree/radian conversions:

```
rad = deg * DEG_TO_RAD;  // a way to convert degrees to radians
deg = rad * RAD_TO_DEG;  // a way to convert radians to degrees
```

Using `deg * DEG_TO_RAD` looks more efficient than `deg * PI / 180`, but it's not, since the Arduino compiler is smart enough to recognize that `PI / 180` is a constant (the value will never change), so it substitutes the result of dividing `PI` by 180, which happens to be the same value as the constant `DEG_TO_RAD` (0.017453292519...). Use whichever approach you prefer.

See Also

Arduino references for `sin` (*http://www.arduino.cc/en/Reference/Sin*), `cos` (*http://arduino.cc/en/Reference/Cos*), and `tan` (*http://arduino.cc/en/Reference/Tan*)

3.11 Generating Random Numbers

Problem

You want to get a random number, either ranging from zero up to a specified maximum or constrained between a minimum and maximum value you provide.

Solution

Use the `random` function to return a random number. Calling `random` with a single parameter sets the upper bound; the values returned will range from zero to one less than the upper bound:

```
random(max);        // returns a random number between 0 and max -1
```

Calling `random` with two parameters sets the lower and upper bounds; the values returned will range from the lower bound (inclusive) to one less than the upper bound:

```
random(min, max); // returns a random number between min and max -1
```

Discussion

Although there appears to be no obvious pattern to the numbers returned, the values are not truly random. Exactly the same sequence will repeat each time the sketch starts. In many applications, this does not matter. But if you need a different sequence each time your sketch starts, use the function **randomSeed(seed)** with a different seed value each time (if you use the same seed value, you'll get the same sequence). This function starts the random number generator at some arbitrary place based on the seed parameter you pass:

```
randomSeed(1234);  // change the starting sequence of random numbers.
```

Here is an example that uses the different forms of random number generation available on Arduino:

```
// Random
// demonstrates generating random numbers

int randNumber;

void setup()
{
  Serial.begin(9600);

  // Print random numbers with no seed value
  Serial.println("Print 20 random numbers between 0 and 9");
  for(int i=0; i < 20; i++)
  {
    randNumber = random(10);
    Serial.print(randNumber);
    Serial.print(" ");
  }
  Serial.println();
  Serial.println("Print 20 random numbers between 2 and 9");
  for(int i=0; i < 20; i++)
  {
    randNumber = random(2,10);
    Serial.print(randNumber);
    Serial.print(" ");
  }

  // Print random numbers with the same seed value each time
  randomSeed(1234);
  Serial.println();
  Serial.println("Print 20 random numbers between 0 and 9 after constant seed ");
  for(int i=0; i < 20; i++)
  {
    randNumber = random(10);
    Serial.print(randNumber);
    Serial.print(" ");
  }

  // Print random numbers with a different seed value each time
  randomSeed(analogRead(0));  // read from an analog port with nothing connected
```

```
  Serial.println();
  Serial.println("Print 20 random numbers between 0 and 9 after floating seed ");
  for(int i=0; i < 20; i++)
  {
    randNumber = random(10);
    Serial.print(randNumber);
    Serial.print(" ");
  }
  Serial.println();
  Serial.println();
}

void loop()
{
}
```

Here is the output from this code:

```
Print 20 random numbers between 0 and 9
7 9 3 8 0 2 4 8 3 9 0 5 2 2 7 3 7 9 0 2
Print 20 random numbers between 2 and 9
9 3 7 7 2 7 5 8 2 9 3 4 2 5 4 3 5 7 5 7
Print 20 random numbers between 0 and 9 after constant seed
8 2 8 7 1 8 0 3 6 5 9 0 3 4 3 1 2 3 9 4
Print 20 random numbers between 0 and 9 after floating seed
0 9 7 4 4 7 7 4 4 9 1 6 0 2 3 1 5 9 1 1
```

If you press the reset button on your Arduino to restart the sketch, the first three lines of random numbers will be unchanged. Only the last line changes each time the sketch starts, because it sets the seed to a different value by reading it from an unconnected analog input port as a seed to the `randomSeed` function. If you are using analog port 0 for something else, change the argument to `analogRead` to an unused analog port.

See Also

Arduino references for `random` (*http://www.arduino.cc/en/Reference/Random*) and `randomSeed` (*http://arduino.cc/en/Reference/RandomSeed*)

3.12 Setting and Reading Bits

Problem

You want to read or set a particular bit in a numeric variable.

Solution

Use the following functions:

`bitSet(x, bitPosition)`
: Sets (writes a 1 to) the given `bitPosition` of variable `x`

`bitClear(x, bitPosition)`
: Clears (writes a 0 to) the given `bitPosition` of variable `x`

`bitRead(x, bitPosition)`
: Returns the value (as 0 or 1) of the bit at the given `bitPosition` of variable `x`

`bitWrite(x, bitPosition, value)`
: Sets the given value (as 0 or 1) of the bit at the given `bitPosition` of variable `x`

`bit(bitPosition)`
: Returns the value of the given bit position: `bit(0)` is 1, `bit(1)` is 2, `bit(2)` is 4, and so on

In all these functions, `bitPosition 0` is the least significant (rightmost) bit.

Here is a sketch that uses these functions to manipulate the bits of an 8-bit variable called `flags`:

```
// bitFunctions
// demonstrates using the bit functions


byte flags = 0; // these examples set, clear or read bits in a variable called flags.

// bitSet example
void setFlag( int flagNumber)
{
   bitSet(flags, flagNumber);
}

// bitClear example
void  clearFlag( int flagNumber)
{
   bitClear(flags, flagNumber);
}

// bitPosition example

int  getFlag( int flagNumber)
{
   return  bitRead(flags, flagNumber);
}

void setup()
{
  Serial.begin(9600);
}

void loop()
{
    showFlags();
    setFlag(2);  // set some flags;
    setFlag(5);
    showFlags();
    clearFlag(2);
```

```
    showFlags();

    delay(10000); // wait a very long time
}

// reports flags that are set
void showFlags()
{
    for(int flag=0; flag < 8; flag++)
    {
      if (getFlag(flag) == true)
         Serial.print("* bit set for flag ");
      else
         Serial.print("bit clear for flag ");

      Serial.println(flag);
    }
    Serial.println();
}
```

This code will print the following:

```
bit clear for flag 0
bit clear for flag 1
bit clear for flag 2
bit clear for flag 3
bit clear for flag 4
bit clear for flag 5
bit clear for flag 6
bit clear for flag 7

bit clear for flag 0
bit clear for flag 1
* bit set for flag 2
bit clear for flag 3
bit clear for flag 4
* bit set for flag 5
bit clear for flag 6
bit clear for flag 7

bit clear for flag 0
bit clear for flag 1
bit clear for flag 2
bit clear for flag 3
bit clear for flag 4
* bit set for flag 5
bit clear for flag 6
bit clear for flag 7
```

Discussion

Reading and setting bits is a common task, and many of the Arduino libraries use this functionality. One of the more common uses of bit operations is to efficiently store and retrieve binary values (on/off, true/false, 1/0, high/low, etc.).

Arduino defines the constants `true` and `HIGH` as 1 and `false` and `LOW` as 0.

The state of eight switches can be packed into a single 8-bit value instead of requiring eight bytes or integers. The example in this recipe's Solution shows how eight values can be individually set or cleared in a single byte.

The term *flag* is a programming term for values that store the state of some aspect of a program. In this sketch, the flag bits are read using `bitRead`, and they are set or cleared using `bitSet` or `bitClear`. These functions take two parameters: the first is the value to read or write (`flags` in this example), and the second is the bit position indicating where the read or write should take place. Bit position 0 is the least significant (rightmost) bit; position 1 is the second position from the right, and so on. So:

```
bitRead(2, 1); // returns 1 : 2 is binary 10 and bit in position 1 is 1
bitRead(4, 1); // returns 0 : 4 is binary 100 and bit in position 1 is 0
```

There is also a function called `bit` that returns the value of each bit position:

```
bit(0)  is equal to 1;
bit(1)  is equal to 2;
bit(2)  is equal to 4;
...
bit(7)  is equal to 128
```

See Also

Arduino references for bit and byte functions:

`lowByte`
: *http://www.arduino.cc/en/Reference/LowByte*

`highByte`
: *http://arduino.cc/en/Reference/HighByte*

`bitRead`
: *http://www.arduino.cc/en/Reference/BitRead*

`bitWrite`
: *http://arduino.cc/en/Reference/BitWrite*

`bitSet`
: *http://arduino.cc/en/Reference/BitSet*

`bitClear`
: *http://arduino.cc/en/Reference/BitClear*

`bit`
: *http://arduino.cc/en/Reference/Bit*

3.13 Shifting Bits

Problem

You need to perform bit operations that shift bits left or right in a `byte`, `int`, or `long`.

Solution

Use the `<<` (bit-shift left) and `>>` (bit-shift right) operators to shift the bits of a value.

Discussion

This fragment sets variable `x` equal to 6. It shifts the bits left by one and prints the new value (`12`). Then that value is shifted right two places (and in this example becomes equal to `3`):

```
int x = 6;
int result = x << 1;  // 6  shifted left 1 is 12
Serial.println(result);
int result = x >> 2;   // 12 shifted right 2 is 3;
Serial.println(result);
```

Here is how this works: 6 shifted left one place equals 12, because the decimal number 6 is 0110 in binary. When the digits are shifted left, the value becomes 1100 (decimal 12). Shifting 1100 right two places becomes 0011 (decimal 3). You may notice that shifting a number left by *n* places is the same as multiplying the value by 2 raised to the power of *n*. Shifting a number right by *n* places is the same as dividing the value by 2 raised to the power of *n*. In other words, the following pairs of expressions are the same:

`x << 1` is the same as `x * 2`.
`x << 2` is the same as `x * 4`.
`x << 3` is the same as `x * 8`.
`x >> 1` is the same as `x / 2`.
`x >> 2` is the same as `x / 4`.
`x >> 3` is the same as `x / 8`.

The Arduino controller chip can shift bits more efficiently than it can multiply and divide, and you may come across code that uses the bit shift to multiply and divide:

```
int c = (a << 1) + (b >> 2); //add (a times 2) plus ( b divided by 4)
```

The expression `(a << 1) + (b >> 2);` does not look much like `(a * 2) + (b / 4);`, but both expressions do the same thing. Indeed, the Arduino compiler is smart enough to recognize that multiplying an integer by a constant that is a power of two is identical to a shift and will produce the same machine code as the version using shift. The source code using arithmetic operators is easier for humans to read, so it is preferred when the intent is to multiply and divide.

See Also

Arduino references for bit and byte functions: `lowByte`, `highByte`, `bitRead`, `bitWrite`, `bitSet`, `bitClear`, and `bit` (see Recipe 3.12)

3.14 Extracting High and Low Bytes in an int or long

Problem

You want to extract the high byte or low byte of an integer; for example, when sending integer values as bytes on a serial or other communication line.

Solution

Use `lowByte(i)` to get the least significant byte from an integer. Use `highByte(i)` to get the most significant byte from an integer.

The following sketch converts an integer value into low and high bytes:

```
//ByteOperators

int intValue = 258; // 258 in hexadecimal notation is 0x102


void setup()
{
  Serial.begin(9600);
}

void loop()
{
  int loWord,hiWord;
  byte loByte, hiByte;


  hiByte = highByte(intValue);
  loByte = lowByte(intValue);

  Serial.println(intValue,DEC);
  Serial.println(intValue,HEX);
  Serial.println(loByte,DEC);
  Serial.println(hiByte,DEC);

  delay(10000); // wait a very long time
}
```

Discussion

The example sketch prints `intValue` followed by the low byte and high byte:

```
258     // the integer value to be converted
102     // the value in hexadecimal notation
2       // the low byte
1       // the high byte
```

To extract the byte values from a `long`, the 32-bit long value first gets broken into two 16-bit *words* that can then be converted into bytes as shown in the earlier code. At the time of this writing, the standard Arduino library did not have a function to perform this operation on a `long`, but you can add the following lines to your sketch to provide this:

```
#define highWord(w) ((w) >> 16)
#define lowWord(w) ((w) & 0xffff)
```

These are *macro expressions*: `hiWord` performs a 16-bit shift operation to produce a 16-bit value, and `lowWord` masks the lower 16 bits using the bitwise And operator (see Recipe 2.20).

The number of bits in an `int` varies on different platforms. On Arduino it is 16 bits, but in other environments it is 32 bits. The term *word* as used here refers to a 16-bit value.

This code converts the 32-bit hex value 0x1020304 to its 16-bit constituent high and low values:

```
loword = lowWord(longValue);
hiword = highWord(longValue);
Serial.println(loword,DEC);
Serial.println(hiword,DEC);
```

This prints the following values:

```
772   // 772 is 0x0304 in hexadecimal
258   // 258 is 0x0102 in hexadecimal
```

Note that 772 in decimal is 0x0304 in hexadecimal, which is the low-order word (16 bits) of the `longValue 0x1020304`. You may recognize 258 from the first part of this recipe as the value produced by combining a high byte of 1 and a low byte of 2 (0x0102 in hexadecimal).

See Also

Arduino references for bit and byte functions: `lowByte`, `highByte`, `bitRead`, `bitWrite`, `bitSet`, `bitClear`, and `bit` (see Recipe 3.12)

3.15 Forming an int or long from High and Low Bytes

Problem

You want to create a 16-bit (`int`) or 32-bit (`long`) integer value from individual bytes; for example, when receiving integers as individual bytes over a serial communication link. This is the inverse operation of Recipe 3.14.

Solution

Use the `word(h,l)` function to convert two bytes into a single Arduino integer. Here is the code from Recipe 3.14 expanded to convert the individual high and low bytes back into an integer:

```
//ByteOperators

int intValue = 0x102;  // 258


void setup()
{
  Serial.begin(9600);
}

void loop()
{
  int loWord,hiWord;
  byte loByte, hiByte;


  hiByte = highByte(intValue);
  loByte = lowByte(intValue);

  Serial.println(intValue,DEC);
  Serial.println(loByte,DEC);
  Serial.println(hiByte,DEC);

  loWord = word(hiByte, loByte);   // convert the bytes back into a word
  Serial.println(loWord,DEC);
  delay(10000); // wait a very long time
}
```

Discussion

The `word(high,low)` expression assembles a high and low byte into a 16-bit value. The code in this recipe's Solution takes the low and high bytes formed as shown in Recipe 3.14, and assembles them back into a word. The output is the integer value, the low byte, the high byte, and the bytes converted back to an integer value:

```
258
2
```

```
1
258
```

Arduino does not have a function to convert a 32-bit long value into two 16-bit words (at the time of this writing), but you can add your own `makeLong()` capability by adding the following line to the top of your sketch:

```
#define makeLong(hi, low)  ((hi) << 16 & (low))
```

This defines a command that will shift the high value 16 bits to the left and add it to the low value:

```
#define makeLong(hi, low)  (((long) hi) << 16 | (low))
#define highWord(w) ((w) >> 16)
#define lowWord(w) ((w) & 0xffff)

// declare a value to test
long longValue = 0x1020304;  // in decimal: 16909060
                             // in binary : 00000001 00000010 00000011 00000100

void setup()
{
  Serial.begin(9600);
}

void loop()
{
  int loWord,hiWord;

  Serial.println(longValue,DEC);  // this prints  16909060
  loWord = lowWord(longValue);    // convert long to two words
  hiWord = highWord(longValue);
  Serial.println(loWord,DEC);     // print the value 772
  Serial.println(hiWord,DEC);     // print the value 258
  longValue = makeLong( hiWord, loWord);  // convert the words back to a long
  Serial.println(longValue,DEC);  // this again prints  16909060

  delay(10000); // wait a very long time
}
```

The output is:

```
16909060
772
258
16909060
```

See Also

Arduino references for bit and byte functions: `lowByte`, `highByte`, `bitRead`, `bitWrite`, `bitSet`, `bitClear`, and `bit` (see Recipe 3.12)

CHAPTER 4
Serial Communications

4.0 Introduction

Serial communications provide an easy and flexible way for your Arduino board to interact with your computer and other devices. This chapter explains how to send and receive information using this capability.

Chapter 1 described how to connect the Arduino serial port to your computer to upload sketches. The upload process sends data from your computer to Arduino and Arduino sends status messages back to the computer to confirm the transfer is working. The recipes here show how you can use this communication link to send and receive any information between Arduino and your computer or another serial device.

Serial communications are also a handy tool for debugging. You can send debug messages from Arduino to the computer and display them on your computer screen or an external LCD display.

The Arduino IDE (described in Recipe 1.3) provides a Serial Monitor (shown in Figure 4-1) to display serial data sent from Arduino.

You can also send data from the Serial Monitor to Arduino by entering text in the text box to the left of the Send button. Baud rate (the speed at which data is transmitted, measured in bits per second) is selected using the drop-down box on the bottom right. You can use the drop down labeled "No line ending" to automatically send a carriage return or a combination of a carriage return and a line at the end of each message sent when clicking the Send button, by changing "No line ending" to your desired option.

Your Arduino sketch can use the serial port to indirectly access (usually via a proxy program written in a language like Processing) all the resources (memory, screen, keyboard, mouse, network connectivity, etc.) that your computer has. Your computer can also use the serial link to interact with sensors or other devices connected to Arduino.

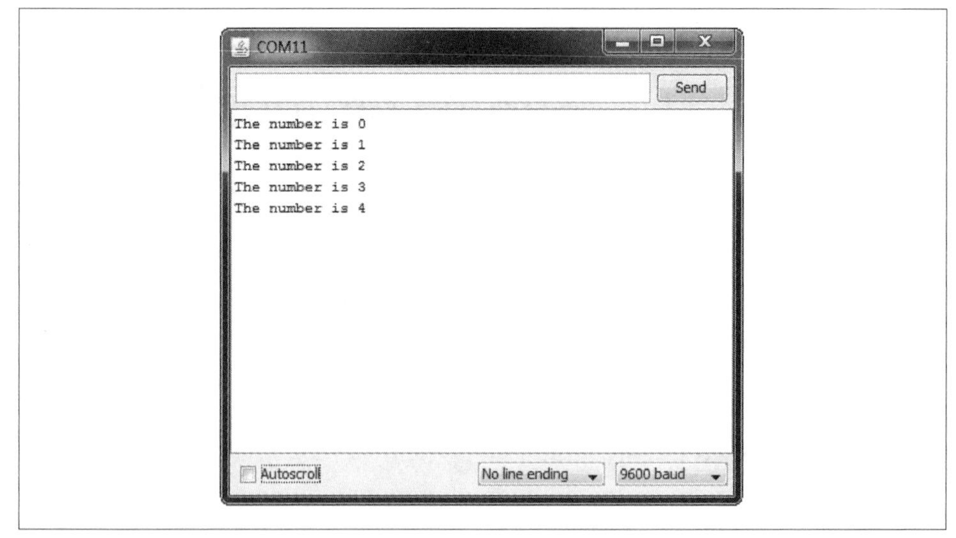

Figure 4-1. Arduino Serial Monitor screen

Implementing serial communications involves hardware and software. The hardware provides the electrical signaling between Arduino and the device it is talking to. The software uses the hardware to send bytes or bits that the connected hardware understands. The Arduino serial libraries insulate you from most of the hardware complexity, but it is helpful for you to understand the basics, especially if you need to troubleshoot any difficulties with serial communications in your projects.

Serial Hardware

Serial hardware sends and receives data as electrical pulses that represent sequential bits. The zeros and ones that carry the information that makes up a byte can be represented in various ways. The scheme used by Arduino is 0 volts to represent a bit value of 0, and 5 volts (or 3.3 volts) to represent a bit value of 1.

Using 0 volts (for 0) and 5 volts (for 1) is very common. This is referred to as the *TTL level* because that was how signals were represented in one of the first implementations of digital logic, called Transistor-Transistor Logic (TTL).

Boards including the Uno, Duemilanove, Diecimila, Nano, and Mega have a chip to convert the hardware serial port on the Arduino chip to Universal Serial Bus (USB) for connection to the hardware serial port. Other boards, such as the Mini, Pro, Pro Mini, Boarduino, Sanguino, and Modern Device Bare Bones Board, do not have USB support and require an adapter for connecting to your computer that converts TTL to USB. See *http://www.arduino.cc/en/Main/Hardware* for more details on these boards.

Some popular USB adapters include:

- Mini USB Adapter (*http://arduino.cc/en/Main/MiniUSB*)
- USB Serial Light Adapter (*http://arduino.cc/en/Main/USBSerial*)
- FTDI USB TTL Adapter (*http://www.ftdichip.com/Products/FT232R.htm*)
- Modern Device USB BUB board (*http://shop.moderndevice.com/products/usb-bub*)
- Seeedstudio UartSBee (*http://www.seeedstudio.com/depot/uartsbee-v31-p-688.html*)

Some serial devices use the RS-232 standard for serial connection. These usually have a nine-pin connector, and an adapter is required to use them with the Arduino. RS-232 is an old and venerated communications protocol that uses voltage levels not compatible with Arduino digital pins.

You can buy Arduino boards that are built for RS-232 signal levels, such as the Freeduino Serial v2.0 (*http://www.nkcelectronics.com/freeduino-serial-v20-board-kit-arduino-diecimila-compatib20.html*).

RS-232 adapters that connect RS-232 signals to Arduino 5V (or 3.3V) pins include the following:

- RS-232 to TTL 3V–5.5V adapter (*http://www.nkcelectronics.com/rs232-to-ttl-converter-board-33v232335.html*)
- P4 RS232 to TTL Serial Adapter Kits (*http://shop.moderndevice.com/products/p4*)
- RS232 Shifter SMD (*http://www.sparkfun.com/commerce/product_info.php?products_id=449*)

A standard Arduino has a single hardware serial port, but serial communication is also possible using software libraries to emulate additional ports (communication channels) to provide connectivity to more than one device. Software serial requires a lot of help from the Arduino controller to send and receive data, so it's not as fast or efficient as hardware serial.

The Arduino Mega has four hardware serial ports that can communicate with up to four different serial devices. Only one of these has a USB adapter built in (you could wire a USB-TTL adapter to any of the other serial ports). Table 4-1 shows the port names and pins used for all of the Mega serial ports.

Table 4-1. Arduino Mega serial ports

Port name	Transmit pin	Receive pin
Serial	1 (also USB)	0 (also USB)
Serial1	18	19
Serial2	16	17
Serial3	14	15

Software Serial

You will usually use the built-in Arduino Serial library to communicate with the hardware serial ports. Serial libraries simplify the use of the serial ports by insulating you from hardware complexities.

Sometimes you need more serial ports than the number of hardware serial ports available. If this is the case, you can use an additional library that uses software to emulate serial hardware. Recipes 4.13 and 4.14 show how to use a software serial library to communicate with multiple devices.

Serial Message Protocol

The hardware or software serial libraries handle sending and receiving information. This information often consists of groups of variables that need to be sent together. For the information to be interpreted correctly, the receiving side needs to recognize where each message begins and ends. Meaningful serial communication, or any kind of machine-to-machine communication, can only be achieved if the sending and receiving sides fully agree how information is organized in the message. The formal organization of information in a message and the range of appropriate responses to requests is called a *communications protocol*.

Messages can contain one or more special characters that identify the start of the message—this is called the *header*. One or more characters can also be used to identify the end of a message—this is called the *footer*. The recipes in this chapter show examples of messages in which the values that make up the body of a message can be sent in either text or binary format.

Sending and receiving messages in text format involves sending commands and numeric values as human-readable letters and words. Numbers are sent as the string of digits that represent the value. For example, if the value is 1234, the characters 1, 2, 3, and 4 are sent as individual characters.

Binary messages comprise the bytes that the computer uses to represent values. Binary data is usually more efficient (requiring fewer bytes to be sent), but the data is not as human-readable as text, which makes it more difficult to debug. For example, Arduino represents 1234 as the bytes 4 and 210 (4 * 256 + 210 = 1234). If the device you are connecting to sends or receives only binary data, that is what you will have to use, but if you have the choice, text messages are easier to implement and debug.

There are many ways to approach software problems, and some of the recipes in this chapter show two or three different ways to achieve a similar result. The differences (e.g., sending text instead of raw binary data) may offer a different balance between simplicity and efficiency. Where choices are offered, pick the solution that you find easiest to understand and adapt—this will probably be the first solution covered. Alternatives may be a little more efficient, or they may be more appropriate for a specific protocol that you want to connect to, but the "right way" is the one you find easiest to get working in your project.

The Processing Development Environment

Some of the examples in this chapter use the Processing language to send and receive serial messages on a computer talking to Arduino.

Processing is a free open source tool that uses a similar development environment to Arduino, but instead of running your sketches on a microcontroller, your Processing sketches run on your computer. You can read more about Processing and download everything you need at the Processing website (*http://processing.org/*).

Processing is based on the Java language, but the Processing code samples in this book should be easy to translate into other environments that support serial communications. Processing comes with some example sketches illustrating communication between Arduino and Processing. SimpleRead is a Processing example that includes Arduino code. In Processing, select File→Examples→Libraries→Serial→SimpleRead to see an example that reads data from the serial port and changes the color of a rectangle when a switch connected to Arduino is pressed and released.

New in Arduino 1.0

Arduino 1.0 introduced a number of Serial enhancements and changes :

- `Serial.flush` now waits for all outgoing data to be sent rather than discarding received data. You can use the following statement to discard all data in the receive buffer: `while(Serial.read() >= 0) ; // flush the receive buffer`
- `Serial.write` and `Serial.print` do not block. Earlier code would wait until all characters were sent before returning. From 1.0, characters sent using `Serial.write` are transmitted in the background (from an interrupt handler) allowing your sketch code to immediately resume processing. This is usually a good thing (it can make the sketch more responsive) but sometimes you want to wait until all characters are sent. You can achieve this by calling `Serial.flush()` immediately following `Serial.write()`.
- Serial print functions return the number of characters printed. This is useful when text output needs to be aligned or for applications that send data that includes the total number of characters sent.
- There is a built-in parsing capability for streams such as Serial to easily extract numbers and find text. See the Discussion section of Recipe 4.5 for more on using this capability with Serial.
- The SoftwareSerial library bundled with Arduino has had significant enhancements; see Recipes 4.13 and 4.14.

- A `Serial.peek` function has been added to let you 'peek' at the next character in the receive buffer. Unlike `Serial.read`, the character is not removed from the buffer with `Serial.peek`.

See Also

An Arduino RS-232 tutorial is available at *http://www.arduino.cc/en/Tutorial/Arduino SoftwareRS232*. Lots of information and links are available at the Serial Port Central website, *http://www.lvr.com/serport.htm*.

In addition, a number of books on Processing are also available:

- *Getting Started with Processing: A Quick, Hands-on Introduction* by Casey Reas and Ben Fry (Make).
- *Processing: A Programming Handbook for Visual Designers and Artists* by Casey Reas and Ben Fry (MIT Press).
- *Visualizing Data* by Ben Fry (O'Reilly; search for it on oreilly.com).
- *Processing: Creative Coding and Computational Art* by Ira Greenberg (Apress).
- *Making Things Talk* by Tom Igoe (Make). This book covers Processing and Arduino and provides many examples of communication code.

4.1 Sending Debug Information from Arduino to Your Computer

Problem

You want to send text and data to be displayed on your PC or Mac using the Arduino IDE or the serial terminal program of your choice.

Solution

This sketch prints sequential numbers on the Serial Monitor:

```
/*
 * SerialOutput sketch
 * Print numbers to the serial port
*/
void setup()
{
  Serial.begin(9600); // send and receive at 9600 baud
}

int number = 0;

void loop()
{
  Serial.print("The number is ");
```

```
    Serial.println(number);     // print the number

    delay(500); // delay half second between numbers
    number++; // to the next number
}
```

Connect Arduino to your computer just as you did in Chapter 1 and upload this sketch. Click the Serial Monitor icon in the IDE and you should see the output displayed as follows:

```
The number is 0
The number is 1
The number is 2
```

Discussion

To display text and numbers from your sketch on a PC or Mac via a serial link, put the `Serial.begin(9600)` statement in `setup()`, and then use `Serial.print()` statements to print the text and values you want to see.

The Arduino Serial Monitor function can display serial data sent from Arduino. To start the Serial Monitor, click the Serial Monitor toolbar icon as shown in Figure 4-2. A new window will open for displaying output from Arduino.

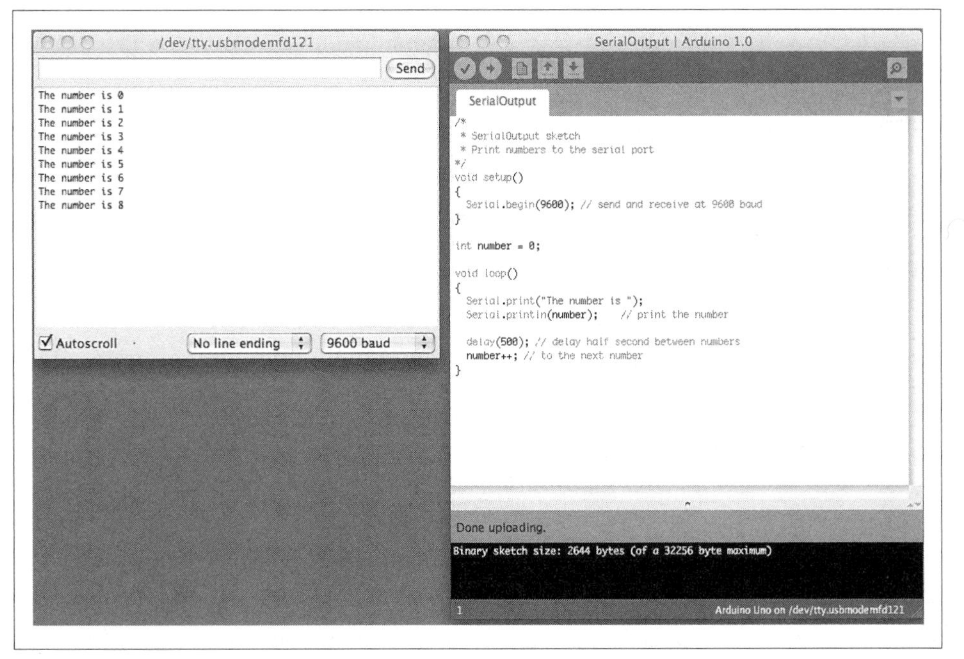

Figure 4-2. Arduino Serial Monitor screen

Your sketch must call the `Serial.begin()` function before it can use serial input or output. The function takes a single parameter: the desired communication speed. You must use the same speed for the sending side and the receiving side, or you will see gobbledygook (or nothing at all) on the screen. This example and most of the others in this book use a speed of 9,600 baud (*baud* is a measure of the number of bits transmitted per second). The 9,600 baud rate is approximately 1,000 characters per second. You can send at lower or higher rates (the range is 300 to 115,200), but make sure both sides use the same speed. The Serial Monitor sets the speed using the baud rate drop down (at the bottom right of the Serial Monitor window in Figure 4-2). If your output looks something like this:

```
`3??f<ÌxÌ□□□ü`³??f<
```

you should check that the selected baud rate on the serial monitor on your computer matches the rate set by `Serial.begin()` in your sketch.

If your send and receive serial speeds are set correctly but you are still getting unreadable text, check that you have the correct board selected in the IDE Tools→Board menu. There are chip speed variants of some boards, if you have selected the wrong one, change it to the correct one and upload to the board again.

You can display text using the `Serial.print()` function. Strings (text within double quotes) will be printed as is (but without the quotes). For example, the following code:

```
Serial.print("The number is ");
```

prints this:

```
The number is
```

The values (numbers) that you print depend on the type of variable; see Recipe 4.2 for more about this. For example, printing an integer will print its numeric value, so if the variable `number` is `1`, the following code:

```
Serial.println(number);
```

will print this:

```
1
```

In the example sketch, the number printed will be `0` when the loop starts and will increase by one each time through the loop. The `ln` at the end of `println` causes the next print statement to start on a new line.

That should get you started printing text and the decimal value of integers. See Recipe 4.2 for more detail on print formatting options.

You may want to consider a third-party terminal program that has more features than Serial Monitor. Displaying data in text or binary format (or both), displaying control characters, and logging to a file are just a few of the additional capabilities available from the many third-party terminal programs. Here are some that have been recommended by Arduino users:

CoolTerm (http://freeware.the-meiers.org/)
: An easy-to-use freeware terminal program for Windows, Mac, and Linux

CuteCom (http://cutecom.sourceforge.net/)
: An open source terminal program for Linux

Bray Terminal (https://sites.google.com/site/terminalbpp/)
: A free executable for the PC

GNU screen (http://www.gnu.org/software/screen/)
: An open source virtual screen management program that supports serial communications; included with Linux and Mac OS X

moserial (http://live.gnome.org/moserial)
: Another open source terminal program for Linux

PuTTY (http://www.chiark.greenend.org.uk/~sgtatham/putty/)
: An open source SSH program for Windows and Linux that supports serial communications

RealTerm (http://realterm.sourceforge.net/)
: An open source terminal program for the PC

ZTerm (http://homepage.mac.com/dalverson/zterm/)
: A shareware program for the Mac

In addition, an article in the Arduino wiki explains how to configure Linux to communicate with Arduino using TTY (see *http://www.arduino.cc/playground/Interfacing/LinuxTTY*).

You can use a liquid crystal display as a serial output device, although it will be very limited in functionality. Check the documentation to see how your display handles carriage returns, as some displays may not automatically advance to a new line after `println` statements.

See Also

The Arduino LiquidCrystal library for text LCDs uses underlying print functionality similar to the Serial library, so you can use many of the suggestions covered in this chapter with that library (see Chapter 11).

4.2 Sending Formatted Text and Numeric Data from Arduino

Problem

You want to send serial data from Arduino displayed as text, decimal values, hexadecimal, or binary.

Solution

You can print data to the serial port in many different formats; here is a sketch that demonstrates all the format options:

```
/*
 * SerialFormatting
 * Print values in various formats to the serial port
 */
char chrValue = 65;  // these are the starting values to print
byte byteValue = 65;
int intValue  = 65;
float floatValue = 65.0;

void setup()
{
  Serial.begin(9600);
}

void loop()
{
  Serial.println("chrValue: ");
  Serial.println(chrValue);
  Serial.write(chrValue);
  Serial.println();
  Serial.println(chrValue,DEC);

  Serial.println("byteValue: ");
  Serial.println(byteValue);
  Serial.write(byteValue);
  Serial.println();
  Serial.println(byteValue,DEC);

  Serial.println("intValue: ");
  Serial.println(intValue);
  Serial.println(intValue,DEC);
  Serial.println(intValue,HEX);
  Serial.println(intValue,OCT);
  Serial.println(intValue,BIN);

  Serial.println("floatValue: ");
  Serial.println(floatValue);

  delay(1000); // delay a second between numbers
  chrValue++;  // to the next value
  byteValue++;
  intValue++;
  floatValue +=1;
}
```

The output (condensed here onto a few lines) is as follows:

```
chrValue:   A  A  65
byteValue:  65 A  65
intValue:   65 65 41 101 1000001
floatValue: 65.00
```

```
chrValue:   B  B  66
byteValue:  66 B  66
intValue:   66 66 42 102 1000010
floatValue: 66.00
```

Discussion

Printing a text string is simple: `Serial.print("hello world");` sends the text string "hello world" to a device at the other end of the serial port. If you want your output to print a new line after the output, use `Serial.println()` instead of `Serial.print()`.

Printing numeric values can be more complicated. The way that byte and integer values are printed depends on the type of variable and an optional formatting parameter. The Arduino language is very easygoing about how you can refer to the value of different data types (see Recipe 2.2 for more on data types). But this flexibility can be confusing, because even when the numeric values are similar, the compiler considers them to be separate types with different behaviors. For example, printing a `char`, `byte`, and `int` of the same value will not necessarily produce the same output.

Here are some specific examples; all of them create variables that have similar values:

```
char asciiValue  = 'A';   // ASCII A has a value of 65
char chrValue    = 65;    // an 8 bit signed character, this also is ASCII 'A'
byte byteValue   = 65;    // an 8 bit unsigned character, this also is ASCII 'A'
int intValue     = 65;    // a 16 bit signed integer set to a value of 65
float floatValue = 65.0;  // float with a value of 65
```

Table 4-2 shows what you will see when you print variables using Arduino routines.

Table 4-2. Output formats using Serial.print

Data type	print (`val`)	print (`val,DEC`)	write (`val`)	print (`val,HEX`)	print (`val,OCT`)	print (`val,BIN`)
`char`	A	65	A	41	101	1000001
`byte`	65	65	A	41	101	1000001
`int`	65	65	A	41	101	1000001
`long`	Format of `long` is the same as `int`					
`float`	65.00	Formatting not supported for floating-point values				
`double`	65.00	`double` is the same as `float`				

The expression `Serial.print(val,BYTE);` is no longer supported in Arduino 1.0.

If your code expects byte variables to behave the same as char variables (that is, for them to print as ASCII), you will need to change this to `Serial.write(val);`.

The sketch in this recipe uses a separate line of source code for each print statement. This can make complex print statements bulky. For example, to print the following line:

```
At 5 seconds: speed = 17, distance = 120
```

you'd typically have to code it like this:

```
Serial.print("At ");
Serial.print(t);
Serial.print(" seconds: speed= ");
Serial.print(s);
Serial.print(", distance= ");
Serial.println(d);
```

That's a lot of code lines for a single line of output. You could combine them like this:

```
Serial.print("At "); Serial.print(t); Serial.print(" seconds, speed= ");
Serial.print(s); Serial.print(", distance= ");Serial.println(d);
```

Or you could use the *insertion-style* capability of the compiler used by Arduino to format your print statements. You can take advantage of some advanced C++ capabilities (streaming insertion syntax and templates) that you can use if you declare a streaming template in your sketch. This is most easily achieved by including the Streaming library developed by Mikal Hart. You can read more about this library and download the code from Mikal's website (*http://arduiniana.org/libraries/streaming/*).

If you use the Streaming library, the following gives the same output as the lines shown earlier:

```
Serial << "At " << t << " seconds, speed= " << s << ", distance = " << d << endl;
```

See Also

Chapter 2 provides more information on data types used by Arduino. The Arduino web reference at *http://arduino.cc/en/Reference/HomePage* covers the serial commands, and the Arduino web reference at *http://www.arduino.cc/playground/Main/StreamingOutput* covers streaming (insertion-style) output.

4.3 Receiving Serial Data in Arduino

Problem

You want to receive data on Arduino from a computer or another serial device; for example, to have Arduino react to commands or data sent from your computer.

Solution

It's easy to receive 8-bit values (chars and bytes), because the `Serial` functions use 8-bit values. This sketch receives a digit (single characters 0 through 9) and blinks the LED on pin 13 at a rate proportional to the received digit value:

```
/*
 * SerialReceive sketch
 * Blink the LED at a rate proportional to the received digit value
*/
const int ledPin = 13; // pin the LED is connected to
int  blinkRate=0;     // blink rate stored in this variable

void setup()
{
  Serial.begin(9600); // Initialize serial port to send and receive at 9600 baud
  pinMode(ledPin, OUTPUT); // set this pin as output
}

void loop()
{
  if ( Serial.available()) // Check to see if at least one character is available
  {
    char ch = Serial.read();
    if( isDigit(ch) ) // is this an ascii digit between 0 and 9?
    {
       blinkRate = (ch - '0');      // ASCII value converted to numeric value
       blinkRate = blinkRate * 100; // actual rate is 100ms times received digit
    }
  }
  blink();
}

// blink the LED with the on and off times determined by blinkRate
void blink()
{
  digitalWrite(ledPin,HIGH);
  delay(blinkRate); // delay depends on blinkrate value
  digitalWrite(ledPin,LOW);
  delay(blinkRate);
}
```

Upload the sketch and send messages using the Serial Monitor. Open the Serial Monitor by clicking the Monitor icon (see Recipe 4.1) and type a digit in the text box at the top of the Serial Monitor window. Clicking the Send button will send the character typed into the text box; if you type a digit, you should see the blink rate change.

Discussion

Converting the received ASCII characters to numeric values may not be obvious if you are not familiar with the way ASCII represents characters. The following converts the character ch to its numeric value:

```
blinkRate = (ch - '0');    // ASCII value converted to numeric value
```

The ASCII characters '0' through '9' have a value of 48 through 57 (see Appendix G). Converting '1' to the numeric value one is done by subtracting '0' because '1' has an ASCII value of 49, so 48 (ASCII '0') must be subtracted to convert this to the number one. For example, if ch is representing the character 1, its ASCII value is 49. The expression 49- '0' is the same as 49-48. This equals 1, which is the numeric value of the character 1.

In other words, the expression (ch - '0') is the same as (ch - 48); this converts the ASCII value of the variable ch to a numeric value.

Receiving numbers with more than one digit involves accumulating characters until a character that is not a valid digit is detected. The following code uses the same setup() and blink() functions as those shown earlier, but it gets digits until the newline character is received. It uses the accumulated value to set the blink rate.

The newline character (ASCII value 10) can be appended automatically each time you click Send. The Serial Monitor has a drop-down box at the bottom of the Serial Monitor screen (see Figure 4-1); change the option from "No line ending" to "Newline."

Change the code as follows:

```
int value;

void loop()
{
  if( Serial.available())
  {
    char ch = Serial.read();
    if( isDigit(ch) )// is this an ascii digit between 0 and 9?
    {
       value = (value * 10) + (ch - '0'); // yes, accumulate the value
    }
    else if (ch == 10)  // is the character the newline character?
    {
       blinkRate = value;  // set blinkrate to the accumulated value
       Serial.println(blinkRate);
       value = 0; // reset val to 0 ready for the next sequence of digits
    }
  }
  blink();
}
```

Enter a value such as 123 into the Monitor text box and click Send, and the blink delay will be set to 123 milliseconds. Each digit is converted from its ASCII value to its numeric value. Because the numbers are decimal numbers (base 10), each successive number is multiplied by 10. For example, the value of the number 234 is 2 * 100 + 3 * 10 + 4. The code to accomplish that is:

```
    if( isDigit(ch) ) // is this an ascii digit between 0 and 9?
    {
       value = (value * 10) + (ch - '0'); // yes, accumulate the value
    }
```

If you want to handle negative numbers, your code needs to recognize a leading minus (`'-'`) sign. In this example, each numeric value must be separated by a character that is not a digit or minus sign:

```
int value = 0;
int sign = 1;

void loop()
{
  if( Serial.available())
  {
    char ch = Serial.read();
    if( isDigit(ch) ) // is this an ascii digit between 0 and 9?
       value = (value * 10) + (ch - '0'); // yes, accumulate the value
    else if( ch == '-')
       sign = -1;
    else // this assumes any char not a digit or minus sign terminates the value
    {
       value = value * sign ;  // set value to the accumulated value
       Serial.println(value);
       value = 0; // reset value to 0 ready for the next sequence of digits
       sign = 1;
    }
  }
}
```

Another approach to converting text strings representing numbers is to use the C language conversion function called **atoi** (for **int** variables) or **atol** (for **long** variables). These obscurely named functions convert a string into integers or long integers. To use them you have to receive and store the entire string in a character array before you can call the conversion function.

This code fragment terminates the incoming digits on any character that is not a digit (or if the buffer is full):

```
const int MaxChars = 5; // an int string contains up to 5 digits and
                        // is terminated by a 0 to indicate end of string
char strValue[MaxChars+1]; // must be big enough for digits and terminating null
int index = 0;         // the index into the array storing the received digits

void loop()
{
  if( Serial.available())
  {
    char ch = Serial.read();
    if( index < MaxChars && isDigit(ch) ){
      strValue[index++] = ch; // add the ASCII character to the string;
    }
    else
    {
      // here when buffer full or on the first non digit
      strValue[index] = 0;        // terminate the string with a 0
      blinkRate = atoi(strValue);  // use atoi to convert the string to an int
      index = 0;
```

```
    }
  }
  blink();
  }
```

strValue is a numeric string built up from characters received from the serial port.

> See Recipe 2.6 for information about character strings.

atoi (short for ASCII to integer) is a function that converts a character string to an integer (atol converts to a long integer).

Arduino 1.0 added the serialEvent function that you can use to handle incoming serial characters. If you have code within a serialEvent function in your sketch, this will be called once each time through the loop function. The following sketch performs the same function as the first sketch in this Recipe but uses serialEvent to handle the incoming characters:

```
/*
 * SerialReceive sketch
 * Blink the LED at a rate proportional to the received digit value
 */
const int ledPin = 13; // pin the LED is connected to
int   blinkRate=0;     // blink rate stored in this variable

void setup()
{
  Serial.begin(9600); // Initialize serial port to send and receive at 9600 baud
  pinMode(ledPin, OUTPUT); // set this pin as output
}

void loop()
{
  blink();
}

void serialEvent()
{
  while(Serial.available())
  {
    char ch = Serial.read();
    Serial.write(ch);
    if( isDigit(ch) ) // is this an ascii digit between 0 and 9?
    {
      blinkRate = (ch - '0');      // ASCII value converted to numeric value
      blinkRate = blinkRate * 100; // actual rate is 100mS times received digit
    }
  }
}
```

```
// blink the LED with the on and off times determined by blinkRate
void blink()
{
  digitalWrite(ledPin,HIGH);
  delay(blinkRate); // delay depends on blinkrate value
  digitalWrite(ledPin,LOW);
  delay(blinkRate);
}
```

Arduino 1.0 also introduced the `parseInt` and `parseFloat` methods that simplify extracting numeric values from Serial (it also works with Ethernet and other objects derived from the `Stream` class; see the introduction to Chapter 15 for more about stream-parsing with the networking objects).

`Serial.parseInt()` and `Serial.parseFloat()` read Serial characters and return their numeric representation. Nonnumeric characters before the number are ignored and the number ends with the first character that is not a numeric digit (or '`.`' if using `parseFloat`.)

See the discussion of Recipe 4.5 for an example showing `parseInt` used to find and extract numbers from Serial data.

See Also

A web search for "atoi" or "atol" provides many references to these functions. Also see the Wikipedia reference at *http://en.wikipedia.org/wiki/Atoi.*

4.4 Sending Multiple Text Fields from Arduino in a Single Message

Problem

You want to send a message that contains more than one piece of information (field). For example, your message may contain values from two or more sensors. You want to use these values in a program such as Processing, running on your PC or Mac.

Solution

The easiest way to do this is to send a text string with all the fields separated by a delimiting (separating) character, such as a comma:

```
// CommaDelimitedOutput sketch

void setup()
{
  Serial.begin(9600);
}

void loop()
```

```
{
  int value1 = 10;     // some hardcoded values to send
  int value2 = 100;
  int value3 = 1000;

  Serial.print('H'); // unique header to identify start of message
  Serial.print(",");
  Serial.print(value1,DEC);
  Serial.print(",");
  Serial.print(value2,DEC);
  Serial.print(",");
  Serial.print(value3,DEC);
  Serial.print(",");  // note that a comma is sent after the last field
  Serial.println();  // send a cr/lf
  delay(100);
}
```

Here is the Processing sketch that reads this data from the serial port:

```
// Processing Sketch to read comma delimited serial
// expects format: H,1,2,3,

import processing.serial.*;

Serial myPort;        // Create object from Serial class
char HEADER = 'H';    // character to identify the start of a message
short LF = 10;        // ASCII linefeed

// WARNING!
// If necessary change the definition below to the correct port
short portIndex = 1;  // select the com port, 0 is the first port

void setup() {
  size(200, 200);
  println(Serial.list());
  println(" Connecting to -> " + Serial.list()[portIndex]);
  myPort = new Serial(this,Serial.list()[portIndex], 9600);
}

void draw() {
}

void serialEvent(Serial p)
{
  String message = myPort.readStringUntil(LF); // read serial data

  if(message != null)
  {
    print(message);
    String [] data = message.split(","); // Split the comma-separated message
    if(data[0].charAt(0) == HEADER && data.length > 3) // check validity
    {
      for( int i = 1; i < data.length-1; i++) // skip the header & end if line
      {
        println("Value " +  i + " = " + data[i]);  // Print the field values
      }
```

```
            println();
        }
    }
}
```

Discussion

The Arduino code in this recipe's Solution will send the following text string to the serial port (`\r` indicates a carriage return and `\n` indicates a line feed):

```
H,10,100,1000,\r\n
```

You must choose a separating character that will never occur within actual data; if your data consists only of numeric values, a comma is a good choice for a delimiter. You may also want to ensure that the receiving side can determine the start of a message to make sure it has all the data for all the fields. You do this by sending a header character to indicate the start of the message. The header character must also be unique; it should not appear within any of the data fields and it must also be different from the separator character. The example here uses an uppercase *H* to indicate the start of the message. The message consists of the header, three comma-separated numeric values as ASCII strings, and a carriage return and line feed.

The carriage return and line-feed characters are sent whenever Arduino prints using the `println()` function, and this is used to help the receiving side know that the full message string has been received. A comma is sent after the last numerical value to aid the receiving side in detecting the end of the value.

The Processing code reads the message as a string and uses the Java `split()` method to create an array from the comma-separated fields.

In most cases, the first serial port will be the one you want when using a Mac and the last serial port will be the one you want when using Windows. The Processing sketch includes code that shows the ports available and the one currently selected—check that this is the port connected to Arduino.

Using Processing to display sensor values can save hours of debugging time by helping you to visualize the data. The following Processing sketch adds real-time visual display of up to 12 values sent from Arduino. This version displays 8-bit values in a range from –127 to +127 and was created to demonstrate the nunchuck sketch in Recipe 13.2:

```
/*
 * ShowSensorData.
 *
 * Displays bar graph of CSV sensor data ranging from -127 to 127
 * expects format as: "Data,s1,s2,...s12\n" (any number of to 12 sensors is supported)
 * labels can be sent as follows: "Labels,label1, label2,...label12\n");
 */
```

```
import processing.serial.*;

Serial myPort;  // Create object from Serial class
String message = null;
PFont fontA;    // font to display servo pin number
int fontSize = 12;

int maxNumberOfLabels = 12;

int rectMargin = 40;
int windowWidth = 600;
int windowHeight = rectMargin + (maxNumberOfLabels + 1) * (fontSize *2);
int rectWidth = windowWidth - rectMargin*2;
int rectHeight = windowHeight - rectMargin;
int rectCenter = rectMargin + rectWidth / 2;

int origin = rectCenter;
int minValue = -127;
int maxValue = 127;

float scale = float(rectWidth) / (maxValue - minValue);

String [] sensorLabels = {"s1", "s2", "s3", "s4", "s5", "s6", "s7", "s8", "s9",
                          "s10", "s11", "s12"};
// this will be changed to the number of labels actually received
int labelCount = maxNumberOfLabels;

void setup() {
  size(windowWidth, windowHeight);
  short portIndex = 1;  // select the com port, 0 is the first port
  String portName = Serial.list()[portIndex];
  println(Serial.list());
  println(" Connecting to -> " + portName) ;
  myPort = new Serial(this, portName, 57600);
  fontA = createFont("Arial.normal", fontSize);
  textFont(fontA);
  labelCount = sensorLabels.length;
}

void drawGrid() {
  fill(0);
  text(minValue, xPos(minValue), rectMargin-fontSize);
  line(xPos(minValue), rectMargin, xPos(minValue), rectHeight + fontSize);
  text((minValue+maxValue)/2, rectCenter, rectMargin-fontSize);
  line(rectCenter, rectMargin, rectCenter, rectHeight + fontSize);
  text(maxValue, xPos(maxValue), rectMargin-fontSize);
  line( xPos(maxValue), rectMargin, xPos(maxValue), rectHeight + fontSize);

  for (int i=0; i < labelCount; i++) {
    text(sensorLabels[i], fontSize, yPos(i));
    text(sensorLabels[i], xPos(maxValue) + fontSize, yPos(i));
  }
}
```

```
int yPos(int index) {
  return rectMargin + fontSize + (index * fontSize*2);
}

int xPos(int value) {
  return origin  + int(scale * value);
}

void drawBar(int yIndex, int value) {
  rect(origin, yPos(yIndex)-fontSize, value * scale, fontSize);   //draw the value
}

void draw() {

  while (myPort.available () > 0) {
    try {
      message = myPort.readStringUntil(10);
      if (message != null) {
        print(message);
        String [] data  = message.split(","); // Split the CSV message
        if ( data[0].equals("Labels") ) { // check for label header
          labelCount = min(data.length-1, maxNumberOfLabels) ;
          arrayCopy(data, 1, sensorLabels, 0, labelCount );
        }
        else if ( data[0].equals("Data"))// check for data header
        {
          background(255);
          drawGrid();
          fill(204);
          println(data.length);
          for ( int i=1; i <= labelCount && i < data.length-1; i++)
          {
              drawBar(i-1, Integer.parseInt(data[i]));
          }
        }
      }
    }
    catch (Exception e) {
      e.printStackTrace(); // Display whatever error we received
    }
  }
}
```

Figure 4-3 shows how nunchuck accelerometer values (`aX,Ay,aZ`) and joystick (`jX,Jy`) values are displayed. Bars will appear when the nunchuck buttons (`bC` and `bZ`) are pressed.

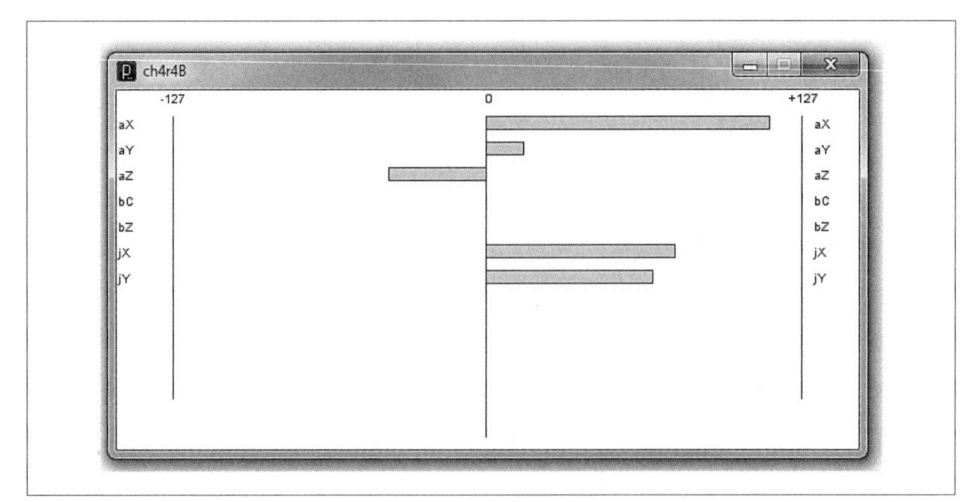

Figure 4-3. Processing screen showing nunchuck sensor data

The range of values and the origin of the graph can be easily changed if desired. For example, to display bars originating at the lefthand axis with values from 0 to 1024, use the following:

```
int origin = rectMargin; // rectMargin is the left edge of the graphing area
int minValue = 0;
int maxValue = 1024;
```

If you don't have a nunchuck, you can generate values with the following simple sketch that displays analog input values. If you don't have any sensors to connect, running your fingers along the bottom of the analog pins will produce levels that can be viewed in the Processing sketch. The values range from 0 to 1023, so change the origin and min and max values in the Processing sketch, as described in the previous paragraph:

```
void setup() {
  Serial.begin(57600);
  delay(1000);
  Serial.println("Labels,A0,A1,A2,A3,A4,A5");
}

void loop() {
  Serial.print("Data,");
  for(int i=0; i < 6; i++)
  {
    Serial.print( analogRead(i) );
    Serial.print(",");
  }
  Serial.print('\n'); // newline character
  delay(100);
}
```

See Also

The Processing website provides more information on installing and using this programming environment. See *http://processing.org/*.

4.5 Receiving Multiple Text Fields in a Single Message in Arduino

Problem

You want to receive a message that contains more than one field. For example, your message may contain an identifier to indicate a particular device (such as a motor or other actuator) and what value (such as speed) to set it to.

Solution

Arduino does not have the `split()` function used in the Processing code in Recipe 4.4, but similar functionality can be implemented as shown in this recipe. The following code receives a message with three numeric fields separated by commas. It uses the technique described in Recipe 4.4 for receiving digits, and it adds code to identify comma-separated fields and store the values into an array:

```
/*
 * SerialReceiveMultipleFields sketch
 * This code expects a message in the format: 12,345,678
 * This code requires a newline character to indicate the end of the data
 * Set the serial monitor to send newline characters
 */

const int NUMBER_OF_FIELDS = 3; // how many comma separated fields we expect
int fieldIndex = 0;            // the current field being received
int values[NUMBER_OF_FIELDS];   // array holding values for all the fields



void setup()
{
  Serial.begin(9600); // Initialize serial port to send and receive at 9600 baud
}

void loop()
{
  if( Serial.available())
  {
    char ch = Serial.read();
    if(ch >= '0' && ch <= '9') // is this an ascii digit between 0 and 9?
    {
      // yes, accumulate the value if the fieldIndex is within range
      // additional fields are not stored
      if(fieldIndex < NUMBER_OF_FIELDS) {
```

```
          values[fieldIndex] = (values[fieldIndex] * 10) + (ch - '0');
        }
      }
      else if (ch == ',')  // comma is our separator, so move on to the next field
      {
          fieldIndex++;    // increment field index
      }
      else
      {
        // any character not a digit or comma ends the acquisition of fields
        // in this example it's the newline character sent by the Serial Monitor

        // print each of the stored fields
        for(int i=0; i < min(NUMBER_OF_FIELDS, fieldIndex+1); i++)
        {
          Serial.println(values[i]);
          values[i] = 0; // set the values to zero, ready for the next message
        }
        fieldIndex = 0;  // ready to start over
      }
    }
  }
```

Discussion

This sketch accumulates values (as explained in Recipe 4.3), but here each value is added to an array (which must be large enough to hold all the fields) when a comma is received. A character other than a digit or comma (such as the newline character; see Recipe 4.3) triggers the printing of all the values that have been stored in the array. You can either type a nondigit, noncomma character before pressing Send, or set the "No line ending" menu at the bottom right of the Serial Monitor to some other option.

Arduino 1.0 introduced the `parseInt` method that makes it easy to extract information from serial and web streams. Here is an example of how to use this capability (Chapter 15 has more examples of stream parsing).

The following sketch uses `parseInt` to provide similar functionality to the previous sketch:

```
// Receive multiple numeric fields using Arduino 1.0 Stream parsing

const int NUMBER_OF_FIELDS = 3; // how many comma-separated fields we expect
int fieldIndex = 0;             // the current field being received
int values[NUMBER_OF_FIELDS];   // array holding values for all the fields


void setup()
{
  Serial.begin(9600); // Initialize serial port to send and receive at 9600 baud
}

void loop()
{
```

```
  if( Serial.available()) {
    for(fieldIndex = 0; fieldIndex  < 3; fieldIndex ++)
    {
      values[fieldIndex] = Serial.parseInt(); // get a numeric value

    }
    Serial.print( fieldIndex);
    Serial.println(" fields received:");
    for(int i=0; i <  fieldIndex; i++)
    {
       Serial.println(values[i]);
    }
    fieldIndex = 0;  // ready to start over
  }
}
```

The stream-parsing functions will time out waiting for a character; the default is one second. If no digits have been received and `parseInt` times out then it will return 0. You can change the timeout by calling `Stream.setTimeout(timeoutPeriod)`. The timeout parameter is a long integer indicating the number of milliseconds, so the timeout range is from 1 millisecond to 2,147,483,647 milliseconds.

`Stream.setTimeout(2147483647);` will change the timeout interval to just under 25 days.

Here is a summary of the methods supported by Arduino 1.0 Stream parsing (not all are used in the preceding example):

`boolean find(char *target);`
: Reads from the stream until the given target is found. It returns `true` if the target string is found. A return of `false` means the data has not been found anywhere in the stream and that there is no more data available. Note that Stream parsing takes a single pass through the stream; there is no way to go back to try to find or get something else (see the `findUntil` method).

`boolean findUntil(char *target, char *terminate);`
: Similar to the `find` method, but the search will stop if the terminate string is found. Returns `true` only if the target is found. This is useful to stop a search on a keyword or terminator. For example:

```
finder.findUntil("target", "\n");
```

will try to seek to the string `"value"`, but will stop at a newline character so that your sketch can do something else if the target is not found.

`long parseInt();`
: Returns the first valid (long) integer value. Leading characters that are not digits or a minus sign are skipped. The integer is terminated by the first nondigit character following the number. If no digits are found, the function returns `0`.

`long parseInt(char skipChar);`
: Same as `parseInt`, but the given `skipChar` within the numeric value is ignored. This can be helpful when parsing a single numeric value that uses a comma between

blocks of digits in large numbers, but bear in mind that text values formatted with commas cannot be parsed as a comma-separated string (for example, 32,767 would be parsed as 32767).

`float parseFloat();`
: The `float` version of `parseInt`.

`size_t readBytes(char *buffer, size_t length);`
: Puts the incoming characters into the given buffer until timeout or length characters have been read. Returns the number of characters placed in the buffer.

`size_t readBytesUntil(char terminator,char *buf,size_t length);`
: Puts the incoming characters into the given buffer until the `terminator` character is detected. Strings longer than the given `length` are truncated to fit. The function returns the number of characters placed in the buffer.

See Also

Chapter 15 provides more examples of Stream parsing used to find and extract data from a stream.

4.6 Sending Binary Data from Arduino

Problem

You need to send data in binary format, because you want to pass information with the fewest number of bytes or because the application you are connecting to only handles binary data.

Solution

This sketch sends a header followed by two integer (16-bit) values as binary data. The values are generated using the Arduino `random` function (see Recipe 3.11):

```
/*
 * SendBinary sketch
 * Sends a header followed by two random integer values as binary data.
*/

int intValue;    // an integer value (16 bits)

void setup()
{
  Serial.begin(9600);
}

void loop()
{
  Serial.print('H'); // send a header character

  // send a random integer
```

```
    intValue = random(599); // generate a random number between 0 and 599
    // send the two bytes that comprise an integer
    Serial.write(lowByte(intValue));  // send the low byte
    Serial.write(highByte(intValue)); // send the high byte

    // send another random integer
    intValue = random(599); // generate a random number between 0 and 599
    // send the two bytes that comprise an integer
    Serial.write(lowByte(intValue));  // send the low byte
    Serial.write(highByte(intValue)); // send the high byte

    delay(1000);
}
```

Discussion

Sending binary data requires careful planning, because you will get gibberish unless the sending side and the receiving side understand and agree exactly how the data will be sent. Unlike text data, where the end of a message can be determined by the presence of the terminating carriage return (or another unique character you pick), it may not be possible to tell when a binary message starts or ends by looking just at the data—data that can have any value can therefore have the value of a header or terminator character.

This can be overcome by designing your messages so that the sending and receiving sides know exactly how many bytes are expected. The end of a message is determined by the number of bytes sent rather than detection of a specific character. This can be implemented by sending an initial value to say how many bytes will follow. Or you can fix the size of the message so that it's big enough to hold the data you want to send. Doing either of these is not always easy, as different platforms and languages can use different sizes for the binary data types—both the number of bytes and their order may be different from Arduino. For example, Arduino defines an `int` as two bytes, but Processing (Java) defines an `int` as four bytes (`short` is the Java type for a 16-bit integer). Sending an `int` value as text (as seen in earlier text recipes) simplifies this problem because each individual digit is sent as a sequential digit (just as the number is written). The receiving side recognizes when the value has been completely received by a carriage return or other nondigit delimiter. Binary transfers can only know about the composition of a message if it is defined in advance or specified in the message.

This recipe's Solution requires an understanding of the data types on the sending and receiving platforms and some careful planning. Recipe 4.7 shows example code using the Processing language to receive these messages.

Sending single bytes is easy; use `Serial.write(byteVal)`. To send an integer from Arduino you need to send the low and high bytes that make up the integer (see Recipe 2.2 for more on data types). You do this using the `lowByte` and `highByte` functions (see Recipe 3.14):

```
Serial.write(lowByte(intValue), BYTE);
Serial.write(highByte(intValue), BYTE);
```

Sending a long integer is done by breaking down the four bytes that comprise a `long` in two steps. The `long` is first broken into two 16-bit integers; each is then sent using the method for sending integers described earlier:

```
int longValue = 1000;
int intValue;
```

First you send the lower 16-bit integer value:

```
intValue = longValue && 0xFFFF;  // get the value of the lower 16 bits
Serial.write(lowByte(intVal));
Serial.writet(highByte(intVal));
```

Then you send the higher 16-bit integer value:

```
intValue = longValue >> 16;  // get the value of the higher 16 bits
Serial.write(lowByte(intVal));
Serial.writet(highByte(intVal));
```

You may find it convenient to create functions to send the data. Here is a function that uses the code shown earlier to print a 16-bit integer to the serial port:

```
// function to send the given integer value to the serial port
void sendBinary(int value)
{
  // send the two bytes that comprise a two byte (16 bit) integer
  Serial.write(lowByte(value));  // send the low byte
  Serial.write(highByte(value)); // send the high byte
}
```

The following function sends the value of a `long` (4-byte) integer by first sending the two low (rightmost) bytes, followed by the high (leftmost) bytes:

```
// function to send the given long integer value to the serial port
void sendBinary(long value)
{
  // first send the low 16 bit integer value
  int temp = value && 0xFFFF;  // get the value of the lower 16 bits
  sendBinary(temp);
  // then send the higher 16 bit integer value:
  temp = value >> 16;  // get the value of the higher 16 bits
  sendBinary(temp);
}
```

These functions to send binary `int` and `long` values have the same name: `sendBinary`. The compiler distinguishes them by the type of value you use for the parameter. If your code calls `printBinary` with a 2-byte value, the version declared as `void sendBinary(int value)` will be called. If the parameter is a `long` value, the version declared as `void sendBinary(long value)` will be called. This behavior is called *function overloading*. Recipe 4.2 provides another illustration of this; the different functionality you saw in `Serial.print` is due to the compiler distinguishing the different variable types used.

You can also send binary data using *structures*. Structures are a mechanism for organizing data, and if you are not already familiar with their use you may be better off sticking with the solutions described earlier. For those who are comfortable with the concept of structure pointers, the following is a function that will send the bytes within a structure to the serial port as binary data:

```
void sendStructure( char *structurePointer, int structureLength)
{
  int i;

  for (i = 0 ; i < structureLength ; i++)
    serial.write(structurePointer[i]);
  }

sendStructure((char *)&myStruct, sizeof(myStruct));
```

Sending data as binary bytes is more efficient than sending data as text, but it will only work reliably if the sending and receiving sides agree exactly on the composition of the data. Here is a summary of the important things to check when writing your code:

Variable size
: Make sure the size of the data being sent is the same on both sides. An integer is 2 bytes on Arduino, 4 bytes on most other platforms. Always check your programming language's documentation on data type size to ensure agreement. There is no problem with receiving a 2-byte Arduino integer as a 4-byte integer in Processing as long as Processing expects to get only two bytes. But be sure that the sending side does not use values that will overflow the type used by the receiving side.

Byte order
: Make sure the bytes within an `int` or `long` are sent in the same order expected by the receiving side.

Synchronization
: Ensure that your receiving side can recognize the beginning and end of a message. If you start listening in the middle of a transmission stream, you will not get valid data. This can be achieved by sending a sequence of bytes that won't occur in the body of a message. For example, if you are sending binary values from `analogRead`, these can only range from 0 to 1,023, so the most significant byte must be less than 4 (the `int` value of 1,023 is stored as the bytes 3 and 255); therefore, there will never be data with two consecutive bytes greater than 3. So, sending two bytes of 4 (or any value greater than 3) cannot be valid data and can be used to indicate the start or end of a message.

Structure packing
: If you send or receive data as structures, check your compiler documentation to make sure the *packing* is the same on both sides. Packing is the padding that a compiler uses to align data elements of different sizes in a structure.

Flow control

Either choose a transmission speed that ensures that the receiving side can keep up with the sending side, or use some kind of *flow control*. Flow control is a handshake that tells the sending side that the receiver is ready to get more data.

See Also

Chapter 2 provides more information on the variable types used in Arduino sketches.

Also, check the Arduino references for `lowByte` at *http://www.arduino.cc/en/Reference/LowByte* and `highByte` at *http://www.arduino.cc/en/Reference/HighByte*.

The Arduino compiler packs structures on byte boundaries; see the documentation for the compiler you use on your computer to set it for the same packing. If you are not clear on how to do this, you may want to avoid using structures to send data.

For more on flow control, see *http://en.wikipedia.org/wiki/Flow_control*.

4.7 Receiving Binary Data from Arduino on a Computer

Problem

You want to respond to binary data sent from Arduino in a programming language such as Processing. For example, you want to respond to Arduino messages sent in Recipe 4.6.

Solution

This recipe's Solution depends on the programming environment you use on your PC or Mac. If you don't already have a favorite programming tool and want one that is easy to learn and works well with Arduino, Processing is an excellent choice.

Here are the two lines of Processing code to read a byte, taken from the Processing `SimpleRead` example (see this chapter's introduction):

```
if ( myPort.available() > 0) {  // If data is available,
  val = myPort.read();          // read it and store it in val
```

As you can see, this is very similar to the Arduino code you saw in earlier recipes.

The following is a Processing sketch that sets the size of a rectangle proportional to the integer values received from the Arduino sketch in Recipe 4.6:

```
/*
 * ReceiveBinaryData_P
 *
 * portIndex must be set to the port connected to the Arduino
 */
import processing.serial.*;

Serial myPort;        // Create object from Serial class
```

```
short portIndex = 1;  // select the com port, 0 is the first port

char HEADER = 'H';
int value1, value2;         // Data received from the serial port

void setup()
{
  size(600, 600);
  // Open whatever serial port is connected to Arduino.
  String portName = Serial.list()[portIndex];
  println(Serial.list());
  println(" Connecting to -> " + Serial.list()[portIndex]);
  myPort = new Serial(this, portName, 9600);
}

void draw()
{
  // read the header and two binary *(16 bit) integers:
  if ( myPort.available() >= 5)  // If at least 5 bytes are available,
  {
    if( myPort.read() == HEADER) // is this the header
    {
      value1 = myPort.read();                 // read the least significant byte
      value1 =  myPort.read() * 256 + value1; // add the most significant byte

      value2 = myPort.read();                 // read the least significant byte
      value2 =  myPort.read() * 256 + value2; // add the most significant byte

      println("Message received: " + value1 + "," + value2);
    }
  }
  background(255);            // Set background to white
  fill(0);                    // set fill to black

  // draw rectangle with coordinates based on the integers received from Arduino
  rect(0, 0, value1,value2);
}
```

Discussion

The Processing language influenced Arduino, and the two are intentionally similar. The `setup` function in Processing is used to handle one-time initialization, just like in Arduino. Processing has a display window, and `setup` sets its size to 600 × 600 pixels with the call to `size(600,600)`.

The line `String portName = Serial.list()[portIndex];` selects the serial port—in Processing, all available serial ports are contained in the `Serial.list` object and this example uses the value of a variable called `portIndex`. `println(Serial.list())` prints all the available ports, and the line `myPort = new Serial(this, portName, 9600);` opens the port selected as `portName`. Ensure that you set `portIndex` to the serial port that is connected to your Arduino (Arduino is usually the first port on a Mac; on Windows, it's usually the last port if Arduino is the most recent serial device installed).

The `draw` function in Processing works like `loop` in Arduino; it is called repeatedly. The code in `draw` checks if data is available on the serial port; if so, bytes are read and converted to the integer value represented by the bytes. A rectangle is drawn based on the integer values received.

See Also

You can read more about Processing on the Processing website (*http://processing.org/*).

4.8 Sending Binary Values from Processing to Arduino

Problem

You want to send binary bytes, integers, or long values from Processing to Arduino. For example, you want to send a message consisting of a message identifier "tag" and two 16-bit values.

Solution

Use this code:

```
// Processing Sketch

/* SendingBinaryToArduino
 * Language: Processing
 */
import processing.serial.*;

Serial myPort;  // Create object from Serial class
public static final char HEADER    = 'H';
public static final char MOUSE_TAG = 'M';

void setup()
{
  size(512, 512);
  String portName = Serial.list()[1];
  myPort = new Serial(this, portName, 9600);
}

void draw(){
}

void serialEvent(Serial p) {
  // handle incoming serial data
  String inString = myPort.readStringUntil('\n');
  if(inString != null) {
    print( inString );   // echo text string from Arduino
  }
}

void mousePressed() {
```

```
  sendMessage(MOUSE_TAG, mouseX, mouseY);
}

void sendMessage(char tag, int x, int y){
  // send the given index and value to the serial port
  myPort.write(HEADER);
  myPort.write(tag);
  myPort.write((char)(x / 256)); // msb
  myPort.write(x & 0xff);  //lsb
  myPort.write((char)(y / 256)); // msb
  myPort.write(y & 0xff);  //lsb
}
```

When the mouse is clicked in the Processing window, `sendMessage` will be called with the 8-bit tag indicating that this is a mouse message and the two 16-bit mouse `x` and `y` coordinates. The `sendMessage` function sends the 16-bit `x` and `y` values as two bytes, with the most significant byte first.

Here is the Arduino code to receive these messages and echo the results back to Processing:

```
// BinaryDataFromProcessing
// These defines must mirror the sending program:
const char HEADER       = 'H';
const char MOUSE_TAG    = 'M';
const int  TOTAL_BYTES  = 6  ; // the total bytes in a message

void setup()
{
  Serial.begin(9600);
}

void loop(){
  if ( Serial.available() >= TOTAL_BYTES)
  {
     if( Serial.read() == HEADER)
    {
      char tag = Serial.read();
      if(tag == MOUSE_TAG)
      {
        int x = Serial.read() * 256;
        x = x + Serial.read();
        int y = Serial.read() * 256;
        y = y + Serial.read();
        Serial.print("Received mouse msg, x = ");
        Serial.print(x);
        Serial.print(", y =  ");
        Serial.println(y);
      }
      else
      {
        Serial.print("got message with unknown tag ");
        Serial.write(tag);
      }
    }
```

```
    }
  }
```

Discussion

The Processing code sends a header byte to indicate that a valid message follows. This is needed so Arduino can synchronize if it starts up in the middle of a message or if the serial connection can lose data, such as with a wireless link. The tag provides an additional check for message validity and it enables any other message types you may want to send to be handled individually. In this example, the function is called with three parameters: a tag and the 16-bit `x` and `y` mouse coordinates.

The Arduino code checks that at least `MESSAGE_BYTES` have been received, ensuring that the message is not processed until all the required data is available. After the header and tag are checked, the 16-bit values are read as two bytes, with the first multiplied by 256 to restore the most significant byte to its original value.

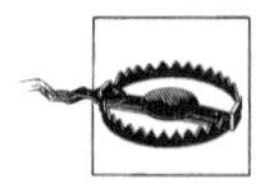

The sending side and receiving side must use the same message size for binary messages to be handled correctly. If you want to increase or decrease the number of bytes to send, change `TOTAL_BYTES` in the Arduino code to match.

4.9 Sending the Value of Multiple Arduino Pins

Problem

You want to send groups of binary bytes, integers, or long values from Arduino. For example, you may want to send the values of the digital and analog pins to Processing.

Solution

This recipe sends a header followed by an integer containing the bit values of digital pins 2 to 13. This is followed by six integers containing the values of analog pins 0 through 5. Chapter 5 has many recipes that set values on the analog and digital pins that you can use to test this sketch:

```
/*
 * SendBinaryFields
 * Sends digital and analog pin values as binary data
 */

const char HEADER = 'H';  // a single character header to indicate
                          // the start of a message

void setup()
{
  Serial.begin(9600);
  for(int i=2; i <= 13; i++)
  {
```

```
    pinMode(i, INPUT);       // set pins 2 through 13 to inputs
    digitalWrite(i, HIGH);   // turn on pull-ups
  }
}

void loop()
{
  Serial.write(HEADER); // send the header
  // put the bit values of the pins into an integer
  int values = 0;
  int bit = 0;

  for(int i=2; i <= 13; i++)
  {
    bitWrite(values, bit, digitalRead(i));  // set the bit to 0 or 1 depending
                                            // on value of the given pin
    bit = bit + 1;                          // increment to the next bit
  }
  sendBinary(values); // send the integer

  for(int i=0; i < 6; i++)
  {
    values = analogRead(i);
    sendBinary(values); // send the integer
  }
  delay(1000); //send every second
}

// function to send the given integer value to the serial port
void sendBinary( int value)
{
  // send the two bytes that comprise an integer
  Serial.write(lowByte(value));  // send the low byte
  Serial.write(highByte(value)); // send the high byte
}
```

Discussion

The code sends a header (the character `H`), followed by an integer holding the digital pin values using the **`bitRead`** function to set a single bit in the integer to correspond to the value of the pin (see Chapter 3). It then sends six integers containing the values read from the six analog ports (see Chapter 5 for more information). All the integer values are sent using **`sendBinary`**, introduced in Recipe 4.6. The message is 15 bytes long—1 byte for the header, 2 bytes for the digital pin values, and 12 bytes for the six analog integers. The code for the digital and analog inputs is explained in Chapter 5.

Assuming analog pins have values of 0 on pin 0, 100 on pin 1, and 200 on pin 2 through 500 on pin 5, and digital pins 2 through 7 are high and 8 through 13 are low, this is the decimal value of each byte that gets sent:

```
72   // the character 'H' - this is the header
     // two bytes in low high order containing bits representing pins 2-13
63   // binary 00111111 :  this indicates that pins 2-7 are high
```

```
0    // this indicates that  8-13 are low

     // two bytes for each pin representing the analog value
0    // pin 0 has an integer value of 0 so this is sent as two bytes
0

100  // pin 1 has a value of 100, sent as a byte of 100 and a byte of 0
0
...
     // pin 5 has a value of 500
244  // the remainder when dividing 500 by 256
1    //  the number of times 500 can be divided by 256
```

This Processing code reads the message and prints the values to the Processing console:

```
// Processing Sketch

/*
 * ReceiveMultipleFieldsBinary_P
 *
 * portIndex must be set to the port connected to the Arduino
*/

import processing.serial.*;

Serial myPort;        // Create object from Serial class
short portIndex = 1;  // select the com port, 0 is the first port

char HEADER = 'H';

void setup()
{
  size(200, 200);
  // Open whatever serial port is connected to Arduino.
  String portName = Serial.list()[portIndex];
  println(Serial.list());
  println(" Connecting to -> " + Serial.list()[portIndex]);
  myPort = new Serial(this, portName, 9600);
}

void draw()
{
int val;

  if ( myPort.available() >= 15)  // wait for the entire message to arrive
  {
    if( myPort.read() == HEADER) // is this the header
    {
      println("Message received:");
      // header found
      // get the integer containing the bit values
      val = readArduinoInt();
      // print the value of each bit
      for(int pin=2, bit=1; pin <= 13; pin++){
        print("digital pin " + pin + " = " );
        int isSet = (val & bit);
```

```
          if( isSet == 0) {
            println("0");
          }
          else{
            println("1");
          }
          bit = bit * 2; //shift the bit to the next higher binary place
        }
        println();
        // print the six analog values
        for(int i=0; i < 6; i ++){
          val = readArduinoInt();
          println("analog port " + i + "= " + val);
        }
        println("----");
      }
    }
  }

  // return integer value from bytes received from serial port (in low,high order)
  int readArduinoInt()
  {
    int val;      // Data received from the serial port

    val = myPort.read();              // read the least significant byte
    val =  myPort.read() * 256 + val; // add the most significant byte
    return val;
  }
```

The Processing code waits for 15 characters to arrive. If the first character is the header, it then calls the function named `readArduinoInt` to read two bytes and transform them back into an integer by doing the complementary mathematical operation that was performed by Arduino to get the individual bits representing the digital pins. The six integers are then representing the analog values.

See Also

To send Arduino values back to the computer or drive the pins from the computer (without making decisions on the board), consider using Firmata (*http://www.firmata.org*). The Firmata library and example sketches (File→Examples→Firmata) are included in the Arduino software distribution, and a library is available to use in Processing. You load the Firmata code onto Arduino, control whether pins are inputs or outputs from the computer, and then set or read those pins.

4.10 How to Move the Mouse Cursor on a PC or Mac

Problem

You want Arduino to interact with an application on your computer by moving the mouse cursor. Perhaps you want to move the mouse position in response to Arduino information. For example, suppose you have connected a Wii nunchuck (see Recipe 13.2) to your Arduino and you want your hand movements to control the position of the mouse cursor in a program running on a PC.

Solution

You can send serial commands that specify the mouse cursor position to a program running on the target computer. Here is a sketch that moves the mouse cursor based on the position of two potentiometers:

```
// SerialMouse sketch
const int buttonPin = 2;  //LOW on digital pin enables mouse

const int potXPin = 4;    // analog pins for pots
const int potYPin = 5;

void setup()
{
  Serial.begin(9600);
  pinMode(buttonPin, INPUT);
  digitalWrite(buttonPin, HIGH); // turn on pull-ups
}

void loop()
{
  int x = (512 - analogRead(potXPin)) / 4;  // range is -127 to +127
  int y = (512 - analogRead(potYPin)) / 4;
  Serial.print("Data,");
  Serial.print(x,DEC);
  Serial.print(",");
  Serial.print(y,DEC);
  Serial.print(",");
  if(digitalRead(buttonPin) == LOW)
    Serial.print(1);  // send 1 when button pressed
  else
    Serial.print(0);
  Serial.println(",");
  delay(50); // send position 20 times a second
}
```

Figure 4-4 illustrates the wiring for two potentiometers (see Chapter 5 for more details). The switch is included so you can enable and disable Arduino mouse control by closing and opening the contacts.

The Processing code is based on the code shown in Recipe 4.4, with code added to control a mouse:

```
// Processing Sketch

/*
 * ArduinoMouse.pde  (Processing sketch)
 */

/* WARNING: This sketch takes over your mouse
```

```
 Press escape to close running sketch */

import java.awt.AWTException;
import java.awt.Robot;
import processing.serial.*;

Serial    myPort;   // Create object from Serial class
arduMouse myMouse;  // create arduino controlled mouse

public static final short LF = 10;          // ASCII linefeed
public static final short portIndex = 1;  // select the com port,
                                            // 0 is the first port

int posX, posY, btn; // data from msg fields will be stored here

void setup() {
  size(200, 200);
  println(Serial.list());
  println(" Connecting to -> " + Serial.list()[portIndex]);
  myPort = new Serial(this,Serial.list()[portIndex], 9600);
  myMouse = new arduMouse();
  btn = 0; // turn mouse off until requested by Arduino message
}

void draw() {
   if ( btn != 0)
      myMouse.move(posX, posY); // move mouse to received x and y position
}

void serialEvent(Serial p) {
  String message = myPort.readStringUntil(LF); // read serial data
  if(message != null)
  {
    //print(message);
    String [] data  = message.split(","); // Split the comma-separated message
    if ( data[0].equals("Data"))// check for data header
    {
      if( data.length > 3 )
      {
        try {
          posX = Integer.parseInt(data[1]);
          posY = Integer.parseInt(data[2]);
          btn  = Integer.parseInt(data[3]);
        }
        catch (Throwable t) {
          println("."); // parse error
          print(message);
        }
      }
    }
  }
}

class arduMouse {
  Robot myRobot;     // create object from Robot class;
```

```
  static final short rate = 4; // multiplier to adjust movement rate
  int centerX, centerY;
  arduMouse() {
    try {
      myRobot = new Robot();
    }
    catch (AWTException e) {
      e.printStackTrace();
    }
    Dimension screen = java.awt.Toolkit.getDefaultToolkit().getScreenSize();
    centerY =  (int)screen.getHeight() / 2 ;
    centerX =  (int)screen.getWidth() / 2;
  }
  // method to move mouse from center of screen by given offset
  void move(int offsetX, int offsetY) {
    myRobot.mouseMove(centerX + (rate* offsetX), centerY - (rate * offsetY));
  }
}
```

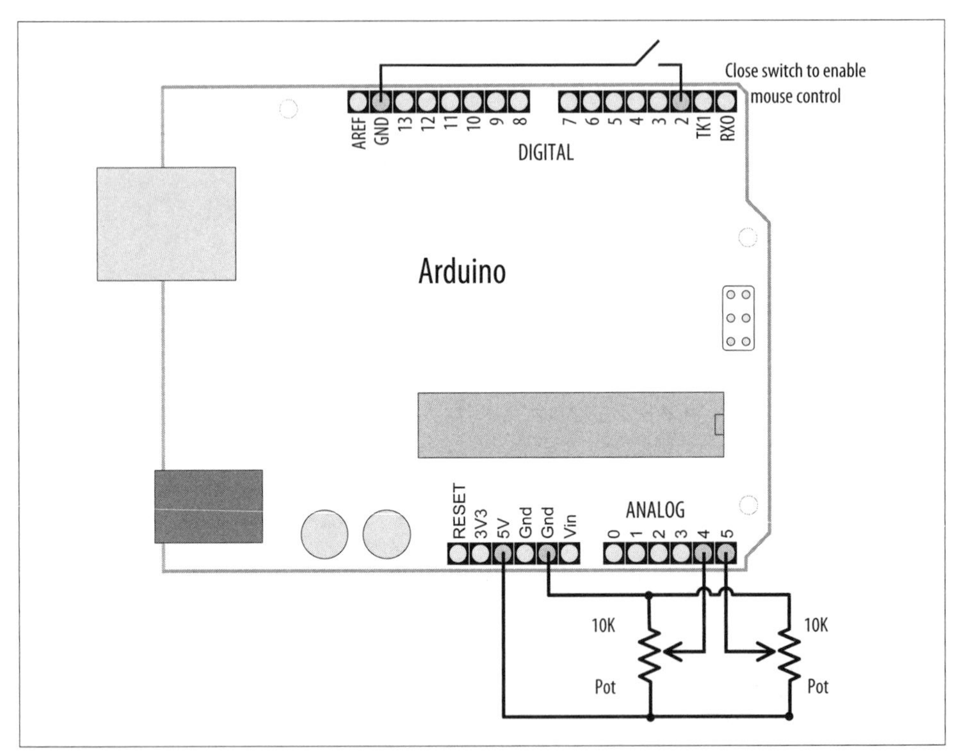

Figure 4-4. Wiring for mouse control using two potentiometers

The Processing code splits the message containing the `x` and `y` coordinates and sends them to the `mouseMove` method of the Java `Robot` class. In this example, the Robot class has a wrapper named `arduMouse` that provides a `move` method that scales to your screen size.

Discussion

This technique for controlling applications running on your computer is easy to implement and should work with any operating system that can run the Processing application. If you need to invert the direction of movement on the X or Y axis you can do this by changing the sign of the axis in the Processing sketch as follows:

```
posX = -Integer.parseInt(data[1]); // minus sign inverts axis
```

Some platforms require special privileges or extensions to access low-level input control. If you can't get control of the mouse, check the documentation for your operating system.

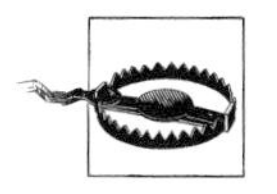

A runaway Robot object has the ability to remove your control over the mouse and keyboard if used in an endless loop. In this recipe a value is sent to Processing to enable and disable control based on the level of digital pin 2.

Boards using the ATmeg32U4 controller chip can directly emulate a USB mouse. The Arduino Leonardo board and the PJRC Teensy come with examples showing how to emulate a USB mouse.

Leonardo board:
: *http://blog.makezine.com/archive/2011/09/arduino-leonardo-opens-doors-to-product-development.html*

Teensy USB mouse example:
: *http://www.pjrc.com/teensy/usb_mouse.html*

See Also

Go to *http://java.sun.com/j2se/1.3/docs/api/java/awt/Robot.html* for more information on the Java Robot class.

An article on using the Robot class is available at *http://www.developer.com/java/other/article.php/10936_2212401_1*.

If you prefer to use a Windows programming language, the low-level Windows API function to insert keyboard and mouse events into the input stream is called SendInput. You can visit *http://msdn.microsoft.com/en-us/library/ms646310(VS.85).aspx* for more information.

Recipe 4.11 that follows shows how to apply this technique to control the Google Earth application.

4.11 Controlling Google Earth Using Arduino

Problem

You want to control movement in an application such as Google Earth using sensors attached to Arduino. For example, you want sensors to detect hand movements to act as the control stick for the flight simulator in Google Earth. The sensors could use a joystick (see Recipe 6.17) or a Wii nunchuck (see Recipe 13.2).

Solution

Google Earth lets you "fly" anywhere on Earth to view satellite imagery, maps, terrain, and 3-D buildings (see Figure 4-5). It contains a flight simulator that can be controlled by a mouse, and this recipe uses techniques described in Recipe 4.10 combined with a sensor connected to Arduino to provide the joystick input.

Figure 4-5. Google Earth flight simulator

The Arduino code sends the horizontal and vertical positions determined by reading an input device such as a joystick. There are many input options, for example you can use the circuit from Recipe 4.10 (this works well if you can find an old analog joystick that uses potentiometers that you can re-purpose).

Discussion

Google Earth is a free download; you can get it from the Google website, *http://earth.google.com/download-earth.html*. Download and run the version for your operating system to install it on your computer. Start Google Earth, and from the Tools menu, select Enter Flight Simulator. Select an aircraft (the SR22 is easier to fly than the F16) and an airport. The Joystick support should be left unchecked—you will be using the Arduino-controlled mouse to fly the aircraft. Click the Start Flight button (if the aircraft is already flying when you start, you can press the space bar to pause the simulator so that you can get the Processing sketch running).

Upload the Arduino sketch from Recipe 4.10 and run the Processing sketch from that recipe on your computer. Make Google Earth the Active window by clicking in the Google Earth window. Activate Arduino mouse control by connecting digital pin 2 to Gnd.

You are now ready to fly. Press Page Up on your keyboard a few times to increase the throttle (and then press the space bar on your keyboard if you had paused the simulator). When the SR22 reaches an air speed that is a little over 100 knots, you can "pull back" on the stick and fly. Information explaining the simulator controls can be found in the Google Help menu.

When you are finished flying you can relinquish Arduino mouse control back to your computer mouse by disconnecting pin 2 from Gnd.

Here is another variation that sends messages to the Processing sketch. This one combines the Wii nunchuck code from Recipe 13.2 with a library discussed in Recipe 16.5. The connections are as shown in Recipe 13.2:

```
/*
 * WiichuckSerial
 *
 * Uses Nunchuck Library discussed in Recipe 16.5
 * sends comma-separated values for data
 * Label string separated by commas can be used by receiving program
 * to identify fields
 */


#include <Wire.h>
#include "Nunchuck.h"

// values to add to the sensor to get zero reading when centered
int offsetX, offsetY, offsetZ;

#include <Wire.h>
#include "Nunchuck.h"
void setup()
{
    Serial.begin(57600);
    nunchuckSetPowerpins();
    nunchuckInit(); // send the initialization handshake
```

```
    nunchuckRead(); // ignore the first time
    delay(50);
}
void loop()
{
  nunchuckRead();
  delay(6);
  nunchuck_get_data();
  boolean btnC = nunchuckGetValue(wii_btnC);
  boolean btnZ = nunchuckGetValue(wii_btnZ);

  if(btnC) {
    offsetX = 127 - nunchuckGetValue(wii_accelX) ;
    offsetY = 127 - nunchuckGetValue(wii_accelY) ;
  }
  Serial.print("Data,");
  printAccel(nunchuckGetValue(wii_accelX),offsetX) ;
  printAccel(nunchuckGetValue(wii_accelY),offsetY) ;
  printButton(nunchuckGetValue(wii_btnZ));

  Serial.println();
}

void printAccel(int value, int offset)
{
  Serial.print(adjReading(value, 127-50, 127+50, offset));
  Serial.print(",");
}

void printJoy(int value)
{
  Serial.print(adjReading(value,0, 255, 0));
  Serial.print(",");
}

void printButton(int value)
{
  if( value != 0)
     value = 127;
  Serial.print(value,DEC);
  Serial.print(",");
}

int adjReading( int value, int min, int max, int offset)
{
   value = constrain(value + offset, min, max);
   value = map(value, min, max, -127, 127);
   return value;
}
```

These sketches use a Serial speed of 57600 to minimize latency. If you want to view the Arduino output on the Serial Monitor, you will need to change its baud rate accordingly. You will need to change the Serial Monitor's baud rate back to 9600 to view the output of most other sketches in this book. If you don't have a Wii nunchuck, you can use the Arduino sketch from Recipe 4.10, but you will need to change that sketch's baud rate to 57600 and upload it to the Arduino.

You can send nunchuck joystick values instead of the accelerometer values by replacing the two lines that begin `printAccel` with the following lines:

```
printJoy(nunchuckGetValue(wii_joyX));
printJoy(nunchuckGetValue(wii_joyY));
```

You can use the Processing sketch from Recipe 4.10, but this enhanced version displays the control position in the Processing window and activates the flight simulator using the nunchuck 'z' button:

```
/**
 * GoogleEarth_FS.pde
 *
 * Drives Google Flight Sim using CSV sensor data
 */

import java.awt.AWTException;
import java.awt.Robot;
import java.awt.event.InputEvent;
import processing.serial.*;
Serial myPort;  // Create object from Serial class

arduMouse myMouse;

String message = null;
int maxDataFields = 7; // 3 axis accel, 2 buttons, 2 joystick axis
boolean isStarted = false;
int accelX, accelY, btnZ; // data from msg fields will be stored here


void setup() {
  size(260, 260);
  PFont fontA = createFont("Arial.normal", 12);
  textFont(fontA);

  short portIndex = 1;  // select the com port, 0 is the first port
  String portName = Serial.list()[portIndex];
  println(Serial.list());
  println(" Connecting to -> " + portName) ;
  myPort = new Serial(this, portName, 57600);
  myMouse = new arduMouse();

  fill(0);
  text("Start Google FS in the center of your screen", 5, 40);
  text("Center the mouse pointer in Google earth", 10, 60);
```

```
  text("Press and release Nunchuck Z button to play", 10, 80);
  text("Press Z button again to pause mouse", 20, 100);
}

void draw() {
  processMessages();
  if (isStarted == false) {
    if ( btnZ != 0) {
      println("Release button to start");
      do{ processMessages();}
         while(btnZ != 0);
      myMouse.mousePress(InputEvent.BUTTON1_MASK); // start the SIM
      isStarted = true;
    }
  }
  else
  {
    if ( btnZ != 0) {
      isStarted = false;
      background(204);
      text("Release Z button to play", 20, 100);
      print("Stopped, ");
    }
    else{
      myMouse.move(accelX, accelY); // move mouse to received x and y position
      fill(0);
      stroke(255, 0, 0);
      background(#8CE7FC);
      ellipse(127+accelX, 127+accelY, 4, 4);
    }
  }
}

void processMessages() {
  while (myPort.available () > 0) {
    message = myPort.readStringUntil(10);
    if (message != null) {
      //print(message);
      String [] data  = message.split(","); // Split the CSV message
      if ( data[0].equals("Data"))// check for data header
      {
        try {
          accelX = Integer.parseInt(data[1]);
          accelY = Integer.parseInt(data[2]);
          btnZ = Integer.parseInt(data[3]);
        }
        catch (Throwable t) {
          println("."); // parse error
        }
      }
    }
  }
}

class arduMouse {
```

```
    Robot myRobot;     // create object from Robot class;
    static final short rate = 4; // pixels to move
    int centerX, centerY;
    arduMouse() {
      try {
        myRobot = new Robot();
      }
      catch (AWTException e) {
        e.printStackTrace();
      }
      Dimension screen = java.awt.Toolkit.getDefaultToolkit().getScreenSize();
      centerY =  (int)screen.getHeight() / 2 ;
      centerX =  (int)screen.getWidth() / 2;
    }
    // method to move mouse from center of screen by given offset
    void move(int offsetX, int offsetY) {
      myRobot.mouseMove(centerX + (rate* offsetX), centerY - (rate * offsetY));
    }
    // method to simulate pressing mouse button
    void mousePress( int button) {
      myRobot.mousePress(button) ;
    }
  }
```

See Also

The Google Earth website contains the downloadable code and instructions needed to get this going on your computer: *http://earth.google.com/*.

4.12 Logging Arduino Data to a File on Your Computer

Problem

You want to create a file containing information received over the serial port from Arduino. For example, you want to save the values of the digital and analog pins at regular intervals to a logfile.

Solution

We covered sending information from Arduino to your computer in previous recipes. This solution uses the same Arduino code explained in Recipe 4.9. The Processing sketch that handles file logging is based on the Processing sketch also described in that recipe.

This Processing sketch creates a file (using the current date and time as the filename) in the same directory as the Processing sketch. Messages received from Arduino are added to the file. Pressing any key saves the file and exits the program:

```
/*
 * ReceiveMultipleFieldsBinaryToFile_P
 *
```

```
 * portIndex must be set to the port connected to the Arduino
 * based on ReceiveMultipleFieldsBinary, this version saves data to file
 * Press any key to stop logging and save file
 */

import processing.serial.*;


PrintWriter output;
DateFormat fnameFormat= new SimpleDateFormat("yyMMdd_HHmm");
DateFormat  timeFormat = new SimpleDateFormat("hh:mm:ss");
String fileName;

Serial myPort;        // Create object from Serial class
short portIndex = 0;  // select the com port, 0 is the first port
char HEADER = 'H';

void setup()
{
  size(200, 200);
  // Open whatever serial port is connected to Arduino.
  String portName = Serial.list()[portIndex];
  println(Serial.list());
  println(" Connecting to -> " + Serial.list()[portIndex]);
  myPort = new Serial(this, portName, 9600);
  Date now = new Date();
  fileName = fnameFormat.format(now);
  output = createWriter(fileName + ".txt"); // save the file in the sketch folder
}

void draw()
{
  int val;
  String time;

  if ( myPort.available() >= 15)  // wait for the entire message to arrive
  {
    if( myPort.read() == HEADER) // is this the header
    {
      String timeString = timeFormat.format(new Date());
      println("Message received at " + timeString);
      output.println(timeString);
      // header found
      // get the integer containing the bit values
      val = readArduinoInt();
      // print the value of each bit
      for(int pin=2, bit=1; pin <= 13; pin++){
        print("digital pin " + pin + " = " );
        output.print("digital pin " + pin + " = " );
        int isSet = (val & bit);
        if( isSet == 0){
           println("0");
           output.println("0");
        }
        else  {
```

```
          println("1");
          output.println("0");
        }
        bit = bit * 2; // shift the bit
      }
      // print the six analog values
      for(int i=0; i < 6; i ++){
        val = readArduinoInt();
        println("analog port " + i + "= " + val);
        output.println("analog port " + i + "= " + val);
      }
      println("----");
      output.println("----");
    }
  }
}

void keyPressed() {
  output.flush(); // Writes the remaining data to the file
  output.close(); // Finishes the file
  exit(); // Stops the program
}

// return the integer value from bytes received on the serial port
// (in low,high order)
int readArduinoInt()
{
  int val;      // Data received from the serial port

  val = myPort.read();          // read the least significant byte
  val =  myPort.read() * 256 + val; // add the most significant byte
  return val;
}
```

Don't forget that you need to set `portIndex` to the serial port connected to Arduino.

Discussion

The base name for the logfile is formed using the `DateFormat` function in Processing:

```
DateFormat fnameFormat= new SimpleDateFormat("yyMMdd_HHmm");
```

The full filename is created with code that adds a directory and file extension:

```
  output = createWriter(fileName + ".txt");
```

The file will be created in the same directory as the Processing sketch (the sketch needs to be saved at least once to ensure that the directory exists). To find this directory, choose Sketch→Show Sketch Folder in Processing. `createWriter` is the Processing function that opens the file; this creates an object (a unit of runtime functionality) called `output` that handles the actual file output. The text written to the file is the same as what is printed to the console in Recipe 4.9, but you can format the file contents as required by using the standard string-handling capabilities of Processing. For example, the following variation on the `draw` routine produces a comma-separated file that can be read by a spreadsheet or database. The rest of the Processing sketch can be the same, although you may want to change the extension from *.txt* to *.csv*:

```
void draw()
{
  int val;
  String time;

  if ( myPort.available() >= 15)  // wait for the entire message to arrive
  {
    if( myPort.read() == HEADER) // is this the header
    {
       String timeString = timeFormat.format(new Date());
       output.print(timeString);
       val = readArduinoInt(); // read but don't output the digital values

      // output the six analog values delimited by a comma
      for(int i=0; i < 6; i ++){
        val = readArduinoInt();
        output.print("," + val);
      }
      output.println();
    }
  }
}
```

See Also

For more on `createWriter`, see *http://processing.org/reference/createWriter_.html*.

4.13 Sending Data to Two Serial Devices at the Same Time

Problem

You want to send data to a serial device such as a serial LCD, but you are already using the built-in serial port to communicate with your computer.

Solution

On a Mega this is not a problem, as it has four hardware serial ports; just create two serial objects and use one for the LCD and one for the computer:

```
void setup() {
  // initialize two serial ports on a Mega
  Serial.begin(9600);  // primary serial port
  Serial1.begin(9600); // Mega can also use Serial1 through Serial3
}
```

On a standard Arduino board (such as the Uno or Duemilanove) that only has one hardware serial port, you will need to create an emulated or "soft" serial port.

You can use the distributed SoftwareSerial library for sending data to multiple devices.

Arduino releases from 1.0 use an improved SoftwareSerial library based on Mikal Hart's NewSoftSerial Library. If you are using an Arduino release prior to 1.0, you can download NewSoftSerial from *http://arduiniana.org/libraries/newsoftserial*.

Select two available digital pins, one each for transmit and receive, and connect your serial device to them. It is convenient to use the hardware serial port for communication with the computer because this has a USB adapter on the board. Connect the device's transmit line to the receive pin and the receive line to the transmit pin. In Figure 4-6, we have selected pin 2 as the receive pin and pin 3 as the transmit pin.

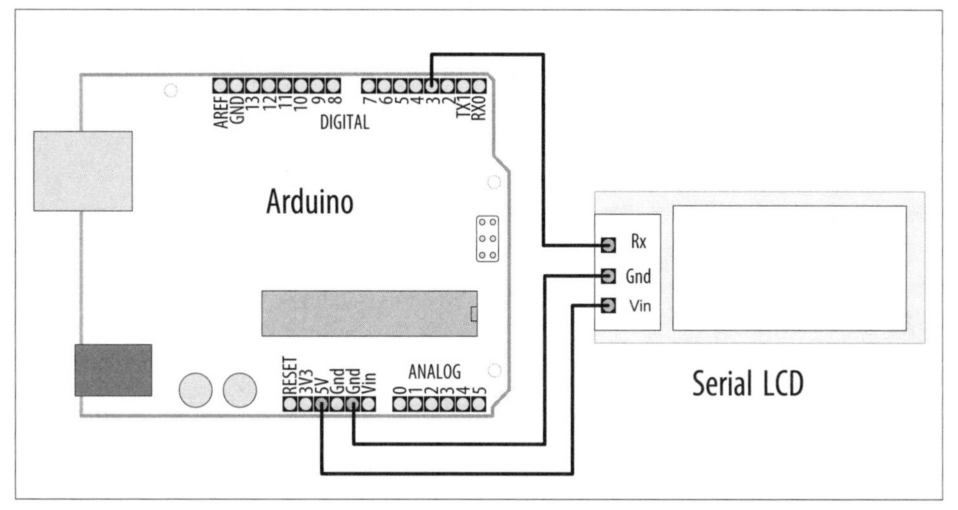

Figure 4-6. Connecting a serial device to a "soft" serial port

In your sketch, create a `SoftwareSerial` object and tell it which pins you chose as your emulated serial port. In this example, we're creating an object named `serial_lcd`, which we instruct to use pins 2 and 3:

```
/*
 * SoftwareSerialOutput sketch
 * Output data to a software serial port
 */

#include <SoftwareSerial.h>

const int rxpin = 2;           // pin used to receive (not used in this version)
const int txpin = 3;           // pin used to send to LCD
SoftwareSerial serial_lcd(rxpin, txpin); // new serial port on pins 2 and 3

void setup()
{
  Serial.begin(9600); // 9600 baud for the built-in serial port
  serial_lcd.begin(9600); //initialize the software serial port also for 9600
```

```
}

int number = 0;

void loop()
{
  serial_lcd.print("The number is "); // send text to the LCD
  serial_lcd.println(number);    // print the number on the LCD
  Serial.print("The number is ");
  Serial.println(number);        // print the number on the PC console

  delay(500); // delay half second between numbers
  number++;   // to the next number
}
```

If you are using Arduino versions prior to 1.0, download the NewSoftSerial library and replace references to SoftwareSerial with NewSoftSerial:

```
// NewSoftSerial version

#include <NewSoftSerial.h>

const int rxpin = 2;           // pin used to receive from LCD
const int txpin = 3;           // pin used to send to LCD
NewSoftSerial serial_lcd(rxpin, txpin); // new serial port on pins 2 + 3
```

This sketch assumes that a serial LCD has been connected to pin 3 as shown in Figure 4-6, and that a serial console is connected to the built-in port. The loop will repeatedly display the same message on each:

```
The number is 0
The number is 1
...
```

Discussion

Every Arduino microcontroller contains at least one built-in serial port. This special piece of hardware is responsible for generating the series of precisely timed pulses its partner device sees as data and for interpreting the similar stream that it receives in return. Although the Mega has four such ports, most Arduino flavors have only one. For projects that require connections to two or more serial devices, you'll need a software library that emulates the additional ports. A "software serial" library effectively turns an arbitrary pair of digital I/O pins into a new serial port.

To build your software serial port, you select a pair of pins that will act as the port's transmit and receive lines in much the same way that pins 1 and 0 are controlled by Arduino's built-in port. In Figure 4-6, pins 3 and 2 are shown, but any available digital pins can be used. It's wise to avoid using 0 and 1, because these are already being driven by the built-in port.

The syntax for writing to the soft port is identical to that for the hardware port. In the example sketch, data is sent to both the "real" and emulated ports using `print()` and `println()`:

```
serial_lcd.print("The number is "); // send text to the LCD
serial_lcd.println(number);         // send the number on the LCD

Serial.print("The number is ");     // send text to the hardware port
Serial.println(number);             // to output on Arduino Serial Monitor
```

If you are using a *unidirectional* serial device—that is, one that only sends or receives—you can conserve resources by specifying a nonexistent pin number in the `SoftwareSerial` constructor for the line you don't need. For example, a serial LCD is fundamentally an output-only device. If you don't expect (or want) to receive data from it, you can tell SoftwareSerial using this syntax:

```
#include <SoftwareSerial.h>
...
const int no_such_pin = 255;
const int txpin = 3;
SoftwareSerial serial_lcd(no_such_pin, txpin); // TX-only on pin 3
```

In this case, we would only physically connect a single pin (3) to the serial LCD's "input" or "RX" line.

See Also

SoftwareSerial for Arduino 1.0 and later releases is based on NewSoftSerial. You can read more about NewSoftSerial on Mikal Hart's website (*http://arduiniana.org/libraries/newsoftserial/*)

4.14 Receiving Serial Data from Two Devices at the Same Time

Problem

You want to receive data from a serial device such as a serial GPS, but you are already using the built-in serial port to communicate with your computer.

Solution

This problem is similar to the one in Recipe 4.13, and indeed the solution is much the same. If your Arduino's serial port is connected to the console and you want to attach a second serial device, you must create an emulated port using a software serial library. In this case, we will be receiving data from the emulated port instead of writing to it, but the basic solution is very similar.

See the previous recipe regarding the NewSoftSerial library if you are using an Arduino release prior to 1.0.

Select two pins to use as your transmit and receive lines.

Connect your GPS as shown in Figure 4-7. Rx (receive) is not used in this example, so you can ignore the Rx connection to pin 3 if your GPS does not have a receive pin.

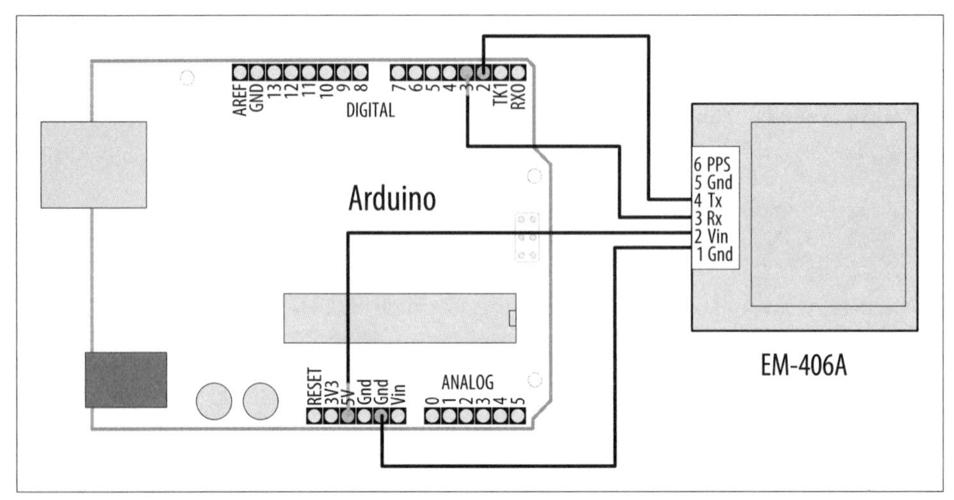

Figure 4-7. Connecting a serial GPS device to a "soft" serial port

As you did in Recipe 4.13, create a `SoftwareSerial` object in your sketch and tell it which pins to control. In the following example, we define a soft serial port called `serial_gps`, using pins 2 and 3 for receive and transmit, respectively:

```
/*
 * SoftwareSerialInput sketch
 * Read data from a software serial port
 */

#include <SoftwareSerial.h>
const int rxpin = 2;                     // pin used to receive from GPS
const int txpin = 3;                     // pin used to send to GPS
SoftwareSerial serial_gps(rxpin, txpin); // new serial port on pins 2 and 3

void setup()
{
  Serial.begin(9600); // 9600 baud for the built-in serial port
  serial_gps.begin(4800); // initialize the port, most GPS devices
                          // use 4800 baud
}

void loop()
```

```
{
  if (serial_gps.available() > 0) // any character arrived yet?
  {
    char c = serial_gps.read();   // if so, read it from the GPS
    Serial.write(c);              // and echo it to the serial console
  }
}
```

If you are using Arduino versions prior to 1.0, download the NewSoftSerial library and replace references to SoftwareSerial with NewSoftSerial:

```
// NewSoftSerial version
#include <NewSoftSerial.h>
const int rxpin = 2;                    // pin used to receive from GPS
const int txpin = 3;                    // pin used to send to GPS
NewSoftSerial serial_gps(rxpin, txpin); // new serial port on pins 2 and 3
```

This short sketch simply forwards all incoming data from the GPS to the Arduino Serial Monitor. If the GPS is functioning and your wiring is correct, you should see GPS data displayed on the Serial Monitor.

Discussion

You initialize an emulated SoftwareSerial port by providing pin numbers for transmit and receive. The following code will set up the port to receive on pin 2 and send on pin 3:

```
const int rxpin = 2;                     // pin used to receive from GPS
const int txpin = 3;                     // pin used to send to GPS
SoftwareSerial serial_gps(rxpin, txpin); // new serial port on pins 2 and 3
```

The `txpin` is not used in this example and can be set to 255 to free up pin 3, as explained in the previous recipe.

The syntax for reading an emulated port is very similar to that for reading from a built-in port. First check to make sure a character has arrived from the GPS with `available()`, and then read it with `read()`.

It's important to remember that software serial ports consume time and resources. An emulated serial port must do everything that a hardware port does, using the same processor your sketch is trying to do "real work" with. Whenever a new character arrives, the processor must interrupt whatever it is doing to handle it. This can be time-consuming. At 4,800 baud, for example, it takes the Arduino about two milliseconds to process a single character. While two milliseconds may not sound like much, consider that if your peer device—say, the GPS unit shown earlier—transmits 200 to 250 characters per second, your sketch is spending 40 to 50 percent of its time trying to keep up with the serial input. This leaves very little time to actually *process* all that data. The lesson is that if you have two serial devices, when possible connect the one with the higher bandwidth consumption to the built-in (hardware) port. If you must connect a high-bandwidth device to a software serial port, make sure the rest of your sketch's loop is very efficient.

Receiving data from multiple SoftwareSerial ports

With the SoftwareSerial library included with Arduino 1.0, it is possible to create multiple "soft" serial ports in the same sketch. This is a useful way to control, say, several XBee radios or serial displays in the same project. The caveat is that at any given time, only one of these ports can actively receive data. Reliable communication on a software port requires the processor's undivided attention. That's why SoftwareSerial can only actively communicate with one port at a given time.

It *is* possible to receive on two different SoftwareSerial ports in the same sketch. You just have to take some care that you aren't trying to receive from both simultaneously. There are many successful designs which, say, monitor a serial GPS device for a while, then later accept input from an XBee. The key is to alternate slowly between them, switching to a second device only when a transmission from the first is complete.

For example, in the sketch that follows, imagine a remote XBee module sending commands. The sketch listens to the command stream through the "xbee" port until it receives the signal to begin gathering data from a GPS module attached to a second SoftwareSerial port. The sketch then monitors the GPS for 10 seconds—long enough to establish a "fix"—before returning to the XBee.

In a system with multiple "soft" ports, only one is actively receiving data. By default, the "active" port is the one for which `begin()` has been called most recently. However, you can change which port is active by calling its `listen()` method. `listen()` instructs the SoftwareSerial system to stop receiving data on one port and begin listening for data on another.

The following code fragment illustrates how you might design a sketch to read first from one port and then another:

```
/*
 * MultiRX sketch
 * Receive data from two software serial ports
 */
#include <SoftwareSerial.h>
const int rxpin1 = 2;
const int txpin1 = 3;
const int rxpin2 = 4;
const int txpin2 = 5;

SoftwareSerial gps(rxpin1, txpin1); // gps device connected to pins 2 and 3
SoftwareSerial xbee(rxpin2, txpin2); // xbee device connected to pins 4 and 5

void setup()
{
  xbee.begin(9600);
  gps.begin(4800);
  xbee.listen(); // Set "xbee" to be the active device
}

void loop()
{
```

```
    if (xbee.available() > 0) // xbee is active. Any characters available?
    {
      if (xbee.read() == 'y') // if xbee received a 'y' character?
      {
        gps.listen(); // now start listening to the gps device

        unsigned long start = millis(); // begin listening to the GPS
        while (start + 100000 > millis())
        // listen for 10 seconds
        {
          if (gps.available() > 0) // now gps device is active
          {
            char c = gps.read();
            // *** process gps data here
          }
        }
        xbee.listen(); // After 10 seconds, go back to listening to the xbee
      }
    }
  }
```

This sketch is designed to treat the XBee radio as the active port until it receives a y character, at which point the GPS becomes the active listening device. After processing GPS data for 10 seconds, the sketch resumes listening on the XBee port. Data that arrives on an inactive port is simply discarded.

Note that the "active port" restriction only applies to multiple *soft* ports. If your design really must receive data from more than one serial device simultaneously, consider attaching one of these to the built-in hardware port. Alternatively, it is perfectly possible to add additional hardware ports to your projects using external chips, devices called *UARTs*.

4.15 Setting Up Processing on Your Computer to Send and Receive Serial Data

Problem

You want to use the Processing development environment to send and receive serial data.

Solution

You can get the Processing application from the Downloads section of the Processing website, *http://processing.org*. Files are available for each major operating system. Download the appropriate one for your operating system and unzip the file to somewhere that you normally store applications. On a Windows computer, this might be a location like *C:\Program Files\Processing*. On a Mac, it might be something like */Applications/Processing.app*.

If you installed Processing on the same computer that is running the Arduino IDE, the only other thing you need to do is identify the serial port in Processing. The following Processing sketch prints the serial ports available:

```
/**
 * GettingStarted
 *
 * A sketch to list the available serial ports
 * and display characters received
 */


import processing.serial.*;

Serial myPort;       // Create object from Serial class
int portIndex = 0;  // set this to the port connected to Arduino
int val;             // Data received from the serial port

void setup()
{
  size(200, 200);
  println(Serial.list()); // print the list of all the ports
  println(" Connecting to -> " + Serial.list()[portIndex]);
  myPort = new Serial(this, Serial.list()[portIndex], 9600);
}

void draw()
{
  if ( myPort.available() > 0) // If data is available,
  {
    val = myPort.read();          // read it and store it in val
    print(val);
  }
}
```

If you are running Processing on a computer that is not running the Arduino development environment, you may need to install the Arduino USB drivers (Chapter 1 describes how to do this).

Set the variable `portIndex` to match the port used by Arduino. You can see the port numbers printed in the Processing text window (the area below the source code, not the separate Display window; see *http://processing.org/reference/environment*). Recipe 1.4 describes how to find out which serial port your Arduino board is using.

CHAPTER 5

Simple Digital and Analog Input

5.0 Introduction

The Arduino's ability to sense digital and analog inputs allows it to respond to you and to the world around you. This chapter introduces techniques you can use to do useful things with these inputs. This is the first of many chapters to come that cover electrical connections to Arduino. If you don't have an electronics background, you may want to look through Appendix A on electronic components, Appendix B on schematic diagrams and data sheets, Appendix C on building and connecting circuits, and Appendix E on hardware troubleshooting. In addition, many good introductory tutorials are available. Two that are particularly relevant to Arduino are *Getting Started with Arduino* by Massimo Banzi and *Making Things Talk* by Tom Igoe (both O'Reilly; search on oreilly.com). Other books offering a background on electronics topics covered in this and the following chapters include *Getting Started in Electronics* by Forrest Mims (Master Publishing) and *Physical Computing* by Tom Igoe (Cengage).

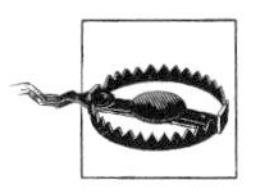

If wiring components to your Arduino is new to you, be careful about how you connect and power the things you attach. Arduino uses a robust controller chip that can take a fair amount of abuse, but you can damage the chip if you connect the wrong voltages or short-circuit an output pin. Most Arduino controller chips are powered by 5 volts, and you must not connect external power to Arduino pins with a higher voltage than this (or 3.3 volts if your Arduino controller runs on this voltage).

Most Arduino boards have the main chip in a socket that can be removed and replaced, so you don't need to replace the whole board if you damage the chip.

Figure 5-1 shows the arrangement of pins on a standard Arduino board. See *http://www.arduino.cc/en/Main/Hardware* for a list of all the official boards along with links to connection information for each. If your board is not on that list, check your board supplier's website for connection information.

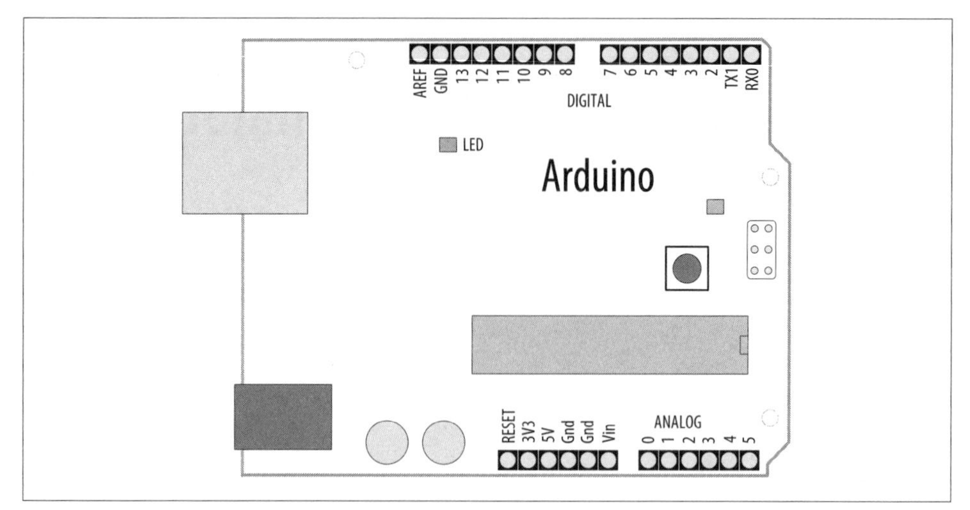

Figure 5-1. Digital and analog pins on a standard Arduino board

This chapter covers the Arduino pins that can sense *digital* and *analog* inputs. Digital input pins sense the presence and absence of voltage on a pin. Analog input pins measure a range of voltages on a pin.

The Arduino function to detect digital input is `digitalRead` and it tells your sketch if a voltage on a pin is `HIGH` (5 volts) or `LOW` (0 volts). The Arduino function to configure a pin for reading input is `pinMode(pin, INPUT)`.

On a typical board, there are 14 digital pins (numbered 0 to 13) as shown at the top of Figure 5-1. Pins 0 and 1 (marked RX and TX) are used for the USB serial connection and should be avoided for other uses. If you need more digital pins on a standard board, you can use the analog pins as digital pins (analog pins 0 through 5 can be used as digital pins 14 through 19).

Arduino 1.0 introduced logical names for many of the pins. The constants in Table 5-1 can be used in all functions that expect a pin number.

Table 5-1. Pin constants introduced in Arduino 1.0

Constant	Pin Number	Constant	Pin Number
A0	Analog input 0 (Digital 14)	LED_BUILTIN	On-board LED (Digital 13)
A1	Analog input 1 (Digital 15)	SDA	I2C Data (Digital 18)
A2	Analog input (Digital 16)	SCL	I2C Clock (Digital 19)
A3	Analog input (Digital 17)	SS	SPI Select (Digital 10)
A4	Analog input (Digital 18)	MOSI	SPI Input (Digital 11)
A5	Analog input (Digital 19)	MISO	SPI Output (Digital 12)
		SCL	SPI Clock (Digital 13)

The Mega board has many more digital and analog pins. Digital pins 0 through 13 and analog pins 0 through 5 are located in the same place as on the standard board, so that hardware shields designed for the standard board can fit onto a Mega. As with the standard board, you can use analog pins as digital pins, but with the Mega, analog pins 0 through 15 are digital pin numbers 54 through 69. Figure 5-2 shows the Mega pin layout.

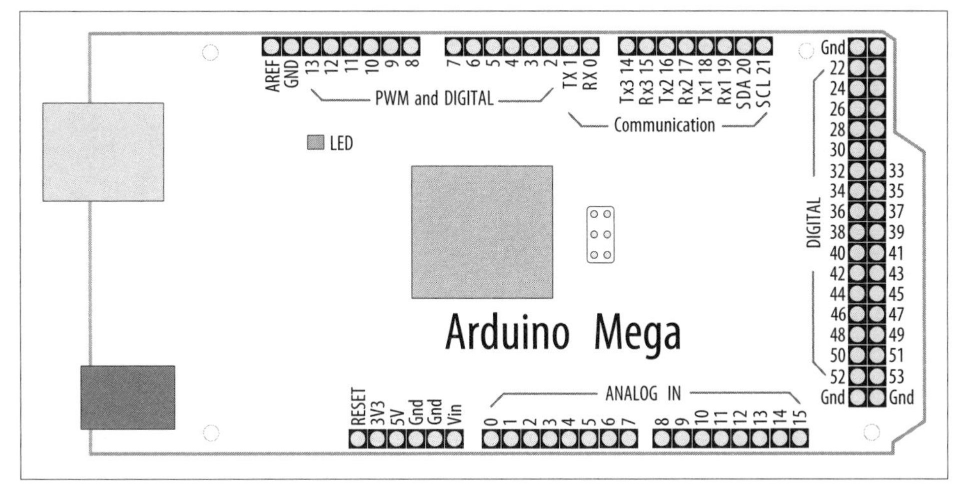

Figure 5-2. Arduino Mega board

Most boards have an LED connected to pin 13, and some of the recipes use this as an output indicator. If your board does not have an LED on pin 13, skip ahead to Recipe 7.1 if you need help connecting an LED to a digital pin.

Recipes covering digital input sometimes use external resistors to provide the voltage that is sensed by `digitalRead`. These resistors are called *pull-up* resistors (so named because the voltage is "pulled up" to the 5V line that the resistor is connected to) or *pull-down* resistors (the voltage is "pulled down" to 0 volts). Although 10K ohms is a commonly used value, anything between 4.7K and 20K or more will work; see Appendix A for more information about the components used in this chapter.

Unlike a digital value, which is only on or off, analog values are continuously variable. The volume setting of a device is a good example; it is not just on or off, but it can have a range of values in between. Many sensors provide information by varying the voltage to correspond to the sensor measurement. Arduino code uses a function called `analogRead` to get a value proportional to the voltage it sees on one of its analog pins. The value will be 0 if there are 0 volts on the pin and 1,023 for 5 volts. The value in between will be proportional to the voltage on the pin, so 2.5 volts (half of 5 volts) will result in a value of roughly 511 (half of 1,023). You can see the six analog input pins (marked 0 to 5) at the bottom of Figure 5-1 (these pins can also be used as digital pins 14 to 19 if they are not needed for analog). Some of the analog recipes use a *potentiometer* (*pot* for short, also called a *variable resistor*) to vary the voltage on a pin. When choosing a potentiometer, a value of 10K is the best option for connecting to analog pins.

Although most of the circuits in this chapter are relatively easy to connect, you may want to consider getting a solderless breadboard to simplify your wiring to external components: some choices are the Jameco 20723 (two bus rows per side); RadioShack 276-174 (one bus row per side); Digi-Key 438-1045-ND; and SparkFun PRT-00137.

Another handy item is an inexpensive multimeter. Almost any will do, as long as it can measure voltage and resistance. Continuity checking and current measurement are nice additional features to have. (The Jameco 220812, RadioShack 22-810, and SparkFun TOL-00078 offer these features.)

5.1 Using a Switch

Problem

You want your sketch to respond to the closing of an electrical contact; for example, a pushbutton or other switch or an external device that makes an electrical connection.

Solution

Use `digitalRead` to determine the state of a switch connected to an Arduino digital pin set as input. The following code lights an LED when a switch is pressed (Figure 5-3 shows how it should be wired up):

```
/*
   Pushbutton sketch
   a switch connected to pin 2 lights the LED on pin 13
*/

const int ledPin = 13;             // choose the pin for the LED
const int inputPin = 2;            // choose the input pin (for a pushbutton)

void setup() {
  pinMode(ledPin, OUTPUT);          // declare LED as output
  pinMode(inputPin, INPUT);         // declare pushbutton as input
}

void loop(){
  int val = digitalRead(inputPin);  // read input value
  if (val == HIGH)                  // check if the input is HIGH
  {
    digitalWrite(ledPin, HIGH);     // turn LED on if switch is pressed
  }
  else
  {
    digitalWrite(ledPin, LOW);      // turn LED off
```

```
    }
  }
```

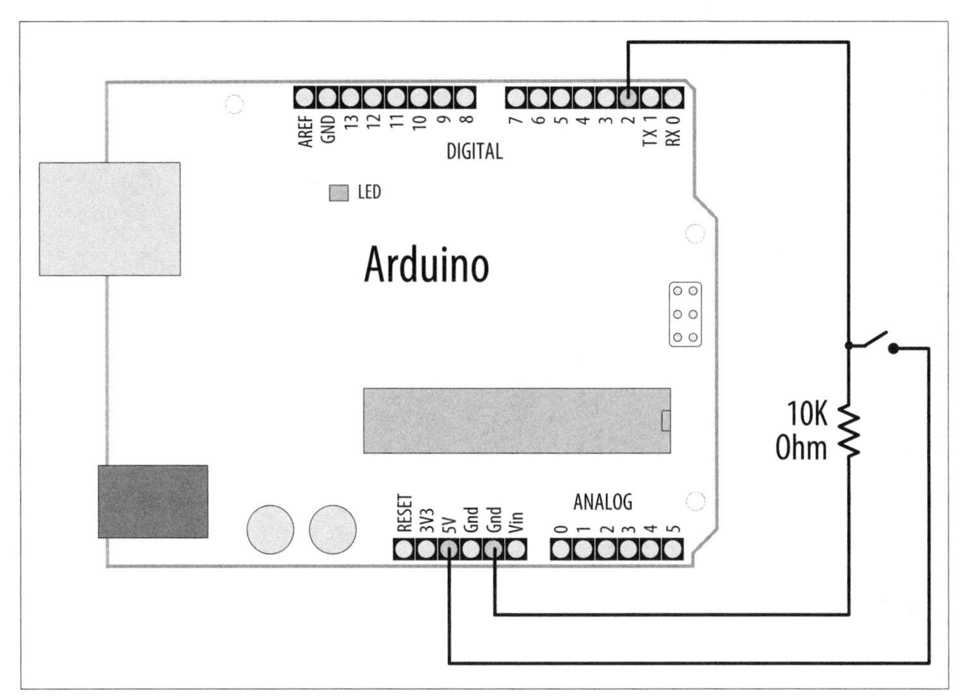

Figure 5-3. Switch connected using pull-down resistor

Standard Arduino boards have a built-in LED connected to pin 13. If your board does not, see Recipe 7.1 for information on connecting an LED to an Arduino pin.

Discussion

The `setup` function configures the LED pin as `OUTPUT` and the switch pin as `INPUT`.

A pin must be set to `OUTPUT` mode for `digitalWrite` to control the pin's output voltage. It must be in `INPUT` mode to read the digital input.

The `digitalRead` function monitors the voltage on the input pin (`inputPin`), and it returns a value of `HIGH` if the voltage is high (5 volts) and `LOW` if the voltage is low (0 volts). Actually, any voltage that is greater than 2.5 volts (half of the voltage powering the chip) is considered `HIGH` and less than this is treated as `LOW`. If the pin is left unconnected (known as *floating*), the value returned from `digitalRead` is indeterminate (it may be `HIGH` or `LOW`, and it cannot be reliably used). The resistor shown in Figure 5-3 ensures that the voltage on the pin will be low when the switch is not pressed, because the resistor "pulls down" the voltage to ground. When the switch is pushed, a connection is made between the pin and +5 volts, so the value on the pin interpreted by `digital Read` changes from `LOW` to `HIGH`.

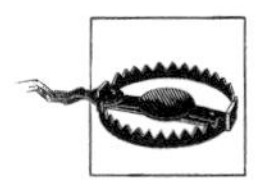

Do not connect a digital or analog pin to a voltage higher than 5 volts (or 3.3 volts on a 3.3V board). This can damage the pin and possibly destroy the entire chip. Also, make sure you don't wire the switch so that it shorts the 5 volts to ground (without a resistor). Although this may not damage the Arduino chip, it is not good for the power supply.

In this example, the value from `digitalRead` is stored in the variable `val`. This will be `HIGH` if the button is pressed, `LOW` otherwise.

The switch used in this example (and almost everywhere else in this book) makes electrical contact when pressed and breaks contact when not pressed. These switches are called Normally Open (NO); see this book's website (*http://shop.oreilly.com/product/0636920022244.do*) for part numbers. The other kind of momentary switch is called Normally Closed (NC).

The output pin connected to the LED is turned on when you set `val` to `HIGH`, illuminating the LED.

Although Arduino sets all digital pins as inputs by default, it is a good practice to set this explicitly in your sketch to remind yourself about the pins you are using.

You may see similar code that uses `true` instead of `HIGH`; these can be used interchangeably (they are also sometimes represented as `1`). Likewise, `false` is the same as `LOW` and `0`. Use the form that best expresses the meaning of the logic in your application.

Almost any switch can be used, although the ones called *momentary tactile switches* are popular because they are inexpensive and can plug directly into a breadboard. See the website for this book (*http://shop.oreilly.com/product/0636920022244.do*) for some supplier part numbers.

Here is another way to implement the logic in the preceding sketch:

```
void loop()
{
    digitalWrite(ledPin, digitalRead(inputPin));  // turn LED ON if input pin is
                                                  // HIGH, else turn OFF
}
```

This doesn't store the button state into a variable. Instead, it sets the LED on or off directly from the value obtained from `digitalRead`. It is a handy shortcut, but if you find it overly terse, there is no practical difference in performance, so pick whichever form you find easier to understand.

The pull-up code is similar to the pull-down version, but the logic is reversed: the value on the pin goes `LOW` when the button is pressed (see Figure 5-4 for a schematic diagram of this). It may help to think of this as pressing the switch `DOWN`, causing the output to go `LOW`:

```
void loop()
{
  int val = digitalRead(inputPin);  // read input value
  if (val == HIGH)                  // check if the input is HIGH
  {
    digitalWrite(ledPin, LOW);      // turn LED OFF
  }
  else
  {
    digitalWrite(ledPin, HIGH);     // turn LED ON
  }
}
```

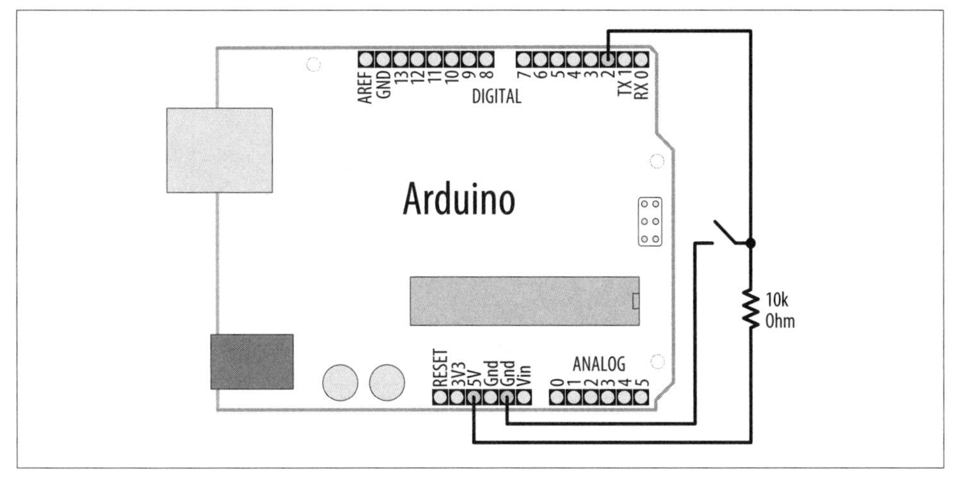

Figure 5-4. Switch connected using pull-up resistor

See Also

The Arduino reference for `digitalRead`: *http://arduino.cc/en/Reference/DigitalRead*

The Arduino reference for `digitalWrite`: *http://arduino.cc/en/Reference/DigitalWrite*

The Arduino reference for `pinMode`: *http://arduino.cc/en/Reference/PinMode*

The Arduino references for constants (`HIGH`, `LOW`, etc.): *http://arduino.cc/en/Reference/Constants*

Arduino tutorial on digital pins: *http://arduino.cc/en/Tutorial/DigitalPins*

5.2 Using a Switch Without External Resistors

Problem

You want to simplify your wiring by eliminating external pull-up resistors when connecting switches.

Solution

As explained in Recipe 5.1, digital inputs must have a resistor to hold the pin to a known value when the switch is not pressed. Arduino has internal pull-up resistors that can be enabled by writing a `HIGH` value to a pin that is in `INPUT` mode (the code for this is shown in Recipe 5.1).

For this example, the switch is wired as shown in Figure 5-5. This is almost exactly the same as Figure 5-4, but without an external resistor.

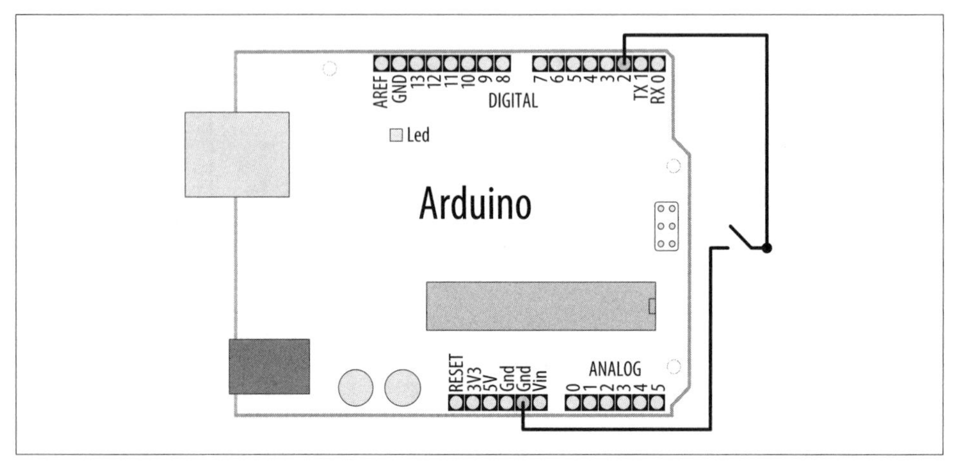

Figure 5-5. Switch wired for use with internal pull-up resistor

The switch is only connected between pin 2 and Gnd. Gnd is short for *ground* and is at 0 volts by definition:

```
/*
  Pullup sketch
  a switch connected to pin 2 lights the LED on pin 13
*/

const int ledPin = 13;          // output pin for the LED
const int inputPin = 2;         // input pin for the switch

void setup() {
  pinMode(ledPin, OUTPUT);
  pinMode(inputPin, INPUT);
```

```
  digitalWrite(inputPin,HIGH);  // turn on internal pull-up on the inputPin
}
void loop(){
  int val = digitalRead(inputPin);  // read input value
  if (val == HIGH)                  // check if the input is HIGH
  {
    digitalWrite(ledPin, HIGH);     // turn LED OFF
  }
  else
  {
    digitalWrite(ledPin, LOW);      // turn LED ON
  }
}
```

> There is more than one Gnd pin on an Arduino board; they are all connected together, so pick whichever is convenient.

Discussion

You enable internal pull-up resistors by writing a `HIGH` value to a pin in input mode. Using `digitalWrite(`*`pin`*`, HIGH)` on a pin in input mode may not be intuitive at first, but you'll soon get used to it. You can turn the pull-up off by writing a `LOW` value to the pin.

If your application switches the pin mode back and forth between input and output, bear in mind that the state of the pin will remain `HIGH` or `LOW` when you change modes. In other words, if you have set an output pin `HIGH` and then change to input mode, the pull-up will be on, and reading the pin will produce a `HIGH`. If you set the pin `LOW` in output mode with `digitalWrite(`*`pin`*`, LOW)` and then change to input mode with `pinMode(`*`pin`*`, INPUT)`, the pull-up will be off. If you turn a pull-up on, changing to output mode will set the pin `HIGH`, which could, for example, unintentionally light an LED connected to it.

The internal pull-up resistors are 20K ohms or more (between 20K and 50K). This is suitable for most applications, but some devices may require lower-value resistors—see the data sheet for external devices you want to connect to Arduino to see if the internal pull-ups are suitable or not.

5.3 Reliably Detecting the Closing of a Switch

Problem

You want to avoid false readings due to *contact bounce* (contact bounce produces spurious signals at the moment the switch contacts close or open). The process of eliminating spurious readings is called *debouncing*.

Solution

There are many ways to solve this problem; here is one using the wiring shown in Figure 5-3 from Recipe 5.1:

```
/*
 * Debounce sketch
 * a switch connected to pin 2 lights the LED on pin 13
 * debounce logic prevents misreading of the switch state
 */

const int inputPin = 2;        // the number of the input pin
const int ledPin = 13;         // the number of the output pin
const int debounceDelay = 10;  // milliseconds to wait until stable

// debounce returns true if the switch in the given pin is closed and stable
boolean debounce(int pin)
{
  boolean state;
  boolean previousState;

  previousState = digitalRead(pin);          // store switch state
  for(int counter=0; counter < debounceDelay; counter++)
  {
      delay(1);                  // wait for 1 millisecond
      state = digitalRead(pin);  // read the pin
      if( state != previousState)
      {
         counter = 0; // reset the counter if the state changes
         previousState = state;  // and save the current state
      }
  }
  // here when the switch state has been stable longer than the debounce period
  return state;
}

void setup()
{
  pinMode(inputPin, INPUT);
  pinMode(ledPin, OUTPUT);
}

void loop()
{
  if (debounce(inputPin))
  {
    digitalWrite(ledPin, HIGH);
  }
}
```

The debounce function is called (used) with the pin number of the switch you want to debounce; the function returns `true` if the switch is pressed and stable. It returns `false` if it is not pressed or not yet stable.

Discussion

The debounce method checks to see if it gets the same reading from the switch after a delay that needs to be long enough for the switch contacts to stop bouncing. You may require longer intervals for "bouncier" switches (some switches can require as much as 50 ms or more). The function works by repeatedly checking the state of the switch for as many milliseconds as defined in the debounce time. If the switch remains stable for this time, the state of the switch will be returned (true if pressed and false if not). If the switch state changes within the debounce period, the counter is reset so that the checks start over until the switch state does not change within the debounce time.

If your wiring uses pull-up resistors instead of pull-down resistors (see Recipe 5.2) you need to invert the value returned from the debounce function, because the state goes LOW when the switch is pressed using pull-ups, but the function should return true (true is the same as HIGH) when the switch is pressed. The debounce code using pull-ups is as follows; only the last four lines (highlighted) are changed from the previous version:

```
boolean debounce(int pin)
{
  boolean state;
  boolean previousState;

  previousState = digitalRead(pin);          // store switch state
  for(int counter=0; counter < debounceDelay; counter++)
  {
      delay(1);                  // wait for 1 millisecond
      state = digitalRead(pin);  // read the pin
      if( state != previousState)
      {
         counter = 0; // reset the counter if the state changes
         previousState = state;  // and save the current state
      }
  }
  // here when the switch state has been stable longer than the debounce period
  if(state == LOW)  // LOW means pressed (because pull-ups are used)
     return true;
  else
    return false;
}
```

For testing, you can add a count variable to display the number of presses. If you view this on the Serial Monitor (see Chapter 4), you can see whether it increments once per press. Increase the value of debounceDelay until the count keeps step with the presses. The following fragment prints the value of count when used with the debounce function shown earlier:

```
int count;   // add this variable to store the number of presses

void setup()
{
  pinMode(inPin, INPUT);
```

```
  pinMode(outPin, OUTPUT);
  Serial.begin(9600); // add this to the setup function
}

void loop()
{
  if(debounce(inPin))
  {
    digitalWrite(outPin, HIGH);
    count++;        // increment count
    Serial.println(count);  // display the count on the Serial Monitor
  }
}
```

This debounce() function will work for any number of switches, but you must ensure that the pins used are in input mode.

A potential disadvantage of this method for some applications is that from the time the debounce function is called, everything waits until the switch is stable. In most cases this doesn't matter, but your sketch may need to be attending to other things while waiting for your switch to stabilize. You can use the code shown in Recipe 5.4 to overcome this problem.

See Also

See the Debounce example sketch distributed with Arduino. From the File menu, select Examples→Digital→Debounce.

5.4 Determining How Long a Switch Is Pressed

Problem

Your application wants to detect the length of time a switch has been in its current state. Or you want to increment a value while a switch is pushed and you want the rate to increase the longer the switch is held (the way many electronic clocks are set). Or you want to know if a switch has been pressed long enough for the reading to be stable (see Recipe 5.3).

Solution

The following sketch demonstrates the setting of a countdown timer. The wiring is the same as in Figure 5-5 from Recipe 5.2. Pressing a switch sets the timer by incrementing the timer count; releasing the switch starts the countdown. The code debounces the switch and accelerates the rate at which the counter increases when the switch is held for longer periods. The timer count is incremented by one when the switch is initially pressed (after debouncing). Holding the switch for more than one second increases the increment rate by four; holding the switch for four seconds increases the rate by ten.

Releasing the switch starts the countdown, and when the count reaches zero, a pin is set `HIGH` (in this example, lighting an LED):

```
/*
 SwitchTime sketch
 Countdown timer that decrements every tenth of a second
 lights an LED when 0
 Pressing button increments count, holding button down increases
 rate of increment

 */
const int ledPin = 13;                // the number of the output pin
const int inPin = 2;                  // the number of the input pin

const int  debounceTime = 20;         // the time in milliseconds required
                                      // for the switch to be stable
const int  fastIncrement = 1000;      // increment faster after this many
                                      // milliseconds
const int  veryFastIncrement = 4000;  // and increment even faster after
                                      // this many milliseconds
int count = 0;                        // count decrements every tenth of a
                                      // second until reaches 0


void setup()
{
  pinMode(inPin, INPUT);
  digitalWrite(inPin, HIGH); // turn on pull-up resistor
  pinMode(ledPin, OUTPUT);
  Serial.begin(9600);
}



void loop()
{
  int duration = switchTime();
  if( duration > veryFastIncrement)
    count = count + 10;
  else if ( duration > fastIncrement)
    count = count + 4;
  else if ( duration > debounceTime)
    count = count + 1;

  else
  {
    // switch not pressed so service the timer
    if( count == 0)
      digitalWrite(ledPin, HIGH);  // turn the LED on if the count is 0
    else
    {
      digitalWrite(ledPin, LOW);   // turn the LED off if the count is not 0
      count = count - 1;           // and decrement the count
    }
  }
```

```
  Serial.println(count);
  delay(100);
}

// return the time in milliseconds that the switch has been in pressed (LOW)
long switchTime()
{
  // these variables are static  - see Discussion for an explanation
  static unsigned long startTime = 0;  // the time the switch state change was
first detected
  static boolean state;                // the current state of the switch

  if(digitalRead(inPin) != state) // check to see if the switch has changed state
  {
    state = ! state;       // yes, invert the state
    startTime = millis();  // store the time
  }
  if( state == LOW)
    return millis() - startTime;   // switch pushed, return time in milliseconds
  else
    return 0; // return 0 if the switch is not pushed (in the HIGH state);
}
```

Discussion

The heart of this recipe is the `switchTime` function. This returns the number of milliseconds that the switch has been pressed. Because this recipe uses internal pull-up resistors (see Recipe 5.2), the `digitalRead` of the switch pin will return `LOW` when the switch is pressed.

The `loop` checks the value returned from `switchTime` to see what should happen. If the time the switch has been held down is long enough for the fastest increment, the counter is incremented by that amount; if not, it checks the `fast` value to see if that should be used; if not, it checks if the switch has been held down long enough to stop bouncing and if so, it increments a small amount. At most, one of those will happen. If none of them are `true`, the switch is not being pressed, or it has not been pressed long enough to have stopped bouncing. The counter value is checked and an LED is turned on if it is zero; if it's not zero, the counter is decremented and the LED is turned off.

You can use the `switchTime` function just for debouncing a switch. The following code handles debounce logic by calling the `switchTime` function:

```
// the time in milliseconds that the switch needs to be stable
const int  debounceTime = 20;

if( switchTime() > debounceTime);
    Serial.print("switch is debounced");
```

This approach to debouncing can be handy if you have more than one switch, because you can peek in and look at the amount of time a switch has been pressed and process other tasks while waiting for a switch to become stable. To implement this, you need to store the current state of the switch (pressed or not) and the time the state last changed. There are many ways to do this—in this example, you will use a separate function for each switch. You could store the variables associated with all the switches at the top of your sketch as *global variables* (called "global" because they are accessible everywhere). But it is more convenient to have the variables for each switch contained with the function.

Retaining values of variables defined in a function is achieved by using *static variables*. Static variables within a function provide permanent storage for values that must be maintained between function calls. A value assigned to a static variable is retained even after the function returns. The last value set will be available the next time the function is called. In that sense, static variables are similar to the global variables (variables declared outside a function, usually at the beginning of a sketch) that you saw in the other recipes. But unlike global variables, static variables declared in a function are only accessible within that function. The benefit of static variables is that they cannot be accidentally modified by some other function.

This sketch shows an example of how you can add separate functions for different switches. The wiring for this is similar to Recipe 5.2, with the second switch wired similarly to the first (as shown in Figure 5-5) but connected between pin 3 and Gnd:

```
/*
 SwitchTimeMultiple sketch
 Prints how long more than one switch has been pressed
 */

const int switchAPin = 2;              // the pin for switch A
const int switchBPin = 3;              // the pin for switch B

// functions with references must be explicitly declared
unsigned long switchTime(int pin, boolean &state, unsigned long  &startTime);

void setup()
{
  pinMode(switchAPin, INPUT);
  digitalWrite(switchAPin, HIGH); // turn on pull-up resistors
  pinMode(switchBPin, INPUT);
  digitalWrite(switchBPin, HIGH); // turn on pull-up resistors
  Serial.begin(9600);
}

void loop()
{
unsigned long time;

  Serial.print("switch A time =");
  time = switchATime();
  Serial.print(time);

  Serial.print(", switch B time =");
  time = switchBTime();
  Serial.println(time);
```

```
  delay(1000);
}

unsigned long switchTime(int pin, boolean &state, unsigned long  &startTime)
{
  if(digitalRead(pin) != state) // check to see if the switch has changed state
  {
    state = ! state;       //yes,  invert the state
    startTime = millis();  // store the time
  }
  if( state == LOW)
    return millis() - startTime;   // return the time in milliseconds
  else
    return 0; // return 0 if the switch is not pushed (in the HIGH state);
}


long switchATime()
{
  // these variables are static  - see text for an explanation
  // the time the switch state change was first detected
  static unsigned long startTime = 0;
  static boolean state;                 // the current state of the switch
  return switchTime(switchAPin, state, startTime);
}

long switchBTime()
{
  // these variables are static  - see text for an explanation
  // the time the switch state change was first detected
  static unsigned long startTime = 0;
  static boolean state;                 // the current state of the switch
  return switchTime(switchBPin, state, startTime);
}
```

The time calculation is performed in a function called `switchTime()`. This function examines and updates the switch state and duration. The function uses references to handle the parameters—references were covered in Recipe 2.11. A function for each switch (`switchATime()` and `switchBTime()`) is used to retain the start time and state for each switch. Because the variables holding the values are declared as static, the values will be retained when the functions exit. Holding the variables within the function ensures that the wrong variable will not be used. The pins used by the switches are declared as global variables because the values are needed by `setup` to configure the pins. But because these variables are declared with the `const` keyword, the compiler will not allow the values to be modified, so there is no chance that these will be accidentally changed by the sketch code.

Limiting the exposure of a variable becomes more important as projects become more complex. The Arduino environment provides a more elegant way to handle this; see Recipe 16.4 for a discussion on how to implement this using classes.

5.5 Reading a Keypad

Problem

You have a matrix keypad and want to read the key presses in your sketch. For example, you have a telephone-style keypad similar to the SparkFun 12-button keypad (SparkFun COM-08653).

Solution

Wire the rows and columns from the keypad connector to the Arduino, as shown in Figure 5-6.

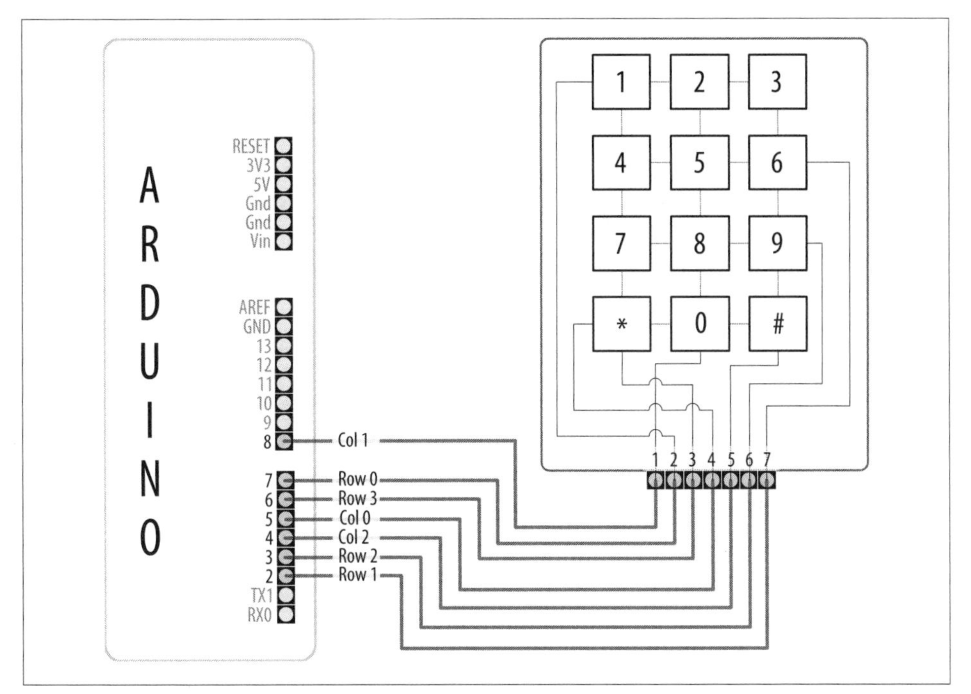

Figure 5-6. Connecting the SparkFun keyboard matrix

If you've wired your Arduino and keypad as shown in Figure 5-6, the following sketch will print key presses to the Serial Monitor:

```
/*
  Keypad sketch
  prints the key pressed on a keypad to the serial port
*/

const int numRows = 4;       // number of rows in the keypad
const int numCols = 3;       // number of columns
const int debounceTime = 20; // number of milliseconds for switch to be stable
```

```
// keymap defines the character returned when the corresponding key is pressed
const char keymap[numRows][numCols] = {
  { '1', '2', '3'  } ,
  { '4', '5', '6'  } ,
  { '7', '8', '9'  } ,
  { '*', '0', '#'  }
};

// this array determines the pins used for rows and columns
const int rowPins[numRows] = { 7, 2, 3, 6 }; // Rows 0 through 3
const int colPins[numCols] = { 5, 8, 4 };    // Columns 0 through 2

void setup()
{
  Serial.begin(9600);
  for (int row = 0; row < numRows; row++)
  {
    pinMode(rowPins[row],INPUT);       // Set row pins as input
    digitalWrite(rowPins[row],HIGH);   // turn on Pull-ups
  }
  for (int column = 0; column < numCols; column++)
  {
    pinMode(colPins[column],OUTPUT);     // Set column pins as outputs
                                         // for writing
    digitalWrite(colPins[column],HIGH);  // Make all columns inactive
  }
}

void loop()
{
  char key = getKey();
  if( key != 0) {       // if the character is not 0 then
                        // it's a valid key press
    Serial.print("Got key ");
    Serial.println(key);
  }
}

// returns with the key pressed, or 0 if no key is pressed
char getKey()
{
  char key = 0;                                  // 0 indicates no key pressed

  for(int column = 0; column < numCols; column++)
  {
    digitalWrite(colPins[column],LOW);         // Activate the current column.
    for(int row = 0; row < numRows; row++)     // Scan all rows for
                                               // a key press.
    {
      if(digitalRead(rowPins[row]) == LOW)     // Is a key pressed?
      {
        delay(debounceTime);                   // debounce
        while(digitalRead(rowPins[row]) == LOW)
            ;                                  // wait for key to be released
```

```
        key = keymap[row][column];           // Remember which key
                                             // was pressed.
      }
    }
    digitalWrite(colPins[column],HIGH);    // De-activate the current column.
  }
  return key;  // returns the key pressed or 0 if none
}
```

This sketch will only work correctly if the wiring agrees with the code. Table 5-2 shows how the rows and columns should be connected to Arduino pins. If you are using a different keypad, check your data sheet to determine the row and column connections. Check carefully, as incorrect wiring can short out the pins, and that could damage your controller chip.

Table 5-2. Mapping of Arduino pins to SparkFun connector and keypad rows and columns

Arduino pin	Keypad connector	Keypad row/column
2	7	Row 1
3	6	Row 2
4	5	Column 2
5	4	Column 0
6	3	Row 3
7	2	Row 0
8	1	Column 1

Discussion

Matrix keypads typically consist of Normally Open switches that connect a row with a column when pressed. (A Normally Open switch only makes electrical connection when pushed.) Figure 5-6 shows how the internal conductors connect the button rows and columns to the keyboard connector. Each of the four rows is connected to an input pin and each column is connected to an output pin. The `setup` function sets the pin modes and enables pull-up resistors on the input pins (see the pull-up recipes in the beginning of this chapter).

The `getkey` function sequentially sets the pin for each column `LOW` and then checks to see if any of the row pins are `LOW`. Because pull-up resistors are used, the rows will be high (pulled up) unless a switch is closed (closing a switch produces a `LOW` signal on the input pin). If they are `LOW`, this indicates that the switch for that row and column is closed. A delay is used to ensure that the switch is not bouncing (see Recipe 5.3); the code waits for the switch to be released, and the character associated with the switch is found in the `keymap` array and returned from the function. A `0` is returned if no switch is pressed.

A library in the Arduino Playground that is similar to the preceding example provides more functionality. The library makes it easier to handle different numbers of keys and it can be made to work while sharing some of the pins with an LCD. You can find the library at *http://www.arduino.cc/playground/Main/KeypadTutorial*.

See Also

For more information on the SparkFun 12-button keypad, go to *http://www.sparkfun.com/commerce/product_info.php?products_id=8653*.

5.6 Reading Analog Values

Problem

You want to read the voltage on an analog pin. Perhaps you want a reading from a potentiometer (pot) or a device or sensor that provides a voltage between 0 and 5 volts.

Solution

This sketch reads the voltage on an analog pin and flashes an LED in a proportional rate to the value returned from the `analogRead` function. The voltage is adjusted by a potentiometer connected as shown in Figure 5-7:

```
/*
 Pot sketch
 blink an LED at a rate set by the position of a potentiometer
*/

const int potPin = 0;    // select the input pin for the potentiometer
const int ledPin = 13;   // select the pin for the LED
int val = 0;             // variable to store the value coming from the sensor

void setup()
{
  pinMode(ledPin, OUTPUT);  // declare the ledPin as an OUTPUT
}

void loop() {
  val = analogRead(potPin);     // read the voltage on the pot
  digitalWrite(ledPin, HIGH);  // turn the ledPin on
  delay(val);                  // blink rate set by pot value (in milliseconds)
  digitalWrite(ledPin, LOW);   // turn the ledPin off
  delay(val);                  // turn led off for same period as it was turned on
}
```

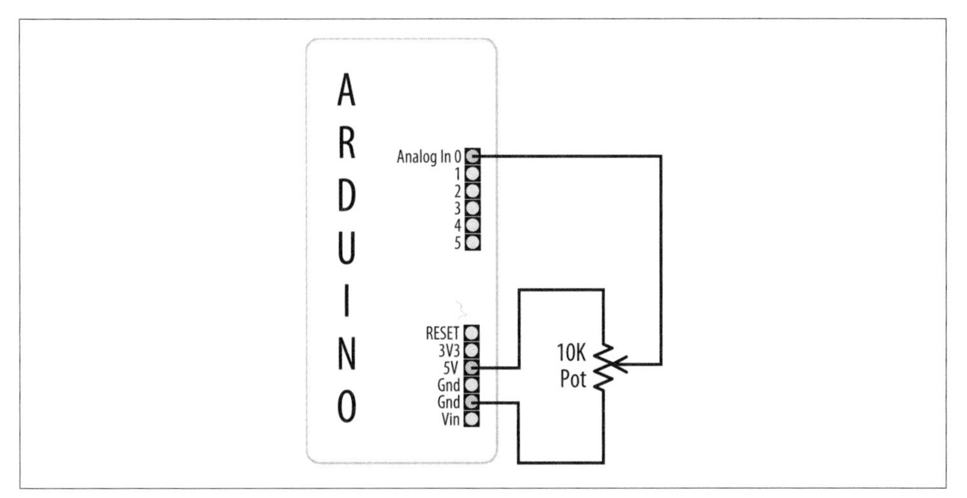

Figure 5-7. Connecting a potentiometer to Arduino

Discussion

This sketch uses the `analogRead` function to read the voltage on the potentiometer's *wiper* (the center pin). A pot has three pins; two are connected to a resistive material and the third pin (usually in the middle) is connected to a wiper that can be rotated to make contact anywhere on the resistive material. As the potentiometer rotates, the resistance between the wiper and one of the pins increases, while the other decreases. The schematic diagram for this recipe (Figure 5-7) may help you visualize how a potentiometer works; as the wiper moves toward the bottom end, the wiper (the line with the arrow) will have lower resistance connecting to Gnd and higher resistance connecting to 5 volts. As the wiper moves down, the voltage on the analog pin will decrease (to a minimum of 0 volts). Moving the wiper upward will have the opposite effect, and the voltage on the pin will increase (up to a maximum of 5 volts).

If the voltage on the pin decreases, rather than increases, as you increase the rotation of the potentiometer, you can reverse the connections to the +5 volts and Gnd pins.

The voltage is measured using `analogRead`, which provides a value proportional to the actual voltage on the analog pin. The value will be 0 when there are 0 volts on the pin and 1,023 when there are 5 volts. A value in between will be proportional to the ratio of the voltage on the pin to 5 volts.

Potentiometers with a value of 10K ohms are the best choice for connecting to analog pins. See this book's website (*http://shop.oreilly.com/product/0636920022244.do*) for recommended part numbers.

potPin does not need to be set as input. (This is done for you automatically each time you call analogRead.)

See Also

Appendix B, for tips on reading schematic diagrams

Arduino reference for analogRead: *http://www.arduino.cc/en/Reference/AnalogRead*

Getting Started with Arduino by Massimo Banzi (Make)

5.7 Changing the Range of Values

Problem

You want to change the range of a value, such as the value from analogRead obtained by connecting a potentiometer (pot) or other device that provides a variable voltage. For example, suppose you want to display the position of a potentiometer knob as a percentage from 0 percent to 100 percent.

Solution

Use the Arduino map function to scale values to the range you want. This sketch reads the voltage on a pot into the variable val and scales this from 0 to 100 as the pot is rotated from one end to the other. It blinks an LED with a rate proportional to the voltage on the pin and prints the scaled range to the serial port (see Recipe 4.2 for instructions on monitoring the serial port). Recipe 5.6 shows how the pot is connected (see Figure 5-7):

```
/*
 * Map sketch
 * map the range of analog values from a pot to scale from 0 to 100
 * resulting in an LED blink rate ranging from 0 to 100 milliseconds.
 * and Pot rotation percent is written to the serial port
 */

const int potPin = 0;          // select the input pin for the potentiometer
int ledPin = 13;               // select the pin for the LED

void setup()
{
  pinMode(ledPin, OUTPUT);     // declare the ledPin as an OUTPUT
  Serial.begin(9600);
}

void loop() {
  int val;                       // The value coming from the sensor
  int percent;                   // The mapped value
```

```
    val = analogRead(potPin);         // read the voltage on the pot (val ranges
                                      // from 0 to 1023)
    percent = map(val,0,1023,0,100); // percent will range from 0 to 100.
    digitalWrite(ledPin, HIGH);       // turn the ledPin on
    delay(percent);                   // On time given by percent value
    digitalWrite(ledPin, LOW);        // turn the ledPin off
    delay(100 - percent);             // Off time is 100 minus On time
    Serial.println(percent);          // show the % of pot rotation on Serial Monitor
}
```

Discussion

Recipe 5.6 describes how the position of a pot is converted to a value. Here you use this value with the map function to scale the value to your desired range. In this example, the value provided by analogRead (0 to 1023) is mapped to a percentage (0 to 100). The values from analogRead will range from 0 to 1023 if the voltage ranges from 0 to 5 volts, but you can use any appropriate values for the source and target ranges. For example, a typical pot only rotates 270 degrees from end to end, and if you wanted to display the angle of the knob on your pot, you could use this code:

```
angle = map(val,0,1023,0,270); // angle of pot derived from analogRead val
```

Range values can also be negative. If you want to display 0 when the pot is centered and negative values when the pot is rotated left and positive values when it is rotated right, you can do this:

```
// show angle of 270 degree pot with center as 0
angle = map(val,0,1023,-135,135);
```

The map function can be handy where the input range you are concerned with does not start at zero. For example, if you have a battery where the available capacity is proportional to a voltage that ranges from 1.1 volts (1,100 millivolts) to 1.5 volts (1,500 millivolts), you can do the following:

```
const int empty  = 5000 / 1100; // the voltage is 1.1 volts (1100mv) when empty
const int full   = 5000 / 1500; // the voltage is 1.5 volts (1500mv) when full

int val = analogRead(potPin);                 // read the analog voltage
int percent = map(val, empty, full, 0,100); // map the actual range of voltage
to a percent
Serial.println(percent);
```

If you are using sensor readings with map then you will need to determine the minimum and maximum values from your sensor. You can monitor the reading on the serial port to determine the lowest and highest values. Enter these as the lower and upper bound into the map function.

If the range can't be determined in advance, you can determine the values by calibrating the sensor. Recipe 8.11 shows one technique for calibration; another can be found in the Calibration examples sketch distributed with Arduino (Examples→Analog→Calibration).

Bear in mind that if you feed values into `map` that are outside the upper and lower limits, the output will also be outside the specified output range. You can prevent this happening by using the `constrain` function; see Recipe 3.5.

`map` uses integer math, so it will only return whole numbers in the range specified. Any fractional element is truncated, not rounded.

(See Recipe 5.9 for more details on how `analogRead` values relate to actual voltage.)

See Also

The Arduino reference for `map`: *http://www.arduino.cc/en/Reference/Map*

5.8 Reading More Than Six Analog Inputs

Problem

You have more analog inputs to monitor than you have available analog pins. A standard Arduino board has six analog inputs (the Mega has 16) and there may not be enough analog inputs available for your application. Perhaps you want to adjust eight parameters in your application by turning knobs on eight potentiometers.

Solution

Use a multiplexer chip to select and connect multiple voltage sources to one analog input. By sequentially selecting from multiple sources, you can read each source in turn. This recipe uses the popular 4051 chip connected to Arduino as shown in Figure 5-8. Your analog inputs get connected to the 4051 pins marked Ch 0 to Ch 7. Make sure the voltage on the channel input pins is never higher than 5 volts:

```
/*
 * multiplexer sketch
 * read 1 of 8 analog values into single analog input pin with 4051 multiplexer
 */

// array of pins used to select 1 of 8 inputs on multiplexer
const int select[] = {2,3,4}; // pins connected to the 4051 input select lines
const int analogPin = 0;      // the analog pin connected to multiplexer output

// this function returns the analog value for the given channel
int getValue( int channel)
{
   // set the selector pins HIGH and LOW to match the binary value of channel
   for(int bit = 0; bit < 3; bit++)
   {
      int pin = select[bit]; // the pin wired to the multiplexer select bit
```

```
      int isBitSet =  bitRead(channel, bit); // true if given bit set in channel
      digitalWrite(pin, isBitSet);
    }
    return analogRead(analogPin);
}

void setup()
{
  for(int bit = 0; bit < 3; bit++)
    pinMode(select[bit], OUTPUT);  // set the three select pins to output
  Serial.begin(9600);
}
void loop () {
  // print the values for each channel once per second
  for(int channel = 0; channel < 8; channel++)
  {
     int value = getValue(channel);
     Serial.print("Channel ");
     Serial.print(channel);
     Serial.print(" = ");
     Serial.println(value);
  }
  delay (1000);
}
```

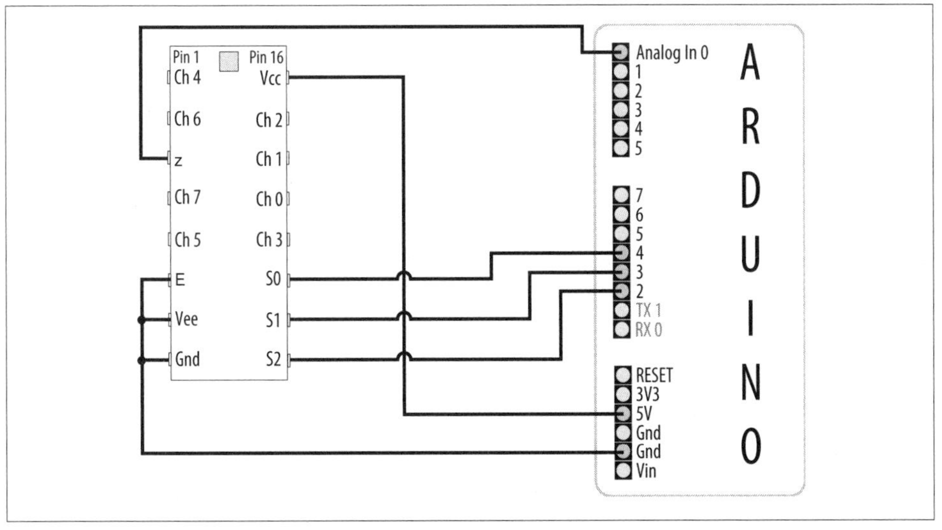

Figure 5-8. The 4051 multiplexer connected to Arduino

Discussion

Analog multiplexers are digitally controlled analog switches. The 4051 selects one of eight inputs through three selector pins (S0, S1, and S2). There are eight different combinations of values for the three selector pins, and the sketch sequentially selects each of the possible bit patterns; see Table 5-3.

Table 5-3. Truth table for 4051 multiplexer

Selector pins			Selected input
S2	S1	S0	
0	0	0	0
0	0	1	1
0	1	0	2
0	1	1	3
1	0	0	4
1	0	1	5
1	1	0	6
1	1	1	7

You may recognize the pattern in Table 5-3 as the binary representation of the decimal values from 0 to 7.

In the preceding sketch, `getValue()` is the function that sets the correct selector bits for the given channel using `digitalWrite(pin, isBitSet)` and reads the analog value from the selected 4051 input with `analogRead(analogPin)`. The code to produce the bit patterns uses the built-in `bitRead` function (see Recipe 3.12).

Don't forget to connect the ground from the devices you are measuring to the ground on the 4051 and Arduino, as shown in Figure 5-8.

Bear in mind that this technique selects and monitors the eight inputs sequentially, so it requires more time between the readings on a given input compared to using `analogRead` directly. If you are reading eight inputs, it will take eight times longer for each input to be read. This may make this method unsuitable for inputs that change value quickly.

See Also

Arduino Playground tutorial for the 4051: *http://www.arduino.cc/playground/Learning/4051*

CD4051 data sheet: *http://www.fairchildsemi.com/ds/CD%2FCD4052BC.pdf*

Analog/digital MUX breakout board data sheet: *http://www.nkcelectronics.com/analog digital-mux-breakout.html*

5.9 Displaying Voltages Up to 5V

Problem

You want to monitor and display the value of a voltage between 0 and 5 volts. For example, suppose you want to display the voltage of a single 1.5V cell on the Serial Monitor.

Solution

Use `AnalogRead` to measure the voltage on an analog pin. Convert the reading to a voltage by using the ratio of the reading to the reference voltage (5 volts), as shown in Figure 5-9.

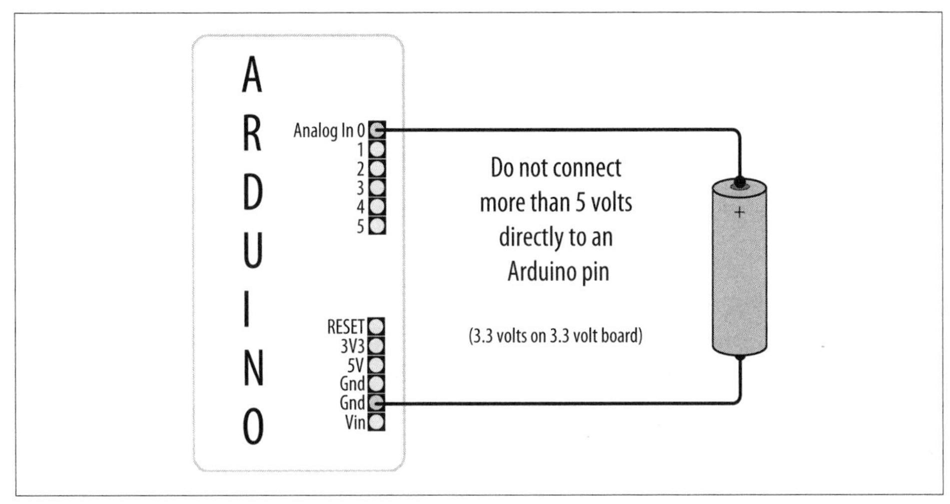

Figure 5-9. Measuring voltages up to 5 volts using 5V board

The simplest solution uses a floating-point calculation to print the voltage; this example sketch calculates and prints the ratio as a voltage:

```
/*
 * Display5vOrless sketch
 * prints the voltage on analog pin to the serial port
 * Warning - do not connect more than 5 volts directly to an Arduino pin.
 */

const float referenceVolts = 5.0;  // the default reference on a 5-volt board
const int batteryPin = 0;          // battery is connected to analog pin 0

void setup()
{
   Serial.begin(9600);
}
```

```
void loop()
{
   int val = analogRead(batteryPin);  // read the value from the sensor
   float volts = (val / 1023.0) * referenceVolts;  // calculate the ratio
   Serial.println(volts);  // print the value in volts
}
```

The formula is: Volts = (analog reading / analog steps) × Reference voltage

Printing a floating-point value to the serial port with `println` will format the value to two decimal places.

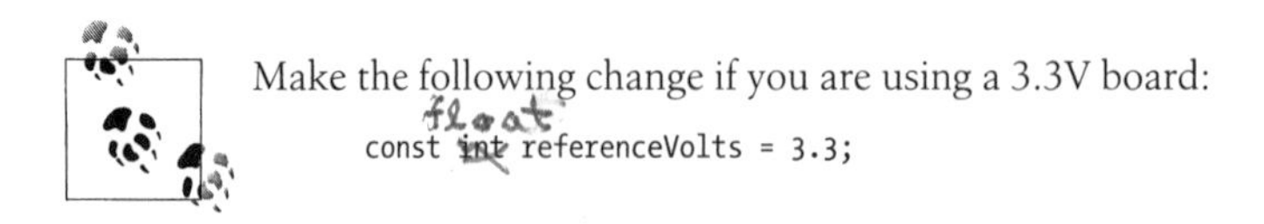

Make the following change if you are using a 3.3V board:

```
const int referenceVolts = 3.3;
```

Floating-point numbers consume lots of memory, so unless you are already using floating point elsewhere in your sketch, it is more efficient to use integer values. The following code looks a little strange at first, but because `analogRead` returns a value of `1023` for 5 volts, each step in value will be 5 divided by 1,023. In units of millivolts, this is 5,000 divided by 1,023.

This code prints the value in millivolts:

```
const int  batteryPin = 0;

void setup()
{
  Serial.begin(9600);
}

void loop()
{
  long val = analogRead(batteryPin);    // read the value from the sensor -
                                        // note val is a long int
  Serial.println( (val * (500000/1023)) / 100);  // print the value in millivolts
}
```

The following code prints the value using decimal points. It prints `1.5` if the voltage is 1.5 volts:

```
const int batteryPin = 0;

void setup()
{
  Serial.begin(9600);
}

void loop()
{
  int val = analogRead(batteryPin); // read the value from the sensor
```

```
  long mv =  (val * (500000/1023L)) / 100; // calculate the value in millivolts
  Serial.print(mv/1000); // print the integer value of the voltage
  Serial.print('.');
  int fraction = (mv % 1000); // calculate the fraction
  if (fraction == 0)
     Serial.print("000");       // add three zero's
  else if (fraction < 10)     // if fractional < 10 the 0 is ignored giving a wrong
                               // time, so add the zeros
     Serial.print("00");        // add two zeros
  else if (fraction < 100)
     Serial.print("0");
  Serial.println(fraction); // print the fraction
}
```

If you are using a 3.3V board, change `(1023/5)` to `(int)(1023/3.3)`.

Discussion

The `analogRead()` function returns a value that is proportional to the ratio of the measured voltage to the reference voltage (5 volts). To avoid the use of floating point, yet maintain precision, the code operates on values as millivolts instead of volts (there are 1,000 millivolts in 1 volt). Because a value of `1023` indicates 5,000 millivolts, each unit represents 5,000 divided by 1,023 millivolts (that is, 4.89 millivolts).

You will see both 1023 and 1024 used for converting `analogRead` values to millivolts. 1024 is commonly used by engineers because there are 1024 possible values between 0 and 1023. However, 1023 is more intuitive for some because the highest possible value is 1023. In practice, the hardware inaccuracy is greater than the difference between the calculations so choose whichever value you feel more comfortable with.

To eliminate the decimal point, the values are multiplied by 100. In other words, 5,000 millivolts times 100 divided by 1,023 gives the number of millivolts times 100. Dividing this by 100 yields the value in millivolts. If multiplying fractional numbers by 100 to enable the compiler to perform the calculation using fixed-point arithmetic seems convoluted, you can stick to the slower and more memory-hungry floating-point method.

This solution assumes you are using a standard Arduino powered from 5 volts. If you are using a 3.3V board, the maximum voltage you can measure is 3.3 volts without using a voltage divider—see Recipe 5.11.

5.10 Responding to Changes in Voltage

Problem

You want to monitor one or more voltages and take some action when the voltage rises or falls below a threshold. For example, you want to flash an LED to indicate a low battery level—perhaps to start flashing when the voltage drops below a warning threshold and increasing in urgency as the voltage drops further.

Solution

You can use the connections shown in Figure 5-7 in Recipe 5.9, but here we'll compare the value from **analogRead** to see if it drops below a threshold. This example starts flashing an LED at 1.2 volts and increases the on-to-off time as the voltage decreases below the threshold. If the voltage drops below a second threshold, the LED stays lit:

```
/*
 RespondingToChanges sketch
 flash an LED to indicate low voltage levels
*/

long warningThreshold  = 1200;  // Warning level in millivolts - LED flashes
long criticalThreshold = 1000;  // Critical voltage level - LED stays on

const int batteryPin = 0;
const int ledPin = 13;

void setup()
{
   pinMode(ledPin, OUTPUT);
}

void loop()
{
  int val = analogRead(batteryPin);     // read the value from the sensor
  if( val < (warningThreshold  * 1023L)/5000) {
    // in the line above, L following a number makes it a 32 bit value
    flash(val) ;
  }
}

// function to flash an LED
// on/off time determined by value passed as percent
void flash(int percent)
{
  digitalWrite(ledPin, HIGH);
  delay(percent + 1);
  digitalWrite(ledPin, LOW);
  delay(100 - percent );  // check delay == 0?
}
```

Discussion

The highlighted line in this sketch calculates the ratio of the value read from the analog port to the value of the threshold voltage. For example, with a warning threshold of 1 volt and a reference voltage of 5 volts, you want to know when the analog reading is one-fifth of the reference voltage. The expression `1023L` tells the compiler that this is a long integer (a 32-bit integer; see Recipe 2.2), so the compiler will promote all the variables in this expression to long integers to prevent overflowing the capacity of an `int` (a normal 16-bit integer).

When reading analog values, you can work in the units that are returned from `analogRead`—ranging from 0 to 1023—or you can work in the actual voltages they represent (see Recipe 5.7). As in this recipe, if you are not displaying voltage, it's simpler and more efficient to use the output of `analogRead` directly.

5.11 Measuring Voltages More Than 5V (Voltage Dividers)

Problem

You want to measure voltages greater than 5 volts. For example, you want to display the voltage of a 9V battery and trigger an alarm LED when the voltage falls below a certain level.

Solution

Use a solution similar to Recipe 5.9, but connect the voltage through a voltage divider (see Figure 5-10). For voltages up to 10 volts, you can use two 4.7K ohm resistors. For higher voltages, you can determine the required resistors using Table 5-4.

Table 5-4. Resistor values

Max voltage	R1	R2	Calculation R2/(R1 + R2)	value of `resistorFactor`
5	Short (+V connected to analog pin)	None (Gnd connected to Gnd)	None	1023
10	1K	1K	1(1 + 1)	511
15	2K	1K	1(2 + 1)	341
20	3K	1K	1(3 + 1)	255
30	4K (3.9K)	1K	1(4 + 1)	170

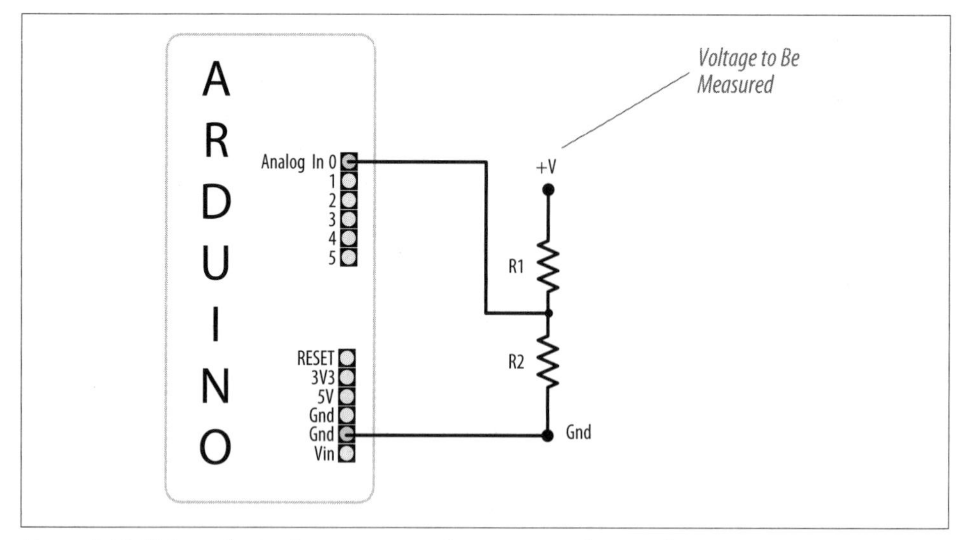

Figure 5-10. Voltage divider for measuring voltages greater than 5 volts

Select the row with the highest voltage you need to measure to find the values for the two resistors:

```
/*
  DisplayMoreThan5V sketch
  prints the voltage on analog pin to the serial port
  Do not connect more than 5 volts directly to an Arduino pin.
*/

const float referenceVolts = 5;        // the default reference on a 5-volt board
//const float referenceVolts = 3.3;  // use this for a 3.3-volt board

const float R1 = 1000; // value for a maximum voltage of 10 volts
const float R2 = 1000;
// determine by voltage divider resistors, see text
const float resistorFactor = 1023.0 * (R2/(R1 + R2));
const int batteryPin = 0;         // +V from battery is connected to analog pin 0

void setup()
{
   Serial.begin(9600);
}

void loop()
{
   int val = analogRead(batteryPin);  // read the value from the sensor
   float volts = (val / resistorFactor) * referenceVolts ; // calculate the ratio
   Serial.println(volts);  // print the value in volts
}
```

Discussion

Like the previous analog recipes, this recipe relies on the fact that the `analogRead` value is a ratio of the measured voltage to the reference. But because the measured voltage is divided by the two dropping resistors, the `analogRead` value needs to be multiplied to get the actual voltage. The code here is similar to that in Recipe 5.7, but the value of `resistorFactor` is selected based on the voltage divider resistors as shown in Table 5-4:

```
const int resistorFactor = 511;     // determine by voltage divider resistors,
see Table 5-3
```

The value read from the analog pin is divided not by 1,023, but by a value determined by the dropping resistors:

```
float volts = (val / resistorFactor) * referenceVolts ;  // calculate the ratio
```

The calculation used to produce the table is based on the following formula: the output voltage is equal to the input voltage times R2 divided by the sum of R1 and R2. In the example where two equal-value resistors are used to drop the voltage from a 9V battery by half, `resistorFactor` is 511 (half of 1,023), so the value of the `volts` variable will be twice the voltage that appears on the input pin. With resistors selected for 10 volts, the analog reading from a 9V battery will be approximately 920.

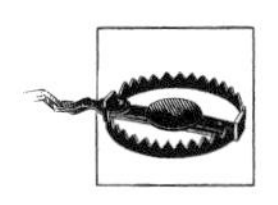

More than 5 volts on the pin can damage the pin and possibly destroy the chip; double-check that you have chosen the right value resistors and wired them correctly before connecting them to an Arduino input pin. If you have a multimeter, measure the voltage before connecting anything that could possibly carry voltages higher than 5 volts.

CHAPTER 6
Getting Input from Sensors

6.0 Introduction

Getting and using input from sensors enables Arduino to respond to or report on the world around it. This is one of the most common tasks you will encounter. This chapter provides simple and practical examples of how to use the most popular input devices and sensors. Wiring diagrams show how to connect and power the devices, and code examples demonstrate how to use data derived from the sensors.

Sensors respond to input from the physical world and convert this into an electrical signal that Arduino can read on an input pin. The nature of the electrical signal provided by a sensor depends on the kind of sensor and how much information it needs to transmit. Some sensors (such as photoresistors and Piezo knock sensors) are constructed from a substance that alters their electrical properties in response to physical change. Others are sophisticated electronic modules that use their own microcontroller to process information before passing a signal on for the Arduino.

Sensors use the following methods to provide information:

Digital on/off
: Some devices, such as the tilt sensor in Recipe 6.1 and the motion sensor in Recipe 6.3, simply switch a voltage on and off. These can be treated like the switch recipes shown in Chapter 5.

Analog
: Other sensors provide an analog signal (a voltage that is proportional to what is being sensed, such as temperature or light level). The recipes for detecting light (Recipe 6.2), motion (Recipes 6.1 and 6.3), vibration (Recipe 6.6), sound (Recipe 6.7), and acceleration (Recipe 6.18) demonstrate how analog sensors can be used. All of them use the `analogRead` command that is discussed in Chapter 5.

Pulse width
: Distance sensors, such as the PING))) in Recipe 6.4, provide data using pulse duration proportional to the distance value. Applications using these sensors measure the duration of a pulse using the `pulseIn` command.

Serial

Some sensors provide values using a serial protocol. For example, the RFID reader in Recipe 6.9 and the GPS in Recipe 6.14 communicate through the Arduino serial port (see Chapter 4 for more on serial). Most Arduino boards only have one hardware serial port, so read Recipe 6.14 for an example of how you can add additional software serial ports if you have multiple serial sensors or the hardware serial port is occupied for some other task.

Synchronous protocols: I2C and SPI

The I2C and SPI digital standards were created for microcontrollers like Arduino to talk to external sensors and modules. Recipe 6.16 shows how a compass module is connected using synchronous digital signaling. These protocols are used extensively for sensors, actuators, and peripherals, and they are covered in detail in Chapter 13.

There is another generic class of sensing devices that you may make use of. These are consumer devices that contain sensors but are sold as devices in their own right, rather than as sensors. Examples of these in this chapter include a PS/2 mouse and a PlayStation game controller. These devices can be very useful; they provide sensors already incorporated into robust and ergonomic devices. They are also inexpensive (often less expensive than buying the raw sensors that they contain), as they are mass-produced. You may have some of these lying around.

If you are using a device that is not specifically covered in a recipe, you may be able to adapt a recipe for a device that produces a similar type of output. Information about a sensor's output signal is usually available from the company from which you bought the device or from a data sheet for your device (which you can find through a Google search of the device part number or description).

Data sheets are aimed at engineers designing products to be manufactured, and they usually provide more detail than you need to just get the product up and running. The information on output signal will usually be in a section referring to data format, interface, output signal, or something similar. Don't forget to check the maximum voltage (usually in a section labeled "Absolute Maximum Ratings") to ensure that you don't damage the component.

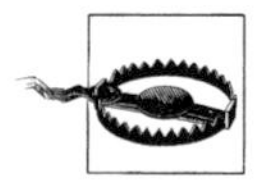

Sensors designed for a maximum of 3.3 volts can be destroyed by connecting them to 5 volts. Check the absolute maximum rating for your device before connecting.

Reading sensors from the messy analog world is a mixture of science, art, and perseverance. You may need to use ingenuity and trial and error to get a successful result. A common problem is that the sensor just tells you a physical condition has occurred, not what caused it. Putting the sensor in the right context (location, range, orientation) and limiting its exposure to things that you don't want to activate it are skills you will acquire with experience.

Another issue concerns separating the desired signal from background noise; Recipe 6.6 shows how you can use a threshold to detect when a signal is above a certain level, and Recipe 6.7 shows how you can take the average of a number of readings to smooth out noise spikes.

See Also

For information on connecting electronic components, see *Make: Electronics* by Charles Platt (Make).

See the introduction to Chapter 5 and Recipe 5.6 for more on reading analog values from sensors.

6.1 Detecting Movement

Problem

You want to detect when something is moved, tilted, or shaken.

Solution

This sketch uses a switch that closes a circuit when tilted, called a *tilt sensor*. The switch recipes in Chapter 5 (Recipes 5.1 and 5.2) will work with a tilt sensor substituted for the switch.

The sketch below (circuit shown in Figure 6-1) will switch on the LED attached to pin 11 when the tilt sensor is tilted one way, and the LED connected to pin 12 when it is tilted the other way:

```
/*
tilt sketch

  a tilt sensor attached to pin 2 lights one of
  the LEDs connected to pins 11 and 12 depending
  on which way the sensor is tilted
*/

const int tiltSensorPin = 2;           //pin the tilt sensor is connected to
const int firstLEDPin = 11;            //pin for one LED
const int secondLEDPin = 12;           //pin for the other

void setup()
{
  pinMode (tiltSensorPin, INPUT);      //the code will read this pin
  digitalWrite (tiltSensorPin, HIGH);  //and use a pull-up resistor

  pinMode (firstLEDPin, OUTPUT);       //the code will control this pin
  pinMode (secondLEDPin, OUTPUT);      //and this one
}
```

```
void loop()
{
  if (digitalRead(tiltSensorPin)){          //check if the pin is high
    digitalWrite(firstLEDPin, HIGH);        //if it is high turn on firstLED
    digitalWrite(secondLEDPin, LOW);        //and turn off secondLED
  }
  else{                                     //if it isn't
    digitalWrite(firstLEDPin, LOW);         //do the opposite
    digitalWrite(secondLEDPin, HIGH);
  }
}
```

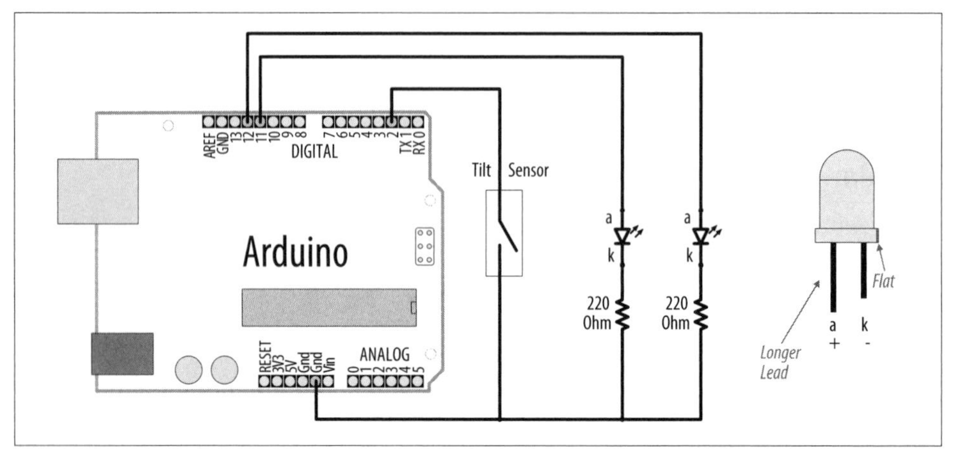

Figure 6-1. Tilt sensor and LEDs

Discussion

The most common tilt sensor is a ball bearing in a box with contacts at one end. When the box is tilted the ball rolls away from the contacts and the connection is broken. When the box is tilted to roll the other way the ball touches the contacts and completes a circuit. Markings, or pin configurations, show which way the sensor should be oriented. Tilt sensors are sensitive to small movements of around 5 to 10 degrees when oriented with the ball just touching the contacts. If you position the sensor so that the ball bearing is directly above (or below) the contacts, the LED state will only change if it is turned right over. This can be used to tell if something is upright or upside down.

To determine if something is being shaken, you need to check how long it's been since the state of the tilt sensor changed (this recipe's Solution just checks if the switch was open or closed). If it hasn't changed for a time you consider significant, the object is not shaking. Changing the orientation of the tilt sensor will change how vigorous the shaking needs to be to trigger it. The following code lights an LED when the sensor is shaken:

```
/*
  shaken sketch
  tilt sensor connected to pin 2
  led connected to pin 13
*/

const int tiltSensorPin = 2;
const int ledPin = 13;
int tiltSensorPreviousValue = 0;
int tiltSensorCurrentValue = 0;
long lastTimeMoved = 0;
int shakeTime=50;

void setup()
{
  pinMode (tiltSensorPin, INPUT);
  digitalWrite (tiltSensorPin, HIGH);
  pinMode (ledPin, OUTPUT);
}

void loop()
{
  tiltSensorCurrentValue=digitalRead(tiltSensorPin);
  if (tiltSensorPreviousValue != tiltSensorCurrentValue){
    lastTimeMoved = millis();
    tiltSensorPreviousValue = tiltSensorCurrentValue;
  }

  if (millis() - lastTimeMoved < shakeTime){
    digitalWrite(ledPin, HIGH);
  }
  else{
    digitalWrite(ledPin, LOW);
  }
}
```

Many mechanical switch sensors can be used in similar ways. A float switch can turn on when the water level in a container rises to a certain level (similar to the way a ball cock works in a toilet cistern). A pressure pad such as the one used in shop entrances can be used to detect when someone stands on it. If your sensor turns a digital signal on and off, something similar to this recipe's sketch should be suitable.

See Also

Chapter 5 contains background information on using switches with Arduino.

Recipe 12.2 has more on using the `millis` function to determine delay.

6.2 Detecting Light

Problem

You want to detect changes in light levels. You may want to detect a change when something passes in front of a light detector or to measure the light level—for example, detecting when a room is getting too dark.

Solution

The easiest way to detect light levels is to use a light dependent resistor (LDR). This changes resistance with changing light levels, and when connected in the circuit shown in Figure 6-2 it produces a change in voltage that the Arduino analog input pins can sense.

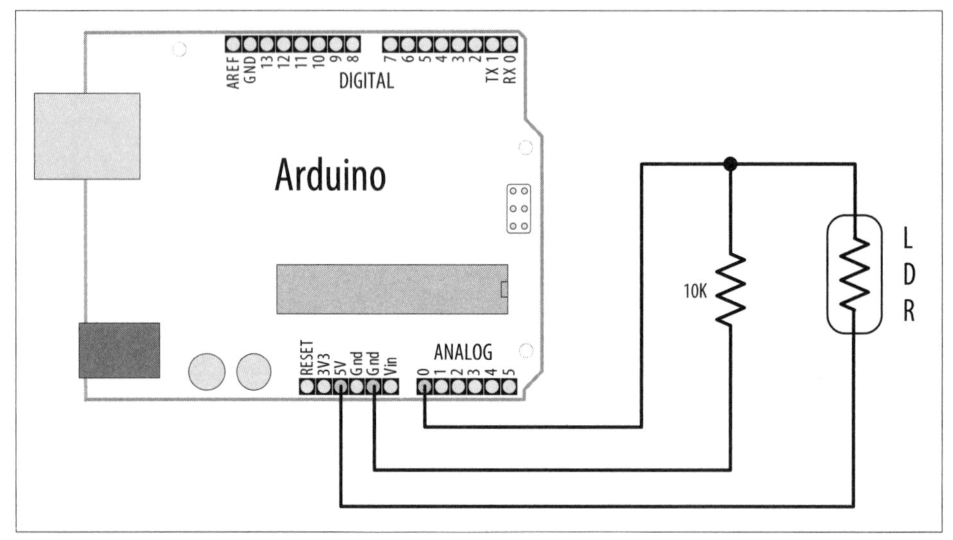

Figure 6-2. Connecting a light dependent resistor

The sketch for this recipe is simple:

```
const int ledPin =  13;     // LED connected to digital pin 13
const int sensorPin = 0;    // connect sensor to analog input 0

void setup()
{
  pinMode(ledPin, OUTPUT); // enable output on the led pin
}

void loop()
{
  int rate = analogRead(sensorPin);    // read the analog input
  digitalWrite(ledPin, HIGH);   // set the LED on
```

```
    delay(rate);                    // wait duration dependent on light level
    digitalWrite(ledPin, LOW);      // set the LED off
    delay(rate);
}
```

Discussion

The circuit for this recipe is the standard way to use any sensor that changes its resistance based on some physical phenomenon (see Chapter 5 for background information on responding to analog signals). With the circuit in Figure 6-2, the voltage on analog pin 0 changes as the resistance of the LDR changes with varying light levels.

A circuit such as this will not give the full range of possible values from the analog input—0 to 1,023—as the voltage will not be swinging from 0 volts to 5 volts. This is because there will always be a voltage drop across each resistance, so the voltage where they meet will never reach the limits of the power supply. When using sensors such as these, it is important to check the actual values the device returns in the situation you will be using it. Then you have to determine how to convert them to the values you need to control whatever you are going to control. See Recipe 5.7 for more details on changing the range of values.

The LDR is a simple kind of sensor called a *resistive sensor*. A range of resistive sensors respond to changes in different physical characteristics. Similar circuits will work for other kinds of simple resistive sensors, although you may need to adjust the resistor to suit the sensor.

Choosing the best resistor value depends on the LDR you are using and the range of light levels you want to monitor. Engineers would use a light meter and consult the data sheet for the LDR, but if you have a multimeter, you can measure the resistance of the LDR at a light level that is approximately midway in the range of illumination you want to monitor. Note the reading and choose the nearest convenient resistor to this value.

See Also

This sketch was introduced in Recipe 1.6; see that Recipe for more on this and variations on this sketch.

6.3 Detecting Motion (Integrating Passive Infrared Detectors)

Problem

You want to detect when people are moving near a sensor.

Solution

Use a motion sensor such as a Passive Infrared (PIR) sensor to change values on a digital pin when someone moves nearby.

Sensors such as the SparkFun PIR Motion Sensor (SEN-08630) and the Parallax PIR Sensor (555-28027) can be easily connected to Arduino pins, as shown in Figure 6-3.

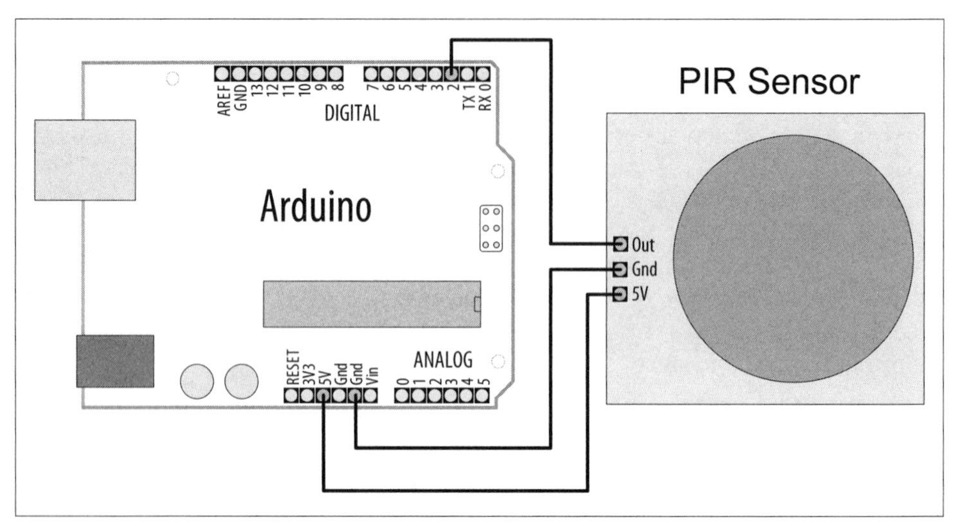

Figure 6-3. Connecting a PIR motion sensor

Check the data sheet for your sensor to identify the correct pins. The Parallax sensor has pins marked "OUT," "-," and "+" (for Output, Gnd, and +5V). The SparkFun sensor is marked with "Alarm," "GND," and "DC" (for Output, Gnd, and +5V).

The following sketch will light the LED on Arduino pin 13 when the sensor detects motion:

```
/*
   PIR sketch
   a Passive Infrared motion sensor connected to pin 2
   lights the LED on pin 13
*/

const int ledPin = 13;              // choose the pin for the LED
const int inputPin = 2;             // choose the input pin (for the PIR sensor)

void setup() {
  pinMode(ledPin, OUTPUT);           // declare LED as output
  pinMode(inputPin, INPUT);          // declare pushbutton as input
}

void loop(){
  int val = digitalRead(inputPin);  // read input value
  if (val == HIGH)                  // check if the input is HIGH
```

```
    {
      digitalWrite(ledPin, HIGH);     // turn LED on if motion detected
      delay(500);
      digitalWrite(ledPin, LOW);      // turn LED off
    }
}
```

Discussion

This code is similar to the pushbutton examples shown in Chapter 5. That's because the sensor acts like a switch when motion is detected. Different kinds of PIR sensors are available, and you should check the information for the one you have connected.

Some sensors, such as the Parallax, have a jumper that determines how the output behaves when motion is detected. In one mode, the output remains `HIGH` while motion is detected, or it can be set so that the output goes `HIGH` briefly and then `LOW` when triggered. The example sketch in this recipe's Solution will work in either mode.

Other sensors may go `LOW` on detecting motion. If your sensor's output pin goes `LOW` when motion is detected, change the line that checks the input value so that the LED is turned on when `LOW`:

```
if (val == LOW)                    // motion when the input is LOW
```

PIR sensors come in a variety of styles and are sensitive over different distances and angles. Careful choice and positioning can make them respond to movement in part of a room, rather than all of it.

PIR sensors respond to heat and can be triggered by animals such as cats and dogs, as well as by people and other heat sources.

6.4 Measuring Distance

Problem

You want to measure the distance to something, such as a wall or someone walking toward the Arduino.

Solution

This recipe uses the popular Parallax PING))) ultrasonic distance sensor to measure the distance of an object ranging from 2 centimeters to around 3 meters. It displays the distance on the Serial Monitor and flashes an LED faster as objects get closer (Figure 6-4 shows the connections):

```
/* Ping))) Sensor
 * prints distance and changes LED flash rate
 * depending on distance from the Ping))) sensor
 */


const int pingPin = 5;
const int ledPin  = 13; // pin connected to LED

void setup()
{
  Serial.begin(9600);
  pinMode(ledPin, OUTPUT);
}

void loop()
{
  int cm = ping(pingPin) ;
  Serial.println(cm);
  digitalWrite(ledPin, HIGH);
  delay(cm * 10 ); // each centimeter adds 10 milliseconds delay
  digitalWrite(ledPin, LOW);
  delay( cm * 10);
}

// following code based on  http://www.arduino.cc/en/Tutorial/Ping
// returns the distance in cm
int ping(int pingPin)
{
  // establish variables for duration of the ping,
  // and the distance result in inches and centimeters:
  long duration, cm;

  // The PING))) is triggered by a HIGH pulse of 2 or more microseconds.
  // Give a short LOW pulse beforehand to ensure a clean HIGH pulse:
  pinMode(pingPin, OUTPUT);
  digitalWrite(pingPin, LOW);
  delayMicroseconds(2);
  digitalWrite(pingPin, HIGH);
  delayMicroseconds(5);
  digitalWrite(pingPin, LOW);

  pinMode(pingPin, INPUT);
  duration = pulseIn(pingPin, HIGH);

  // convert the time into a distance
  cm = microsecondsToCentimeters(duration);
  return cm ;
}
```

```
long microsecondsToCentimeters(long microseconds)
{
  // The speed of sound is 340 m/s or 29 microseconds per centimeter.
  // The ping travels out and back, so to find the distance of the
  // object we take half of the distance travelled.
  return microseconds / 29 / 2;
}
```

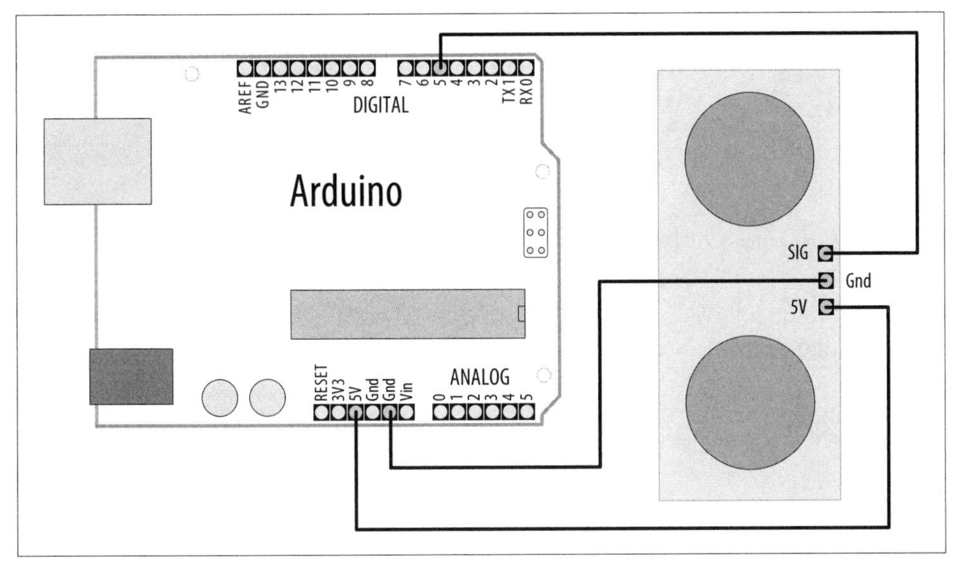

Figure 6-4. Ping))) sensor connections

Discussion

Ultrasonic sensors provide a measurement of the time it takes for sound to bounce off an object and return to the sensor.

The "ping" sound pulse is generated when the `pingPin` level goes `HIGH` for two microseconds. The sensor will then generate a pulse that terminates when the sound returns. The width of the pulse is proportional to the distance the sound traveled and the sketch then uses the `pulseIn` function to measure that duration. The speed of sound is 340 meters per second, which is 29 microseconds per centimeter. The formula for the distance of the round trip is: RoundTrip = microseconds / 29

So, the formula for the one-way distance in centimeters is: microseconds / 29 / 2

The MaxBotix EZ1 is another ultrasonic sensor that can be used to measure distance. It is easier to integrate than the Ping))) because it does not need to be "pinged." It can provide continuous distance information, either as an analog voltage or proportional to pulse width. Figure 6-5 shows the connections.

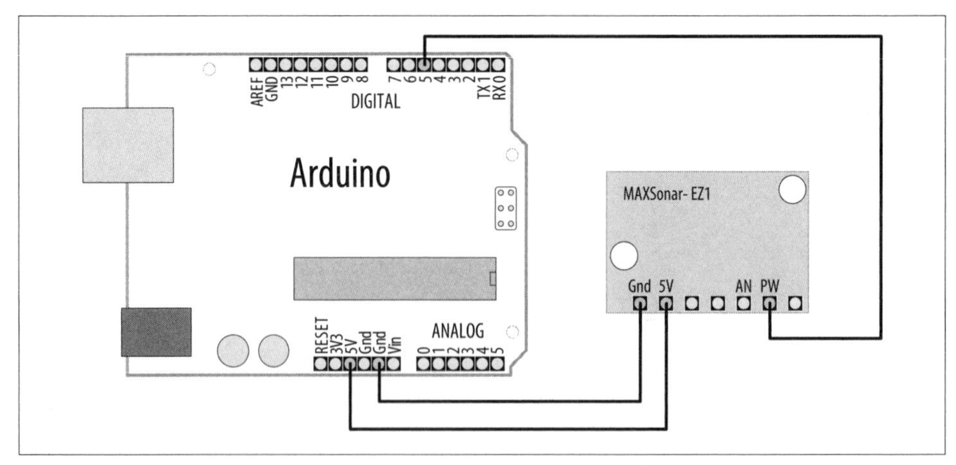

Figure 6-5. Connecting EZ1 PW output to a digital input pin

The sketch that follows uses the EZ1 pulse width (PW) output to produce output similar to that of the previous sketch:

```
/*
 * EZ1Rangefinder Distance Sensor
 * prints distance and changes LED flash rate
 * depending on distance from the Ping))) sensor
 */

const int sensorPin = 5;
const int ledPin  = 13; // pin connected to LED

long value = 0;
int cm = 0;
int inches = 0;

void setup()
{
  Serial.begin(9600);
  pinMode(ledPin, OUTPUT);
}

void loop()
{
  value = pulseIn(sensorPin, HIGH) ;
  cm = value / 58;        // pulse width is 58 microseconds per cm
  inches = value / 147;   // which is 147 microseconds per inch
  Serial.print(cm);
  Serial.print(',');
  Serial.println(inches);

  digitalWrite(ledPin, HIGH);
  delay(cm * 10 ); // each centimeter adds 10 milliseconds delay
```

```
    digitalWrite(ledPin, LOW);
    delay( cm * 10);

    delay(20);
}
```

The EZ1 is powered through +5V and ground pins and these are connected to the respective Arduino pins. Connect the EZ1 PW pin to Arduino digital pin 5. The sketch measures the width of the pulse with the `pulseIn` command. The width of the pulse is 58 microseconds per centimeter, or 147 microseconds per inch.

You may need to add a capacitor across the +5V and Gnd lines to stabilize the power supply to the sensor if you are using long connecting leads. If you get erratic readings, connect a 10 uF capacitor at the sensor (see Appendix C for more on using decoupling capacitors).

You can also obtain a distance reading from the EZ1 through its analog output—connect the AN pin to an analog input and read the value with `analogRead`. The following code prints the analog input converted to inches:

```
value = analogRead(0);
inches = value / 2;    // each digit of analog read is around 5mv
Serial.println(inches);
```

The analog output is around 9.8mV per inch. The value from `analogRead` is around 4.8mV per unit (see Recipe 5.6 for more on `analogRead`) and the preceding code rounds these so that each group of two units is one inch. The rounding error is small compared to the accuracy of the device, but if you want a more precise calculation you can use floating point as follows:

```
value = analogRead(0);
float mv = (value /1024.0) * 5000 ;
float inches  =  mv / 9.8; // 9.8mv per inch
Serial.println(inches) ;
```

See Also

Recipe 5.6 explains how to convert readings from `analogInput` into voltage values.

The Arduino reference for `pulseIn`: *http://www.arduino.cc/en/Reference/PulseIn*

6.5 Measuring Distance Accurately

Problem

You want to measure how far objects are from the Arduino with more accuracy than in Recipe 6.4.

Solution

Infrared (IR) sensors generally provide an analog output that can be measured using analogRead. They can have greater accuracy than ultrasonic sensors, albeit with a smaller range (a range of 10 centimeters to 1 or 2 meters is typical for IR sensors). This sketch provides similar functionality to Recipe 6.4, but it uses an infrared sensor—the Sharp GP2Y0A02YK0F (Figure 6-6 shows the connections):

```
/* ir-distance sketch
 * prints distance and changes LED flash rate based on distance from IR sensor
 */

const int ledPin    = 13; // the pin connected to the LED to flash
const int sensorPin = 0;  // the analog pin connected to the sensor

const long referenceMv = 5000; // long int to prevent overflow when multiplied

void setup()
{
  Serial.begin(9600);
  pinMode(ledPin, OUTPUT);
}

void loop()
{
  int val = analogRead(sensorPin);
  int mV = (val * referenceMv) / 1023;

  Serial.print(mV);
  Serial.print(",");
  int cm = getDistance(mV);
  Serial.println(cm);

  digitalWrite(ledPin, HIGH);
  delay(cm * 10 ); // each centimeter adds 10 milliseconds delay
  digitalWrite(ledPin, LOW);
  delay( cm * 10);

  delay(100);
}

// the following is used to interpolate the distance from a table
// table entries are distances in steps of 250 millivolts
const int TABLE_ENTRIES = 12;
const int firstElement = 250; // first entry is 250 mV
const int INTERVAL  = 250; // millivolts between each element
static int distance[TABLE_ENTRIES] = {150,140,130,100,60,50,40,35,30,25,20,15};

int getDistance(int mV)
{
   if( mV >  INTERVAL * TABLE_ENTRIES-1 )
      return distance[TABLE_ENTRIES-1];
   else
   {
```

```
        int index = mV / INTERVAL;
        float frac = (mV % 250) / (float)INTERVAL;
        return distance[index] - ((distance[index] - distance[index+1]) * frac);
    }
}
```

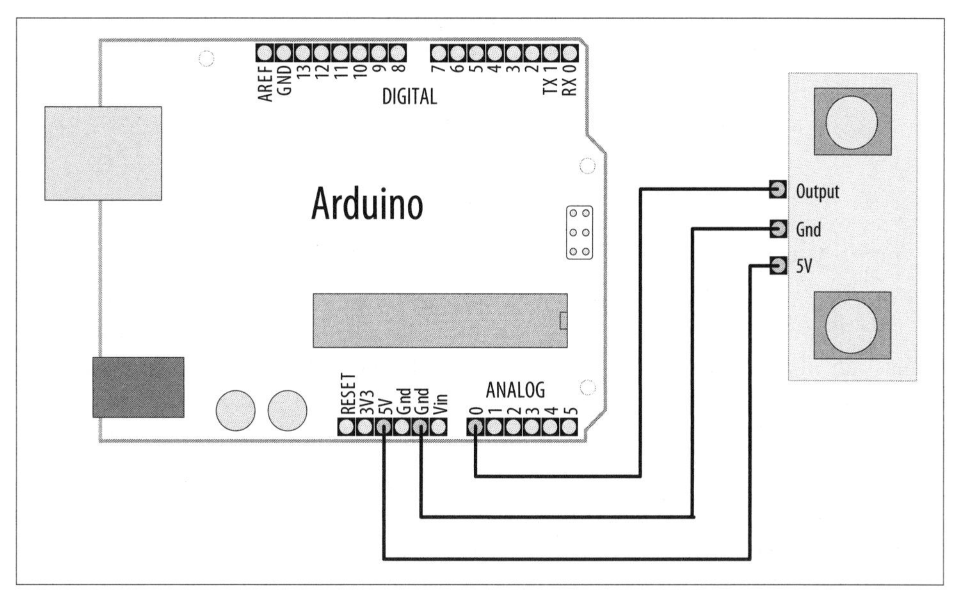

Figure 6-6. Connecting the Sharp IR distance sensor

Discussion

The output from the IR sensor is not linear—in other words, the value read from `analogRead` is not proportional to distance. So, the calculation is more complicated than the one used in Recipe 6.4. The sketch in this recipe's Solution uses a table to interpolate the actual distance by finding the nearest entry in the table and adjusting it based on the ratio of the measured value to the next table entry (this technique is called *interpolating*). You may need to adjust the table values for your sensor—you can do this with information from your data sheet or through trial and error.

As values for the table can be found by trial and error (measuring the voltage until it changes by the required amount, and then measuring the distance), this technique can also be used when you don't have an equation to interpret the values—for example, when you don't have a data sheet for the device you are using.

The conversion from voltage to distance is done in this function:

```
int getDistance(int mV)
```

The function first checks if the value is within the range given in the table. The shortest valid distance is returned if the value is not within range:

```
if( mV >  INTERVAL * TABLE_ENTRIES )
   return distance[TABLE_ENTRIES-1]; //TABLE_ENTRIES-1 is last valid entry
```

If the value is within the table range, integer division calculates which entry is closest but is lower than the reading:

```
int index = mV / INTERVAL ;
```

The modulo operator (see Chapter 3) is used to calculate a fractional value when a reading falls between two entries:

```
float frac = (mV % 250) / (float)INTERVAL;

return distance[index] + (distance[index]* (frac / interval));
```

The last line in the `getDistance` function uses the index and fraction to calculate and return a distance value. It reads the value from the table, and then adds a proportion of that value based on the `frac` value. This final element is an approximation, but as it is for a small range of the result, it gives acceptable results. If it is not accurate enough for you, you need to produce a table with more values closer together.

A table can also be used to improve performance if the calculation takes significant time to complete, or is done repeatedly with a limited number of values. Calculations, particularly with floating point, can be slow. Replacing the calculation with a table can speed things up.

The values can either be hardcoded into the sketch, like this one, or be calculated in `setup()`. This may make the sketch take longer to start, but as this only happens once each time the Arduino gets power, you will then get a speed gain every time around the main `loop()`. The trade-off for the speed is that the table consumes memory—the bigger the table, the more RAM memory used. See Chapter 17 for help using `Progmem` to store data in program memory.

You may need to add a capacitor across the +5V and Gnd lines to stabilize the power supply to the sensor if you are using long connecting leads. If you get erratic readings, connect a 10 uF capacitor at the sensor (see Appendix C for more on using decoupling capacitors).

See Also

A detailed explanation of the Sharp IR sensor is available at *http://www.societyofrobots.com/sensors_sharpirrange.shtml*.

6.6 Detecting Vibration

Problem

You want to respond to vibration; for example, when a door is knocked on.

Solution

A Piezo sensor responds to vibration. It works best when connected to a larger surface that vibrates. Figure 6-7 shows the connections:

```
/* piezo sketch
 * lights an LED when the Piezo is tapped
 */

const int sensorPin = 0;  // the analog pin connected to the sensor
const int ledPin  = 13;     // pin connected to LED
const int THRESHOLD = 100;


void setup()
{
   pinMode(ledPin, OUTPUT);
}

void loop()
{
  int val = analogRead(sensorPin);
  if (val >= THRESHOLD)
  {
    digitalWrite(ledPin, HIGH);
    delay(100);  // to make the LED visible
  }
  else
    digitalWrite(ledPin, LOW);
}
```

Discussion

A Piezo sensor, also known as a knock sensor, produces a voltage in response to physical stress. The more it is stressed, the higher the voltage. The Piezo is polarized and the positive side (usually a red wire or a wire marked with a "+") is connected to the analog input; the negative wire (usually black or marked with a "–") is connected to ground. A high-value resistor (1 megohm) is connected across the sensor.

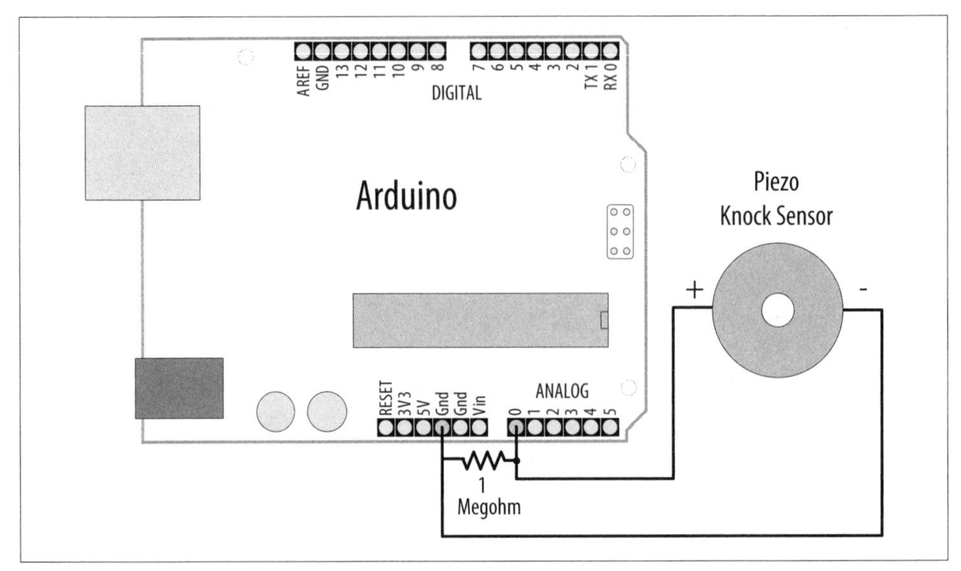

Figure 6-7. Knock sensor connections

The voltage is detected by Arduino `analogRead` to turn on an LED (see Chapter 5 for more about the `analogRead` function). The `THRESHOLD` value determines the level from the sensor that will turn on the LED, and you can decrease or increase this value to make the sketch more or less sensitive.

Piezo sensors can be bought in plastic cases or as bare metal disks with two wires attached. The components are the same; use whichever fits your project best.

Some sensors, such as the Piezo, can be driven by the Arduino to produce the thing that they can sense. Chapter 9 has more about using a Piezo to generate sound.

6.7 Detecting Sound

Problem

You want to detect sounds such as clapping, talking, or shouting.

Solution

This recipe uses the BOB-08669 breakout board for the Electret Microphone (SparkFun). Connect the board as shown in Figure 6-8 and load the code to the board.

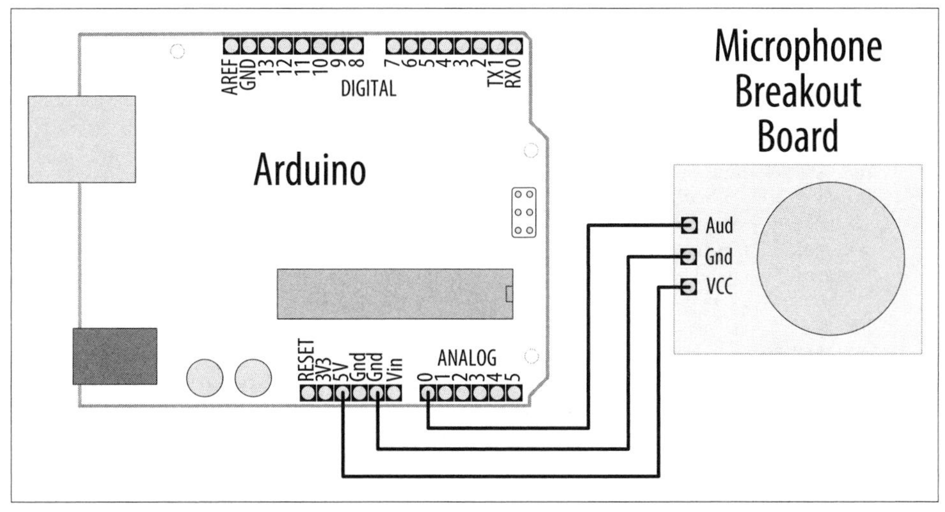

Figure 6-8. Microphone board connections

The built-in LED on Arduino pin 13 will turn on when you clap, shout, or play loud music near the microphone. You may need to adjust the threshold—use the Serial Monitor to view the high and low values, and change the threshold value so that it is between the high values you get when noise is present and the low values when there is little or no noise. Upload the changed code to the board and try again:

```
/*
microphone sketch

SparkFun breakout board for Electret Microphone is connected to analog pin 0
*/

const int ledPin = 13;            //the code will flash the LED in pin 13
const int middleValue = 512;      //the middle of the range of analog values
const int numberOfSamples = 128;  //how many readings will be taken each time

int sample;                       //the value read from microphone each time
long signal;                      //the reading once you have removed DC offset
long averageReading;              //the average of that loop of readings

long runningAverage=0;          //the running average of calculated values
const int averagedOver= 16;     //how quickly new values affect running average
                                //bigger numbers mean slower

const int threshold=400;        //at what level the light turns on
```

```
void setup() {
  pinMode(ledPin, OUTPUT);
  Serial.begin(9600);
}

void loop() {
  long sumOfSquares = 0;
  for (int i=0; i<numberOfSamples; i++) { //take many readings and average them
    sample = analogRead(0);               //take a reading
    signal = (sample - middleValue);      //work out its offset from the center
    signal *= signal;                     //square it to make all values positive
    sumOfSquares += signal;               //add to the total
  }
  averageReading = sumOfSquares/numberOfSamples;     //calculate running average
  runningAverage=(((averagedOver-1)*runningAverage)+averageReading)/averagedOver;

  if (runningAverage>threshold){         //is average more than the threshold ?
    digitalWrite(ledPin, HIGH);          //if it is turn on the LED
  }else{
    digitalWrite(ledPin, LOW);           //if it isn't turn the LED off
  }
  Serial.println(runningAverage);        //print the value so you can check it
}
```

Discussion

A microphone produces very small electrical signals. If you connected it straight to the pin of an Arduino, you would not get any detectable change. The signal needs to be amplified first to make it usable by Arduino. The SparkFun board has the microphone with an amplifier circuit built in to amplify the signal to a level readable by Arduino.

Because you are reading an audio signal in this recipe, you will need to do some additional calculations to get useful information. An audio signal is changing fairly quickly, and the value returned by `analogRead` will depend on what point in the undulating signal you take a reading. If you are unfamiliar with using `analogRead`, see Chapter 5 and Recipe 6.2. An example waveform for an audio tone is shown in Figure 6-9. As time changes from left to right, the voltage goes up and down in a regular pattern. If you take readings at the three different times marked on it, you will get three different values. If you used this to make decisions, you might incorrectly conclude that the signal got louder in the middle.

An accurate measurement requires multiple readings taken close together. The peaks and troughs increase as the signal gets bigger. The difference between the bottom of a trough and the top of a peak is called the *amplitude* of the signal, and this increases as the signal gets louder.

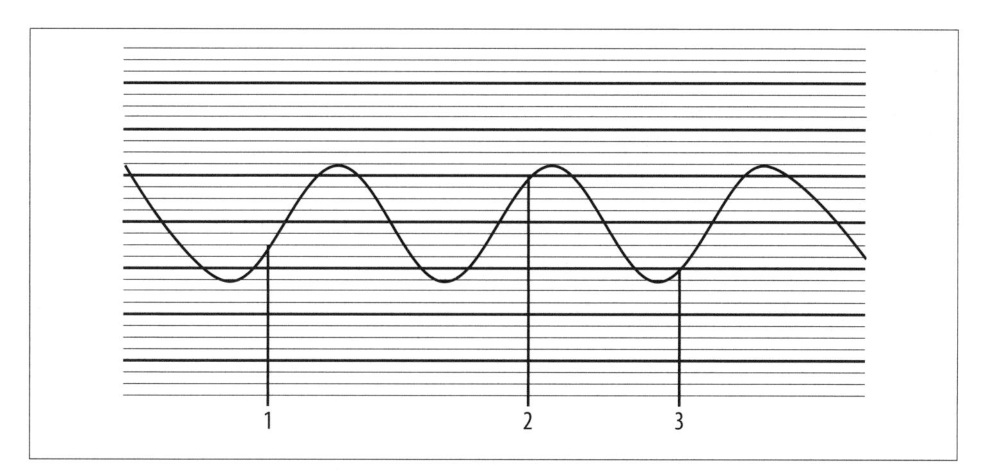

Figure 6-9. Audio signal measured in three places

To measure the size of the peaks and troughs, you measure the difference between the midpoint voltage and the levels of the peaks and troughs. You can visualize this midpoint value as a line running midway between the highest peak and the lowest trough, as shown in Figure 6-10. The line represents the DC offset of the signal (it's the DC value when there are no peaks or troughs). If you subtract the DC offset value from your `analogRead` values, you get the correct reading for the signal amplitude.

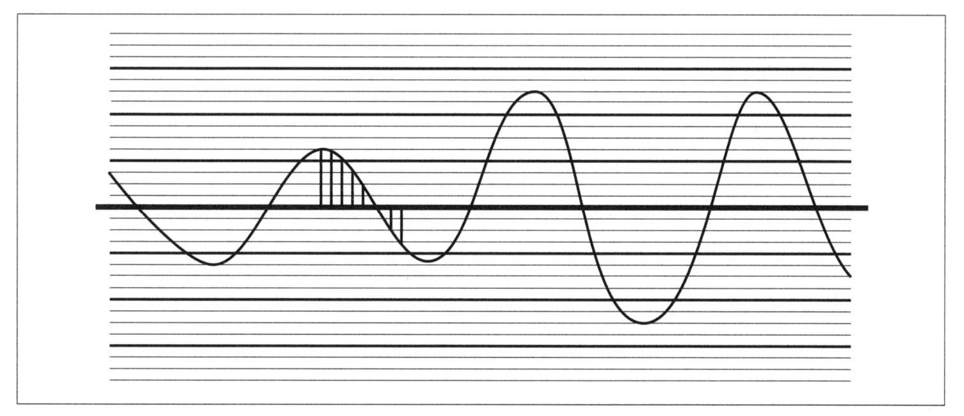

Figure 6-10. Audio signal showing DC offset (signal midpoint)

As the signal gets louder, the average size of these values will increase, but as some of them are negative (where the signal has dropped below the DC offset), they will cancel each other out, and the average will tend to be zero. To fix that, we square each value (multiply it by itself). This will make all the values positive, and it will increase the difference between small changes, which helps you evaluate changes as well. The average value will now go up and down as the signal amplitude does.

To do the calculation, we need to know what value to use for the DC offset. To get a clean signal, the amplifier circuit for the microphone will have been designed to have a DC offset as close as possible to the middle of the possible range of voltage so that the signal can get as big as possible without distorting. The code assumes this and uses the value 512 (right in the middle of the analog input range of 0 to 1,023).

The values of variables at the top of the sketch can be varied if the sketch does not trigger well for the level of sound you want.

The `numberOfSamples` is set at 128—if it is set too small, the average may not adequately cover complete cycles of the waveform and you will get erratic readings. If the value is set too high, you will be averaging over too long a time, and a very short sound might be missed as it does not produce enough change once a large number of readings are averaged. It could also start to introduce a noticeable delay between a sound and the light going on. Constants used in calculations, such as `numberOfSamples` and `averagedOver`, are set to powers of 2 (128 and 16, respectively). Try to use values evenly divisible by two for these to give you the fastest performance (see Chapter 3 for more on math functions).

6.8 Measuring Temperature

Problem

You want to display the temperature or use the value to control a device; for example, to switch something on when the temperature reaches a threshold.

Solution

This recipe displays the temperature in Fahrenheit and Celsius (Centigrade) using the popular LM35 heat detection sensor. The sensor looks similar to a transistor and is connected as shown in Figure 6-11:

```
/*
 lm35 sketch
 prints the temperature to the Serial Monitor
 */

const int inPin = 0; // analog pin


void setup()
{
  Serial.begin(9600);
}


void loop()
{
  int value = analogRead(inPin);
```

```
    Serial.print(value); Serial.print(" > ");
    float millivolts = (value / 1024.0) * 5000;
    float celsius = millivolts / 10;  // sensor output is 10mV per degree Celsius
    Serial.print(celsius);
    Serial.print(" degrees Celsius, ");

    Serial.print( (celsius * 9)/ 5 + 32 );  //  converts to fahrenheit
    Serial.println(" degrees Fahrenheit");

    delay(1000); // wait for one second

}
```

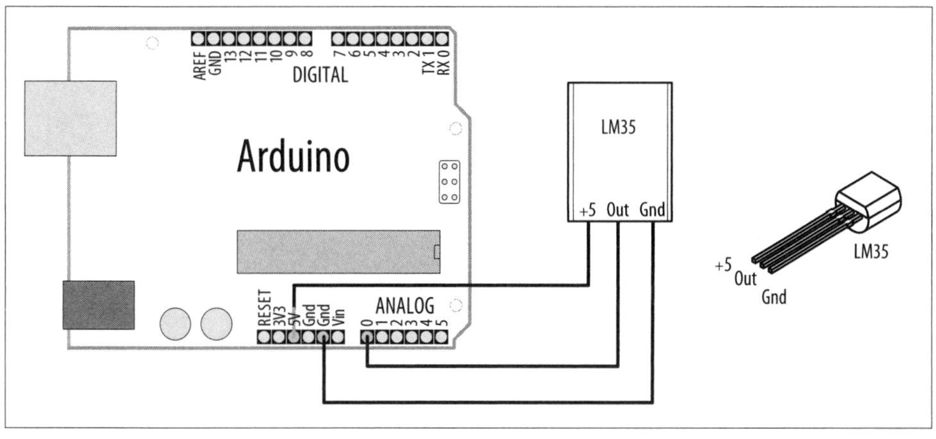

Figure 6-11. Connecting the LM35 temperature sensor

Discussion

The LM35 temperature sensor produces an analog voltage directly proportional to temperature with an output of 1 millivolt per 0.1°C (10mV per degree).

The sketch converts the `analogRead` values into millivolts (see Chapter 5) and divides this by 10 to get degrees.

The sensor accuracy is around 0.5°C, and in many cases you can use integer math instead of floating point.

The following sketch triggers pin 13 when the temperature is above a threshold:

```
const int inPin = 0;  // sensor connected to this analog pin
const int outPin = 13;  // digital output pin

const int threshold = 25; // the degrees celsius that will trigger the output pin


void setup()
{
  Serial.begin(9600);
  pinMode(outPin, OUTPUT);
```

```
}

void loop()
{
  int value = analogRead(inPin);
  long celsius =  (value * 500L) /1024;     // 10 mV per degree c, see text
  Serial.print(celsius);
  Serial.print(" degrees Celsius: ");
  if(celsius > threshold)
  {
     digitalWrite(outPin, HIGH);
     Serial.println("pin is on");
  }
  else
  {
     digitalWrite(outPin, LOW);
     Serial.println("pin is off");
  }
  delay(1000); // wait for one second
}
```

The sketch uses long (32-bit) integers to calculate the value. The letter *L* after the number causes the calculation to be performed using long integer math, so the multiplication of the maximum temperature (500 on a 5V Arduino) and the value read from the analog input does not overflow. See the recipes in Chapter 5 for more about converting analog levels into voltage values.

If you need the values in Fahrenheit, you could use the LM34 sensor, as this produces an output in Fahrenheit, or you can convert the values in this recipe using the following formula:

```
float  f = (celsius * 9)/ 5 + 32 );
```

An alternative sensor for measuring temperature is the LM335. The device looks similar to the LM35 but it is wired and used differently.

The LM335 output is 10mV per degree Kelvin, so zero degrees Celsius results in 2.731 volts. A series resistor is required to set the operating current. A 2K ohm resistor is often used, but 2.2K ohms can also be used. Here is a sketch that displays temperature using the LM335 (Figure 6-12 shows the connections):

```
/*
  lm335 sketch
  prints the temperature to the Serial Monitor
 */

const int inPin = 0; // analog pin


void setup()
{
  Serial.begin(9600);
}
```

```
void loop()
{
  int value = analogRead(inPin);
  Serial.print(value); Serial.print(" > ");
  float millivolts = (value / 1024.0) * 5000;
  // sensor output is 10mV per degree Kelvin, 0 Celsius is 273.15
  float celsius = (millivolts / 10) - 273.15 ;

  Serial.print(celsius);
  Serial.print(" degrees Celsius, ");

  Serial.print( (celsius * 9)/ 5 + 32 );  //  converts to fahrenheit
  Serial.println(" degrees Fahrenheit");

  delay(1000); // wait for one second
}
```

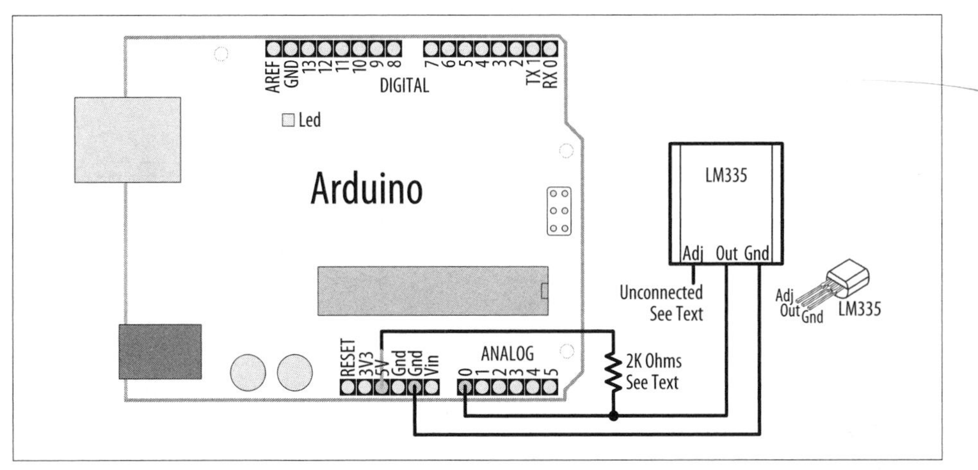

Figure 6-12. Connecting the LM335 temperature sensor

You can improve the accuracy by wiring the unconnected *adj* pin to the slider of a 10K potentiometer with the other leads connected to +5V and Gnd. Adjust the pot to get a reading to match a known accurate thermometer.

See Also

LM35 data sheet: *http://www.national.com/ds/LM/LM35.pdf*

LM335 data sheet: *http://www.national.com/ds/LM/LM135.pdf*

6.9 Reading RFID Tags

Problem

You want to read an RFID tag and respond to specific IDs.

Solution

Figure 6-13 shows a Parallax RFID (radio frequency identification) reader connected to the Arduino serial port. (You may need to disconnect the reader from the serial port when uploading the sketch.)

This reader works with 125kHz tags. If you are using a different reader, check the documentation to ensure correct connections and usage.

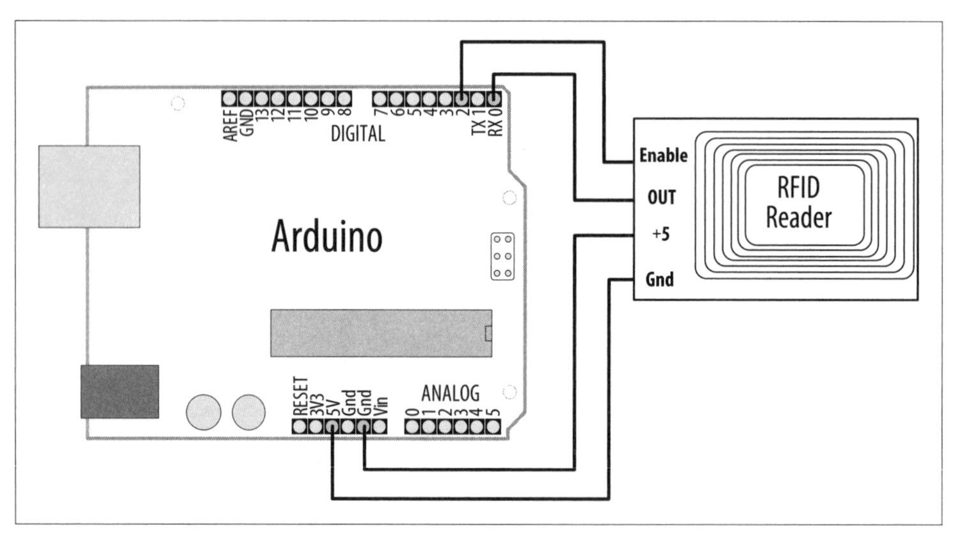

Figure 6-13. Serial RFID reader connected to Arduino

The sketch reads and displays the value of an RFID tag:

```
/*
 RFID sketch
 Displays the value read from an RFID tag
 */

const int startByte   = 10;  // ASCII line feed precedes each tag
const int endByte     = 13;  // ASCII carriage return terminates each tag
const int tagLength   = 10;  // the number of digits in tag
const int totalLength = tagLength + 2; //tag length + start and end bytes
```

```
char tag[tagLength + 1];  // holds the tag and a terminating null

int bytesread = 0;

void setup()
{
  Serial.begin(2400);      // set this to the baud rate of your RFID reader
  pinMode(2,OUTPUT);       // connected to the RFID ENABLE pin
  digitalWrite(2, LOW);  // enable the RFID reader
}

void loop()
{
  if(Serial.available() >= totalLength)  // check if there's enough data
  {
    if(Serial.read() == startByte)
    {
      bytesread = 0;                // start of tag so reset count to 0
      while(bytesread < tagLength)  // read 10 digit code
      {
        int val = Serial.read();
        if((val == startByte)||(val == endByte))  // check for end of code
          break;
        tag[bytesread] = val;
         bytesread = bytesread + 1; // ready to read next digit
      }
      if( Serial.read() == endByte)  // check for the correct end character
      {
        tag[bytesread] = 0; // terminate the string
        Serial.print("RFID tag is: ");
        Serial.println(tag);
      }
    }
  }
}
```

Discussion

A tag consists of a start character followed by a 10-digit tag and is terminated by an end character. The sketch waits for a complete tag message to be available and displays the tag if it is valid. The tag is received as ASCII digits (see Recipe 4.4 for more on receiving ASCII digits). You may want to convert this into a number if you want to store or compare the values received. To do this, change the last few lines as follows:

```
      if( Serial.read() == endByte)  // check for the correct end character
      {
        tag[bytesread] = 0; // terminate the string
        long tagValue = atol(tag);  // convert the ASCII tag to a long integer
        Serial.print("RFID tag is: ");
        Serial.println(tagValue);
      }
```

RFID stands for radio frequency identification, and as the name implies, it is sensitive to radio frequencies and can be prone to interference. The code in this recipe's Solution will only use code of the correct length that contains the correct start and end bits, which should eliminate most errors. But you can make the code more resilient by reading the tag more than once and only using the data if it's the same each time. (RFID readers such as the Parallax will repeat the code while a valid card is near the reader.) To do this, add the following lines to the last few lines in the preceding code snippet:

```
if( Serial.read() == endByte)  // check for the correct end character
{
  tag[bytesread] = 0; // terminate the string
  long tagValue = atol(tag);  // convert the ASCII tag to a long integer
  if (tagValue == lastTagValue)
  {
    Serial.print("RFID tag is: ");
    Serial.println(tagValue);
    lastTagValue = tagValue;
  }
}
```

You will need to add the declaration for `lastTagValue` at the top of the sketch:

```
long lastTagValue=0;
```

This approach is similar to the code from Recipe 5.3. It means you will only get confirmation of a card if it is presented long enough for two readings to be taken, but false readings will be less likely. You can avoid accidental triggering by making it necessary for the card to be present for a certain amount of time before the number is reported.

6.10 Tracking Rotary Movement

Problem

You want to measure and display the rotation of something to track its speed and/or direction.

Solution

To sense rotary motion you can use a rotary encoder that is attached to the object you want to track. Connect the encoder as shown in Figure 6-14:

```
/*
 Read a rotary encoder
 This simple version polls the encoder pins
 The position is displayed on the Serial Monitor
 */
```

```
const int encoderPinA = 4;
const int encoderPinB = 2;
const int encoderStepsPerRevolution=16;
int angle = 0;

int val;

int encoderPos = 0;
boolean encoderALast = LOW;  // remembers the previous pin state


void setup()
{
  pinMode(encoderPinA, INPUT);
  pinMode(encoderPinB, INPUT);
  digitalWrite(encoderPinA, HIGH);
  digitalWrite(encoderPinB, HIGH);
  Serial.begin (9600);
}

void loop()
{
  boolean encoderA = digitalRead(encoderPinA);

  if ((encoderALast == HIGH) && (encoderA == LOW))
  {
    if (digitalRead(encoderPinB) == LOW)
    {
      encoderPos--;
    }
    else
    {
      encoderPos++;
    }
    angle=(encoderPos % encoderStepsPerRevolution)*360/encoderStepsPerRevolution;
    Serial.print (encoderPos);
    Serial.print (" ");
    Serial.println (angle);
  }

  encoderALast = encoderA;
}
```

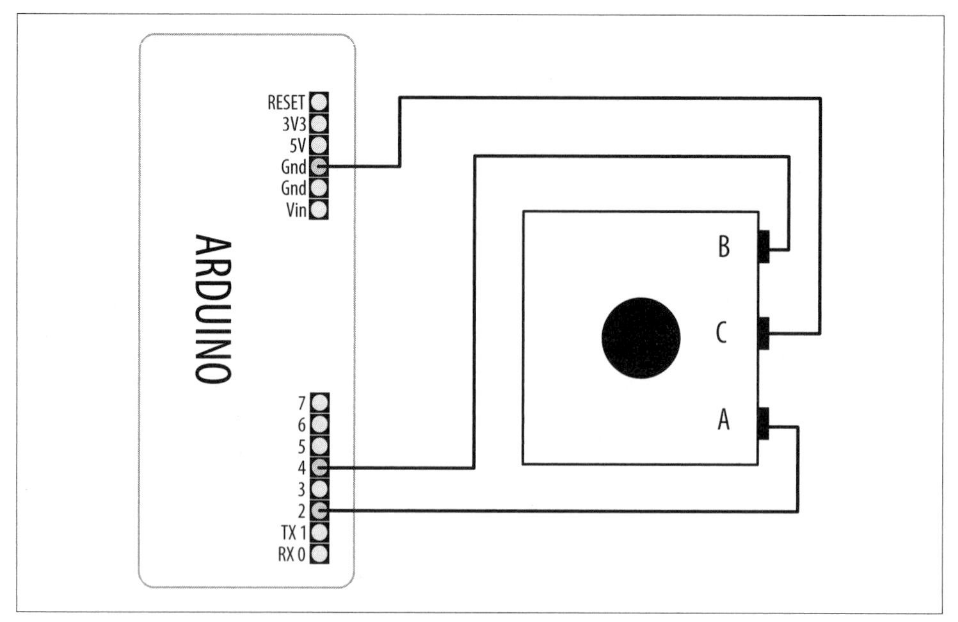

Figure 6-14. Rotary encoder

Discussion

A rotary encoder produces two signals as it is turned. Both signals alternate between `HIGH` and `LOW` as the shaft is turned, but the signals are slightly out of phase with each other. If you detect the point where one of the signals changes from `HIGH` to `LOW`, the state of the other pin (whether it is `HIGH` or `LOW`) will tell you which way the shaft is rotating.

So, the first line of code in the `loop` function reads one of the encoder pins:

```
int encoderA = digitalRead(encoderPinA);
```

Then it checks this value and the previous one to see if the value has just changed to `LOW`:

```
if ((encoderALast == HIGH) && (encoderA == LOW))
```

If it has not, the code doesn't execute the following block; it goes to the bottom of `loop`, saves the value it has just read in `encoderALast`, and goes back around to take a fresh reading.

When the following expression is `true`:

```
if ((encoderALast == HIGH) && (encoderA == LOW))
```

the code reads the other encoder pin and increments or decrements `encoderPos` depending on the value returned. It calculates the angle of the shaft (taking 0 to be the point the shaft was at when the code started running). It then sends the values down the serial port so that you can see it in the Serial Monitor.

Encoders come in different resolutions, quoted as *steps per revolution*. This indicates how many times the signals alternate between HIGH and LOW for one revolution of the shaft. Values can vary from 16 to 1,000. The higher values can detect smaller movements, and these encoders cost much more money. The value for the encoder is hard-coded in the code in the following line:

```
const int encoderStepsPerRevolution=16;
```

If your encoder is different, you need to change that to get the correct angle values.

If you get values out that don't go up and down, but increase regardless of the direction you turn the encoder, try changing the test to look for a rising edge rather than a falling one. Swap the LOW and HIGH values in the line that checks the values so that it looks like this:

```
if ((encoderALast == LOW) && (encoderA == HIGH))
```

Rotary encoders just produce an increment/decrement signal; they cannot directly tell you the shaft angle. The code calculates this, but it will be relative to the start position each time the code runs. The code monitors the pins by *polling* (continuously checking the value of) them. There is no guarantee that the pins have not changed a few times since the last time the code looked, so if the code does lots of other things as well, and the encoder is turned very quickly, it is possible that some of the steps will be missed. For high-resolution encoders this is more likely, as they will send signals much more often as they are turned.

To work out the speed, you need to count how many steps are registered in one direction in a set time.

6.11 Tracking the Movement of More Than One Rotary Encoder

Problem

You have two or more rotary encoders and you want to measure and display rotation.

Solution

The circuit uses two encoders, connected as shown in Figure 6-15. You can read more about rotary encoders in Recipe 6.10:

```
/*
 RotaryEncoderMultiPoll
 This sketch has two encoders connected.
 One is connected to pins 2 and 3
 The other is connected to pins 4 and 5
 */

const int  ENCODERS = 2; // the number of encoders
```

```
const int encoderPinA[ENCODERS] = {2,4};      // encoderA pins on 2 and 4
const int encoderPinB[ENCODERS] = {3,5};      // encoderB pins on 3 and 5
int encoderPos[ ENCODERS] = { 0,0};           // initialize the positions to 0
boolean encoderALast[ENCODERS] = { LOW,LOW}; // holds last state of encoderA pin


void setup()
{
  for (int i=2; i<6; i++){
    pinMode(i, HIGH);
    digitalWrite(i, HIGH);
  }
  Serial.begin (9600);
}

int updatePosition( int encoderIndex)
{
  boolean encoderA = digitalRead(encoderPinA[encoderIndex]);
  if ((encoderALast[encoderIndex] == HIGH) && (encoderA == LOW))
  {
    if (digitalRead(encoderPinB[encoderIndex]) == LOW)
    {
      encoderPos[encoderIndex]--;
    }
    else
    {
      encoderPos[encoderIndex]++;
    }
    Serial.print("Encoder ");
    Serial.print(encoderIndex,DEC);
    Serial.print("=");
    Serial.print (encoderPos[encoderIndex]);
    Serial.println ("/");
  }
  encoderALast[encoderIndex] = encoderA;
}


void loop()
{
  for(int i=0; i < ENCODERS;i++)
  {
    updatePosition(i);
  }
}
```

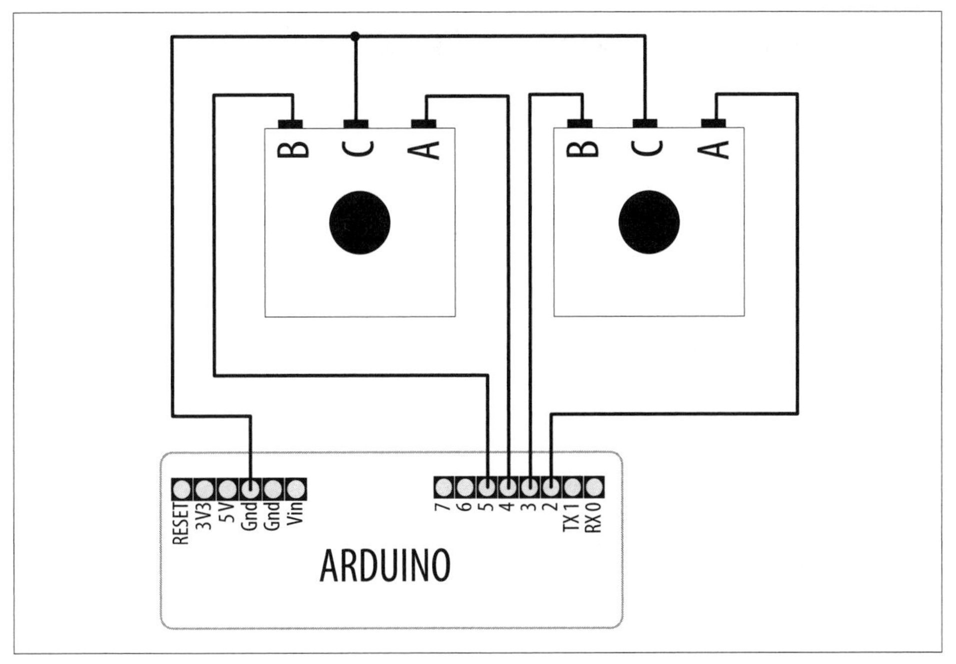

Figure 6-15. Connecting two rotary encoders

Discussion

This recipe uses the same code logic as Recipe 6.10, which was reading one encoder, but it uses arrays for all the variables that must be remembered separately for each encoder. You can then use a `for` loop to go through each one and read it and calculate its rotation. To use more encoders, set the `ENCODERS` values to the number of encoders you have and extend the arrays and the definitions to say which pins they are attached to.

If you get values out that don't go up and down, but increase regardless of the direction you turn the encoder, try changing the test to look for a rising edge rather than a falling one. Swap the `LOW` and `HIGH` values in the line that checks the values from this:

```
if ((encoderALast[encoderIndex] == HIGH) && (encoderA == LOW))
```

to this:

```
if ((encoderALast[encoderIndex] == LOW) && (encoderA == HIGH))
```

If one of the encoders works but the other just counts up, switch over the A and B connections for the one that just counts up.

6.12 Tracking Rotary Movement in a Busy Sketch

Problem

As you extend your code and it is doing other things in addition to reading the encoder, reading the encoder starts to get unreliable. This problem is particularly bad if the shaft rotates quickly.

Solution

The circuit is the same as the one for Recipe 6.11. We will use an interrupt on the Arduino to make sure that every time a step happens, the code responds to it:

```
/*
  RotaryEncoderInterrupt sketch
 */

const int encoderPinA = 2;
const int encoderPinB = 4;
int Pos, oldPos;
volatile int encoderPos = 0; // variables changed within interrupts are volatile


void setup()
{
  pinMode(encoderPinA, INPUT);
  pinMode(encoderPinB, INPUT);
  digitalWrite(encoderPinA, HIGH);
  digitalWrite(encoderPinB, HIGH);
  Serial.begin(9600);

  attachInterrupt(0, doEncoder, FALLING); // encoder pin on interrupt 0 (pin 2)
}

void loop()
{
  uint8_t oldSREG = SREG;

  cli();
  Pos = encoderPos;
  SREG = oldSREG;
  if(Pos != oldPos)
  {
    Serial.println(Pos,DEC);
    oldPos = Pos;
  }
   delay(1000);
}


void doEncoder()
{
  if (digitalRead(encoderPinA) == digitalRead(encoderPinB))
```

```
      encoderPos++;     // count up if both encoder pins are the same
    else
      encoderPos--;     // count down if pins are different
}
```

This code will only report the Pos value on the serial port, at most once every second (because of the delay), but the values reported will take into account any movement that may have happened while it was delaying.

Discussion

As your code has more things to do, the encoder pins will be checked less often. If the pins go through a whole step change before getting read, the Arduino will not detect that step. Moving the shaft quickly will cause this to happen more often, as the steps will be happening more quickly.

To make sure the code responds every time a step happens, you need to use interrupts. When the interrupt condition happens, the code jumps from wherever it is, does what needs to happen, and then returns to where it was and carries on.

On a standard Arduino board, two pins can be used as interrupts: pins 2 and 3. The interrupt is enabled through the following line:

```
attachInterrupt(0, doEncoder, FALLING);
```

The three parameters needed are the interrupt pin identifier (0 for pin 2, 1 for pin 3); the function to jump to when the interrupt happens, in this case `doEncoder`; and finally, the pin behavior to trigger the interrupt, in this case when the voltage falls from 5 to 0 volts. The other options are `RISING` (voltage rises from 0 to 5 volts) and `CHANGE` (voltage falls or rises).

The `doEncoder` function checks the encoder pins to see which way the shaft turned, and changes `encoderPos` to reflect this.

If the values reported only increase regardless of the direction of rotation, try changing the interrupt to look for `RISING` rather than `FALLING`.

Because `encoderPos` is changed in the function that is called when the interrupt happens, it needs to be declared as `volatile` when it is created. This tells the compiler that it could change at any time; don't optimize the code by assuming it won't have changed, as the interrupt can happen at any time.

The Arduino build process optimizes the code by removing code and variables that are not used by your sketch code. Variables that are only modified in an interrupt handler should be declared as `volatile` to tell the compiler not to remove these variables.

To read this variable in the main loop, you should take special precautions to make sure the interrupt does not happen in the middle of reading it. This chunk of code does that:

```
uint8_t oldSREG = SREG;

cli();
Pos = encoderPos;
SREG = oldSREG;
```

First you save the state of SREG (the interrupt registers), and then `cli` turns the interrupt off. The value is read, and then restoring SREG turns the interrupt back on and sets everything back as it was. Any interrupt that occurs when interrupts are turned off will wait until interrupts are turned back on. This period is so short that interrupts will not be missed (as long as you keep the code in the interrupt handler as short as possible).

6.13 Using a Mouse

Problem

You want to detect movements of a PS/2-compatible mouse and respond to changes in the *x* and *y* coordinates.

Solution

This solution uses LEDs to indicate mouse movement. The brightness of the LEDs changes in response to mouse movement in the *x* (left and right) and *y* (nearer and farther) directions. Clicking the mouse buttons sets the current position as the reference point (Figure 6-16 shows the connections):

```
/*
   Mouse
   an arduino sketch using ps2 mouse library
   see: http://www.arduino.cc/playground/ComponentLib/Ps2mouse
 */

// PS2 mouse library from :  http://www.arduino.cc/playground/ComponentLib/Ps2mouse
#define WProgram.h Arduino.h
#include <ps2.h>

const int dataPin =  5;
const int clockPin = 6;

const int xLedPin  = 9;
const int yLedPin  = 11;

const int mouseRange = 255;  // the maximum range of x/y values
```

```
char x;                  // values read from the mouse
char y;
byte status;

int xPosition = 0;       // values incremented and decremented when mouse moves
int yPosition = 0;
int xBrightness = 128;  // values increased and decreased based on mouse position
int yBrightness = 128;

const byte REQUEST_DATA = 0xeb; // command to get data from the mouse

PS2 mouse(clockPin, dataPin);

void setup()
{
  mouseBegin();
}

void loop()
{
  // get a reading from the mouse
  mouse.write(REQUEST_DATA); // ask the mouse for data
  mouse.read();       // ignore ack
  status = mouse.read(); // read the mouse buttons
  if(status & 1) // this bit is set if the left mouse btn pressed
    xPosition = 0; // center the mouse x position
  if(status & 2) // this bit is set if the right mouse btn pressed
    yPosition = 0; // center the mouse y position

  x = mouse.read();
  y = mouse.read();
  if( x != 0 || y != 0)
  {
    // here if there is mouse movement

    xPosition = xPosition + x; // accumulate the position
    xPosition = constrain(xPosition,-mouseRange,mouseRange);

    xBrightness = map(xPosition, -mouseRange, mouseRange, 0,255);
    analogWrite(xLedPin, xBrightness);

    yPosition = constrain(yPosition + y, -mouseRange,mouseRange);
    yBrightness = map(yPosition, -mouseRange, mouseRange, 0,255);
    analogWrite(yLedPin, yBrightness);
  }
}

void mouseBegin()
{
  // reset and initialize the mouse
  mouse.write(0xff);          // reset
  delayMicroseconds(100);
  mouse.read();               // ack byte
  mouse.read();               // blank
  mouse.read();               // blank
```

```
    mouse.write(0xf0);          // remote mode
    mouse.read();               // ack
    delayMicroseconds(100);
}
```

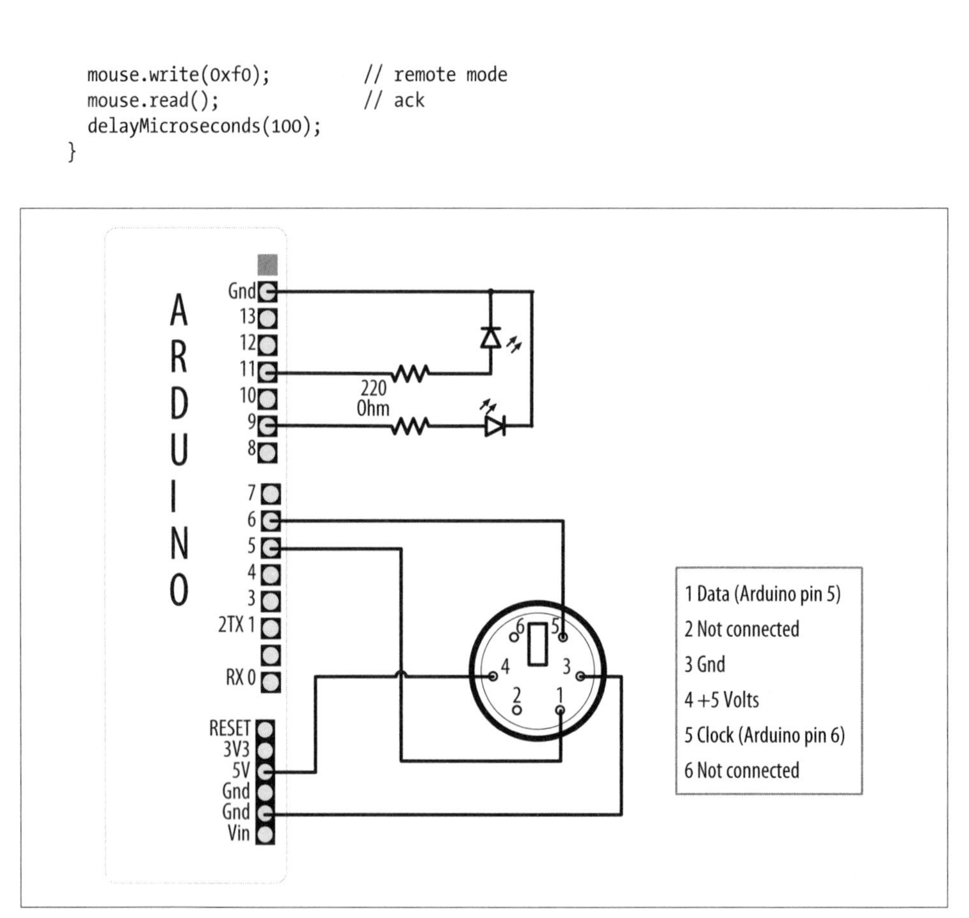

Figure 6-16. Connecting a mouse to indicate position and light LEDs

Figure 6-16 shows a female PS/2 connector from the front. If you don't have a female connector and don't mind chopping the end off your mouse, you can note which wires connect to each of these pins and solder to pin headers that plug directly into the correct Arduino pins.

Discussion

Connect the mouse signal (clock and data) and power leads to Arduino, as shown in Figure 6-16. This solution only works with PS/2-compatible devices, so you may need to find an older mouse—most mice with the round PS/2 connector should work.

The `mouseBegin` function initializes the mouse to respond to requests for movement and button status. The PS/2 library from *http://www.arduino.cc/playground/ComponentLib/Ps2mouse* handles the low-level communication. The `mouse.write` command is used to instruct the mouse that data will be requested. The first call to `mouse.read` gets an acknowledgment (which is ignored in this example). The next call to `mouse.read` gets the button status, and the last two `mouse.read` calls get the *x* and *y* movement that has taken place since the previous request.

The sketch tests to see which bits are `HIGH` in the `status` value to determine if the left or right mouse button was pressed. The two rightmost bits will be `HIGH` when the left and right buttons are pressed, and these are checked in the following lines:

```
status = mouse.read(); // read the mouse buttons
if(status & 1) // rightmost bit is set if the left mouse btn pressed
  xPosition = 0; // center the mouse x position
if(status & 2) // this bit is set if the right mouse btn pressed
  yPosition = 0; // center the mouse y position
```

The `x` and `y` values read from the mouse represent the movement since the previous request, and these values are accumulated in the variables `xPosition` and `yPosition`.

The values of `x` and `y` will be positive if the mouse moves right or away from you, and negative if it moves left or toward you.

The sketch ensures that the accumulated value does not exceed the defined range (`mouseRange`) using the `constrain` function:

```
xPosition = xPosition + x; // accumulate the position
xPosition = constrain(xPosition,-mouseRange,mouseRange);
```

The `yPosition` calculation shows a shorthand way to do the same thing; here the calculation for the `y` value is done within the call to `constrain`:

```
yPosition = constrain(yPosition + y,-mouseRange,mouseRange);
```

The `xPosition` and `yPosition` variables are reset to zero if the left and right mouse buttons are pressed.

LEDs are illuminated to correspond to position using `analogWrite`—half brightness in the center, and increasing and decreasing in brightness as the mouse position increases and decreases.

The position can be displayed on the Serial Monitor by adding the following line just after the second call to `analogWrite()`:

```
printValues(); // show button and x and y values on Serial Monitor
```

You'll also need to add this line to `setup()`:

```
Serial.begin(9600);
```

Add the following function to the end of the sketch to print the values received from the mouse:

```
void printValues()
{
    Serial.println(status, BIN);

    Serial.print("X=");
    Serial.print(x,DEC);
    Serial.print(", position= ");
```

```
        Serial.print(xPosition);
        Serial.print(", brightness= ");
        Serial.println(xBrightness);

        Serial.print("Y=");
        Serial.print(y,DEC);
        Serial.print(", position= ");
        Serial.print(yPosition);
        Serial.print(", brightness= ");
        Serial.println(yBrightness);
        Serial.println();
    }
```

See Also

For a suitable PS/2 connector and breakout, see *http://www.sparkfun.com/products/8509* and *http://www.sparkfun.com/products/8651*.

6.14 Getting Location from a GPS

Problem

You want to determine location using a GPS module.

Solution

A number of fine Arduino-compatible GPS units are available today. Most use a familiar serial interface to communicate with their host microcontroller using a protocol known as *NMEA 0183*. This industry standard provides for GPS data to be delivered to "listener" devices such as Arduino as human-readable ASCII "sentences." For example, the following NMEA sentence:

```
$GPGLL,4916.45,N,12311.12,W,225444,A,*1D
```

describes, among other things, a location on the globe at 49 16.45' north latitude by 123 11.12' west longitude.

To establish location, your Arduino sketch must parse these strings and convert the relevant text to numeric form. Writing code to manually extract data from NMEA sentences can be tricky and cumbersome in the Arduino's limited address space, but fortunately there is a useful library that does this work for you: Mikal Hart's TinyGPS. Download it from *http://arduiniana.org/* and install it. (For instructions on installing third-party libraries, see Chapter 16.)

The general strategy for using a GPS is as follows:

1. Physically connect the GPS device to the Arduino.
2. Read serial NMEA data from the GPS device.
3. Process the data to determine location.

Using TinyGPS, you do the following:

1. Physically connect the GPS device to the Arduino.
2. Create a TinyGPS object.
3. Read serial NMEA data from the GPS device.
4. Process each byte with TinyGPS's `encode()` method.
5. Periodically query TinyGPS's `get_position()` method to determine location.

The following sketch illustrates how you can acquire data from a GPS attached to Arduino's serial port. It lights the built-in LED connected to pin 13 whenever the device is in the Southern Hemisphere:

```
// A simple sketch to detect the Southern Hemisphere
// Assumes: LED on pin 13, GPS connected to Hardware Serial pins 0/1
#include "TinyGPS.h"

TinyGPS gps; // create a TinyGPS object

#define HEMISPHERE_PIN 13

void setup()
{
  Serial.begin(4800); // GPS devices frequently operate at 4800 baud
  pinMode(HEMISPHERE_PIN, OUTPUT);
  digitalWrite(HEMISPHERE_PIN, LOW); // turn off LED to start
}
void loop()
{
  while (Serial.available())
  {
    int c = Serial.read();
    // Encode() each byte
    // Check for new position if encode() returns "True"
    if (gps.encode(c))
    {
      long lat, lon;
      gps.get_position(&lat, &lon);
      if (lat < 0) // Southern Hemisphere?
        digitalWrite(HEMISPHERE_PIN, HIGH);
      else
        digitalWrite(HEMISPHERE_PIN, LOW);
    }
  }
}
```

Start serial communications using the rate required by your GPS. See Chapter 4 if you need more information on using Arduino serial communications.

A 4,800 baud connection is established with the GPS. Once bytes begin flowing, they are processed by `encode()`, which parses the NMEA data. A `true` return from `encode()` indicates that TinyGPS has successfully parsed a complete sentence and that fresh position data may be available. This is a good time to check the device's current location with a call to `get_position()`.

TinyGPS's `get_position()` returns the most recently observed latitude and longitude. The example examines latitude; if it is less than zero, that is, south of the equator, the LED is illuminated.

Discussion

Attaching a GPS unit to an Arduino is usually as simple as connecting two or three data lines from the GPS to input pins on the Arduino. Using the popular USGlobalSat EM-406A GPS module as an example, you can connect the lines as shown in Table 6-1.

Table 6-1. EM-406A GPS pin connections

EM-406A line	Arduino pin
GND	Gnd
VIN	+Vcc
RX	TX (pin 1)
TX	RX (pin 0)
GND	Gnd

Some GPS modules use RS-232 voltage levels, which are incompatible with Arduino's TTL logic and can permanently damage the board. If your GPS uses RS-232 levels then you need some kind of intermediate logic conversion device like the MAX232 integrated circuit.

The code in this recipe's Solution assumes that the GPS is connected directly to Arduino's built-in serial port, but this is not usually the most convenient design. In many projects, the hardware serial port is needed to communicate with a host PC or other peripheral and cannot be used by the GPS. In cases like this, select another pair of digital pins and use a serial port emulation ("soft serial") library to talk to the GPS instead.

SoftwareSerial is the serial emulation library that currently ships with the Arduino IDE. If you are using a version prior to Arduino 1.0, you will need to use a third-party library called NewSoftSerial, also published at *http://arduiniana.org/*. For a more detailed discussion on software serial, see Recipes 4.13 and 4.14.

You can move the GPS's TX line to Arduino pin 2 and RX line to pin 3 to free up the hardware serial port for debugging (see Figure 4-7). Leaving the USB cable connected to the host PC, modify the preceding sketch to use SoftwareSerial to get a detailed glimpse of TinyGPS in action through the Arduino's Serial Monitor:

```
// Another simple sketch to detect the Southern Hemisphere
// Assumes: LED on pin 13, GPS connected to pins 2/3
// (Optional) Serial debug console attached to hardware serial port 0/1
```

```
#include "TinyGPS.h"
#include "SoftwareSerial.h"

#define HEMISPHERE_PIN 13
#define GPS_RX_PIN 2
#define GPS_TX_PIN 3

TinyGPS gps; // create a TinyGPS object
SoftwareSerial ss(GPS_RX_PIN, GPS_TX_PIN); // create soft serial object

void setup()
{
  Serial.begin(9600); // for debugging
  ss.begin(4800); // Use Soft Serial object to talk to GPS
  pinMode(HEMISPHERE_PIN, OUTPUT);
  digitalWrite(HEMISPHERE_PIN, LOW); // turn off LED to start
}
void loop()
{
  while (ss.available())
  {
    int c = ss.read();
    Serial.write(c); // display NMEA data for debug
    // Send each byte to encode()
    // Check for new position if encode() returns "True"
    if (gps.encode(c))
    {
      long lat, lon;
     unsigned long fix_age;
    gps.get_position(&lat, &lon, &fix_age);
    if (fix_age == TinyGPS::GPS_INVALID_AGE )
      Serial.println("No fix ever detected!");
    else if (fix_age > 2000)
      Serial.println("Data is getting STALE!");
    else
      Serial.println("Latitude and longitude valid!");

      Serial.print("Lat: ");
      Serial.print(lat);
      Serial.print(" Lon: ");
      Serial.println(lon);
      if (lat < 0) // Southern Hemisphere?
        digitalWrite(HEMISPHERE_PIN, HIGH);
      else
        digitalWrite(HEMISPHERE_PIN, LOW);
    }
  }
}
```

Note that you can use a different baud rate for connection to the Serial Monitor and the GPS.

This new sketch behaves the same as the earlier example but is much easier to debug. At any time, you can simply hook a monitor up to the built-in serial port to watch the NMEA sentences and TinyGPS data scrolling by.

When power is turned on, a GPS unit begins transmitting NMEA sentences. However, the sentences containing valid location data are only transmitted after the GPS establishes a fix, which requires the GPS antenna to have visibility of the sky and can take up to two minutes or more. Stormy weather or the presence of buildings or other obstacles may also interfere with the GPS's ability to pinpoint location. So, how does the sketch know whether TinyGPS is delivering valid position data? The answer lies in the third parameter to `get_position()`, the optional `fix_age`.

If you supply a pointer to an unsigned `long` variable as `get_position()`'s third parameter, TinyGPS sets it to the number of milliseconds since the last valid position data was acquired; see also Recipe 2.11. A value of `0xFFFFFFFF` here (symbolically, `GPS_INVALID_AGE`) means TinyGPS has not yet parsed any valid sentences containing position data. In this case, the returned latitude and longitude are invalid as well (`GPS_INVALID_ANGLE`).

Under normal operation, you can expect to see quite low values for `fix_age`. Modern GPS devices are capable of reporting position data as frequently as one to five times per second or more, so a `fix_age` in excess of 2,000 ms or so suggests that there may be a problem. Perhaps the GPS is traveling through a tunnel or a wiring flaw is corrupting the NMEA data stream, invalidating the checksum (a calculation to check that the data is not corrupted). In any case, a large `fix_age` indicates that the coordinates returned by `get_position()` are stale. The following code is an example of how `fix_age` can be used to ensure that the position data is fresh:

```
long lat, lon;
unsigned long fix_age;
gps.get_position(&lat, &lon, &fix_age);
if (fix_age == TinyGPS::GPS_INVALID_AGE)
  Serial.println("No fix ever detected!");
else if (fix_age > 2000)
  Serial.println("Data is getting STALE!");
else
  Serial.println("Latitude and longitude valid!");
```

See Also

TinyGPS is available for download at *http://arduiniana.org/libraries/tinygps*.

For a deeper understanding of the NMEA protocol, read the Wikipedia article at *http://en.wikipedia.org/wiki/NMEA*.

Several shops sell GPS modules that interface well with TinyGPS and Arduino. These differ mostly in power consumption, voltage, accuracy, physical interface, and whether they support serial NMEA. SparkFun (*http://www.sparkfun.com*) carries a large range of GPS modules and has an excellent buyer's guide.

GPS technology has inspired lots of creative Arduino projects. A very popular example is the GPS data logger, in which a moving device records location data at regular intervals to the Arduino EEPROM or other on-board storage. See the breadcrumbs project at *http://code.google.com/p/breadcrumbs/wiki/UserDocument* for an example. Ladyada makes a popular GPS data logging shield; see *http://www.ladyada.net/make/gpsshield/*.

Other interesting GPS projects include hobby airplanes and helicopters that maneuver themselves to preprogrammed destinations under Arduino software control. Mikal Hart built a GPS-enabled "treasure chest" with an internal latch that cannot be opened until the box is physically moved to a certain location. See *http://arduiniana.org*.

6.15 Detecting Rotation Using a Gyroscope

Problem

You want to respond to the rate of rotation. This can be used to keep a vehicle or robot moving in a straight line or turning at a desired rate.

Solution

Gyroscopes provide an output related to rotation rate (as opposed to an accelerometer, which indicates rate of change of velocity). Most low-cost gyroscopes use an analog voltage proportional to rotation rate, although some also provide output using I2C (see Chapter 13 for more on using I2C to communicate with devices). This recipe works with a gyro with an analog output proportional to rotation rate. Figure 6-17 shows an LY530AL breakout board from SparkFun. Many low-cost gyros, such as the one used here, are 3.3V devices and must not be plugged in to the 5V power pin.

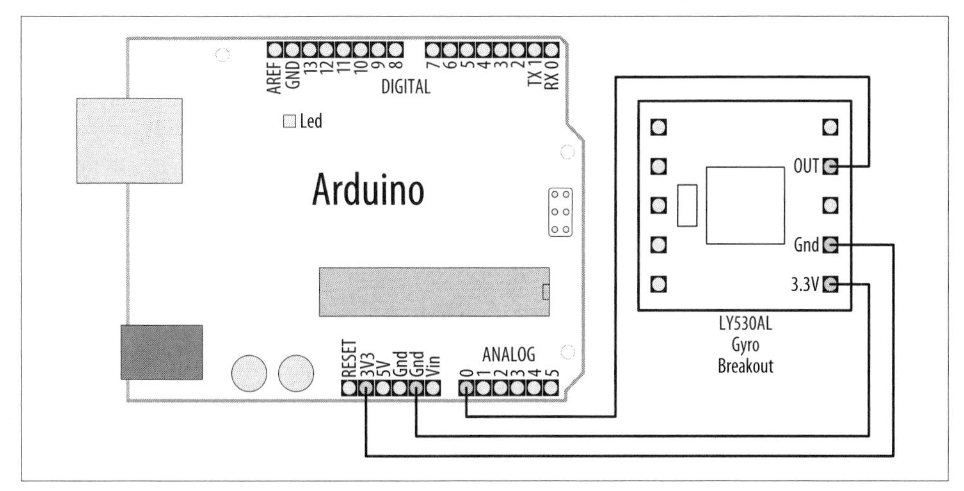

Figure 6-17. LY530AL gyro connected using 3.3V pin

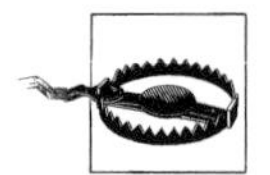

Check the maximum voltage of your gyro before connecting power. Plugging a 3.3V gyro into 5V can permanently damage the device.

The Gyro `OUT` connection is the analog output and is connected to Arduino analog input 0:

```
/*
  gyro sketch
  displays the rotation rate on the Serial Monitor
 */


const int inputPin = 0;  // analog input 0
int rotationRate = 0;

void setup()
{
  Serial.begin(9600);   // sets the serial port to 9600
}

void loop()
{
  rotationRate = analogRead(inputPin); // read the gyro output
  Serial.print("rotation rate is ");
  Serial.println(rotationRate);
  delay(100);                          // wait 100ms for next reading
}
```

Discussion

The `loop` code reads the gyro value on analog pin 0 and displays this on the Serial Monitor.

Using the older LISY300AL gyro

The previous edition covered the LISY300AL gyro, which may now be difficult to obtain. But if you have one, you can use the same sketch if you connect the Power Down (PD) pin to Gnd. Better yet, you can wire the PD pin to an Arduino pin so you can turn the gyro on and off from your sketch. Figure 6-18 shows the connections for the LISY3000AL.

The PD connection enables the gyro to be switched into low power mode and is connected to analog pin 1 (in this sketch, it is used as a digital output pin). You can connect PD to any digital pin; the pin used here was chosen to keep the wiring neater. The sketch above can be modified to control the PD pin as follows:

```
const int inputPin = 0;      // analog input 0
const int powerDownPin = 15; // analog input 1 is digital input 15

int rotationRate = 0;
```

```
void setup()
{
  Serial.begin(9600);            // sets the serial port to 9600
  pinMode(powerDownPin, OUTPUT);
  digitalWrite(powerDownPin, LOW);  // gyro not in power down mode
}

// loop code is same as above
```

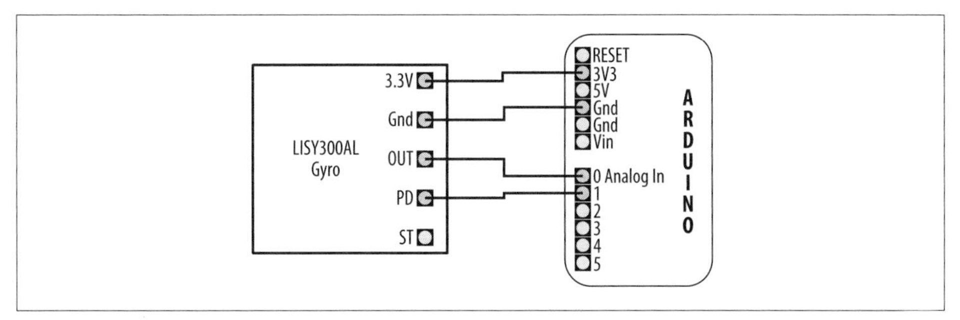

Figure 6-18. LISY3000AL gyro connections

If you don't need to switch the gyro into low-power mode, you can connect the PD line to Gnd (PD `LOW` is on, PD `HIGH` is power down mode).

Analog input pins can be used as digital pins (but not the other way around). Analog input 0 is digital pin 14; analog input 1 is digital pin 15, and so on. Arduino 1.0 introduced new definitions that enable you to refer to Analog input 0 as `A0`, Analog input 1 as `A1`, etc.

Measuring rotation in three dimensions using the ITG-3200 sensor

The ITG-3200 is a 3-axis gyroscope with excellent performance for the price. Even if you only require 2-axis measurements, it is a better choice than the LY530ALH for applications that need accurate measurements or have a high rotation rate (up to 2000° per second). It is a 3.3V I2C device, so if you are not using a 3.3V Arduino board you will need a logic-level converter to protect the gyro's SCL and SDA pins. See the introduction to Chapter 13 for more on I2C and using 3.3V devices.

The breakout board from SparkFun (SEN-09801) makes it easy to connect this up (see Figure 6-19), but don't forget to solder the CLK jumper on the underside of the board that enables the internal clock.

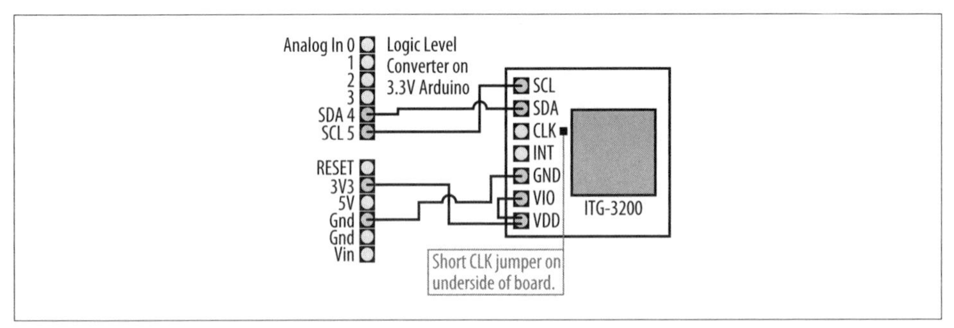

Figure 6-19. Connecting the ITG-3200 to a 3.3-volt board

This sketch below prints the values of each of the x,y and z axis separated by commas:

```
/*
 ITG-3200 example sketch
 Based on the SparkFun quick start guide: http://www.sparkfun.com/tutorials/265
 */
#include <Wire.h>

const int itgAddress = 0x69;

// ITG-3200 constants - see data sheet
const byte SMPLRT_DIV= 0x15;
const byte DLPF_FS   = 0x16;
const byte INT_CFG   = 0x17;
const byte PWR_MGM   = 0x3E;
const byte GYRO_X_ADDRESS = 0x1D; // GYRO_XOUT_H
const byte GYRO_Y_ADDRESS = 0x1F; // GYRO_YOUT_H
const byte GYRO_Z_ADDRESS = 0x21; // GYRO_ZOUT_H

// Configuration settings, see data sheet for details
const byte DLPF_CFG_0     = 0x1;
const byte DLPF_CFG_1     = 0x2;
const byte DLPF_CFG_2     = 0x4;
const byte DLPF_FS_SEL_0 = 0x8;
const byte DLPF_FS_SEL_1 = 0x10;

void setup()
{
  Serial.begin(9600);
  Wire.begin();

  //Configure the gyroscope
  //Set the gyroscope scale for the outputs to +/-2000 degrees per second
  itgWrite(DLPF_FS, (DLPF_FS_SEL_0|DLPF_FS_SEL_1|DLPF_CFG_0));
  //Set the sample rate to 100 hz
  itgWrite(SMPLRT_DIV, 9);
}

//read and output X,Y and Z rates to Serial Monitor
void loop()
{
```

```
  //Create variables to hold the output rates.
  int xRate, yRate, zRate;

  //Read the x,y and z output rates from the gyroscope.
  xRate = readAxis(GYRO_X_ADDRESS);
  yRate = readAxis(GYRO_Y_ADDRESS);
  zRate = readAxis(GYRO_Z_ADDRESS);

  //Print the output rates to the Serial Monitor
  int temperature = 22;
  Serial.print(temperature);
  Serial.print(',');
  Serial.print(xRate);
  Serial.print(',');
  Serial.print(yRate);
  Serial.print(',');
  Serial.println(zRate);

  //Wait 10ms before reading the values again.
  delay(10);
}

//Write the given data to the given itg-3200 register
void itgWrite(char registerAddress, char data)
{

  Wire.beginTransmission(itgAddress); // initiate the send sequence
  Wire.write(registerAddress);        // the register address to write
  Wire.write(data);                   // the data to be written
  Wire.endTransmission();             // this actually sends the data
}

//Read data from the specified register on the ITG-3200 and return the value.
unsigned char itgRead(char registerAddress)
{
  //This variable will hold the contents read from the i2c device.
  unsigned char data=0;

  Wire.beginTransmission(itgAddress);
  Wire.write(registerAddress); //Send the Register Address
  Wire.endTransmission();      //End the communication sequence.

  Wire.beginTransmission(itgAddress);
  Wire.requestFrom(itgAddress, 1);    //Ask the device for data

  if(Wire.available()){  // Wait for a response from device
    data = Wire.read();  // read the data
  }

  Wire.endTransmission();   //End the communication sequence
  return data;   //Return the read data
}

// Read X,Y or Z Axis rate of the gyroscope.
// axisRegAddress argument selects the axis to be read.
```

```
int readAxis(byte axisRegAddress)
{
  int data=0;
  data = itgRead(axisRegAddress)<<8;
  data |= itgRead(axisRegAddress + 1);
  return data;
}
```

See Also

See Chapter 13 for more about I2C.

See "Using 3.3 Volt Devices with 5 Volt Boards" on page 423 for more about that topic.

A SparkFun tutorial for the ITG-3200 is at *http://www.sparkfun.com/tutorials/265*.

6.16 Detecting Direction

Problem

You want your sketch to determine direction from an electronic compass.

Solution

This recipe uses the HM55B Compass Module from Parallax (#29123); Figure 6-20 shows the connections:

```
/*
  HM55bCompass sketch
  uses 'software SPI' serial protocol implemented using Arduino bit operators
  (see Recipe 3.13)
  prints compass angle to Serial Monitor
 */

const int enablePin = 2;
const int clockPin  = 3;
const int dataPin   = 4;

// command codes (from HM55B data sheet)
const byte COMMAND_LENGTH = 4;        // the number of bits in a command
const byte RESET_COMMAND = B0000;     // reset the chip
const byte MEASURE_COMMAND = B1000;   // start a measurement
const byte READ_DATA_COMMAND = B1100; // read data and end flag
const byte MEASUREMENT_READY = B1100; // value returned when measurement complete


int angle;

void setup()
{
  Serial.begin(9600);
  pinMode(enablePin, OUTPUT);
  pinMode(clockPin, OUTPUT);
```

```
  pinMode(dataPin, INPUT);
  reset(); // reset the compass module
}

void loop()
{
  startMeasurement();
  delay(40); // wait for the data to be ready
  if (readStatus()==MEASUREMENT_READY); // check if the data is ready
  {
    angle = readMeasurement();          //read measurement and calculate angle
    Serial.print("Angle = ");
    Serial.println(angle); // print angle
  }
  delay(100);
}

void reset()
{
  pinMode(dataPin, OUTPUT);
  digitalWrite(enablePin, LOW);
  serialOut(RESET_COMMAND, COMMAND_LENGTH);
  digitalWrite(enablePin, HIGH);
}

void startMeasurement()
{
  pinMode(dataPin, OUTPUT);
  digitalWrite(enablePin, LOW);
  serialOut(MEASURE_COMMAND, COMMAND_LENGTH);
  digitalWrite(enablePin, HIGH);
}

int readStatus()
{
  int result = 0;
  pinMode(dataPin, OUTPUT);
  digitalWrite(enablePin, LOW);
  serialOut(READ_DATA_COMMAND, COMMAND_LENGTH);
  result = serialIn(4);
  return result;  // returns the status
}

int readMeasurement()
{
  int X_Data = 0;
  int Y_Data = 0;
  int calcAngle = 0;
  X_Data = serialIn(11); // Field strength in X
  Y_Data = serialIn(11); // and Y direction
  digitalWrite(enablePin, HIGH); // deselect chip
  calcAngle = atan2(-Y_Data , X_Data) / M_PI * 180; // angle is atan(-y/x)
```

```
  if(calcAngle < 0)
    calcAngle = calcAngle + 360; // angle from 0 to 259 instead of plus/minus 180
  return calcAngle;
}

void serialOut(int value, int numberOfBits)
{
  for(int i = numberOfBits; i > 0; i--) // shift the MSB first
  {
    digitalWrite(clockPin, LOW);
    if(bitRead(value, i-1) == 1)
      digitalWrite(dataPin, HIGH);
    else
      digitalWrite(dataPin, LOW);
    digitalWrite(clockPin, HIGH);
  }
}

int serialIn(int numberOfBits)
{
  int result = 0;

  pinMode(dataPin, INPUT);
  for(int i = numberOfBits; i > 0; i--) // get the MSB first
  {
    digitalWrite(clockPin, HIGH);
    if (digitalRead(dataPin) == HIGH)
      result = (result << 1) + 1;
    else
      result = (result << 1) + 0;
    digitalWrite(clockPin, LOW);
  }

  // the following converts the result to a twos-complement negative number
  // if the most significant bit in the 11 bit data is 1
  if(bitRead(result, 11) == 1)
    result = (B11111000 << 8) | result; // twos complement negation

  return result;
}
```

Discussion

The compass module provides magnetic field intensities on two perpendicular axes (x and y). These values vary as the compass orientation is changed with respect to the Earth's magnetic field (magnetic north).

The data sheet for the device tells you what values to send to reset the compass. Check if a valid reading is ready (if so, it will transmit it).

The sketch uses the functions `serialIn()` and `serialOut()` to handle the pin manipulations that send and receive messages.

The compass module is initialized into a known state in the `reset()` function called from `setup()`. The `startMeasurement()` function initiates the measurement, and after a brief delay, the `readStatus()` function indicates if the data is ready. A value of `0` is returned if the measurement is not ready, or `12` (binary 1100) if the compass is ready to transmit data.

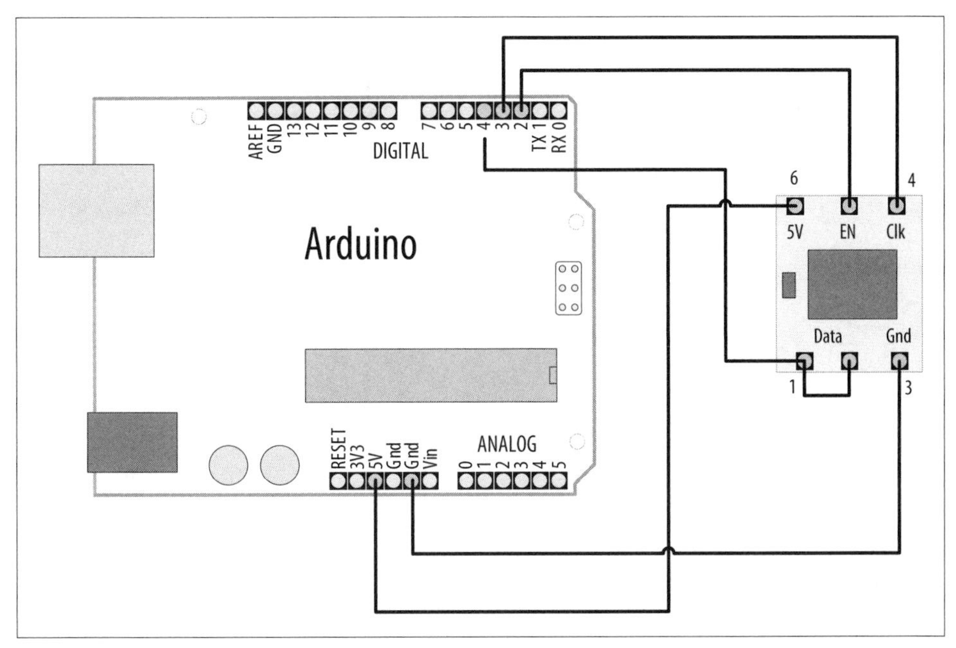

Figure 6-20. HM55B compass connections

Eleven bits of data are read into the `X_Data` and `Y_Data` variables. If you use a different device, you will need to check the data sheet to see how many bits and in what format the data is sent. `X_Data` and `Y_Data` store the magnetic field readings, and the angle to magnetic north is calculated as follows: Radians = arctan(–y/x)

This is implemented in the sketch in the line:

```
calcAngle = atan2(-Y_Data , X_Data) / M_PI * 180; // angle is atan(-y/x)
```

To make a servo follow the compass direction over the first 180 degrees, add the following:

```
#include <Servo.h>
Servo myservo;
```

in `setup`:

```
  myservo.attach(8);
```

and in loop after the angle is calculated:

```
// the servo is driven only up to 180 degrees
    angle = constrain(angle, 0,180);
myservo.write(angle);
```

Direction sensors are increasingly being used in smartphones. Consequently, high performance and low-cost devices are becoming more available. The sketch that follows is for one such device, the 3.3 volt HMC5883L I2C magnetometer chip. Breakout boards are available for this part, for example the SEN-10530 from SparkFun. Connect the GND and VCC pins to Ground and the 3.3V power pin. The SDA and SCL pins are connected to Arduino pins 4 and 5 (see Chapter 13 for more on using I2C devices with Arduino). If you want to use the HMC5883L with a 5 volt Arduino board, see "Using 3.3 Volt Devices with 5 Volt Boards" on page 423 for details on how to use a level shifter.

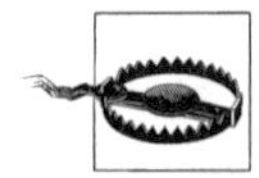

Connecting the HMC5883L directly to Arduino pins on a standard 5V board can permanently damage the HMC5883L chip.

```
/*
 Uses HMC5883L to get earths magnetic field in x,y and z axis
 Displays direction as angle between 0 and 359 degrees
*/

#include <Wire.h> //I2C Arduino Library

const int hmc5883Address = 0x1E; //0011110b, I2C 7bit address of HMC5883
const byte hmc5883ModeRegister      = 0x02;
const byte hmcContinuousMode        = 0x00;
const byte hmcDataOutputXMSBAddress = 0x03;


void setup(){
  //Initialize Serial and I2C communications
  Serial.begin(9600);
  Wire.begin();

  //Put the HMC5883 IC into the correct operating mode
  Wire.beginTransmission(hmc5883Address); //open communication with HMC5883
  Wire.write(hmc5883ModeRegister); //select mode register
  Wire.write(hmcContinuousMode);   //continuous measurement mode
  Wire.endTransmission();
}

void loop(){

  int x,y,z; //triple axis data
```

```
  //Tell the HMC5883 where to begin reading data
  Wire.beginTransmission(hmc5883Address);
  Wire.write(hmcDataOutputXMSBAddress); //select register 3, X MSB register
  Wire.endTransmission();

  //Read data from each axis, 2 registers per axis
  Wire.requestFrom(hmc5883Address, 6);
  if(6<=Wire.available()){
    x = Wire.read()<<8; //X msb
    x |= Wire.read(); //X lsb
    z = Wire.read()<<8; //Z msb
    z |= Wire.read(); //Z lsb
    y = Wire.read()<<8; //Y msb
    y |= Wire.read(); //Y lsb
  }

  //Print out values of each axis
  Serial.print("x: ");
  Serial.print(x);
  Serial.print("  y: ");
  Serial.print(y);
  Serial.print("  z: ");
  Serial.print(z);

  int angle = atan2(-y , x) / M_PI * 180; // angle is atan(-y/x)
  if(angle < 0)
     angle = angle  + 360; // angle from 0 to 359 instead of plus/minus 180
  Serial.print(" Direction = ");
  Serial.println(angle);

  delay(250);
}
```

6.17 Getting Input from a Game Control Pad (PlayStation)

Problem

You want to respond to joystick positions or button presses from a game control pad.

Solution

This recipe uses a Sony PlayStation 2–style controller with the PSX library at *http://www.arduino.cc/playground/Main/PSXLibrary*. Figure 6-21 shows the connections.

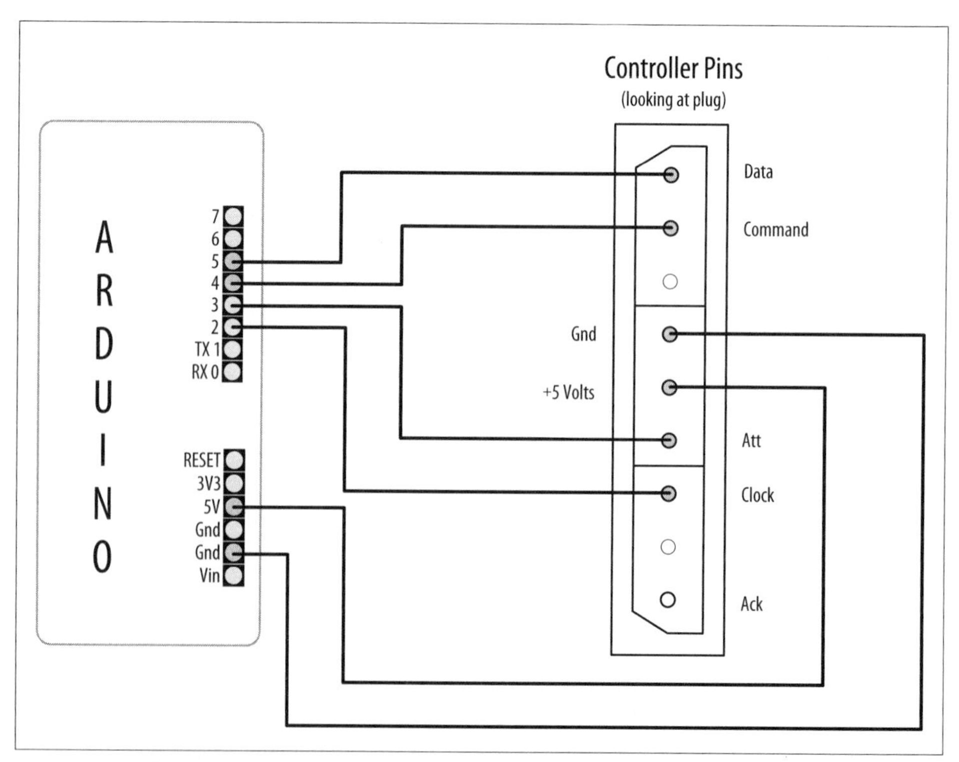

Figure 6-21. PlayStation controller plug connected to Arduino

The sketch uses the Serial Monitor to show which button is pressed:

```
/*
 * PSX sketch
 *
 * Display joystick and button values
 * uses PSX library written by Kevin Ahrendt
 * http://www.arduino.cc/playground/Main/PSXLibrary
 */

#include <Psx.h>                          // Includes the Psx Library

Psx Psx;                                  // Create an instance of the Psx library
const int dataPin  = 5;
const int cmndPin  = 4;
const int attPin   = 3;
const int clockPin = 2;
const int psxDelay = 50;          // determine the clock delay in microseconds

unsigned int data = 0;            // data stores the controller response

void setup()
{
  // initialize the Psx library
```

```
  Psx.setupPins(dataPin, cmndPin, attPin, clockPin, psxDelay);
  Serial.begin(9600); // results will be displayed on the Serial Monitor
}


void loop()
{
  data = Psx.read();        // get the psx controller button data

  // check the button bits to see if a button is pressed
  if(data & psxLeft)
    Serial.println("left button");
  if(data & psxDown)
    Serial.println("down button");
  if(data & psxRight)
    Serial.println("right button");
  if(data & psxUp)
    Serial.println("up button");
  if(data & psxStrt)
    Serial.println("start button");
  if(data & psxSlct)
    Serial.println("select button");

  delay(100);
}
```

Discussion

Game controllers provide information in many different ways. Most recent controllers contain chips that read the switches and joystick in the controller and communicate the information using a protocol depending on the game platform. Older controllers are more likely to give direct access to switches and joysticks using connectors with many connections. The latest wave of game platforms uses USB as the connection and these require hardware support such as a USB host shield.

See Also

Recipe 4.1; Recipe 4.11

PlayStation controller protocol: *http://www.gamesx.com/controldata/psxcont/psxcont.htm*

6.18 Reading Acceleration

Problem

You want to respond to acceleration; for example, to detect when something starts or stops moving. Or you want to detect how something is oriented with respect to the Earth's surface (measure acceleration due to gravity).

Solution

Like many of the sensors discussed in this chapter, there is a wide choice of devices and methods of connection. Recipe 4.11 gave an example of a virtual joystick using the accelerometer in the Wii nunchuck to follow hand movements. Recipe 13.2 has more information on using the Wii nunchuck accelerometer. The recipe here uses analog output proportional to acceleration. Suitable devices include the ADXL203CE (SF SEN-00844), ADXL320 (SF SEN 00847), and MMA7260Q (SF SEN00252)—check the SparkFun accelerometer selection guide (*http://www.sparkfun.com/tutorials/167*) on the SparkFun website for more information.

Figure 6-22 shows the connections for the x- and y-axes of an analog accelerometer.

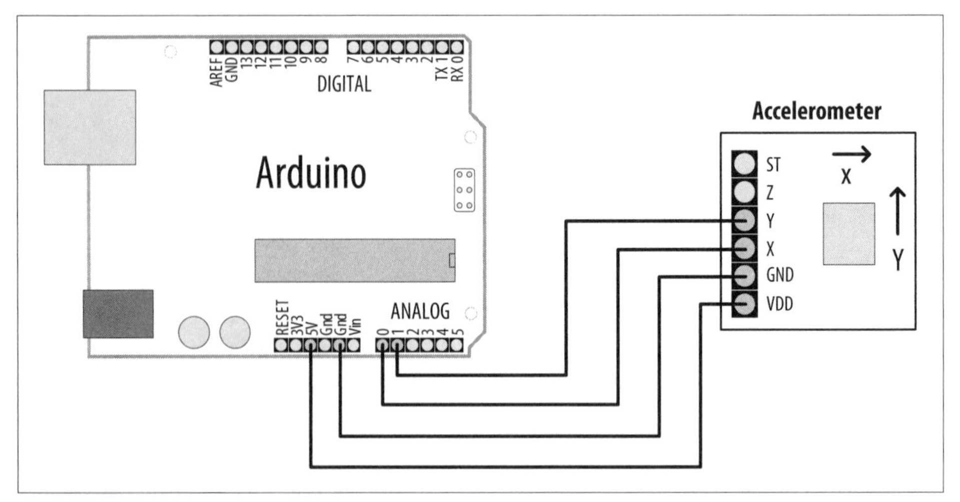

Figure 6-22. Connections for x- and y-axes of an analog accelerometer

Check the data sheet for your device to ensure that you don't exceed the maximum voltage. Many accelerometers are designed for 3.3V operation and can be damaged if connected to the 5V power connection on an Arduino board.

The simple sketch here uses the ADXL320 to display the acceleration in the x- and y-axes:

```
/*
  accel sketch
  simple sketch to output values on the x- and y-axes
 */

const int xPin = 0;  // analog input pins
const int yPin = 1;

void setup()
```

```
{
  Serial.begin(9600);  // note the higher than usual serial speed
}

void loop()
{
int xValue;  // values from accelerometer stored here
int yValue;

    xValue = analogRead(xPin);
    yValue = analogRead(yPin);

    Serial.print("X value = ");
    Serial.println(xValue);

    Serial.print("Y value = ");
    Serial.println(yValue);
    delay(100);
}
```

Discussion

You can use techniques from the previous recipes to extract information from the accelerometer readings. You might need to check for a threshold to work out movement (see Recipe 6.6 for an example of threshold detection). You may need to average values like Recipe 6.7 to get values that are of use. If the accelerometer is reading horizontally, you can use the values directly to work out movement. If it is reading vertically, you will need to take into account the effects of gravity on the values. This is similar to the DC offset in Recipe 6.7, but it can be complicated, as the accelerometer may be changing orientation so that the effect of gravity is not a constant value for each reading.

See Also

SparkFun selection guide: *http://www.sparkfun.com/commerce/tutorial_info.php?tutorials_id=167*

CHAPTER 7

Visual Output

7.0 Introduction

Visual output lets the Arduino show off, and toward that end, the Arduino supports a broad range of LED devices. Before delving into the recipes in this chapter, we'll discuss Arduino digital and analog output. This introduction will be a good starting point if you are not yet familiar with using digital and analog outputs (`digitalWrite` and `analogWrite`).

Digital Output

All the pins that can be used for digital input can also be used for digital output. Chapter 5 provided an overview of the Arduino pin layout; you may want to look through the introduction section in that chapter if you are unfamiliar with connecting things to Arduino pins.

Digital output causes the voltage on a pin to be either high (5 volts) or low (0 volts). Use the `digitalWrite(outputPin, value)` function to turn something on or off. The function has two parameters: `outputPin` is the pin to control, and `value` is either `HIGH` (5 volts) or `LOW` (0 volts).

For the pin voltage to respond to this command, the pin must have been set in *output* mode using the `pinMode(outputPin, OUTPUT)` command. The sketch in Recipe 7.1 provides an example of how to use digital output.

Analog Output

Analog refers to levels that can be gradually varied up to their maximum level (think of light dimmers and volume controls). Arduino has an `analogWrite` function that can be used to control such things as the intensity of an LED connected to the Arduino.

The `analogWrite` function is not truly analog, although it can behave like analog, as you will see. `analogWrite` uses a technique called Pulse Width Modulation (PWM) that emulates an analog signal using digital pulses.

PWM works by varying the proportion of the pulses' on time to off time, as shown in Figure 7-1. Low-level output is emulated by producing pulses that are on for only a short period of time. Higher level output is emulated with pulses that are on more than they are off. When the pulses are repeated quickly enough (almost 500 times per second on Arduino), the pulsing cannot be detected by human senses, and the output from things such as LEDs looks like it is being smoothly varied as the pulse rate is changed.

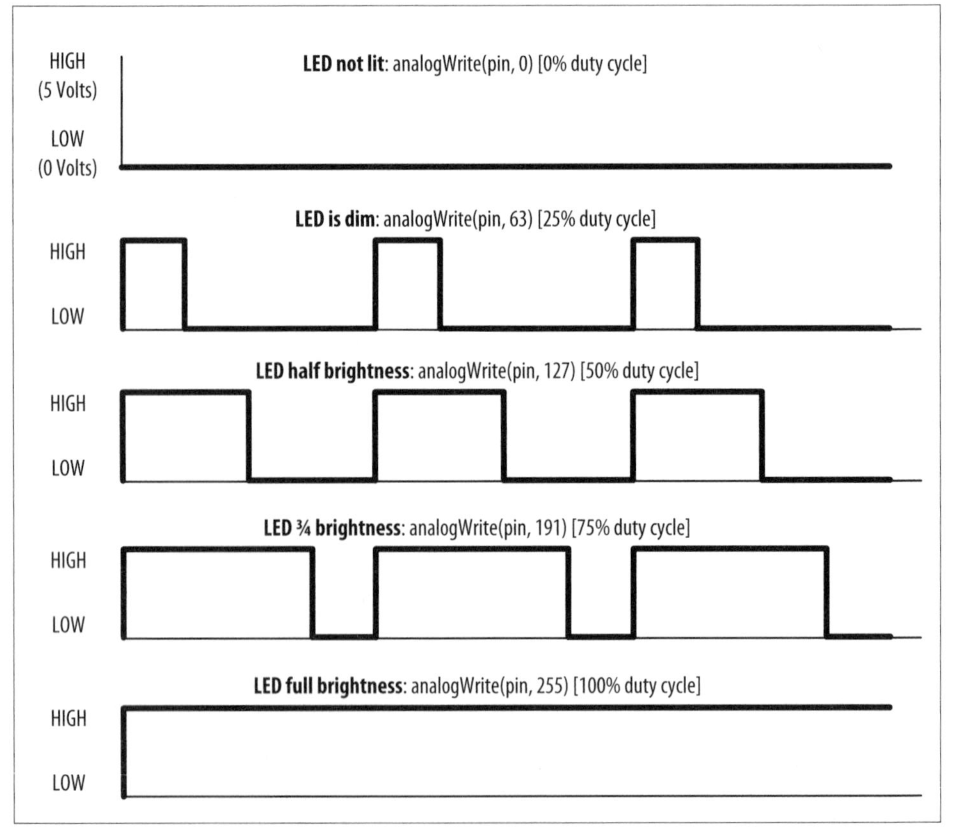

Figure 7-1. PWM output for various analogWrite values

Arduino has a limited number of pins that can be used for analog output. On a standard board, you can use pins 3, 5, 6, 9, 10, and 11. On the Arduino Mega board, you can use pins 2 through 13 for analog output. Many of the recipes that follow use pins that can be used for both digital and analog to minimize rewiring if you want to try out different recipes. If you want to select different pins for analog output, remember to choose one of the supported `analogWrite` pins (other pins will not give any output).

Controlling Light

Controlling light using digital or analog output is a versatile, effective, and widely used method for providing user interaction. Single LEDs, arrays, and numeric displays are covered extensively in the recipes in this chapter. LCD text and graphical displays require different techniques and are covered in Chapter 11.

LED specifications

An LED is a semiconductor device (diode) with two leads, an *anode* and a *cathode*. When the voltage on the anode is more positive than that on the cathode (by an amount called the *forward voltage*) the device emits light (photons). The anode is usually the longer lead, and there is often a flat spot on the housing to indicate the cathode (see Figure 7-2). The LED color and the exact value of the forward voltage depend on the construction of the diode.

A typical red LED has a forward voltage of around 1.8 volts. If the voltage on the anode is not 1.8 volts more positive than the cathode, no current will flow through the LED and no light will be produced. When the voltage on the anode becomes 1.8 volts more positive than that on the cathode, the LED "turns on" (conducts) and effectively becomes a short circuit. You must limit the current with a resistor, or the LED will (sooner or later) burn out. Recipe 7.1 shows you how to calculate values for current-limiting resistors.

You may need to consult an LED data sheet to select the correct LED for your application, particularly to determine values for forward voltage and maximum current. Tables 7-1 and 7-2 show the most important fields you should look for on an LED data sheet.

Table 7-1. Key data sheet specifications: absolute maximum ratings

Parameter	Symbol	Rating	Units	Comment
Forward current	If	25	mA	The maximum continuous current for this LED
Peak forward current (1/10 duty @ 1 kHz)	If	160	mA	The maximum pulsed current (given here for a pulse that is 1/10 on and 9/10 off)

Table 7-2. Key data sheet specifications: electro-optical characteristics

Parameter	Symbol	Rating	Units	Comment
Luminous intensity	Iv	2	mcd	If = 2 mA – brightness with 2 mA current
	Iv	40	mcd	If = 20 mA – brightness with 20 mA current
Viewing angle		120	degrees	The beam angle
Wavelength		620	nm	The dominant or peak wavelength (color)
Forward voltage	Vf	1.8	volts	The voltage across the LED when on

Arduino pins can supply up to 40 mA of current. This is plenty for a typical medium-intensity LED, but not enough to drive the higher brightness LEDs or multiple LEDs connected to a single pin. Recipe 7.3 shows how to use a transistor to increase the current through the LED.

Multicolor LEDs consist of two or more LEDs in one physical package. These may have more than two leads to enable separate control of the different colors. There are many package variants, so you should check the data sheet for your LED to determine how to connect the leads.

Self-color-changing, multicolor LEDs with an integrated chip cannot be controlled in any way; you can't change their colors from Arduino. Because PWM rapidly cycles the power on and off, you are effectively rebooting the integrated chip many times each second, so these LEDs are unsuitable for PWM applications as well.

Multiplexing

Applications that need to control many LEDs can use a technique called *multiplexing*. Multiplexing works by switching groups of LEDs (usually arranged in rows or columns) in sequence. Recipe 7.11 shows how 32 individual LEDs (eight LEDs per digit, including decimal point) with four digits can be driven with just 12 pins. Eight pins drive a digit segment for all the digits and four pins select which digit is active. Scanning through the digits quickly enough (at least 25 times per second) creates the impression that the lights remain on rather than pulsing, through the phenomenon of *persistence of vision*.

Charlieplexing uses multiplexing along with the fact that LEDs have *polarity* (they only illuminate when the anode is more positive than the cathode) to switch between two LEDs by reversing the polarity.

Maximum pin current

LEDs can draw more power than the Arduino chip is designed to handle. The data sheet gives the absolute maximum ratings for the Arduino chip (ATmega328P) as 40 mA per pin. The chip is capable of sourcing and sinking 200 mA overall, so you must also ensure that the total current is less than this. For example, five pins providing a `HIGH` output (sourcing) and five `LOW` (sinking) with each pin at 40 mA. It is good practice to design your applications to operate well within the absolute maximum ratings for best reliability, so best to keep current at or below 30 mA to provide a large comfort margin. For hobby use where more pin current is wanted and reduced reliability is acceptable, you can drive a pin with up to 40 mA as long as the 200 mA source and 200 mA sink limits per chip are not exceeded.

See the discussion section of Recipe 7.3 for a tip on how to get increased current without using external transistors.

The data sheet refers to 40 mA as the absolute maximum rating and some engineers may be hesitant to operate anywhere near this value. However, the 40 mA figure is already de-rated by Atmel and they say the pins can safely handle this current. Recipes that follow refer to the 40 mA maximum rating; however, if you are building anything where reliability is important, de-rating this to 30 mA to provide an added comfort margin is prudent.

7.1 Connecting and Using LEDs

Problem

You want to control one or more LEDs and select the correct current-limiting resistor so that you do not damage the LEDs.

Solution

Turning an LED on and off is easy to do with Arduino, and some of the recipes in previous chapters have included this capability (see Recipe 5.1 for an example that controls the built-in LED on pin 13). The recipe here provides guidance on choosing and using external LEDs. Figure 7-2 shows the wiring for three LEDs, but you can run this sketch with just one or two.

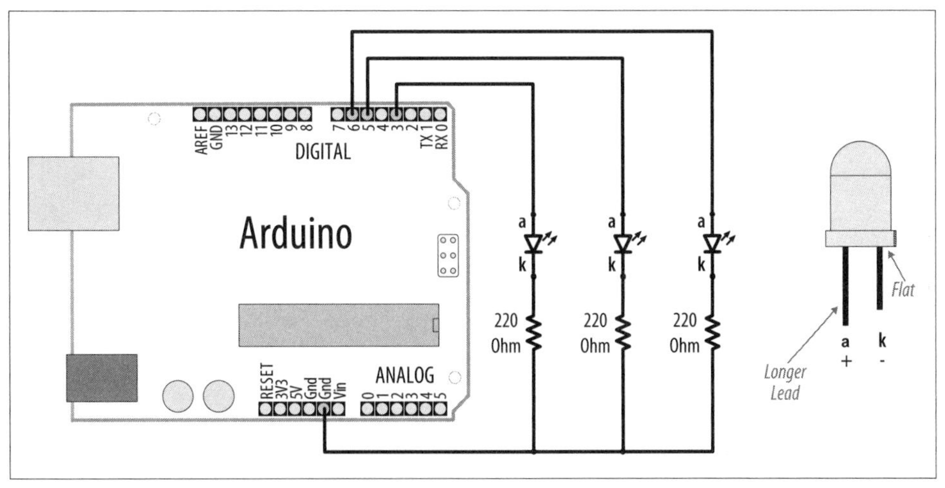

Figure 7-2. Connecting external LEDs

The schematic symbol for the cathode (the negative pin) is *k*, not *c*. The schematic symbol *c* is used for a capacitor.

The following sketch lights up three LEDs connected to pins 3, 5, and 6 in sequence for one second:

```
/*
 LEDs sketch
 Blink three LEDs each connected to a different digital pin
 */

const int firstLedPin  = 3;        // choose the pin for each of the LEDs
const int secondLedPin = 5;
const int thirdLedPin  = 6;

void setup()
{
  pinMode(firstLedPin, OUTPUT);     // declare LED pins as output
  pinMode(secondLedPin, OUTPUT);    // declare LED pins as output
  pinMode(thirdLedPin, OUTPUT);     // declare LED pins as output
}

void loop()
{
  // flash each of the LEDs for 1000 milliseconds (1 second)
  blinkLED(firstLedPin, 1000);
  blinkLED(secondLedPin, 1000);
  blinkLED(thirdLedPin, 1000);
}


// blink the LED on the given pin for the duration in milliseconds
void blinkLED(int pin, int duration)
{
  digitalWrite(pin, HIGH);     // turn LED on
  delay(duration);
  digitalWrite(pin, LOW);      // turn LED off
  delay(duration);
}
```

The sketch sets the pins connected to LEDs as output in the `setup` function. The `loop` function calls `blinkLED` to flash the LED for each of the three pins. `blinkLED` sets the indicated pin `HIGH` for one second (1,000 milliseconds).

Discussion

Because the anodes are connected to Arduino pins and the cathodes are connected to ground, the LEDs will light when the pin goes `HIGH` and will be off when the pin is `LOW`. You can illuminate the LED when the pin is `LOW` by connecting the cathodes to the pins and the anodes to ground (the resistors can be used on either side of the LED).

When LEDs are connected with the anode connected to +5V, as shown in Figure 7-3, the LEDs light when the pin goes `LOW` (the visual effect would reverse—one of the LEDs would turn off for a second while the other two would be lit).

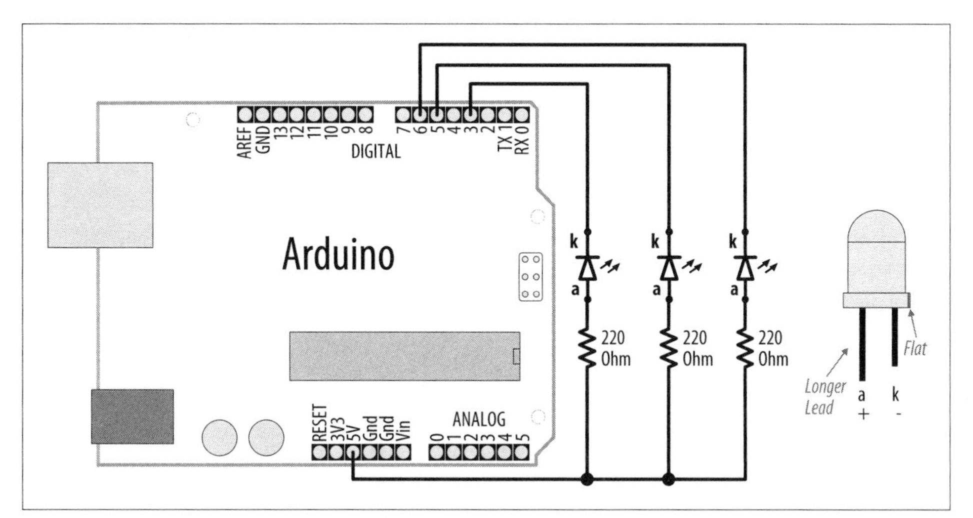

Figure 7-3. Connecting external LEDs with the cathode connected to pins

LEDs require a series resistor to control the current or they can quickly burn out. The built-in LED on pin 13 has a resistor on the circuit board. External LEDs need to be connected through a series resistor on either the anode or the cathode.

A resistor in series with the LED is used to control the amount of current that will flow when the LED conducts. To calculate the resistor value, you need to know the input power supply voltage (Vs, usually 5 volts), the LED forward voltage (Vf), and the amount of current (I) that you want to flow through the LED.

The formula for the resistance in ohms (known as Ohm's law) is

$$R = (Vs - Vf) / I$$

For example, driving an LED with a forward voltage of 1.8 volts with 15 mA of current using an input supply voltage of 5 volts would use the following values:

Vs = 5 (for a 5V Arduino board)
Vf = 1.8 (the forward voltage of the LED)
I = 0.015 (1 milliamp [mA] is one one-thousandth of an amp, so 15 mA is 0.015 amps)

The voltage across the LED when it is on (Vs – Vf) is 5 – 1.8, which is 3.2 volts.

Therefore, the calculation for the series resistor is 3.2 / 0.015, which is 213 ohms.

The value of 213 ohms is not a standard resistor value, so you can round this up to 220 ohms.

The resistor is shown in Figure 7-2 connected between the cathode and ground, but it can be connected to the other side of the LED instead (between the voltage supply and the anode).

Arduino pins have a specified maximum current of 40 mA. If your LED needs more current than this, see Recipe 7.3.

See Also

Recipe 7.3

7.2 Adjusting the Brightness of an LED

Problem

You want to control the intensity of one or more LEDs from your sketch.

Solution

Connect each LED to an analog (PWM) output. Use the wiring shown in Figure 7-2. The sketch will fade the LED(s) from off to maximum intensity and back to off, with each cycle taking around five seconds:

```
/*
 * LedBrightness sketch
 * controls the brightness of LEDs on analog output ports
*/

const int firstLed   = 3;        // specify the pin for each of the LEDs
const int secondLed  = 5;
const int thirdLed   = 6;

int brightness = 0;
int increment = 1;

void setup()
{
  // pins driven by analogWrite do not need to be declared as outputs
}

void loop()
{
  if(brightness > 255)
  {
```

```
    increment = -1; // count down after reaching 255
  }
  else if(brightness < 1)
  {
    increment =  1; // count up after dropping back down to 0
  }
  brightness = brightness + increment; // increment (or decrement sign is minus)

  // write the brightness value to the LEDs
  analogWrite(firstLed, brightness);
  analogWrite(secondLed, brightness);
  analogWrite(thirdLed, brightness );

  delay(10); // 10ms for each step change means 2.55 secs to fade up or down
}
```

Discussion

This uses the same wiring as the previous sketch, but here the pins are controlled using `analogWrite` instead of `digitalWrite`. `analogWrite` uses PWM to control the power to the LED; see this chapter's introduction section for more on analog output.

The sketch fades the light level up and down by increasing (on fade up) or decreasing (on fade down) the value of the `brightness` variable in each pass through the loop. This value is given to the `analogWrite` function for the three connected LEDs. The minimum value for `analogWrite` is `0`—this keeps the voltage on the pin at 0. The maximum value is 255, and this keeps the pin at 5 volts.

When the `brightness` variable reaches the maximum value, it will start to decrease, because the sign of the `increment` is changed from +1 to –1 (adding –1 to a value is the same as subtracting 1 from that value).

See Also

This chapter's introduction describes how Arduino analog output works.

7.3 Driving High-Power LEDs

Problem

You need to switch or control the intensity of LEDs that need more power than the Arduino pins can provide. Arduino chips can only handle current up to 40 mA per pin.

Solution

Use a transistor to switch on and off the current flowing through the LEDs. Connect the LED as shown in Figure 7-4. You can use the same code as shown in the previous recipes (just make sure the pins connected to the transistor base match the pin number used in your sketch).

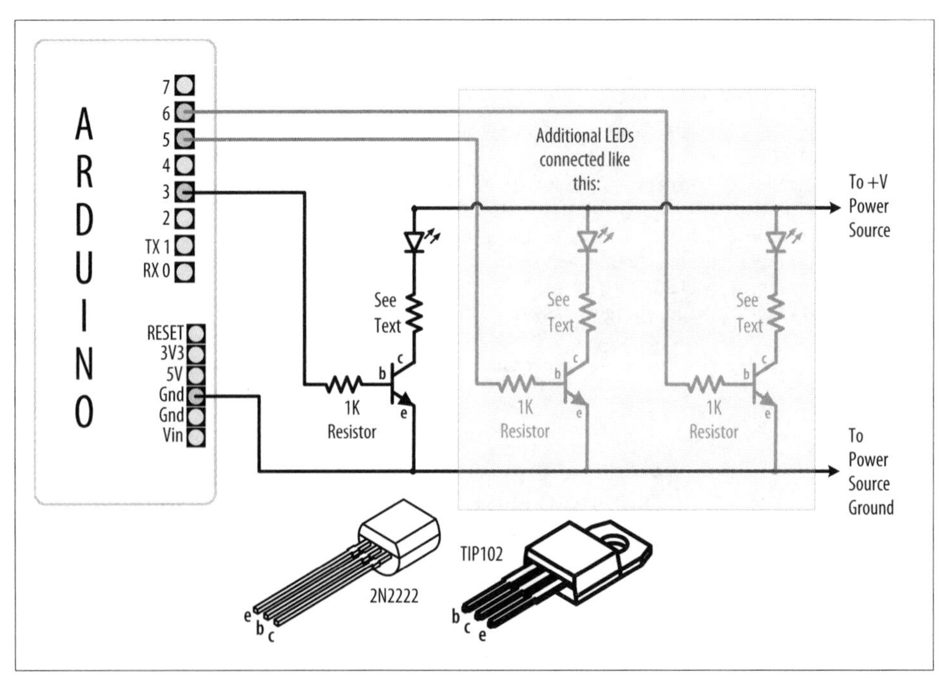

Figure 7-4. Using transistors to drive high-current LEDs

Discussion

Figure 7-4 has an arrow indicating a +V power source. This can be the Arduino +5V power pin, which can supply up to 400 mA or so if powered from USB. The available current when powered through the external power socket is dependent on the current rating and voltage of your DC power supply (the regulator dissipates excess voltage as heat—check that the on-board regulator, a 3-pin chip usually near the DC input socket, is not too hot to the touch). If more current is required than the Arduino +5V can provide, you need a power source separate from the Arduino to drive the LEDs. See Appendix C for information on using an external power supply.

If you're using an external power supply, remember to connect the ground of the external supply to the Arduino ground.

Current is allowed to flow from the collector to the emitter when the transistor is switched on. No significant current flows when the transistor is off. The Arduino can turn a transistor on by making the voltage on a pin `HIGH` with `digitalWrite`. A resistor is necessary between the pin and the transistor base to prevent too much current from flowing—1K ohms is a typical value (this provides 5 mA of current to the base of the transistor). See Appendix B for advice on how to read a data sheet and pick and use a transistor. You can also use specialized integrated circuits such as the ULN2003A for driving multiple outputs. These contain seven high-current (0.5 amp) output drivers.

The resistor used to limit the current flow through the LED is calculated using the technique given in Recipe 7.1, but you may need to take into account that the source voltage will be reduced slightly because of the small voltage drop through the transistor. This will usually be less than three-fourths of a volt (the actual value can be found by looking at collector-emitter saturation voltage; see Appendix B). High-current LEDs (1 watt or more) are best driven using a constant current source (a circuit that actively controls the current) to manage the current through the LED.

How to Exceed 40 mA per Pin

You can also connect multiple pins in parallel to increase current beyond the 40 mA per pin rating (see "Maximum pin current" on page 244).

Figure 7-5 shows how to connect an LED that can be driven with 60 mA through two pins. This shows the LED connecting the resistors to ground through pins 2 and 7—both pins need to be `LOW` for the full 60 mA to flow through the LED. The separate resistors are needed; don't try to use a single resistor to connect the two pins.

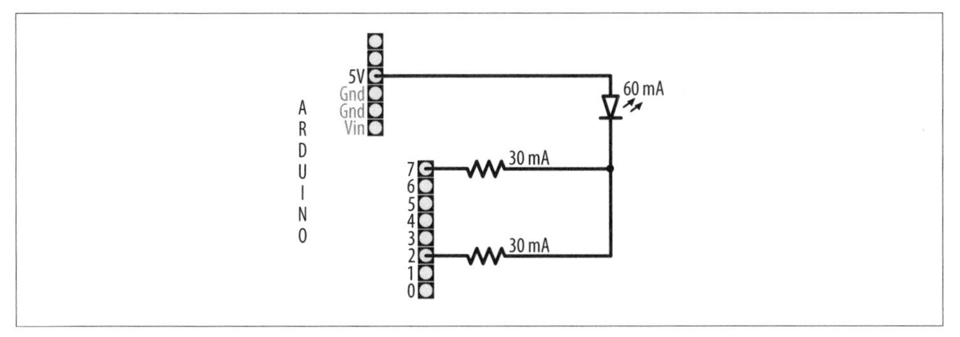

Figure 7-5. How to exceed 40 mA per pin

This technique can also be used to source current. For example, flip the LED around—connect the lead that was going to the resistors (cathode) to GND and the other end (anode) to the resistors—and you illuminate the LED by setting both pins to `HIGH`.

It is best if you use pins that are not adjacent to minimize stress on the chip. This technique works for any pin using `digitalWrite`; it does not work with `analogWrite`—if you need more current for analog outputs (PWM), you will need to use transistors as explained above.

See Also

Web reference for constant current drivers: *http://blog.makezine.com/archive/2009/08/constant_current_led_driver.html*

7.4 Adjusting the Color of an LED

Problem

You want to control the color of an RGB LED under program control.

Solution

RGB LEDs have red, green, and blue elements in a single package, with either the anodes connected together (known as *common anode*) or the cathodes connected together (known as *common cathode*). Use the wiring in Figure 7-6 for common anode (the anodes are connected to +5 volts and the cathodes are connected to pins). Use Figure 7-2 if your RGB LEDs are common cathode.

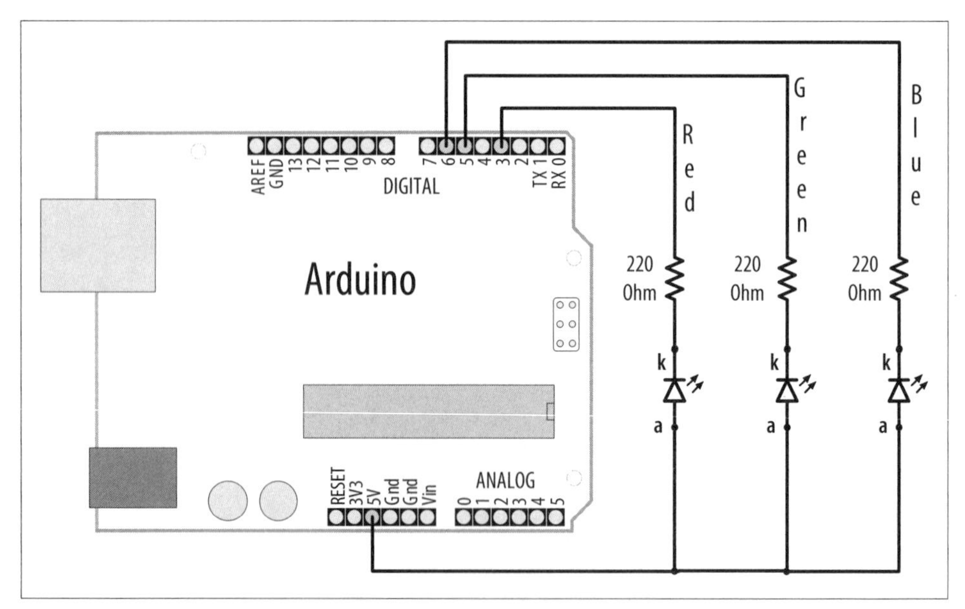

Figure 7-6. RGB connections (common anode)

This sketch continuously fades through the color spectrum by varying the intensity of the red, green, and blue elements:

```
/*
 * RGB_LEDs sketch
 * RGB LEDs driven from analog output ports
```

```
*/

const int redPin   = 3;        // choose the pin for each of the LEDs
const int greenPin = 5;
const int bluePin  = 6;
const boolean invert = true; // set true if common anode, false if common cathode

int color = 0; // a value from 0 to 255 representing the hue
int R, G, B;  // the Red Green and Blue color components

void setup()
{
  // pins driven by analogWrite do not need to be declared as outputs
}

void loop()
{
  int brightness = 255; // 255 is maximum brightness
  hueToRGB( color, brightness);  // call function to convert hue to RGB
  // write the RGB values to the pins
  analogWrite(redPin, R);
  analogWrite(greenPin, G);
  analogWrite(bluePin, B );

  color++;             // increment the color
  if(color > 255)      //
     color = 0;
       delay(10);
}

// function to convert a color to its Red, Green, and Blue components.

void hueToRGB( int hue, int brightness)
{
    unsigned int scaledHue = (hue * 6);
    // segment 0 to 5 around the color wheel
    unsigned int segment = scaledHue / 256;
    // position within the segment
    unsigned int segmentOffset = scaledHue - (segment * 256);

    unsigned int complement = 0;
    unsigned int prev = (brightness * ( 255 -  segmentOffset)) / 256;
    unsigned int next = (brightness *  segmentOffset) / 256;
    if(invert)
    {
      brightness = 255-brightness;
      complement = 255;
      prev = 255-prev;
      next = 255-next;
    }

    switch(segment ) {
    case 0:      // red
     R = brightness;
         G = next;
```

```
            B = complement;
            break;
            case 1:     // yellow
            R = prev;
            G = brightness;
            B = complement;
            break;
            case 2:     // green
            R = complement;
            G = brightness;
            B = next;
            break;
            case 3:     // cyan
            R = complement;
            G = prev;
            B = brightness;
            break;
            case 4:     // blue
            R = next;
            G = complement;
            B = brightness;
            break;
           case 5:      // magenta
            default:
            R = brightness;
            G = complement;
            B = prev;
            break;
            }
    }
```

Discussion

The color of an RGB LED is determined by the relative intensity of its red, green, and blue elements. The core function in the sketch (`hueToRGB`) handles the conversion of a hue value ranging from 0 to 255 into a corresponding color ranging from red to blue. The spectrum of visible colors is often represented using a color wheel consisting of the primary and secondary colors with their intermediate gradients. The spokes of the color wheel representing the six primary and secondary colors are handled by six `case` statements. The code in a `case` statement is executed if the segment variable matches the case number, and if so, the RGB values are set as appropriate for each. Segment 0 is red, segment 1 is yellow, segment 2 is green, and so on.

If you also want to adjust the brightness, you can reduce the value of the `brightness` variable. The following shows how to adjust the brightness with a variable resistor or sensor connected as shown in Figure 7-13 or Figure 7-17:

```
int brightness = map( analogRead(0),0,1023, 0, 255); // get brightness from sensor
```

The `brightness` variable will range in value from 0 to 255 as the analog input ranges from 0 to 1,023, causing the LED to increase brightness as the value increases.

See Also

Recipe 2.16; Recipe 13.1

7.5 Sequencing Multiple LEDs: Creating a Bar Graph

Problem

You want an LED bar graph that lights LEDs in proportion to a value in your sketch or a value read from a sensor.

Solution

You can connect the LEDs as shown in Figure 7-2 (using additional pins if you want more LEDs). Figure 7-7 shows six LEDs connected on consecutive pins.

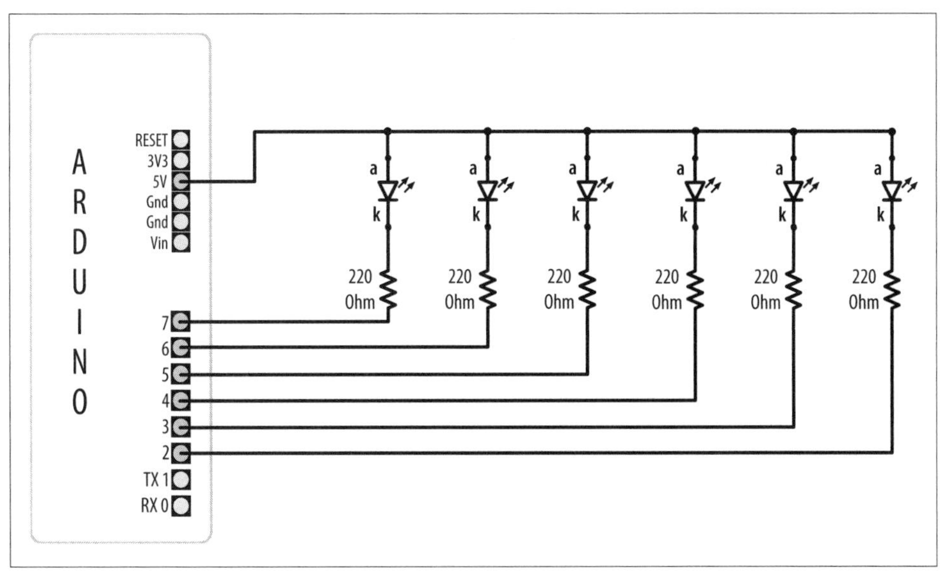

Figure 7-7. Six LEDs with cathodes connected to Arduino pins

The following sketch turns on a series of LEDs, with the number being proportional to the value of a sensor connected to an analog input port (see Figure 7-13 or Figure 7-17 to see how a sensor is connected):

```
/*
  Bargraph sketch

 Turns on a series of LEDs proportional to a value of an analog sensor.
 Six LEDs are controlled but you can change the number of LEDs by changing
 the value of NbrLEDs and adding the pins to the ledPins array
 */
```

```
const int NbrLEDs = 6;
const int ledPins[] = { 2, 3, 4, 5, 6, 7};
const int analogInPin = 0;  // Analog input pin connected to variable resistor
const int wait = 30;

// Swap values of the following two constants if cathodes are connected to Gnd
const boolean LED_ON = LOW;
const boolean LED_OFF = HIGH;

int sensorValue = 0;          // value read from the sensor
int ledLevel = 0;             // sensor value converted into LED 'bars'

void setup() {
  for (int led = 0; led < NbrLEDs; led++)
  {
    pinMode(ledPins[led], OUTPUT);  // make all the LED pins outputs
  }
}

void loop() {
  sensorValue = analogRead(analogInPin);              // read the analog in value
  ledLevel = map(sensorValue, 0, 1023, 0, NbrLEDs);  // map to the number of LEDs
  for (int led = 0; led < NbrLEDs; led++)
  {
    if (led < ledLevel ) {
      digitalWrite(ledPins[led], LED_ON);      // turn on pins less than the level
    }
    else {
      digitalWrite(ledPins[led], LED_OFF); // turn off pins higher than the level
    }
  }
}
```

Discussion

The pins connected to LEDs are held in the array `ledPins`. To change the number of LEDs, you can add (or remove) elements from this array, but make sure the variable `NbrLEDs` is the same as the number of elements (which should be the same as the number of pins). You can have the compiler calculate the value of `NbrLEDs` for you by replacing this line:

```
const int NbrLEDs = 6;
```

with this line:

```
const int NbrLEDs = sizeof(ledPins) / sizof(ledPins[0];
```

The `sizeof` function returns the size (number of bytes) of a variable—in this case, the number of bytes in the `ledPins` array. Because it is an array of integers (with two bytes per element), the total number of bytes in the array is divided by the size of one element (`sizeof(ledPins[0])`) and this gives the number of elements.

The Arduino map function is used to calculate the number of LEDs that should be lit as a proportion of the sensor value. The code loops through each LED, turning it on if the proportional value of the sensor is greater than the LED number. For example, if the sensor value is 0, no pins are lit; if the sensor is at half value, half are lit. When the sensor is at maximum value, all the LEDs are lit.

Figure 7-7 shows all the anodes connected together (known as *common anode*) and the cathodes connected to the pins; the pins need to be LOW for the LED to light. If the LEDs have the anodes connected to pins (as shown in Figure 7-2) and the cathodes are connected together (known as *common cathode*), the LED is lit when the pin goes HIGH. The sketch in this recipe uses the constant names LED_ON and LED_OFF to make it easy to select common anode or common cathode connections. To change the sketch for common cathode connection, swap the values of these constants as follows:

```
const boolean LED_ON = HIGH;  // HIGH is on when using common cathode connection
const boolean LED_OFF = LOW;
```

You may want to slow down the *decay* (rate of change) in the lights; for example, to emulate the movement of the indicator of a sound volume meter. Here is a variation on the sketch that slowly decays the LED bars when the level drops:

```
/*
   LED bar graph - decay version
*/

const int ledPins[] = { 2, 3, 4, 5, 6, 7};
const int NbrLEDs = sizeof(ledPins) / sizof(ledPins[0];
const int analogInPin = 0; // Analog input pin connected to variable resistor
const int decay = 10;      // increasing this reduces decay rate of storedValue

int sensorValue = 0;       // value read from the sensor
int storedValue = 0;       // the stored (decaying) sensor value
int ledLevel = 0;          // value converted into LED 'bars'

void setup() {
  for (int led = 0; led < NbrLEDs; led++)
  {
    pinMode(ledPins[led], OUTPUT);  // make all the LED pins outputs
  }
}


void loop() {
  sensorValue = analogRead(analogInPin);        // read the analog in value
  storedValue = max(sensorValue, storedValue);  // use sensor value if higher
  ledLevel = map(storedValue, 0, 1023, 0, NbrLEDs); // map to number of LEDs
  for (int led = 0; led < NbrLEDs; led++)
  {
    if (led < ledLevel ) {
      digitalWrite(ledPins[led], HIGH);  // turn on pins less than the level
    }
    else {
      digitalWrite(ledPins[led], LOW);   // turn off pins higher than the level
```

```
    }
  }
  storedValue = storedValue - decay;      // decay the value
  delay(10);                              // wait 10 ms before next loop
}
```

The decay is handled by the line that uses the `max` function. This returns either the sensor value or the stored decayed value, whichever is higher. If the sensor is higher than the decayed value, this is saved in `storedValue`. Otherwise, the level of `storedValue` is reduced by the constant `decay` each time through the loop (set to 10 milliseconds by the `delay` function). Increasing the value of the `decay` constant will reduce the time for the LEDs to fade to all off.

See Also

Recipe 3.6 explains the `max` function.

Recipe 5.6 has more on reading a sensor with the `analogRead` function.

Recipe 5.7 describes the `map` function.

See Recipes 12.1 and 12.2 if you need greater precision in your decay times. The total time through the loop is actually greater than 10 milliseconds because it takes an additional millisecond or so to execute the rest of the loop code.

7.6 Sequencing Multiple LEDs: Making a Chase Sequence (Knight Rider)

Problem

You want to light LEDs in a "chasing lights" sequence (as seen on the TV show *Knight Rider*).

Solution

You can use the same connection as shown in Figure 7-7:

```
/* KnightRider
 */

const int NbrLEDs = 6;
const int ledPins[] = {2, 3, 4, 5, 6, 7};
const int wait = 30;

void setup(){
  for (int led = 0; led < NbrLEDs; led++)
  {
    pinMode(ledPins[led], OUTPUT);
  }
```

```
}

void loop() {
  for (int led = 0; led < NbrLEDs-1; led++)
  {
    digitalWrite(ledPins[led], HIGH);
    delay(wait);
    digitalWrite(ledPins[led + 1], HIGH);
    delay(wait);
    digitalWrite(ledPins[led], LOW);
    delay(wait*2);
  }
    for (int led = NbrLEDs-1; led > 0; led--) {
    digitalWrite(ledPins[led], HIGH);
    delay(wait);
    digitalWrite(ledPins[led - 1], HIGH);
    delay(wait);
    digitalWrite(ledPins[led], LOW);
    delay(wait*2);
  }
}
```

Discussion

This code is similar to the code in Recipe 7.5, except the pins are turned on and off in a fixed sequence rather than depending on a sensor level. There are two `for` loops; the first produces the left-to-right pattern by lighting up LEDs from left to right. This loop starts with the first (leftmost) LED and steps through adjacent LEDs until it reaches and illuminates the rightmost LED. The second `for` loop lights the LEDs from right to left by starting at the rightmost LED and decrementing (decreasing by one) the LED that is lit until it gets to the first (rightmost) LED. The delay period is set by the `wait` variable and can be chosen to provide the most pleasing appearance.

7.7 Controlling an LED Matrix Using Multiplexing

Problem

You have a matrix of LEDs and want to minimize the number of Arduino pins needed to turn LEDs on and off.

Solution

This sketch uses an LED matrix of 64 LEDs, with anodes connected in rows and cathodes in columns (as in the Jameco 2132349). Dual-color LED displays may be easier to obtain, and you can drive just one of the colors if that is all you need (Figure 7-8 shows the connections):

```
/*
  matrixMpx sketch

 Sequence LEDs starting from first column and row until all LEDS are lit
 Multiplexing is used to control 64 LEDs with 16 pins
 */

const int columnPins[] = { 2, 3, 4, 5, 6, 7, 8, 9};
const int rowPins[]   = { 10,11,12,15,16,17,18,19};

int pixel       = 0;        // 0 to 63 LEDs in the matrix
int columnLevel = 0;        // pixel value converted into LED column
int rowLevel    = 0;        // pixel value converted into LED row

void setup() {
  for (int i = 0; i < 8; i++)
  {
    pinMode(columnPins[i], OUTPUT);  // make all the LED pins outputs
    pinMode(rowPins[i], OUTPUT);
  }
}

void loop() {
  pixel = pixel + 1;
  if(pixel > 63)
     pixel = 0;

  columnLevel = pixel / 8;                    // map to the number of columns
  rowLevel = pixel % 8;                       // get the fractional value
  for (int column = 0; column < 8; column++)
  {
    digitalWrite(columnPins[column], LOW);     // connect this column to Ground
    for(int row = 0; row < 8; row++)
    {
      if (columnLevel > column)
      {
        digitalWrite(rowPins[row], HIGH);  // connect all LEDs in row to +5 volts
      }
      else if (columnLevel == column && rowLevel >= row)
      {
          digitalWrite(rowPins[row], HIGH);
      }
      else
      {
        digitalWrite(columnPins[column], LOW); // turn off all LEDs in this row
      }
      delayMicroseconds(300);     // delay gives frame time of 20ms for 64 LEDs
      digitalWrite(rowPins[row], LOW);         // turn off LED
    }

    // disconnect this column from Ground
    digitalWrite(columnPins[column], HIGH);
  }
}
```

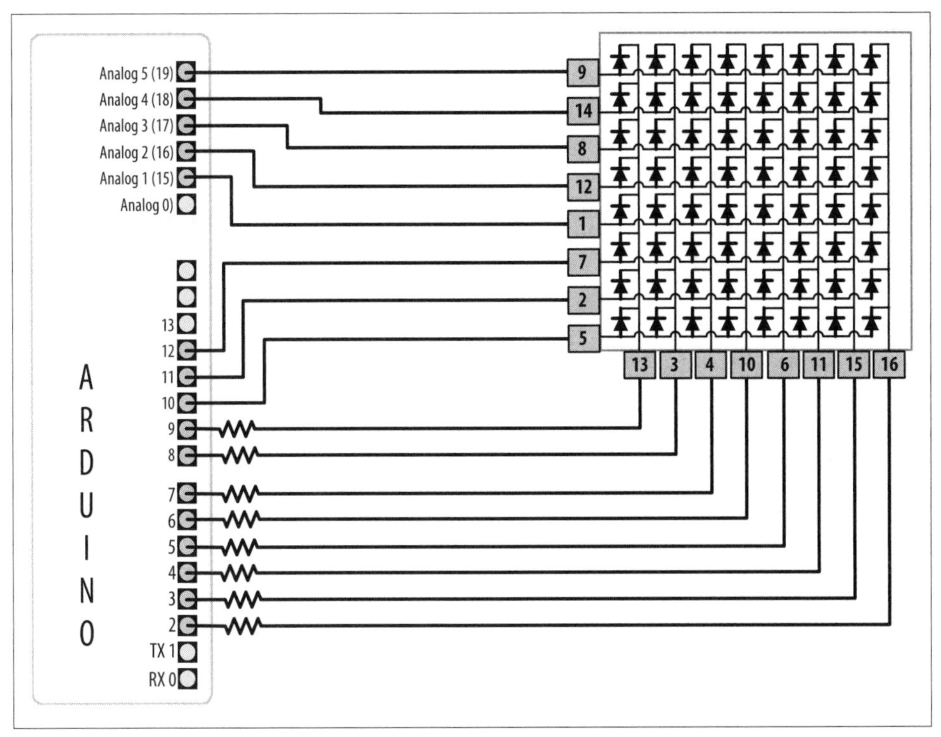

Figure 7-8. An LED matrix connected to 16 digital pins

LED matrix displays do not have a standard pinout, so you must check the data sheet for your display. Wire the rows of anodes and columns of cathodes as shown in Figure 7-15 or Figure 7-16, but use the LED pin numbers shown in your data sheet.

Discussion

The resistor's value must be chosen to ensure that the maximum current through a pin does not exceed 40 mA. Because the current for up to eight LEDs can flow through each column pin, the maximum current for each LED must be one-eighth of 40 mA, or 5 mA. Each LED in a typical small red matrix has a forward voltage of around 1.8 volts. Calculating the resistor that results in 5 mA with a forward voltage of 1.8 volts gives a value of 680 ohms. Check your data sheet to find the forward voltage of the matrix you want to use. Each column of the matrix is connected through the series resistor to a digital pin. When the column pin goes low and a row pin goes high, the corresponding LED will light. For all LEDs where the column pin is high or its row pin is low, no current will flow through the LED and it will not light.

The `for` loop scans through each row and column and turns on sequential LEDs until all LEDs are lit. The loop starts with the first column and row and increments the row counter until all LEDs in that row are lit; it then moves to the next column, and so on, lighting another LED with each pass through the loop until all the LEDs are lit.

You can control the number of lit LEDs in proportion to the value from a sensor (see Recipe 5.6 for connecting a sensor to the analog port) by making the following changes to the sketch.

Comment out or remove these three lines from the beginning of the loop:

```
pixel = pixel + 1;
if(pixel > 63)
   pixel = 0;
```

Replace them with the following lines that read the value of a sensor on pin 0 and map this to a number of pixels ranging from 0 to 63:

```
int sensorValue = analogRead(0);               // read the analog in value
pixel =   map(sensorValue, 0, 1023, 0, 63);    // map sensor value to pixel (LED)
```

You can test this with a variable resistor connected to analog input pin 0 connected as shown in Figure 5-7 in Chapter 5. The number of LEDs lit will be proportional to the value of the sensor.

7.8 Displaying Images on an LED Matrix

Problem

You want to display one or more images on an LED matrix, perhaps creating an animation effect by quickly alternating multiple images.

Solution

This Solution can use the same wiring as in Recipe 7.7. The sketch creates the effect of a heart beating by briefly lighting LEDs arranged in the shape of a heart. A small heart followed by a larger heart is flashed for each heartbeat (the images look like Figure 7-9):

```
/*
 * matrixMpxAnimation sketch
 * animates two heart images to show a beating heart
 */

// the heart images are stored as bitmaps - each bit corresponds to an LED
// a 0 indicates the LED is off, 1 is on
byte bigHeart[] = {
  B01100110,
  B11111111,
  B11111111,
  B11111111,
  B01111110,
  B00111100,
  B00011000,
  B00000000};
```

```
byte smallHeart[] = {
  B00000000,
  B00000000,
  B00010100,
  B00111110,
  B00111110,
  B00011100,
  B00001000,
  B00000000};


const int columnPins[] = {  2, 3, 4, 5, 6, 7, 8, 9};
const int rowPins[]    = { 10,11,12,15,16,17,18,19};

void setup() {
  for (int i = 0; i < 8; i++)
  {
    pinMode(rowPins[i], OUTPUT);        // make all the LED pins outputs
    pinMode(columnPins[i], OUTPUT);
    digitalWrite(columnPins[i], HIGH);  // disconnect column pins from Ground
  }
}

void loop() {
  int pulseDelay = 800 ;          // milliseconds to wait between beats

  show(smallHeart, 80);           // show the small heart image for 100 ms
  show(bigHeart, 160);            // followed by the big heart for 200ms
  delay(pulseDelay);              // show nothing between beats
}


// routine to show a frame of an image stored in the array pointed to by the
// image parameter.
// the frame is repeated for the given duration in milliseconds
void show( byte * image, unsigned long duration)
{
 unsigned long start = millis();            // begin timing the animation
 while (start + duration > millis())        // loop until the duration period
has passed
  {
    for(int row = 0; row < 8; row++)
    {
      digitalWrite(rowPins[row], HIGH);          // connect row to +5 volts
      for(int column = 0; column < 8; column++)
      {
        boolean pixel = bitRead(image[row],column);
        if(pixel == 1)
        {
          digitalWrite(columnPins[column], LOW);  // connect column to Gnd
        }
        delayMicroseconds(300);                   // a small delay for each LED
        digitalWrite(columnPins[column], HIGH);   // disconnect column from Gnd
      }
```

```
        digitalWrite(rowPins[row], LOW);            // disconnect LEDs
      }
    }
  }
```

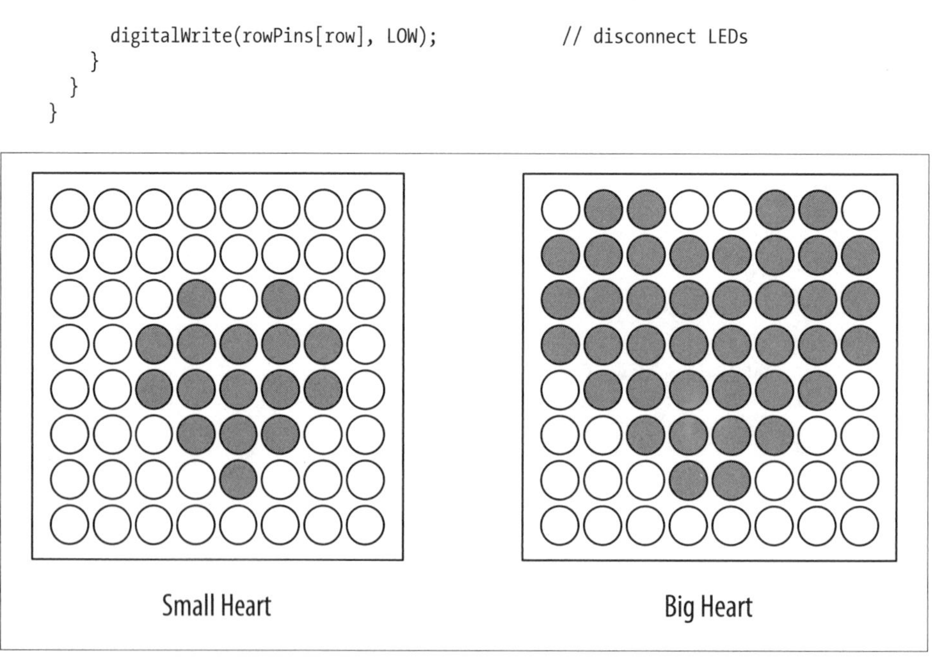

Figure 7-9. The two heart images displayed on each beat

Discussion

Columns and rows are multiplexed (switched) similar to Recipe 7.7, but here the value written to the LED is based on images stored in the `bigHeart` and `smallHeart` arrays. Each element in the array represents a *pixel* (a single LED) and each array row represents a row in the matrix. A row consists of eight bits represented using binary format (as designated by the capital *B* at the start of each row). A bit with a value of `1` indicates that the corresponding LED should be on; a `0` means off. The animation effect is created by rapidly switching between the arrays.

The `loop` function waits a short time (800 milliseconds) between beats and then calls the `show` function, first with the `smallHeart` array and then followed by the `bigHeart` array. The `show` function steps through each element in all the rows and columns, lighting the LED if the corresponding bit is 1. The `bitRead` function (see Recipe 2.20) is used to determine the value of each bit.

A short delay of 300 microseconds between each pixel allows the eye enough time to perceive the LED. The timing is chosen to allow each image to repeat quickly enough (50 times per second) so that blinking is not visible.

Here is a variation that changes the rate at which the heart beats, based on the value from a sensor. You can test this using a variable resistor connected to analog input pin 0, as shown in Recipe 5.6. Use the wiring and code shown earlier, except replace the `loop` function with this code:

```
void loop() {
  sensorValue = analogRead(analogInPin);            // read the analog in value
  int pulseRate =  map(sensorValue,0,1023,40,240); // convert to beats / minute
  int pulseDelay = (60000 / pulseRate );  // milliseconds to wait between beats

  show(smallHeart, 80);             // show the small heart image for 100 ms
  show(bigHeart, 160);              // followed by the big heart for 200ms
  delay(pulseDelay);                // show nothing between beats
}
```

This version calculates the delay between pulses using the `map` function (see Recipe 5.7) to convert the sensor value into beats per minute. The calculation does not account for the time it takes to display the heart, but you can subtract 240 milliseconds (80 ms plus 160 ms for the two images) if you want more accurate timing.

See Also

See Recipes 7.12 and 7.13 for information on how to use shift registers to drive LEDs if you want to reduce the number of Arduino pins needed for driving an LED matrix.

Recipes 12.1 and 12.2 provide more information on how to manage time using the `millis` function.

7.9 Controlling a Matrix of LEDs: Charlieplexing

Problem

You have a matrix of LEDs and you want to minimize the number of pins needed to turn any of them on and off.

Solution

Charlieplexing is a special kind of multiplexing that increases the number of LEDs that can be driven by a group of pins. This sketch sequences through six LEDs using just three pins (Figure 7-10 shows the connections):

```
/*
 * Charlieplexing sketch
 * light six LEDs in sequence that are connected to pins 2, 3, and 4
 */

byte pins[] = {2,3,4};  // the pins that are connected to LEDs

// the next two lines infer number of pins and LEDs from the above array
const int NUMBER_OF_PINS = sizeof(pins)/ sizeof(pins[0]);
```

```
const int NUMBER_OF_LEDS = NUMBER_OF_PINS * (NUMBER_OF_PINS-1);

byte pairs[NUMBER_OF_LEDS/2][2] = { {0,1}, {1,2}, {0,2} }; // maps pins to LEDs


void setup()
{
   // nothing needed here
}

void loop(){
    for(int i=0; i < NUMBER_OF_LEDS; i++)
    {
       lightLed(i);  // light each LED in turn
       delay(1000);
    }
}


// this function lights the given LED, the first LED is 0
void lightLed(int led)
{
  // the following four lines convert LED number to pin numbers
  int indexA = pairs[led/2][0];
  int indexB = pairs[led/2][1];
  int pinA = pins[indexA];
  int pinB = pins[indexB];

  // turn off all pins not connected to the given LED
  for(int i=0; i < NUMBER_OF_PINS; i++)
    if( pins[i] != pinA && pins[i] != pinB)
    {  // if this pin is not one of our pins
        pinMode(pins[i], INPUT);   // set the mode to input
        digitalWrite(pins[i],LOW); // make sure pull-up is off
    }
  // now turn on the pins for the given LED
  pinMode(pinA, OUTPUT);
  pinMode(pinB, OUTPUT);
  if( led % 2 == 0)
  {
     digitalWrite(pinA,LOW);
     digitalWrite(pinB,HIGH);
  }
  else
  {
     digitalWrite(pinB,LOW);
     digitalWrite(pinA,HIGH);
  }
}
```

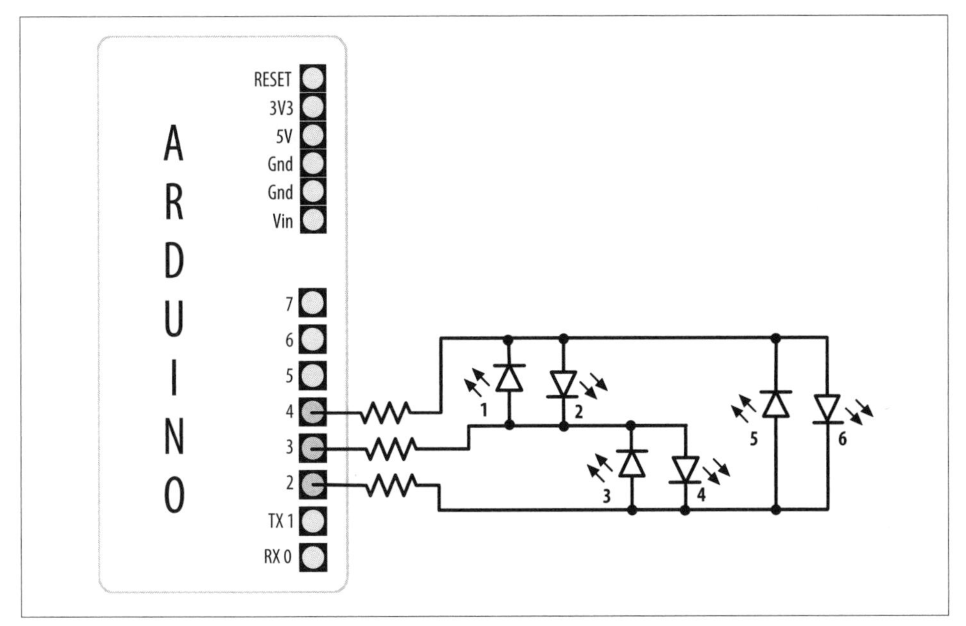

Figure 7-10. Six LEDs driven through three pins using Charlieplexing

Discussion

The term *Charlieplexing* comes from Charlie Allen (of Microchip Technology, Inc.), who published the method. The technique is based on the fact that LEDs only turn on when connected the "right way" around (with the anode more positive than the cathode). Here is the table showing the LED number (see Figure 7-8) that is lit for the valid combinations of the three pins. `L` is `LOW`, `H` is `HIGH`, and `i` is `INPUT` mode. Setting a pin in `INPUT` mode effectively disconnects it from the circuit:

```
  Pins      LEDs
4 3 2    1 2 3 4 5 6
L L L    0 0 0 0 0 0
L H i    1 0 0 0 0 0
H L i    0 1 0 0 0 0
i L H    0 0 1 0 0 0
i H L    0 0 0 1 0 0
L i H    0 0 0 0 1 0
H i L    0 0 0 0 0 1
```

You can double the number of LEDs to 12 using just one more pin. The first six LEDs are connected in the same way as in the preceding example; add the additional six LEDs so that the connections look like Figure 7-11.

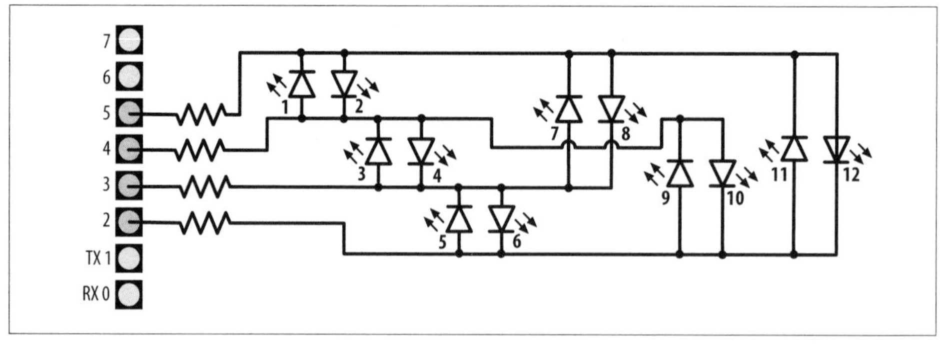

Figure 7-11. Charlieplexing using four pins to drive 12 LEDs

Modify the preceding sketch by adding the extra pin to the `pins` array:

```
byte pins[] = {2,3,4,5};  // the pins that are connected to LEDs
```

Add the extra entries to the `pairs` array so that it reads as follows:

```
byte pairs[NUMBER_OF_LEDS/2][2] = { {0,1}, {1,2}, {0,2}, {2,3}, {1,3}, {0,3} };
```

Everything else can remain the same, so the loop will sequence through all 12 LEDs because the code determines the number of LEDs from the number of entries in the `pins` array.

Because Charlieplexing works by controlling the Arduino pins so that only a single LED is turned on at a time, it is more complicated to create the impression of lighting multiple LEDs. But you can light multiple LEDs using a multiplexing technique modified for Charlieplexing.

This sketch creates a bar graph by lighting a sequence of LEDs based on the value of a sensor connected to analog pin 0:

```
byte pins[] = {2,3,4};
const int NUMBER_OF_PINS = sizeof(pins)/ sizeof(pins[0]);
const int NUMBER_OF_LEDS = NUMBER_OF_PINS * (NUMBER_OF_PINS-1);

byte pairs[NUMBER_OF_LEDS/2][2] = { {0,1}, {1,2}, {0,2} };

int ledStates = 0; //holds states for up to 15 LEDs
int refreshedLed;  // the LED that gets refreshed

void setup()
{
   // nothing here
}

void loop()
{
const int analogInPin = 0; // Analog input pin connected to the variable resistor
```

```
  // here is the code from the bargraph recipe
  int sensorValue = analogRead(analogInPin);          // read the analog in value
  // map to the number of LEDs
  int ledLevel = map(sensorValue, 0, 1023, 0, NUMBER_OF_LEDS);
  for (int led = 0; led < NUMBER_OF_LEDS; led++)
  {
    if (led < ledLevel ) {
      setState(led, HIGH);     // turn on pins less than the level
    }
    else {
      setState(led, LOW);      // turn off pins higher than the level
    }
  }
  ledRefresh();

}

void setState( int led, boolean state)
{
   bitWrite(ledStates,led, state);
}

void ledRefresh()
{
   // refresh a different LED each time this is called.
   if( refreshedLed++ > NUMBER_OF_LEDS) // increment to the next LED
      refreshedLed = 0; // repeat from the first LED if all have been refreshed

   if( bitRead(ledStates, refreshedLed ) == HIGH)
         lightLed( refreshedLed );
}

// this function is identical to the sketch above
// it lights the given LED, the first LED is 0
void lightLed(int led)
{
  // the following four lines convert LED number to pin numbers
  int indexA = pairs[led/2][0];
  int indexB = pairs[led/2][1];
  int pinA = pins[indexA];
  int pinB = pins[indexB];

  // turn off all pins not connected to the given LED
  for(int i=0; i < NUMBER_OF_PINS; i++)
    if( pins[i] != pinA && pins[i] != pinB)
    {  // if this pin is not one of our pins
        pinMode(pins[i], INPUT);    // set the mode to input
        digitalWrite(pins[i],LOW); // make sure pull-up is off
    }
  // now turn on the pins for the given LED
  pinMode(pinA, OUTPUT);
  pinMode(pinB, OUTPUT);
```

```
    if( led % 2 == 0)
    {
      digitalWrite(pinA,LOW);
      digitalWrite(pinB,HIGH);
    }
    else
    {
      digitalWrite(pinB,LOW);
      digitalWrite(pinA,HIGH);
    }
  }
```

This sketch uses the value of the bits in the variable `ledStates` to represent the state of the LEDs (0 if off, 1 if on). The `refresh` function checks each bit and lights the LEDs for each bit that is set to 1. The `refresh` function must be called quickly and repeatedly, or the LEDs will appear to blink.

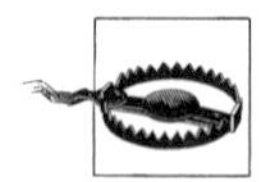

Adding delays into your code can interfere with the "persistence of vision" effect that creates the illusion that hides the flashing of the LEDs.

You can use an interrupt to service the `refresh` function in the background (without needing to explicitly call the function in `loop`). Timer interrupts are covered in Chapter 18, but here is a preview of one approach for using an interrupt to service your LED refreshes. This uses a third-party library called FrequencyTimer2 (available from the Arduino Playground) to create the interrupt:

```
#include <FrequencyTimer2.h>

byte pins[] = {2,3,4,5};
const int NUMBER_OF_PINS = sizeof(pins)/ sizeof(pins[0]);
const int NUMBER_OF_LEDS = NUMBER_OF_PINS * (NUMBER_OF_PINS-1);

byte pairs[NUMBER_OF_LEDS/2][2] = { {0,1}, {1,2}, {0,2} };

int ledStates = 0; //holds states for up to 15 LEDs
int refreshedLed;  // the LED that gets refreshed

---
#include <FrequencyTimer2.h>  // include this library to handle the refresh


byte pins[] = {2,3,4};
const int NUMBER_OF_PINS = sizeof(pins)/ sizeof(pins[0]);
const int NUMBER_OF_LEDS = NUMBER_OF_PINS * (NUMBER_OF_PINS-1);

byte pairs[NUMBER_OF_LEDS/2][2] = { {0,1}, {1,2}, {0,2} };

int ledStates = 0; //holds states for up to 15 LEDs
int refreshedLed;  // the LED that gets refreshed
```

```
void setup()
{
  FrequencyTimer2::setPeriod(20000/ NUMBER_OF_LEDS); // set the period
  // the next line tells FrequencyTimer2 the function to call (ledRefresh)
  FrequencyTimer2::setOnOverflow(ledRefresh);
  FrequencyTimer2::enable();
}

void loop()
{
const int analogInPin = 0; // Analog input pin connected to the variable resistor

  // here is the code from the bargraph recipe
  int sensorValue = analogRead(analogInPin);          // read the analog in value
  // map to the number of LEDs
  int ledLevel = map(sensorValue, 0, 1023, 0, NUMBER_OF_LEDS);
  for (int led = 0; led < NUMBER_OF_LEDS; led++)
  {
    if (led < ledLevel ) {
      setState(led, HIGH);     // turn on pins less than the level
    }
    else {
      setState(led, LOW);      // turn off pins higher than the level
    }
  }
   // the LED is no longer refreshed in loop, it's handled by FrequencyTimer2
}

// the remaining code is the same as the previous example
```

The FrequencyTimer2 library has the period set to 1,666 microseconds (20 ms divided by 12, the number of LEDs). The `FrequencyTimer2setOnOverflow` method gets the function to call (`ledRefresh`) each time the timer "triggers."

See Also

The LOL board is an Arduino shield that drives a 9×14 matrix (126 LEDs) using Charlieplexing. It is an inspiration for what can be achieved using this technique when hardware and software is stretched beyond conventional design constraints: *http://jimmieprodgers.com/kits/lolshield/makelolshield/*.

Chapter 18 provides more information on timer interrupts.

7.10 Driving a 7-Segment LED Display

Problem

You want to display numerals using a 7-segment numeric display.

Solution

The following sketch displays numerals from 0 to 9 on a single-digit, 7-segment display. Figure 7-12 shows the connections. The output is produced by turning on combinations of segments that represent the numerals:

```
/*
 * SevenSegment sketch
 * Shows numerals ranging from 0 through 9 on a single-digit display
 * This example counts seconds from 0 to 9
 */

// bits representing segments A through G (and decimal point) for numerals 0-9
const byte numeral[10] = {
  //ABCDEFG /dp
  B11111100,  // 0
  B01100000,  // 1
  B11011010,  // 2
  B11110010,  // 3
  B01100110,  // 4
  B10110110,  // 5
  B00111110,  // 6
  B11100000,  // 7
  B11111110,  // 8
  B11100110,  // 9
};

// pins for decimal point and each segment
//                          dp,G,F,E,D,C,B,A
const int segmentPins[8] = { 5,9,8,7,6,4,3,2};

void setup()
{
  for(int i=0; i < 8; i++)
  {
    pinMode(segmentPins[i], OUTPUT); // set segment and DP pins to output
  }
}


void loop()
{
  for(int i=0; i <= 10; i++)
  {
    showDigit(i);
    delay(1000);
  }
  // the last value if i is 10 and this will turn the display off
  delay(2000);  // pause two seconds with the display off
}

// Displays a number from 0 through 9 on a 7-segment display
// any value not within the range of 0-9 turns the display off
void showDigit( int number)
{
```

```
    boolean isBitSet;

    for(int segment = 1; segment < 8; segment++)
    {
      if( number < 0 || number > 9){
        isBitSet = 0;   // turn off all segments
      }
      else{
        // isBitSet will be true if given bit is 1
        isBitSet = bitRead(numeral[number], segment);
      }
      isBitSet = ! isBitSet; // remove this line if common cathode display
      digitalWrite( segmentPins[segment], isBitSet);
    }
  }
```

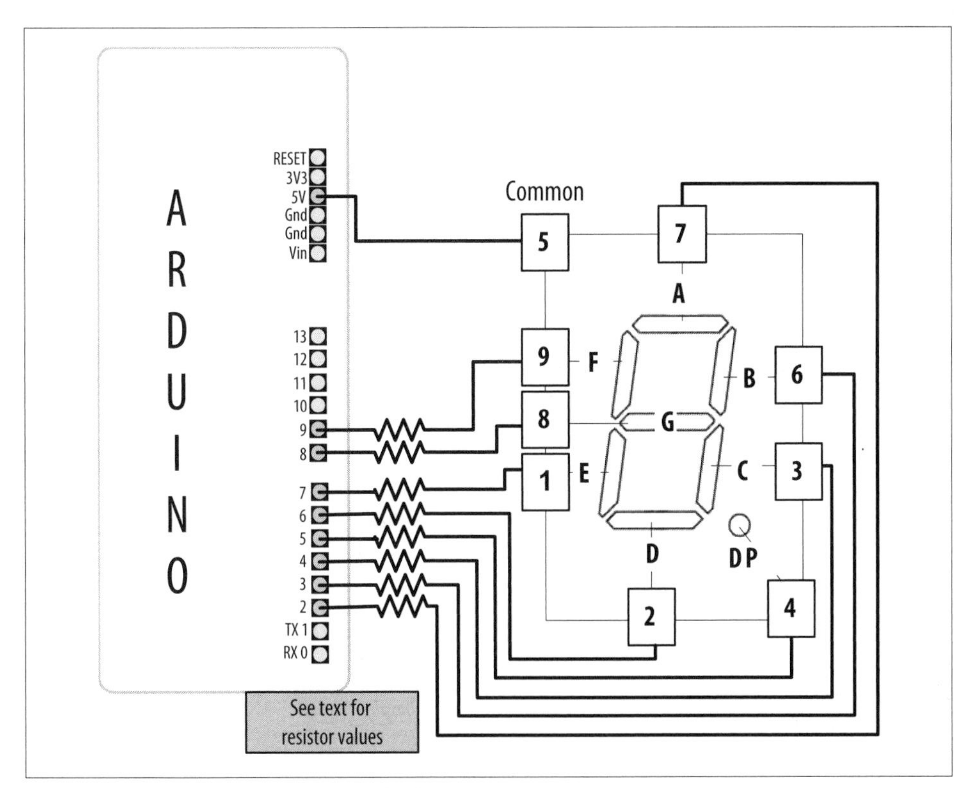

Figure 7-12. Connecting a 7-segment display

Discussion

The segments to be lit for each numeral are held in the array called `numeral`. There is one byte per numeral where each bit in the byte represents one of seven segments (or the decimal point).

The array called segmentPins holds the pins associated with each segment. The showDigit function checks that the number ranges from 0 to 9, and if valid, looks at each segment bit and turns on the segment if the bit is set (equal to 1). See Recipe 3.12 for more on the bitRead function.

As mentioned in Recipe 7.4, a pin is set HIGH when turning on a segment on a common cathode display, and it's set LOW when turning on a segment on a common anode display. The code here is for a common anode display, so it inverts the value (sets 0 to 1 and 1 to 0) as follows:

```
isBitSet = ! isBitSet; // remove this line if common cathode display
```

The ! is the negation operator—see Recipe 2.20. If your display is a common cathode display (all the cathodes are connected together; see the data sheet if you are not sure), you can remove that line.

7.11 Driving Multidigit, 7-Segment LED Displays: Multiplexing

Problem

You want to display numbers using a 7-segment display that shows two or more digits.

Solution

Multidigit, 7-segment displays usually use multiplexing. In earlier recipes, multiplexed rows and columns of LEDs were connected together to form an array; here, corresponding segments from each digit are connected together (see Figure 7-13):

```
/*
 * SevenSegmentMpx sketch
 * Shows numbers ranging from 0 through 9999 on a four-digit display
 * This example displays the value of a sensor connected to an analog input
*/

// bits representing segments A through G (and decimal point) for numerals 0-9
const int numeral[10] = {
  //ABCDEFG /dp
  B11111100,  // 0
  B01100000,  // 1
  B11011010,  // 2
  B11110010,  // 3
  B01100110,  // 4
  B10110110,  // 5
  B00111110,  // 6
  B11100000,  // 7
  B11111110,  // 8
  B11100110,  // 9
};

// pins for decimal point and each segment
                          // dp,G,F,E,D,C,B,A
```

```
const int segmentPins[] = { 4,7,8,6,5,3,2,9};

const int nbrDigits= 4;  // the number of digits in the LED display

                              //dig  1  2  3  4
const int digitPins[nbrDigits] = { 10,11,12,13};

void setup()
{
  for(int i=0; i < 8; i++)
    pinMode(segmentPins[i], OUTPUT); // set segment and DP pins to output

  for(int i=0; i < nbrDigits; i++)
    pinMode(digitPins[i], OUTPUT);
}


void loop()
{
  int value = analogRead(0);
  showNumber(value);
}

void showNumber( int number)
{
  if(number == 0)
    showDigit( 0, nbrDigits-1) ; // display 0 in the rightmost digit
  else
  {
    // display the value corresponding to each digit
    // leftmost digit is 0, rightmost is one less than the number of places
    for( int digit = nbrDigits-1; digit >= 0; digit--)
    {
      if(number > 0)
      {
         showDigit( number % 10, digit)  ;
         number = number / 10;
      }
    }
  }
}

// Displays given number on a 7-segment display at the given digit position
void showDigit( int number, int digit)
{
  digitalWrite( digitPins[digit], HIGH );
  for(int segment = 1; segment < 8; segment++)
  {
    boolean isBitSet = bitRead(numeral[number], segment);
    // isBitSet will be true if given bit is 1
    isBitSet = ! isBitSet; // remove this line if common cathode display
    digitalWrite( segmentPins[segment], isBitSet);
  }
  delay(5);
```

```
    digitalWrite( digitPins[digit], LOW );
  }
```

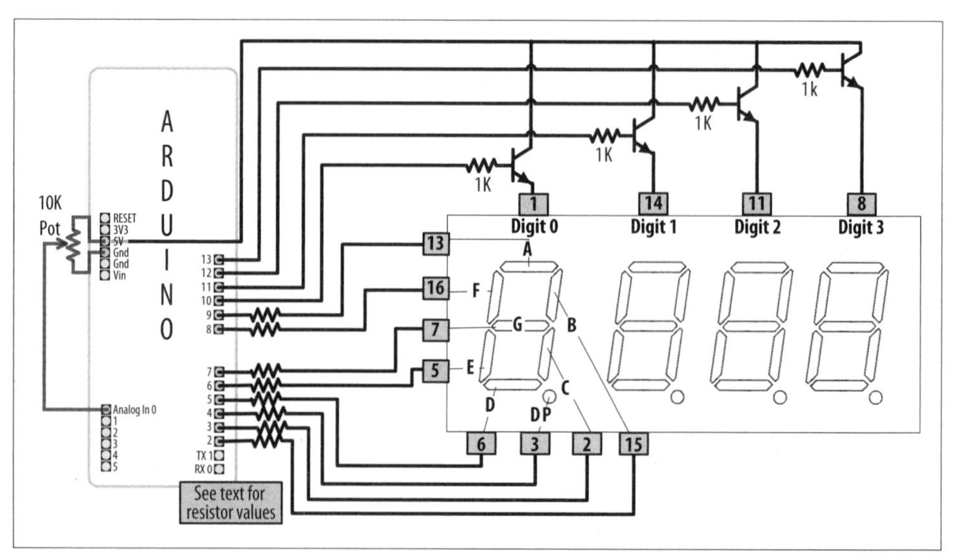

Figure 7-13. Connecting a multidigit, 7-segment display (LTC-2623)

Discussion

This sketch has a `showDigit` function similar to that discussed in Recipe 7.10. Here the function is given the numeral and the digit place. The logic to light the segments to correspond to the numeral is the same, but in addition, the code sets the pin corresponding to the digit place `HIGH`, so only that digit will be written (see the earlier multiplexing explanations).

7.12 Driving Multidigit, 7-Segment LED Displays Using MAX7221 Shift Registers

Problem

You want to control multiple 7-segment displays, but you want to minimize the number of required Arduino pins.

Solution

This Solution uses the popular MAX7221 LED driver chip to control four-digit common cathode displays, such as the Lite-On LTC-4727JR (Digi-Key 160-1551-5-ND). The MAX7221 provides a simpler solution than Recipe 7.11, because it handles multiplexing and digit decoding in hardware.

This sketch will display a number between 0 and 9,999 (Figure 7-14 shows the connections):

```
/*
  Max7221_digits
 */

#include <SPI.h>  // Arduino SPI library introduced in Arduino version 0019

const int slaveSelect = 10; //pin used to enable the active slave

const int numberOfDigits = 2; // change these to match the number of digits
wired up
const int maxCount       = 99;

int number = 0;

void setup()
{
  Serial.begin(9600);
  SPI.begin();   // initialize SPI
  pinMode(slaveSelect, OUTPUT);
  digitalWrite(slaveSelect,LOW);  //select slave
  // prepare the 7221 to display 7-segment data - see data sheet
  sendCommand(12,1);  // normal mode (default is shutdown mode);
  sendCommand(15,0);  // Display test off
  sendCommand(10,8);  // set medium intensity (range is 0-15)
  sendCommand(11,numberOfDigits);  // 7221 digit scan limit command
  sendCommand(9,255); // decode command, use standard 7-segment digits
  digitalWrite(slaveSelect,HIGH);  //deselect slave
}

void loop()
{
  // display a number from serial port terminated by end of line character
  if(Serial.available())
  {
     char ch  = Serial.read();
     if( ch == '\n')
     {
       displayNumber(number);
       number = 0;
     }
     else
     number = (number * 10) + ch - '0'; // see Chapter 4 for details
  }
}

// function to display up to four digits on a 7-segment display
void displayNumber( int number)
{
  for(int i = 0; i < numberOfDigits; i++)
  {
    byte character = number % 10;  // get the value of the rightmost decade
    if(number == 0 && i > 0)
```

```
      character = 0xf;  // the 7221 will blank the segments when receiving value
    // send digit number as command, first digit is command 1
    sendCommand(numberOfDigits-i, character);
    number = number / 10;
  }
}

void sendCommand( int command, int value)
{
  digitalWrite(slaveSelect,LOW); //chip select is active low
  //2-byte data transfer to the 7221
  SPI.transfer(command);
  SPI.transfer(value);
  digitalWrite(slaveSelect,HIGH); //release chip, signal end transfer
}
```

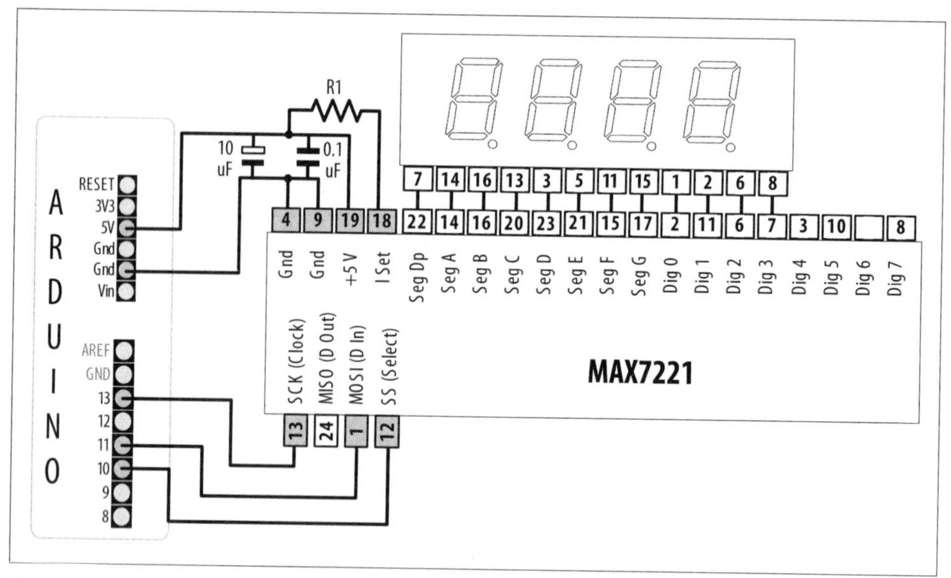

Figure 7-14. MAX7221 driving a multidigit common cathode 7-segment display

Solution

This recipe uses Arduino SPI communication to talk to the MAX7221 chip. Chapter 13 covers SPI in more detail, and Recipe 13.8 explains the SPI-specific code used.

This sketch displays a number if up to four digits are received on the serial port—see Chapter 4 for an explanation of the serial code in `loop`. The `displayNumber` function extracts the value of each digit, starting from the rightmost digit, to the MAX7221, using the `sendCommand` function that sends the values to the MAX7221.

The wiring shown uses a four-digit, 7-segment display, but you can use single- or dual-digit displays for up to eight digits. When combining multiple displays, each corresponding segment pin should be connected together. (Recipe 13.8 shows the connections for a common dual-digit display.)

The MAX72xx chips are designed for common cathode displays. The anode of each segment is available on a separate pin, and the cathodes of all the segments for each digit are connected together.

7.13 Controlling an Array of LEDs by Using MAX72xx Shift Registers

Problem

You have an 8×8 array of LEDs to control, and you want to minimize the number of required Arduino pins.

Solution

As in Recipe 7.12, you can use a shift register to reduce the number of pins needed to control an LED matrix. This Solution uses the popular MAX7219 or MAX7221 LED driver chip to provide this capability. Connect your Arduino, matrix, and MAX72xx as shown in Figure 7-15.

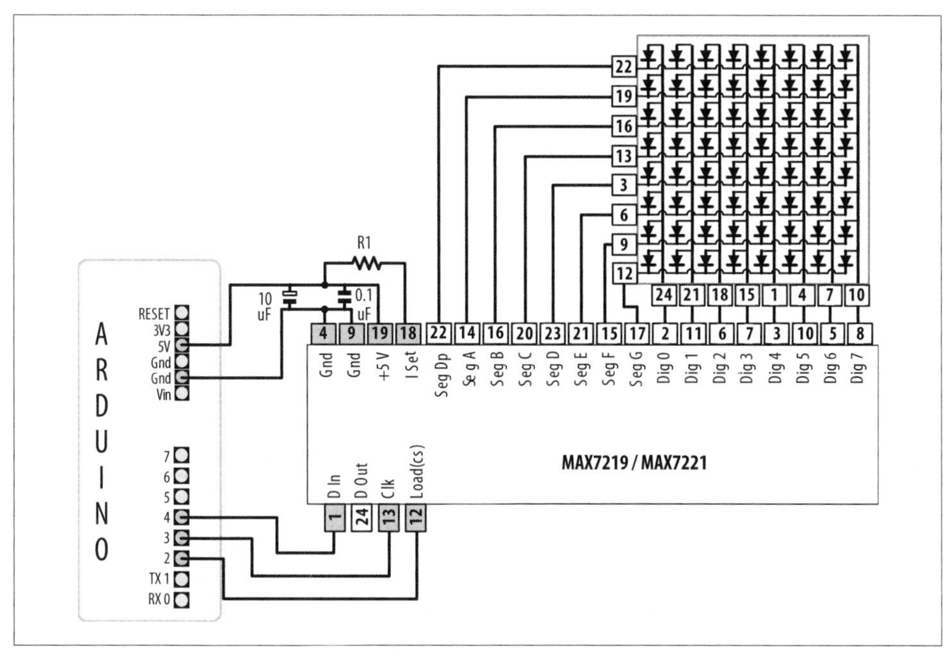

Figure 7-15. MAX72xx driving an 8×8 LED array

This sketch is based on the Arduino hello_matrix library by Nicholas Zambetti, with the pin numbers changed to be consistent with the wiring used elsewhere in this chapter. It uses the Sprite and Matrix libraries that were distributed with Arduino releases prior to 1.0. If you are using Arduino 1.0, and you can't find the libraries on the Arduino Playground, you can obtain the libraries from the 0022 release from *http://arduino.cc/en/Main/Software*.

```
#include <Sprite.h>
#include <Matrix.h>

// Hello Matrix
// by Nicholas Zambetti <http://www.zambetti.com>

// Demonstrates the use of the Matrix library
// For MAX7219 LED Matrix Controllers
// Blinks welcoming face on screen

const int loadPin  = 2;
const int clockPin = 3;
const int dataPin  = 4;

Matrix myMatrix = Matrix(dataPin, clockPin, loadPin); // create a new Matrix

void setup()
{
}

void loop()
{
  myMatrix.clear(); // clear display

  delay(1000);

  // turn some pixels on
  myMatrix.write(1, 5, HIGH);
  myMatrix.write(2, 2, HIGH);
  myMatrix.write(2, 6, HIGH);
  myMatrix.write(3, 6, HIGH);
  myMatrix.write(4, 6, HIGH);
  myMatrix.write(5, 2, HIGH);
  myMatrix.write(5, 6, HIGH);
  myMatrix.write(6, 5, HIGH);

  delay(1000);
}
```

Discussion

A matrix is created by passing pin numbers for the data, load, and clock pins. `loop` uses the `write` method to turn pixels on; the `clear` method turns the pixels off. `write` has three parameters: the first two identify the column and row (`x` and `y`) of an LED and the third parameter (`HIGH` or `LOW`) turns the LED on or off.

The pin numbers shown here are for the green LEDs in the dual-color 8×8 matrix, available from these suppliers:

SparkFun: COM-00681
NKC Electronics Item #: COM-0006

The resistor (marked R1 in Figure 7-15) is used to control the maximum current that will be used to drive an LED. The MAX72xx data sheet has a table that shows a range of values (see Table 7-3).

Table 7-3. Table of resistor values (from MAX72xx data sheet)

LED forward voltage					
Current	**1.5V**	**2.0V**	**2.5V**	**3.0V**	**3.5V**
40 mA	12 kΩ	12 kΩ	11 kΩ	10 kΩ	10 kΩ
30 mA	18 kΩ	17 kΩ	16 kΩ	15 kΩ	14 kΩ
20 mA	30 kΩ	28 kΩ	26 kΩ	24 kΩ	22 kΩ
10 mA	68 kΩ	64 kΩ	60 kΩ	56 kΩ	51 kΩ

The green LED in the LED matrix shown in Figure 7-15 has a forward voltage of 2.0 volts and a forward current of 20 mA. Table 7-3 indicates 28K ohms, but to add a little safety margin, a resistor of 30K or 33K would be a suitable choice. The capacitors (0.1 uf and 10 uf) are required to prevent noise spikes from being generated when the LEDs are switched on and off—see "Using Capacitors for Decoupling" on page 653 in Appendix C if you are not familiar with connecting decoupling capacitors.

See Also

Documentation for the Matrix library: *http://wiring.org.co/reference/libraries/Matrix/index.html*

Documentation for the Sprite library: *http://wiring.org.co/reference/libraries/Sprite/index.html*

MAX72xx data sheet: *http://pdfserv.maxim-ic.com/en/ds/MAX7219-MAX7221.pdf*

7.14 Increasing the Number of Analog Outputs Using PWM Extender Chips (TLC5940)

Problem

You want to have individual control of the intensity of more LEDs than Arduino can support (6 on a standard board and 12 on the Mega).

Solution

The TLC5940 chip drives up to 16 LEDs using only five data pins. Figure 7-16 shows the connections. This sketch is based on the excellent Tlc5940 library written by Alex Leone (*acleone@gmail.com*). You can download the library from *http://code.google.com/p/tlc5940arduino/*:

```
/*
 * TLC sketch
 * Create a Knight Rider-like effect on LEDs plugged into all the TLC outputs
 * this version assumes one TLC with 16 LEDs
 */

#include "Tlc5940.h"

void setup()
{
  Tlc.init();  // initialize the TLC library
}

void loop()
{
  int direction = 1;
  int intensity = 4095; // an intensity from 0 to 4095, full brightness is 4095
  int dim = intensity / 4;  //  1/4 the value dims the LED
  for (int channel = 0; channel < 16; channel += direction) {
  // the following TLC commands set values to be written by the update method
    Tlc.clear();  // turn off all LEDs
    if (channel == 0) {
      direction = 1;
    }
    else {
      Tlc.set(channel - 1, dim);  // set intensity for prev LED
    }
    Tlc.set(channel, intensity); //  full intensity on this LED
    if (channel < 16){
      Tlc.set(channel + 1, dim);  // set the next LED to dim
    }
    else {
      direction = -1;
    }

    Tlc.update();  // this method sends data to the TLC chips to change the LEDs
    delay(75);
  }
}
```

Discussion

This sketch loops through each channel (LED), setting the previous LED to dim, the current channel to full intensity, and the next channel to dim. The LEDs are controlled through a few core methods.

The `Tlc.init` method initializes `Tlc` functions prior to any other function.

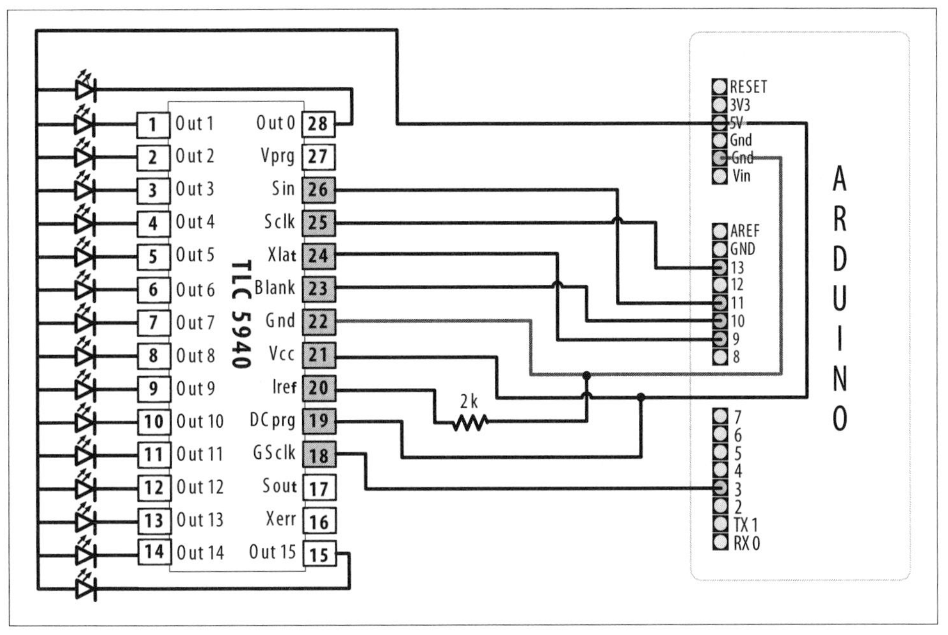

Figure 7-16. Sixteen LEDs driven using external PWM

The following functions only take effect after calling the `update()` method:

`Tlc.clear`
: Turns off all channels

`Tlc.set`
: Sets the intensity for the given channel to a given value

`Tlc.setAll`
: Sets all channels to a given value

`Tlc.update`
: Sends the changes from any of the preceding commands to the TLC chip

More functions are available in the library; see the link to the reference at the end of this recipe.

The 2K resistor between TLC pin 20 (Iref) and Gnd will let around 20 mA through each LED. You can calculate the resistor value R for a different current (in milliamperes) using the formula R = 40,000 / mA. R is 1 ohm, and the calculation does not depend on the LED driving voltage.

If you want the LEDs to turn off when the Arduino is reset, put a pull-up resistor (10K) between +5V and BLANK (pin 23 of the TLC and Arduino pin 10).

Here is a variation that uses a sensor value to set the maximum LED intensity. You can test this using a variable resistor connected as shown in Figure 7-13 or Figure 7-17:

```
#include "Tlc5940.h"

const int sensorPin = 0; // connect sensor to analog input 0

void setup()
{
  Tlc.init();  // initialize the TLC library
}

void loop()
{
  int direction = 1;
  int sensorValue = analogRead(0);  // get the sensor value
  int intensity = map(sensorValue, 0,1023, 0, 4095); // map to TLC range
  int dim = intensity / 4;  //  1/4 the value dims the LED
  for (int channel = 0; channel < NUM_TLCS * 16; channel += direction) {
  // the following TLC commands set values to be written by the update method
    Tlc.clear();  // turn off all LEDs
    if (channel == 0) {
      direction = 1;
    }
    else {
      Tlc.set(channel - 1, dim);  // set intensity for prev LED
    }
    Tlc.set(channel, intensity); //  full intensity on this LED
    if (channel != NUM_TLCS * 16 - 1) {
      Tlc.set(channel + 1, dim);  // set the next LED to dim
    }
    else {
      direction = -1;
    }

    Tlc.update();  // this method sends data to the TLC chips to change the LEDs
    delay(75);
  }
}
```

This version also allows for multiple TLC chips if you want to drive more than 16 LEDs. You do this by "daisy-chaining" the TLC chips—connect the Sout (pin 17) of the first TLC to the Sin (pin 26) of the next. The Sin (pin 26) of the first TLC chip is connected to Arduino pin 11, as shown in Figure 7-16.

The following pins should be connected together when daisy-chaining TLC chips:

- Arduino pin 9 to XLAT (pin 24) of each TLC
- Arduino pin 10 to BLANK (pin 23) of each TLC
- Arduino pin 13 to SCLK (pin 25) of each TLC

Each TLC needs its own resistor between Iref (pin 20) and Gnd.

You must change the value of the NUM_TLCS constant defined in the Tlc5940 library to match the number of chips you have wired.

See Also

Go to *http://code.google.com/p/tlc5940arduino/* to download this library and access its documentation.

7.15 Using an Analog Panel Meter as a Display

Problem

You would like to control the pointer of an analog panel meter from your sketch. Fluctuating readings are easier to interpret on an analog meter, and analog meters add a cool retro look to a project.

Solution

Connect the meter through a series resistor (5K ohms for the typical 1 mA meter) and connect to an analog (PWM) output (see Figure 7-17).

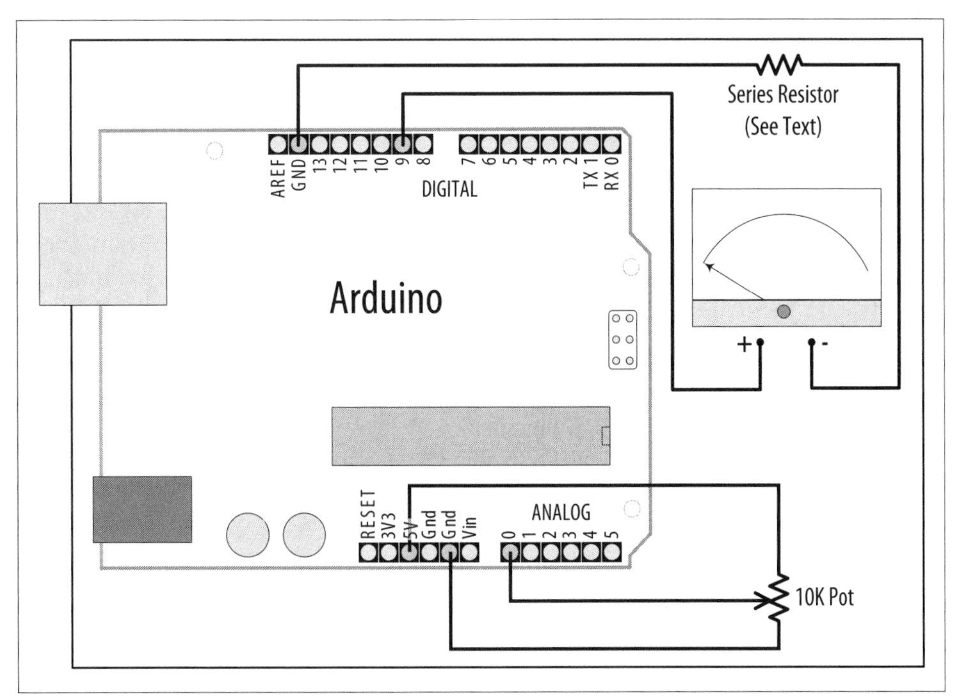

Figure 7-17. Driving an analog meter

The pointer movement corresponds to the position of a pot (variable resistor):

```
/*
 * AnalogMeter sketch
 * Drives an analog meter through an Arduino PWM pin
 * The meter level is controlled by a variable resistor on an analog input pin
 */

const int analogInPin = 0; // Analog input pin connected to the variable resistor
const int analogMeterPin = 9; // Analog output pin connecting to the meter

int sensorValue = 0;        // value read from the pot
int outputValue = 0;        // value output to the PWM (analog out)

void setup()
{
  // nothing in setup
}

void loop()
{
  sensorValue = analogRead(analogInPin);            // read the analog in value
  outputValue = map(sensorValue, 0, 1023, 0, 255);  // scale for analog out
  analogWrite(analogMeterPin, outputValue);         // write the analog out value
}
```

Discussion

In this variation on Recipe 7.2, the Arduino `analogWrite` output drives a panel meter. Panel meters are usually much more sensitive than LEDs; a resistor must be connected between the Arduino output and the meter to drop the current to the level for the meter.

The value of the series resistor depends on the sensitivity of the meter; 5K ohms give full-scale deflection with a 1 mA meter. You can use 4.7K resistors, as they are easier to obtain than 5K, although you will probably need to reduce the maximum value given to `analogWrite` to 240 or so. Here is how you can change the range in the `map` function if you use a 4.7K ohm resistor with a 1 mA meter:

```
outputValue = map(sensorValue, 0, 1023, 0, 240);  // map to meter's range
```

If your meter has a different sensitivity than 1 mA, you will need to use a different value series resistor. The resistor value in ohms is

resistor = 5,000 / mA

So, a 500 microamp meter (0.5 mA) is 5,000 / 0.5, which is 10,000 (10 K) ohms. A 10 mA meter requires 500 ohms, 20 mA 250 ohms.

Some surplus meters already have an internal series resistor—you may need to experiment to determine the correct value of the resistor, but be careful not to apply too much voltage to your meter.

See Also

Recipe 7.2

CHAPTER 8
Physical Output

8.0 Introduction

You can make things move by controlling motors with Arduino. Different types of motors suit different applications, and this chapter shows how Arduino can drive many different kinds of motors.

Motion Control Using Servos

Servos enable you to accurately control physical movement because they generally move to a position instead of continuously rotating. They are ideal for making something rotate over a range of 0 to 180 degrees. Servos are easy to connect and control because the motor driver is built into the servo.

Servos contain a small motor connected through gears to an output shaft. The output shaft drives a servo arm and is also connected to a potentiometer to provide position feedback to an internal control circuit (see Figure 8-1).

You can get continuous rotation servos that have the positional feedback disconnected so that you can instruct the servo to rotate continuously clockwise and counterclockwise with some control over the speed. These function a little like the brushed motors covered in Recipe 8.9, except that continuous rotation servos use the servo library code instead of `analogWrite` and don't require a motor shield.

Continuous rotation servos are easy to use because they don't need a motor shield—the motor drivers are inside the servo. The disadvantages are that the speed and power choices are limited compared to external motors, and the precision of speed control is usually not as good as with a motor shield (the electronics is designed for accurate positioning, not linear speed control). See Recipe 8.3 for more on using continuous rotation servos.

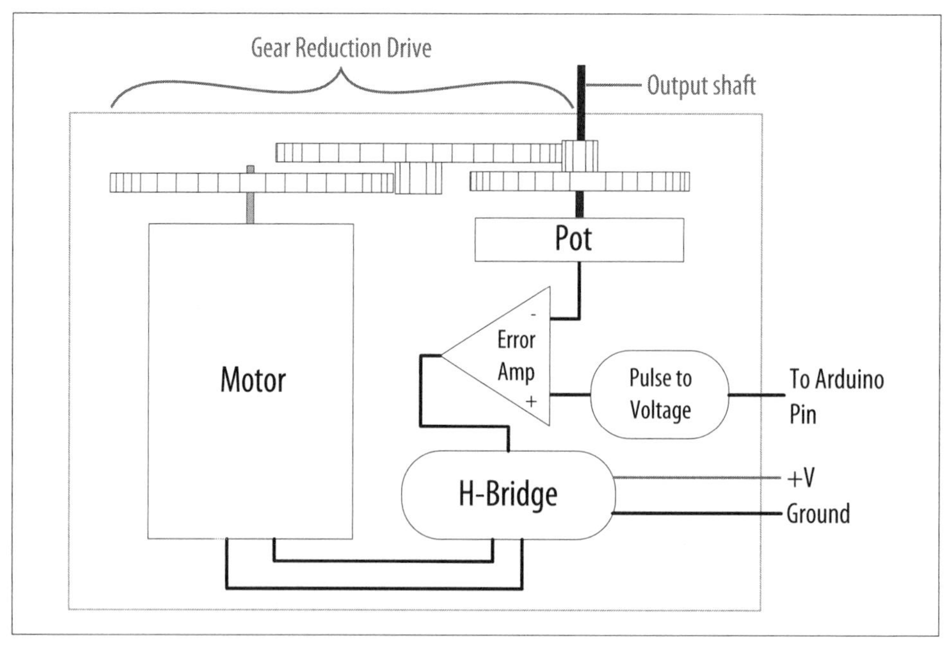

Figure 8-1. Elements inside a hobby servo

Servos respond to changes in the duration of a pulse. A short pulse of 1 ms or less will cause the servo to rotate to one extreme; a pulse duration of 2 ms or so will rotate the servo to the other extreme (see Figure 8-2). Pulses ranging between these values will rotate the servo to a position proportional to the pulse width. There is no standard for the exact relationship between pulses and position, and you may need to tinker with the commands in your sketch to adjust for the range of your servos.

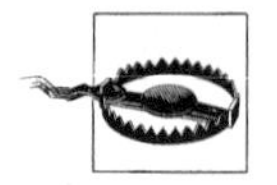

Although the duration of the pulse is *modulated* (controlled), servos require pulses that are different from the Pulse Width Modulation (PWM) output from `analogWrite`. You can damage a hobby servo by connecting it to the output from `analogWrite`—use the Servo library instead.

Solenoids and Relays

Although most motors produce rotary motion, a solenoid produces linear movement when powered. A solenoid has a metallic core that is moved by a magnetic field created when current is passed through a coil. A mechanical relay is a type of solenoid that connects or disconnects electrical contacts (it's a solenoid operating a switch). Relays are controlled just like solenoids. Relays and solenoids, like most motors, require more current than an Arduino pin can safely provide, and the recipes in this chapter show how you can use a transistor or external circuit to drive these devices.

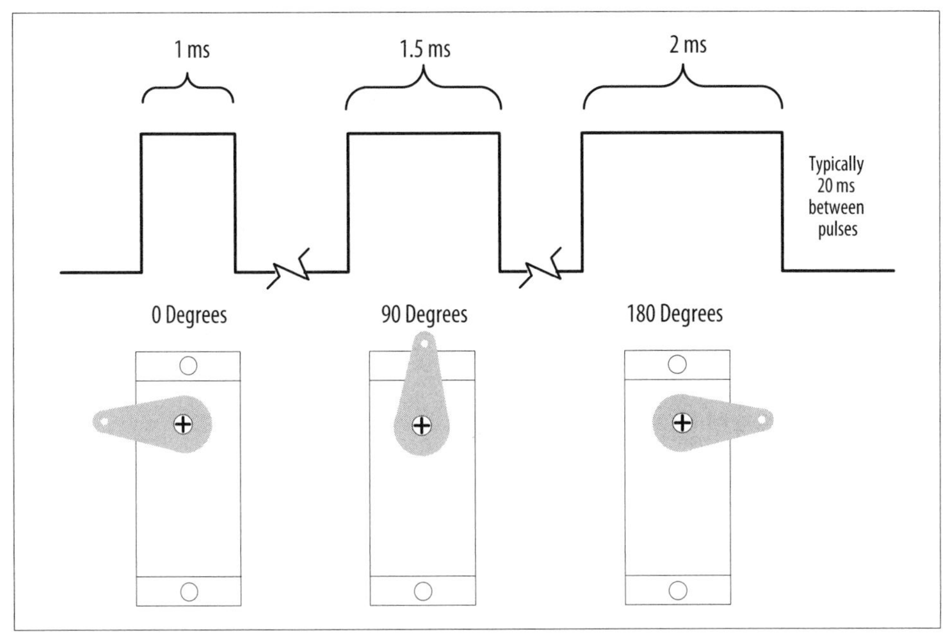

Figure 8-2. Relationship between the pulse width and the servo angle; the servo output arm moves proportionally as the pulse width increases from 1 ms to 2 ms

Brushed and Brushless Motors

Most low-cost direct current (DC) motors are simple devices with two leads connected to brushes (contacts) that control the magnetic field of the coils that drives a metallic core (armature). The direction of rotation can be reversed by reversing the polarity of the voltage on the contacts. DC motors are available in many different sizes, but even the smallest (such as vibration motors used in cell phones) require a transistor or other external control to provide adequate current. The recipes that follow show how to control motors using a transistor or an external control circuit called an H-Bridge.

The primary characteristic in selecting a motor is *torque*. Torque determines how much work the motor can do. Typically, higher torque motors are larger and heavier and draw more current than lower torque motors.

Brushless motors usually are more powerful and efficient for a given size than brushed motors, but they require more complicated electronic control. Where the performance benefit of a brushless motor is desired, components called *electronics speed controllers* intended for hobby radio control use can be easily controlled by Arduino because they are controlled much like a servo motor.

Stepper Motors

Steppers are motors that rotate a specific number of degrees in response to control pulses. The number of degrees in each step is motor-dependent, ranging from one or two degrees per step to 30 degrees or more.

Two types of steppers are commonly used with Arduino: bipolar (typically with four leads attached to two coils) and unipolar (five or six leads attached to two coils). The additional wires in a unipolar stepper are internally connected to the center of the coils (in the five-lead version, each coil has a center tap and both center taps are connected together). The recipes covering bipolar and unipolar steppers have diagrams illustrating these connections.

Troubleshooting Motors

The most common cause of problems when connecting devices that require external power is neglecting to connect all the grounds together. Your Arduino ground must be connected to the external power supply ground and the grounds of external devices being powered.

8.1 Controlling the Position of a Servo

Problem

You want to control the position of a servo using an angle calculated in your sketch. For example, you want a sensor on a robot to swing through an arc or move to a position you select.

Solution

Use the Servo library distributed with Arduino. Connect the servo power and ground to a suitable power supply (a single hobby servo can usually be powered from the Arduino 5V line). Recent versions of the library enable you to connect the servo signal leads to any Arduino digital pin.

Here is the example Sweep sketch distributed with Arduino; Figure 8-3 shows the connections:

```
#include <Servo.h>

Servo myservo;  // create servo object to control a servo

int angle = 0;    // variable to store the servo position

void setup()
{
  myservo.attach(9);  // attaches the servo on pin 9 to the servo object
}
```

```
void loop()
{
  for(angle = 0; angle < 180; angle += 1)  // goes from 0 degrees to 180 degrees
  {                                         // in steps of 1 degree
    myservo.write(angle);     // tell servo to go to position in variable 'angle'
    delay(20);                // waits 20ms between servo commands
  }
  for(angle = 180; angle >= 1; angle -= 1) // goes from 180 degrees to 0 degrees
  {
    myservo.write(angle);    // move servo in opposite direction
    delay(20);               // waits 20ms between servo commands
  }
}
```

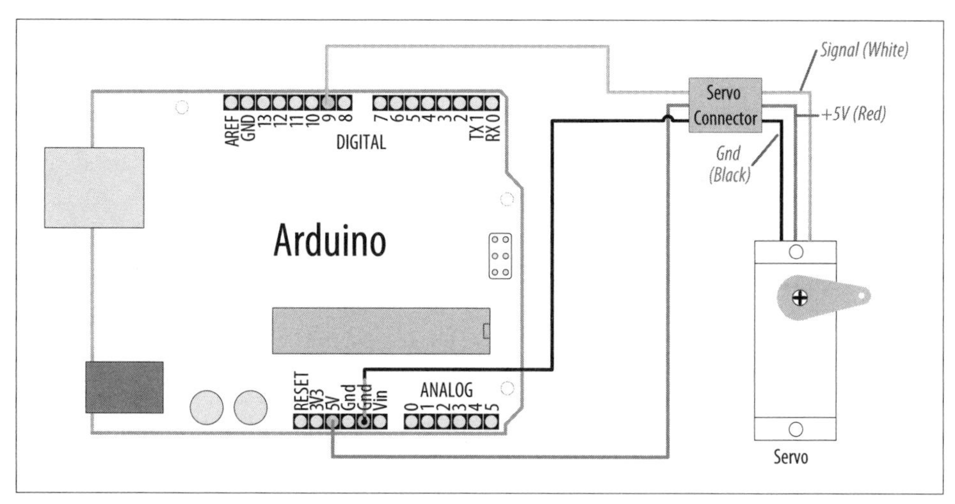

Figure 8-3. Connecting a servo for testing with the example Sweep sketch

Discussion

This example sweeps the servo between 0 and 180 degrees. You may need to tell the library to adjust the minimum and maximum positions so that you get the range of movement you want. Calling `Servo.attach` with optional arguments for minimum and maximum positions will adjust the movement:

```
myservo.attach(9,1000,2000 );  // use pin 9, set min to 1000us, max to 2000us
```

Because typical servos respond to pulses measured in microseconds and not degrees, the arguments following the pin number inform the Servo library how many microseconds to use when 0 degrees or 180 degrees are requested. Not all servos will move over a full 180-degree range, so you may need to experiment with yours to get the range you want.

The parameters for `servo.attach(pin, min, max)` are the following:

`pin`
: The pin number that the servo is attached to (you can use any digital pin)

`min` *(optional)*
: The pulse width, in microseconds, corresponding to the minimum (0-degree) angle on the servo (defaults to 544)

`max` *(optional)*
: The pulse width, in microseconds, corresponding to the maximum (180-degree) angle on the servo (defaults to 2,400)

The Servo library supports up to 12 servos on most Arduino boards and 48 on the Arduino Mega. On standard boards such as the Uno, use of the library disables `analogWrite()` (PWM) functionality on pins 9 and 10, whether or not there is a servo on those pins. See the Servo library reference for more information: *http://arduino.cc/en/Reference/Servo*.

Power requirements vary depending on the servo and how much torque is needed to rotate the shaft.

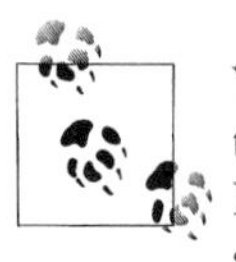

You may need an external source of 5 or 6 volts when connecting multiple servos. Four AA cells work well if you want to use battery power. Remember that you must connect the ground of the external power source to Arduino ground.

8.2 Controlling One or Two Servos with a Potentiometer or Sensor

Problem

You want to control the rotational direction and speed of one or two servos with a potentiometer. For example, you want to control the pan and tilt of a camera or sensor connected to the servos. This recipe can work with any variable voltage from a sensor that can be read from an analog input.

Solution

The same library can be used as in Recipe 8.1, with the addition of code to read the voltage on a potentiometer. This value is scaled so that the position of the pot (from `0` to `1023`) is mapped to a range of 0 to 180 degrees. The only difference in the wiring is the addition of the potentiometer; see Figure 8-4:

```
#include <Servo.h>

Servo myservo;  // create servo object to control a servo

int potpin = 0;  // analog pin used to connect the potentiometer
int val;    // variable to read the value from the analog pin

void setup()
{
  myservo.attach(9);  // attaches the servo on pin 9 to the servo object
}

void loop()
{
  val = analogRead(potpin);            // reads the value of the potentiometer
  val = map(val, 0, 1023, 0, 180);     // scale it to use it with the servo
  myservo.write(val);                  // sets position to the scaled value
  delay(15);                           // waits for the servo to get there
}
```

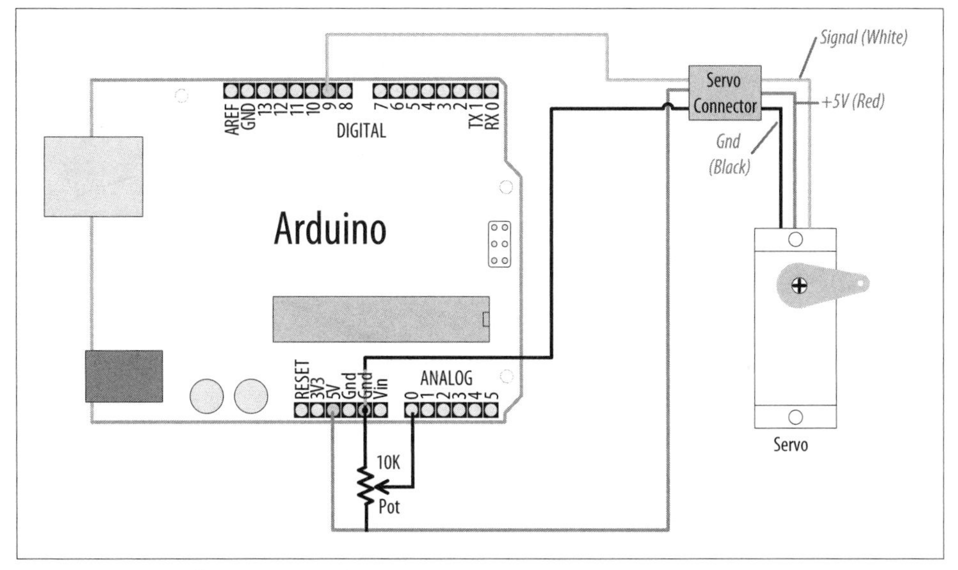

Figure 8-4. Controlling a servo with a potentiometer

Hobby servos have a cable with a 3-pin female connector that can be directly plugged in to a "servo" header fitted to some shields, such as the Adafruit Motor Shield. The physical connector is compatible with the Arduino connectors so you can use the same wire jumpers to those used to connect Arduino pins. Bear in mind that the color of the signal lead is not standardized; yellow is sometimes used instead of white. Red is always in the middle and the ground lead is usually black or brown.

Discussion

Anything that can be read from `analogRead` (see Chapter 5 and Chapter 6) can be used—for example, the gyro and accelerometer recipes in Chapter 6 can be used, so that the angle of the servo is controlled by the yaw of the gyro or angle of the accelerometer.

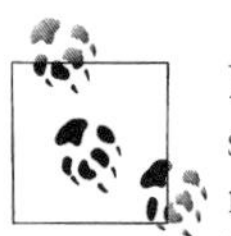

Not all servos will rotate over the full range of the Servo library. If your servo buzzes due to hitting an end stop at an extreme of movement, try reducing the output range in the map function until the buzzing stops. For example:

```
val=map(val,0,1023,10,170); // most function over this range
```

8.3 Controlling the Speed of Continuous Rotation Servos

Problem

You want to control the rotational direction and speed of servos modified for continuous rotation. For example, you are using two continuous rotation servos to power a robot and you want the speed and direction to be controlled by your sketch.

Solution

Continuous rotation servos are a form of gear-reduced motor with forward and backward speed adjustment. Control of continuous rotation servos is similar to normal servos. The servo rotates in one direction as the angle is increased from 90 degrees; it rotates in the other direction when the angle is decreased from 90 degrees. The actual direction forward or backward depends on how you have the servos attached. Figure 8-5 shows the connections for controlling two servos.

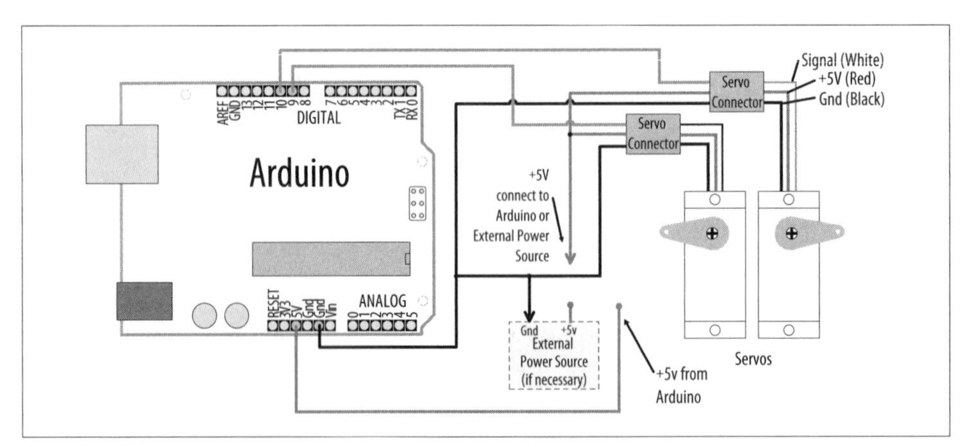

Figure 8-5. Controlling two servos

Servos are usually powered from a 4.8V to 6V source. Heavier duty servos may require more current than the Arduino board can provide through the +5V pin and these will require an external power source. Four 1.2V rechargeable batteries can be used to power Arduino and the servos. Bear in mind that fresh alkaline cells can have a voltage higher than 1.5 volts, so if using alkaline batteries, check with your multimeter that the total voltage does not exceed 6 volts—the absolute maximum operating voltage for Arduino chips.

The sketch sweeps the servos from 90 to 180 degrees, so if the servos were connected to wheels, the vehicle would move forward at a slowly increasing pace and then slow down to a stop. Because the servo control code is in `loop`, this will continue for as long as there is power:

```
#include <Servo.h>

Servo myservoLeft;   // create servo object to control a servo
Servo myservoRight;  // create servo object to control a servo

int angle = 0;     // variable to store the servo position

void setup()
{
  myservoLeft.attach(9);   // attaches left servo on pin 9 to servo object
  myservoRight.attach(10); // attaches right servo on pin 10 to servo object
}

void loop()
{
  for(angle = 90; angle < 180; angle += 1)  // goes from 90 to 180 degrees
  {                                         // in steps of 1 degree.
                                            // 90  degrees is stopped.

    myservoLeft.write(angle);        // rotate servo at speed given by 'angle'
    myservoRight.write(180-angle);   // go in the opposite direction

    delay(20);                       // waits 20ms between servo commands
  }
  for(angle = 180; angle >= 90; angle -= 1) // goes from 180 to 90 degrees
  {
    myservoLeft.write(angle);        // rotate at a speed given by 'angle'
    myservoRight.write(180-angle);   // other servo goes in opposite direction
  }
}
```

Discussion

You can use similar code for continuous rotation and normal servos, but be aware that continuous rotation servos may not stop rotating when writing exactly 90 degrees. Some servos have a small potentiometer you can trim to adjust for this, or you can add or subtract a few degrees to stop the servo. For example, if the left servo stops rotating at 92 degrees, you can change the lines that write to the servos as follows:

```
myservoLeft.write(angle+TRIM); // declare int TRIM=2; at beginning of sketch
```

8.4 Controlling Servos Using Computer Commands

Problem

You want to provide commands to control servos from the serial port. Perhaps you want to control servos from a program running on your computer.

Solution

You can use software to control the servos. This has the advantage that any number of servos can be supported. However, your sketch needs to constantly attend to refreshing the servo position, so the logic can get complicated as the number of servos increases if your project needs to perform a lot of other tasks.

This recipe drives four servos according to commands received on the serial port. The commands are of the following form:

- `180a` writes 180 to servo a
- `90b` writes 90 to servo b
- `0c` writes 0 to servo c
- `17d` writes 17 to servo d

Here is the sketch that drives four servos connected on pins 7 through 10:

```
#include <Servo.h> // the servo library

#define SERVOS 4 // the number of servos
int servoPins[SERVOS] = {7,8,9,10}; // servos on pins 7 through 10

Servo myservo[SERVOS];

void setup()
{
  Serial.begin(9600);
  for(int i=0; i < SERVOS; i++)
    myservo[i].attach(servoPins[i]);
}

void loop()
{
  serviceSerial();
}

// serviceSerial checks the serial port and updates position with received data
// it expects servo data in the form:
//
//  "180a" writes 180 to servo a
//  "90b writes 90 to servo b
//
```

```
void serviceSerial()
{
  static int pos = 0;

  if ( Serial.available()) {
    char ch = Serial.read();

    if( isDigit(ch) )                       // If ch is a number:
      pos = pos * 10 + ch - '0';            // accumulate the value
    else if(ch >= 'a' && ch <= 'a'+ SERVOS) // If ch is a letter for our servos:
      myservo[ch - 'a'].write(pos);         // save the position in position array
  }
}
```

Discussion

Connecting the servos is similar to the previous recipes. Each servo line wire gets connected to a digital pin. All servo grounds are connected to Arduino ground. The servo power lines are connected together, and you may need an external 5V or 6V power source if your servos require more current than the Arduino power supply can provide.

An array named `myservo` (see Recipe 2.4) is used to hold references for the four servos. A `for` loop in `setup` attaches each servo in the array to consecutive pins defined in the `servoPins` array.

If the character received from serial is a digit (the character will be greater than or equal to 0 and less than or equal to 9), its value is accumulated in the variable `pos`. If the character is the letter *a*, the position is written to the first servo in the array (the servo connected to pin 7). The letters *b*, *c*, and *d* control the subsequent servos.

See Also

See Chapter 4 for more on handling values received over serial.

8.5 Driving a Brushless Motor (Using a Hobby Speed Controller)

Problem

You have a hobby brushless motor and you want to control its speed.

Solution

This sketch uses the same code as Recipe 8.2. The wiring is similar, except for the speed controller and motor. A hobby electronic speed controller (ESC) is a device used to control brushless motors in radio-controlled vehicles. Because these items are mass produced, they are a cost-effective way to drive brushless motors. You can find a selection by typing "esc" into the search field of your favorite hobby store website or typing "speed controller esc" into Google.

Brushless motors have three windings and these should be connected following the instructions for your speed controller (see Figure 8-6).

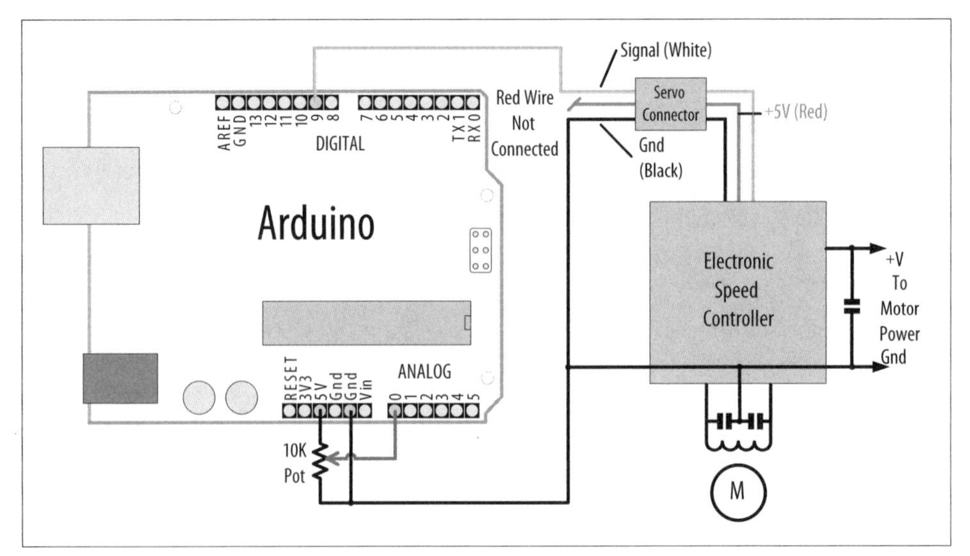

Figure 8-6. Connecting an electronic speed controller

Discussion

Consult the documentation for your speed controller to confirm that it is suitable for your brushless motor and to verify the wiring. Brushless motors have three connections for the three motor wires and two connections for power. Many speed controllers provide power on the center pin of the servo connector. Unless you want to power the Arduino board from the speed controller, you must disconnect or cut this center wire.

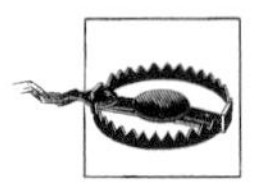

If your speed controller has a feature that provides 5V power to servos and other devices (called a *battery eliminator circuit* or *BEC* for short), you must disconnect this wire when attaching the Arduino to the speed controller (see Figure 8-6).

8.6 Controlling Solenoids and Relays

Problem

You want to activate a solenoid or relay under program control. Solenoids are electromagnets that convert electrical energy into mechanical movement. An electromagnetic relay is a switch that is activated by a solenoid.

Solution

Most solenoids require more power than an Arduino pin can provide, so a transistor is used to switch the current needed to activate a solenoid. Activating the solenoid is achieved by using `digitalWrite` to set the pin `HIGH`.

This sketch turns on a transistor connected as shown in Figure 8-7. The solenoid will be activated for one second every hour:

```
int solenoidPin = 2;           // Solenoid connected to transistor on pin 2

void setup()
{
  pinMode(solenoidPin, OUTPUT);
}

void loop()
{
  long interval = 1000 * 60 * 60 ;   // interval = 60 minutes

  digitalWrite(solenoidPin, HIGH); // activates the solenoid
  delay(1000);                     // waits for a second
  digitalWrite(solenoidPin, LOW);  // deactivates the solenoid
  delay(interval);                 // waits one hour
}
```

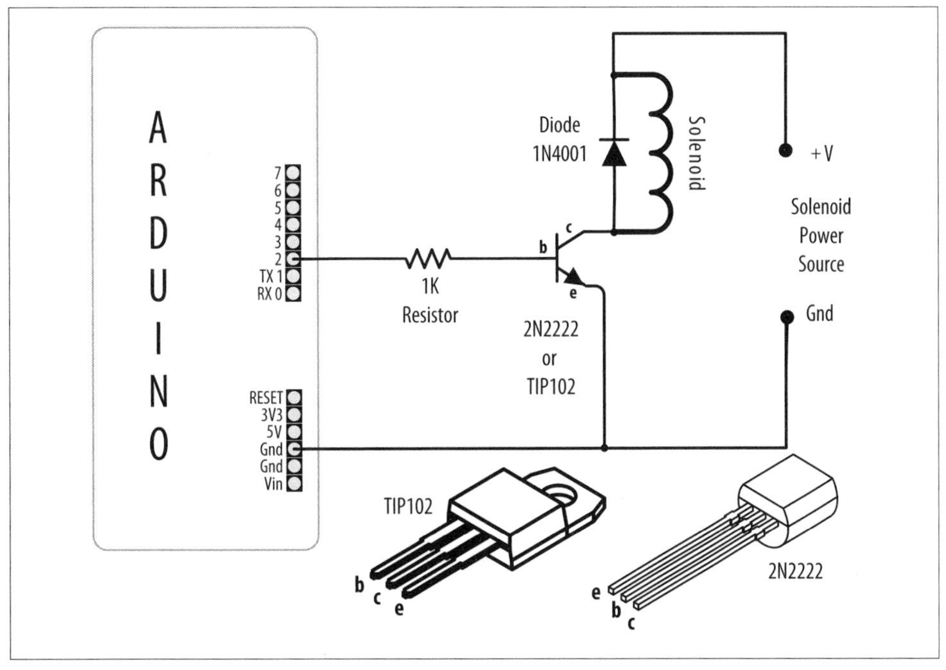

Figure 8-7. Driving a solenoid with a transistor

Discussion

The choice of transistor is dependent on the amount of current required to activate the solenoid or relay. The data sheet may specify this in milliamperes (mA) or as the resistance of the coil. To find the current needed by your solenoid or relay, divide the voltage of the coil by its resistance in ohms. For example, a 12V relay with a coil of 185 ohms draws 65 mA: 12 (volts) / 185 (ohms) = 0.065 amps, which is 65 mA.

Small transistors such as the 2N2222 are sufficient for solenoids requiring up to a few hundred milliamps. Larger solenoids will require a higher power transistor, like the TIP102/TIP120 or similar. There are many suitable transistor alternatives; see Appendix B for help reading a data sheet and choosing transistors.

The purpose of the diode is to prevent reverse EMF from the coil from damaging the transistor (*reverse EMF* is a voltage produced when current through a coil is switched off). The polarity of the diode is important; there is a colored band indicating the cathode—this should be connected to the solenoid positive power supply.

Electromagnetic relays are activated just like solenoids. A special relay called a *solid state relay* (SSR) has internal electronics that can be driven directly from an Arduino pin without the need for the transistor. Check the data sheet for your relay to see what voltage and current it requires; anything more than 40 mA at 5 volts will require a circuit such as the one shown in Figure 8-7.

8.7 Making an Object Vibrate

Problem

You want something to vibrate under Arduino control. For example, you want your project to shake for one second every minute.

Solution

Connect a vibration motor as shown in Figure 8-8.

The following sketch will turn on the vibration motor for one second each minute:

```
/*
 * Vibrate sketch
 * Vibrate for one second every minute
 *
 */

const int motorPin  = 3;  // vibration motor transistor is connected to pin 3

void setup()
{
  pinMode(motorPin, OUTPUT);
}
```

```
void loop()
{
  digitalWrite(motorPin,  HIGH);  // vibrate
  delay(1000); // delay one second
  digitalWrite(motorPin,  LOW);   // stop vibrating
  delay(59000); // wait 59 seconds.
}
```

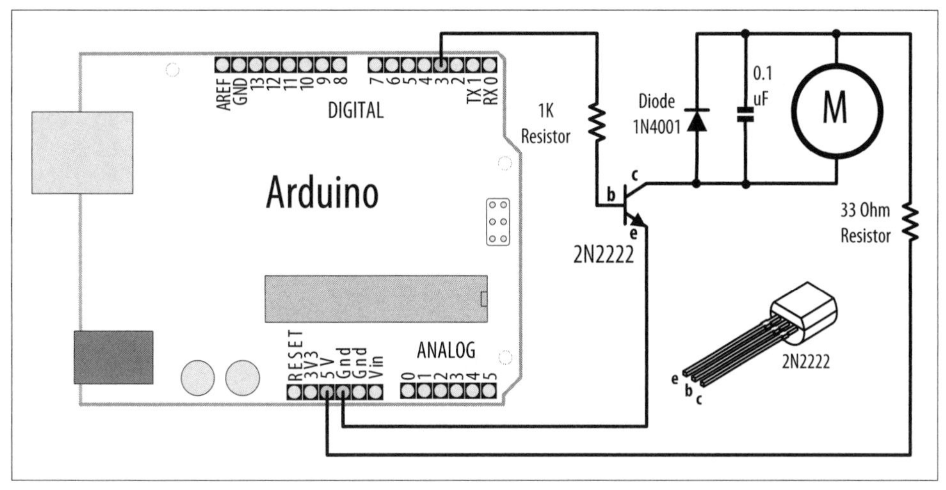

Figure 8-8. Connecting a vibration motor

Discussion

This recipe uses a motor designed to vibrate, such as the SparkFun ROB-08449. If you have an old cell phone you no longer need, it may contain tiny vibration motors that would be suitable. Vibration motors require more power than an Arduino pin can provide, so a transistor is used to switch the motor current on and off. Almost any NPN transistor can be used; Figure 8-3 shows the common 2N2222. See this book's website (*http://shop.oreilly.com/product/0636920022244.do*) for supplier information on this and the other components used. A 1 kilohm resistor connects the output pin to the transistor base; the value is not critical, and you can use values up to 4.7 kilohm or so (the resistor prevents too much current flowing through the output pin). The diode absorbs (or *snubs*—it's sometimes called a *snubber diode*) voltages produced by the motor windings as it rotates. The capacitor absorbs voltage spikes produced when the *brushes* (contacts connecting electric current to the motor windings) open and close. The 33 ohm resistor is needed to limit the amount of current flowing through the motor.

This sketch sets the output pin `HIGH` for one second (1,000 milliseconds) and then waits for 59 seconds. The transistor will turn on (conduct) when the pin is `HIGH`, allowing current to flow through the motor.

Here is a variation of this sketch that uses a sensor to make the motor vibrate. The wiring is similar to that shown in Figure 8-8, with the addition of a photocell connected to analog pin 0 (see Recipe 6.2):

```
/*
 * Vibrate_Photocell sketch
 * Vibrate when photosensor detects light above ambient level
 *
 */

const int motorPin  = 3;  // vibration motor transistor is connected to pin 3
const int sensorPin = 0;  // Photodetector connected to analog input 0
int sensorAmbient = 0;              // ambient light level (calibrated in setup)
const int thresholdMargin = 100;    // how much above ambient needed to vibrate

void setup()
{
  pinMode(motorPin, OUTPUT);
  sensorAmbient = analogRead(sensorPin); // get startup light level;
}

void loop()
{
  int sensorValue = analogRead(sensorPin);
  if( sensorValue > sensorAmbient + thresholdMargin)
  {
     digitalWrite(motorPin,  HIGH); //vibrate
  }
  else
   {
     digitalWrite(motorPin,  LOW);   // stop vibrating
  }
}
```

Here the output pin is turned on when a light shines on the photocell. When the sketch starts, the background light level on the sensor is read and stored in the variable `sensorAmbient`. Light levels read in `loop` that are higher than this will turn on the vibration motor.

8.8 Driving a Brushed Motor Using a Transistor

Problem

You want to turn a motor on and off. You may want to control its speed. The motor only needs to turn in one direction.

Solution

This sketch turns the motor on and off and controls its speed from commands received on the serial port (Figure 8-9 shows the connections):

```
/*
 * SimpleBrushed sketch
 * commands from serial port control motor speed
 * digits '0' through '9' are valid where '0' is off, '9' is max speed
 */

const int motorPin = 3; // motor driver is connected to pin 3

void setup()
{
  Serial.begin(9600);
}

void loop()
{
  if ( Serial.available()) {
    char ch = Serial.read();

    if(isDigit(ch))              // is ch a number?
    {
      int speed = map(ch, '0', '9', 0, 255);
      analogWrite(motorPin, speed);
      Serial.println(speed);
    }
    else
    {
      Serial.print("Unexpected character ");
      Serial.println(ch);
    }
  }
}
```

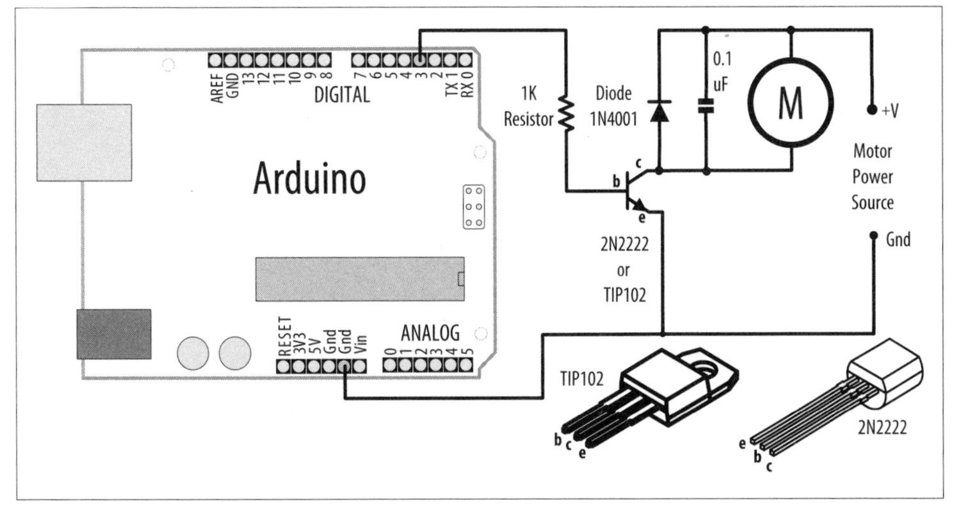

Figure 8-9. Driving a brushed motor

Discussion

This recipe is similar to Recipe 8.7; the difference is that `analogWrite` is used to control the speed of the motor. See "Analog Output" on page 241 for more on `analogWrite` and Pulse Width Modulation (PWM).

8.9 Controlling the Direction of a Brushed Motor with an H-Bridge

Problem

You want to control the direction of a brushed motor—for example, you want to cause a motor to rotate in one direction or the other from serial port commands.

Solution

An H-Bridge can control two brushed motors. Figure 8-10 shows the connections for the L293D H-Bridge IC; you can also use the SN754410, which has the same pin layout:

```
/*
 * Brushed_H_Bridge_simple sketch
 * commands from serial port control motor direction
 * + or - set the direction, any other key stops the motor
 */

const int in1Pin = 5;  // H-Bridge input pins
const int in2Pin = 4;

void setup()
{
  Serial.begin(9600);
  pinMode(in1Pin, OUTPUT);
  pinMode(in2Pin, OUTPUT);
  Serial.println("+ - to set direction, any other key stops motor");
}
void loop()
{
  if ( Serial.available()) {
    char ch = Serial.read();
    if (ch == '+')
    {
      Serial.println("CW");
      digitalWrite(in1Pin,LOW);
      digitalWrite(in2Pin,HIGH);
    }
    else if (ch == '-')
    {
      Serial.println("CCW");
      digitalWrite(in1Pin,HIGH);
      digitalWrite(in2Pin,LOW);
    }
```

```
      else
      {
        Serial.print("Stop motor");
        digitalWrite(in1Pin,LOW);
        digitalWrite(in2Pin,LOW);
      }
    }
  }
```

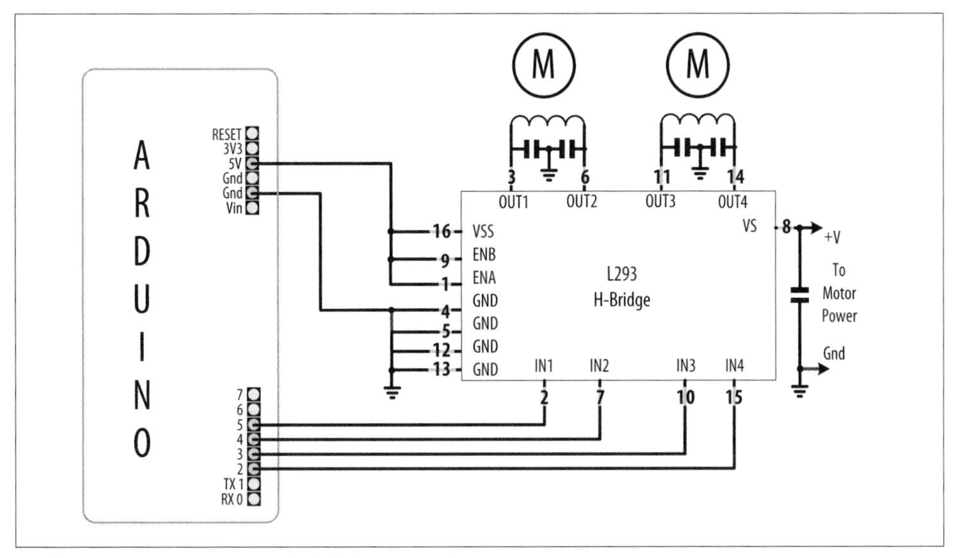

Figure 8-10. Connecting two brushed motors using an L293D H-Bridge

Discussion

Table 8-1 shows how the values on the H-Bridge input affect the motor. In the sketch in this recipe's Solution, a single motor is controlled using the IN1 and IN2 pins; the EN pin is permanently `HIGH` because it is connected to +5V.

Table 8-1. Logic table for H-Bridge

EN	IN1	IN2	Function
`HIGH`	`LOW`	`HIGH`	Turn clockwise
`HIGH`	`HIGH`	`LOW`	Turn counterclockwise
`HIGH`	`LOW`	`LOW`	Motor stop
`HIGH`	`HIGH`	`HIGH`	Motor stop
`LOW`	Ignored	Ignored	Motor stop

Figure 8-10 shows how a second motor can be connected. The following sketch controls both motors together:

```
/*
 * Brushed_H_Bridge_simple2 sketch
 * commands from serial port control motor direction
 * + or - set the direction, any other key stops the motors
 */

const int in1Pin = 5;  // H-Bridge input pins
const int in2Pin = 4;

const int in3Pin = 3;  // H-Bridge pins for second motor
const int in4Pin = 2;

void setup()
{
  Serial.begin(9600);
  pinMode(in1Pin, OUTPUT);
  pinMode(in2Pin, OUTPUT);
  pinMode(in3Pin, OUTPUT);
  pinMode(in4Pin, OUTPUT);
  Serial.println("+ - sets direction of motors, any other key stops motors");
}

void loop()
{
  if ( Serial.available()) {
    char ch = Serial.read();
    if (ch == '+')
    {
      Serial.println("CW");
      // first motor
      digitalWrite(in1Pin,LOW);
      digitalWrite(in2Pin,HIGH);
      //second motor
      digitalWrite(in3Pin,LOW);
      digitalWrite(in4Pin,HIGH);
    }
    else if (ch == '-')
    {
      Serial.println("CCW");
      digitalWrite(in1Pin,HIGH);
      digitalWrite(in2Pin,LOW);

      digitalWrite(in3Pin,HIGH);
      digitalWrite(in4Pin,LOW);
    }
    else
    {
      Serial.print("Stop motors");
      digitalWrite(in1Pin,LOW);
      digitalWrite(in2Pin,LOW);
      digitalWrite(in3Pin,LOW);
      digitalWrite(in4Pin,LOW);
    }
  }
}
```

8.10 Controlling the Direction and Speed of a Brushed Motor with an H-Bridge

Problem

You want to control the direction and speed of a brushed motor. This extends the functionality of Recipe 8.9 by controlling both motor direction and speed through commands from the serial port.

Solution

Connect a brushed motor to the output pins of the H-Bridge as shown in Figure 8-11.

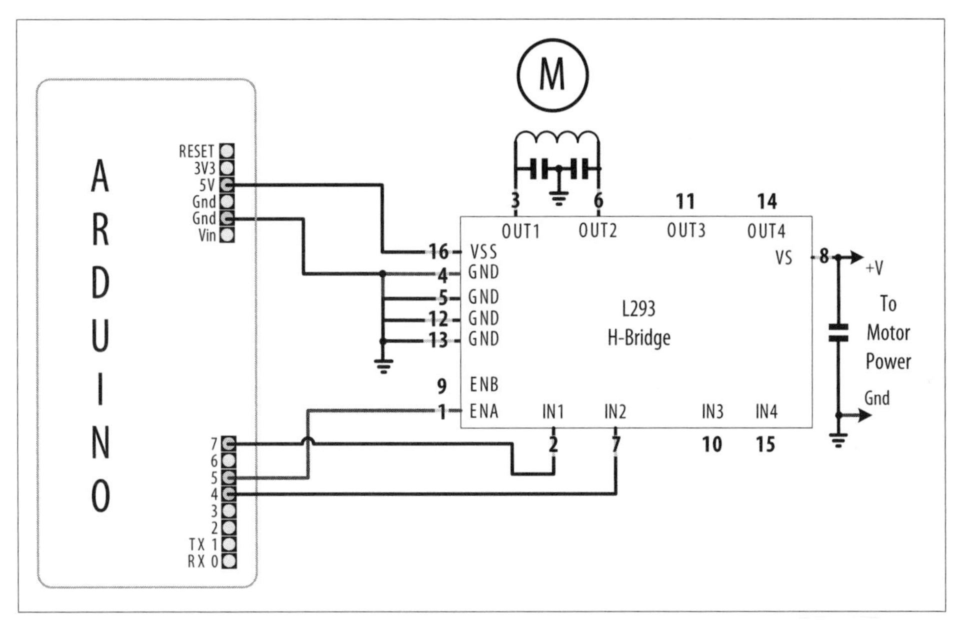

Figure 8-11. Connecting a brushed motor using analogWrite for speed control

This sketch uses commands from the Serial Monitor to control the speed and direction of the motor. Sending 0 will stop the motor, and the digits 1 through 9 will control the speed. Sending "+" and "-" will set the motor direction:

```
/*
 * Brushed_H_Bridge sketch
 * commands from serial port control motor speed and direction
 * digits '0' through '9' are valid where '0' is off, '9' is max speed
 * + or - set the direction
 */

const int enPin  = 5;  // H-Bridge enable pin
```

```
const int in1Pin = 7;  // H-Bridge input pins
const int in2Pin = 4;

void setup()
{
  Serial.begin(9600);
  pinMode(in1Pin, OUTPUT);
  pinMode(in2Pin, OUTPUT);
  Serial.println("Speed (0-9) or + - to set direction");
}

void loop()
{
  if ( Serial.available()) {
    char ch = Serial.read();
    if(isDigit(ch))              // is ch a number?
    {
      int speed = map(ch, '0', '9', 0, 255);
      analogWrite(enPin, speed);
      Serial.println(speed);
    }
    else if (ch == '+')
    {
      Serial.println("CW");
      digitalWrite(in1Pin,LOW);
      digitalWrite(in2Pin,HIGH);
    }
    else if (ch == '-')
    {
      Serial.println("CCW");
      digitalWrite(in1Pin,HIGH);
      digitalWrite(in2Pin,LOW);
    }
    else
    {
      Serial.print("Unexpected character ");
      Serial.println(ch);
    }
  }
}
```

Discussion

This recipe is similar to Recipe 8.9, in which motor direction is controlled by the levels on the IN1 and IN2 pins. But in addition, speed is controlled by the `analogWrite` value on the EN pin (see Chapter 7 for more on PWM). Writing a value of `0` will stop the motor; writing `255` will run the motor at full speed. The motor speed will vary in proportion to values within this range.

8.11 Using Sensors to Control the Direction and Speed of Brushed Motors (L293 H-Bridge)

Problem

You want to control the direction and speed of brushed motors with feedback from sensors. For example, you want two photo sensors to control motor speed and direction to cause a robot to move toward a beam of light.

Solution

This Solution uses similar motor connections to those shown in Figure 8-10, but with the addition of two light-dependent resistors, as shown in Figure 8-12.

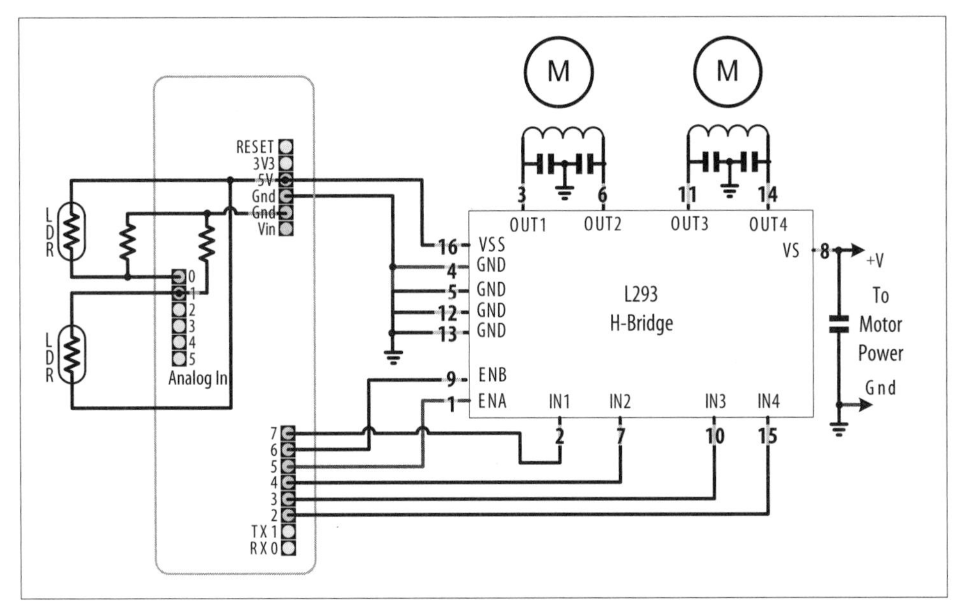

Figure 8-12. Two motors controlled using sensors

The sketch monitors the light level on the sensors and drives the motors to steer toward the sensor detecting the brighter light level:

```
/*
 * Brushed_H_Bridge_Direction sketch
 * uses photo sensors to control motor direction
 * robot moves in the direction of a light
 */

int leftPins[]  = {5,7,4};  // on pin for PWM, two pins for motor direction
int rightPins[] = {6,3,2};
```

```
const int MIN_PWM        = 64;  // this can range from 0 to MAX_PWM;
const int MAX_PWM        = 128; // this can range from around 50 to 255;
const int leftSensorPin = 0;  // analog pins with sensors
const int rightSensorPin = 1;

int sensorThreshold = 0;      // must have this much light on a sensor to move

void setup()
{
  for(int i=1; i < 3; i++)
  {
    pinMode(leftPins[i], OUTPUT);
    pinMode(rightPins[i], OUTPUT);
  }
}

void loop()
{
  int leftVal = analogRead(leftSensorPin);
  int rightVal = analogRead(rightSensorPin);

  if(sensorThreshold == 0){  // have the sensors been calibrated ?
  // if not, calibrate sensors to something above the ambient average
     sensorThreshold = ((leftVal + rightVal) / 2) + 100 ;
  }

  if( leftVal > sensorThreshold || rightVal > sensorThreshold)
  {
    // if there is adequate light to move ahead
    setSpeed(rightPins, map(rightVal,0,1023, MIN_PWM, MAX_PWM));
    setSpeed(leftPins,  map(leftVal ,0,1023, MIN_PWM, MAX_PWM));
  }
}

void setSpeed(int pins[], int speed )
{
  if(speed < 0)
  {
    digitalWrite(pins[1],HIGH);
    digitalWrite(pins[2],LOW);
    speed = -speed;
  }
  else
  {
    digitalWrite(pins[1],LOW);
    digitalWrite(pins[2],HIGH);
  }
  analogWrite(pins[0], speed);
}
```

Discussion

This sketch controls the speed of two motors in response to the amount of light detected by two photocells. The photocells are arranged so that an increase in light on one side will increase the speed of the motor on the other side. This causes the robot to turn toward the side with the brighter light. Light shining equally on both cells makes the robot move forward in a straight line. Insufficient light causes the robot to stop.

Light is sensed through analog inputs 0 and 1 using `analogRead` (see Recipe 6.2). When the program starts, the ambient light is measured and this threshold is used to determine the minimum light level needed to move the robot. A margin of 100 is added to the average level of the two sensors so the robot won't move for small changes in ambient light level. Light level as measured with `analogRead` is converted into a PWM value using the `map` function. Set `MIN_PWM` to the approximate value that enables your robot to move (low values will not provide sufficient torque; find this through trial and error with your robot). Set `MAX_PWM` to a value (up to 255) to determine the fastest speed you want the robot to move.

Motor speed is controlled in the `setSpeed` function. Two pins are used to control the direction for each motor, with another pin to control speed. The pin numbers are held in the `leftPins` and `rightPins` arrays. The first pin in each array is the speed pin; the other two pins are for direction.

An alternative to the L293 is the Toshiba FB6612FNG. This can be used in any of the recipes showing the L293D. Figure 8-13 shows the wiring for the FB6612 as used on the Pololu breakout board (SparkFun ROB-09402).

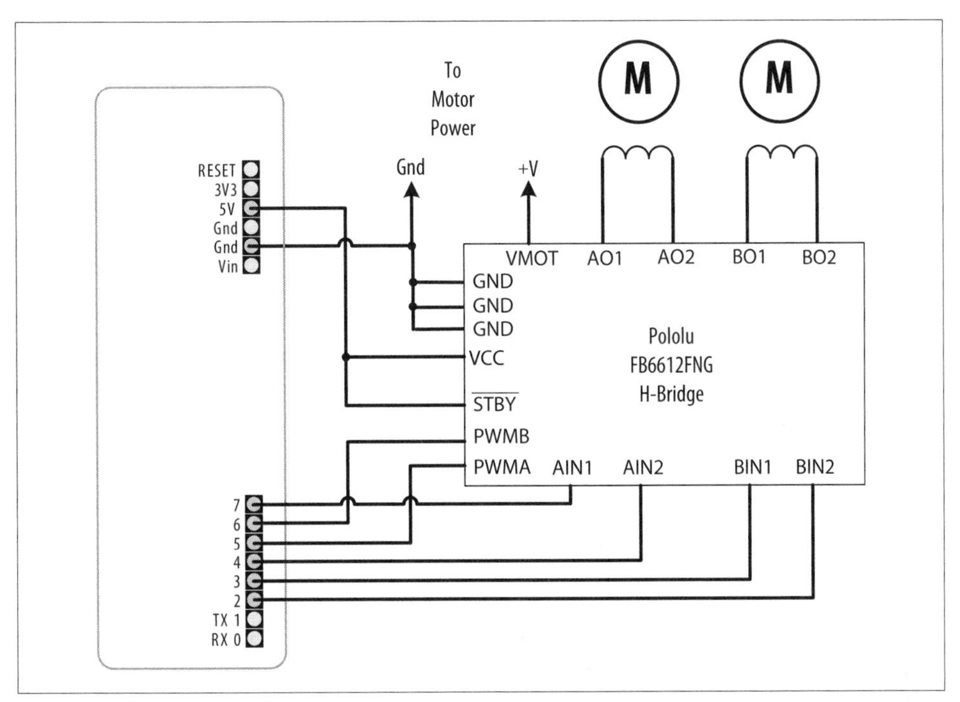

Figure 8-13. H-Bridge wiring for the Pololu breakout board

You can reduce the number of pins needed by adding additional hardware to control the direction pins. This is done by using only one pin per motor for direction, with a transistor or logic gate to invert the level on the other H-Bridge input. You can find circuit diagrams for this in the Arduino wiki, but if you want something already wired up, you can use an H-Bridge shield such as the Freeduino motor control shield (NKC Electronics ARD-0015) or the Ardumoto from SparkFun (DEV-09213). These shields plug directly into Arduino and only require connections to the motor power supply and windings.

Here is the sketch revised for the Ardumoto shield:

```
/*
 * Brushed_H_Bridge_Direction sketch for Ardumotor shield
 * uses photo sensors to control motor direction
 * robot moves in the direction of a light
 */

 int leftPins[]  = {10,12};    // one pin for PWM, one pin for motor direction
 int rightPins[] = {11,13};

const int MIN_PWM        = 64;  // this can range from 0 to MAX_PWM;
const int MAX_PWM        = 128; // this can range from around 50 to 255;
const int leftSensorPin = 0;  // analog pins with sensors
const int rightSensorPin = 1;

int sensorThreshold = 0;      // must have this much light on a sensor to move

void setup()
{
  pinMode(leftPins[1], OUTPUT);
  pinMode(rightPins[1], OUTPUT);
}

void loop()
{
  int leftVal = analogRead(leftSensorPin);
  int rightVal = analogRead(rightSensorPin);
  if(sensorThreshold == 0){  // have the sensors been calibrated ?
    // if not, calibrate sensors to something above the ambient average
     sensorThreshold = ((leftVal + rightVal) / 2) + 100 ;
  }

  if( leftVal > sensorThreshold || rightVal > sensorThreshold)
  {
    // if there is adequate light to move ahead
    setSpeed(rightPins, map(rightVal,0,1023, MIN_PWM, MAX_PWM));
    setSpeed(leftPins,  map(leftVal, 0,1023, MIN_PWM, MAX_PWM));
  }
}

void setSpeed(int pins[], int speed )
{
  if(speed < 0)
  {
```

```
      digitalWrite(pins[1],HIGH);
      speed = -speed;
    }
    else
    {
      digitalWrite(pins[1],LOW);
    }
    analogWrite(pins[0], speed);
  }
```

The `loop` function is identical to the preceding sketch. `setSpeed` has less code because hardware on the shield allows a single pin to control motor direction.

The pin assignments for the Freeduino shield are as follows:

```
int leftPins[]  = {10,13};  // PWM, Direction
int rightPins[] = {9,12};   // PWM, Direction
```

Here is the same functionality implemented using the Adafruit Motor Shield (*http://www.ladyada.net/make/mshield/*); see Figure 8-14. This uses a library named `AFMotor` that can be downloaded from the Adafruit website.

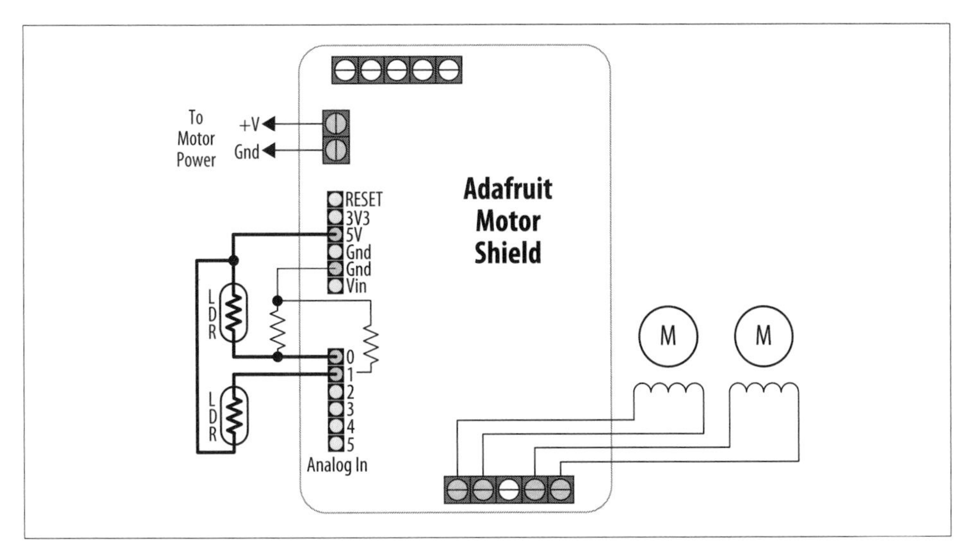

Figure 8-14. Using the Adafruit Motor Shield

The Adafruit shield supports four connections for motor windings; the sketch that follows has the motors connected to connectors 3 and 4:

```
/*
 * Brushed_H_Bridge_Direction sketch for Adafruit Motor shield
 * uses photo sensors to control motor direction
 * robot moves in the direction of a light
 */

#include "AFMotor.h"  // adafruit motor shield library
```

```
AF_DCMotor leftMotor(3, MOTOR12_1KHZ); // motor #3, 1 KHz pwm uses pin 5
AF_DCMotor rightMotor(4, MOTOR12_1KHZ); // motor #4, 1 KHz pwm uses pin 6

const int MIN_PWM        = 64;  // this can range from 0 to MAX_PWM;
const int MAX_PWM        = 128; // this can range from around 50 to 255;
const int leftSensorPin  = 0;   // analog pins with sensors
const int rightSensorPin = 1;

int sensorThreshold = 0;  // must be more light than this on sensors to move

void setup()
{
}

void loop()
{
  int leftVal  = analogRead(leftSensorPin);
  int rightVal = analogRead(rightSensorPin);

  if(sensorThreshold == 0){  // have the sensors been calibrated ?
  // if not, calibrate sensors to something above the ambient average
     sensorThreshold = ((leftVal + rightVal) / 2) + 100 ;
  }

  if( leftVal > sensorThreshold || rightVal > sensorThreshold)
  {
    // if there is adequate light to move ahead
    setSpeed(rightMotor, map(rightVal,0,1023, MIN_PWM, MAX_PWM));
    setSpeed(leftMotor,  map(leftVal ,0,1023, MIN_PWM, MAX_PWM));
  }
}

void setSpeed(AF_DCMotor &motor, int speed )
{
  if(speed < 0)
  {
    motor.run(BACKWARD);
    speed = -speed;
  }
  else
  {
    motor.run(FORWARD);
  }
  motor.setSpeed(speed);
}
```

If you have a different shield than the ones mentioned above, you will need to refer to the data sheet and make sure the values in the sketch match the pins used for PWM and direction.

See Also

The data sheet for the Pololu board: *http://www.pololu.com/file/0J86/TB6612FNG.pdf*

The product page for the Freeduino shield: *http://www.nkcelectronics.com/freeduino-arduino-motor-control-shield-kit.html*

The product page for the Ardumoto shield: *http://www.sparkfun.com/commerce/product_info.php?products_id=9213*

The Adafruit Motor Shield documentation and library can be found here: *http://www.ladyada.net/make/mshield/*

8.12 Driving a Bipolar Stepper Motor

Problem

You have a bipolar (four-wire) stepper motor and you want to step it under program control using an H-Bridge.

Solution

This sketch steps the motor in response to serial commands. A numeric value followed by a + steps in one direction; a - steps in the other. For example, "24+" steps a 24-step motor through one complete revolution in one direction, and "12-" steps half a revolution in the other direction (Figure 8-15 shows the connections to a four-wire bipolar stepper using the L293 H-Bridge):

```
/*
 * Stepper_bipolar sketch
 *
 * stepper is controlled from the serial port.
 * a numeric value followed by '+' or '-' steps the motor
 *
 *
 * http://www.arduino.cc/en/Reference/Stepper
 */

#include <Stepper.h>

// change this to the number of steps on your motor
#define STEPS 24

// create an instance of the stepper class, specifying
// the number of steps of the motor and the pins it's
// attached to
Stepper stepper(STEPS, 2, 3, 4, 5);

int steps = 0;
```

```
void setup()
{
  // set the speed of the motor to 30 RPM
  stepper.setSpeed(30);
  Serial.begin(9600);
}

void loop()
{
  if ( Serial.available()) {
    char ch = Serial.read();

    if(isDigit(ch)){ // is ch a number?
      steps = steps * 10 + ch - '0';  // yes, accumulate the value
    }
    else if(ch == '+'){
      stepper.step(steps);
      steps = 0;
    }
    else if(ch == '-'){
      stepper.step(steps * -1);
      steps = 0;
    }
  }
}
```

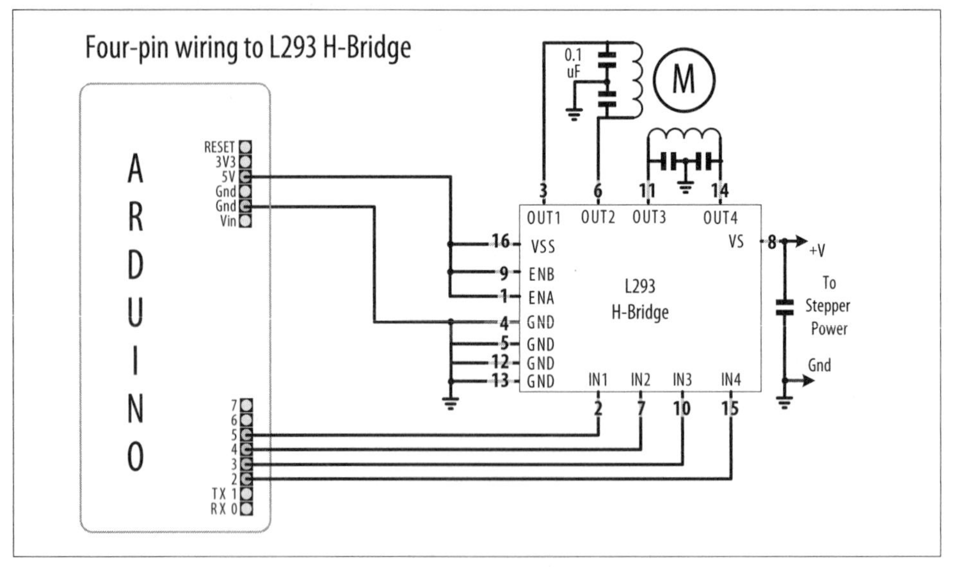

Figure 8-15. Four-wire bipolar stepper using L293 H-Bridge

Discussion

If your stepper requires a higher current than the L293 can provide (600 mA for the L293D), you can use the SN754410 chip for up to 1 amp with the same wiring and code as the L293. For current up to 2 amps, you can use the L298 chip. The L298 can use the same sketch as shown in this recipe's Solution, and it should be connected as shown in Figure 8-16.

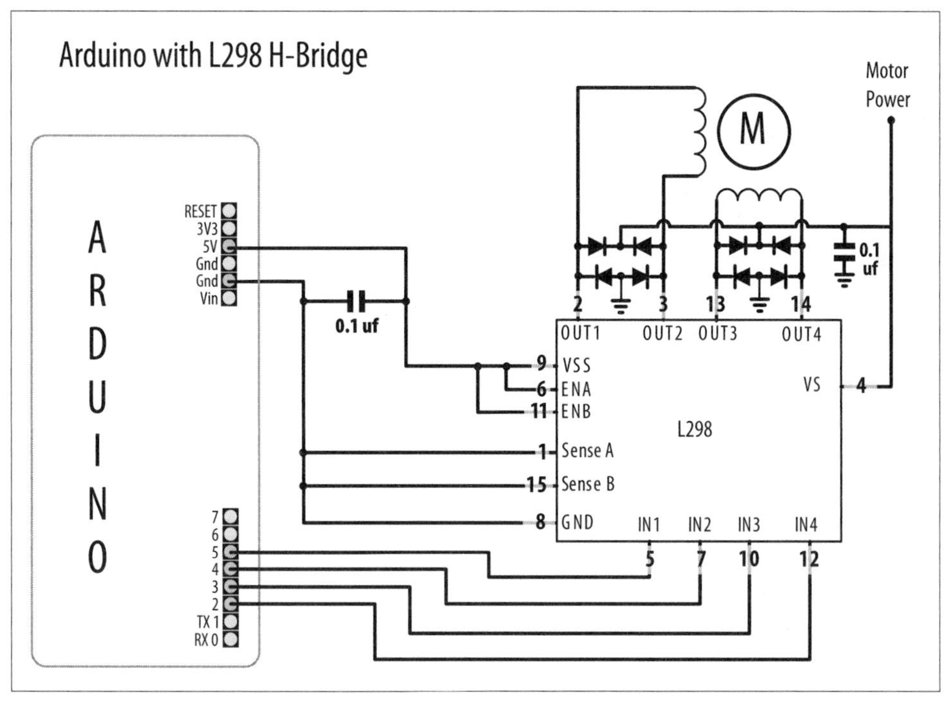

Figure 8-16. Unipolar stepper with L298

A simple way to connect an L298 to Arduino is to use the SparkFun Ardumoto shield (DEV-09213). This plugs on top of an Arduino board and only requires external connection to the motor windings; the motor power comes from the Arduino Vin (external Voltage Input) pin. In1/2 is controlled by pin 12, and ENA is pin 10. In3/4 is connected to pin 13, and ENB is on pin 11. Make the following changes to the code to use the preceding sketch with Ardumoto:

```
Stepper stepper(STEPS, 12,13);
```

Replace all the code inside of `setup()` with the following:

```
pinMode(10, OUTPUT);
digitalWrite(10, LOW);    // enable A

pinMode(11, OUTPUT);
digitalWrite(11, LOW);    // enable B

stepper.setSpeed(30);   // set the speed of the motor to 30 rpm

Serial.begin(9600);
```

The `loop` code is the same as the previous sketch.

See Also

For more on stepper motor wiring, see Tom Igoe's stepper motor notes: *http://www.tigoe.net/pcomp/code/circuits/motors*.

8.13 Driving a Bipolar Stepper Motor (Using the EasyDriver Board)

Problem

You have a bipolar (four-wire) stepper motor and you want to step it under program control using the EasyDriver board.

Solution

This Solution is similar to Recipe 8.12, and uses the same serial command protocol described there, but it uses the popular EasyDriver board. Figure 8-17 shows the connections.

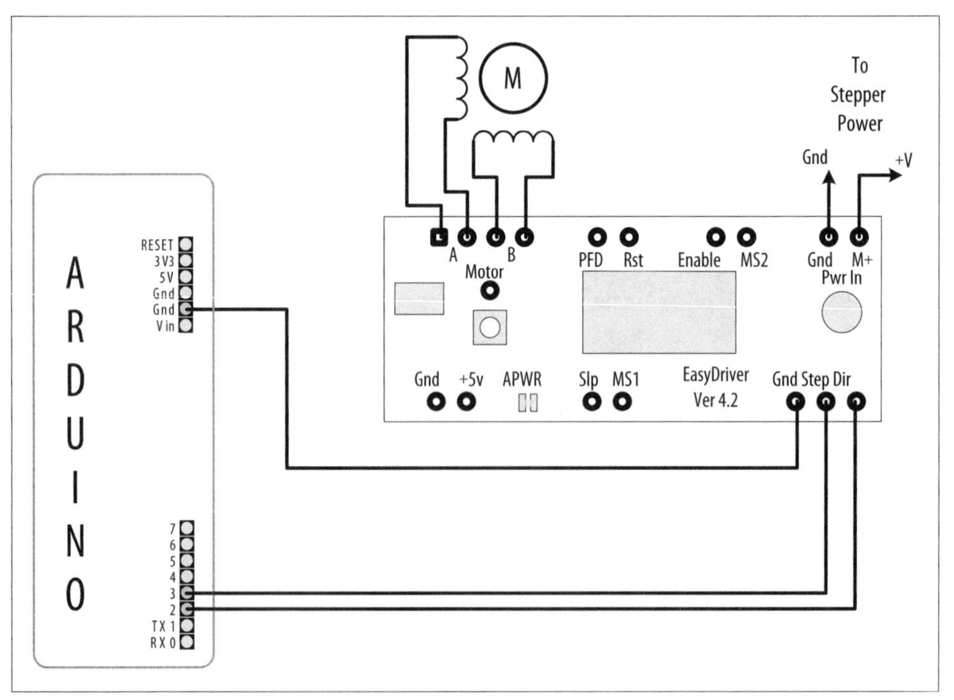

Figure 8-17. Connecting the EasyDriver board

The following sketch controls the step direction and count from the serial port. Unlike the code in Recipe 8.12, it does not require the Stepper library, because the EasyDriver board handles the control of the motor coils in hardware:

```
/*
 * Stepper_Easystepper sketch
 *
 * stepper is controlled from the serial port.
 * a numeric value followed by '+' or '-' steps the motor
 *
 */

const int dirPin = 2;
const int stepPin = 3;

int speed = 100;     // desired speed in steps per second
int steps = 0;       // the number of steps to make

void setup()
{
  pinMode(dirPin, OUTPUT);
  pinMode(stepPin, OUTPUT);
  Serial.begin(9600);
}

void loop()
{
  if ( Serial.available()) {
    char ch = Serial.read();

    if(isDigit(ch)){               // is ch a number?
      steps = steps * 10 + ch - '0';        // yes, accumulate the value
    }
    else if(ch == '+'){
      step(steps);
      steps = 0;
    }
    else if(ch == '-'){
      step(-steps);
      steps = 0;
    }
    else if(ch == 's'){
      speed = steps;
      Serial.print("Setting speed to ");
      Serial.println(steps);
      steps = 0;
    }
  }
}

void step(int steps)
{
  int stepDelay = 1000 / speed;  //delay in ms for speed given as steps per sec
  int stepsLeft;
```

```
    // determine direction based on whether steps_to_mode is + or -
    if (steps > 0)
    {
      digitalWrite(dirPin, HIGH);
      stepsLeft = steps;
    }
    if (steps < 0)
    {
      digitalWrite(dirPin, LOW);
      stepsLeft = -steps;
    }
    // decrement the number of steps, moving one step each time
    while(stepsLeft > 0)
    {
      digitalWrite(stepPin,HIGH);
      delayMicroseconds(1);
      digitalWrite(stepPin,LOW);
      delay(stepDelay);
      stepsLeft--;      // decrement the steps left
    }
  }
```

Discussion

The EasyDriver board is powered through the pins marked M+ and Gnd (shown in the upper right of Figure 8-17). The board operates with voltages between 8 volts and 30 volts; check the specifications of your stepper motor for the correct operating voltage. If you are using a 5V stepper, you must provide 5 volts to the pins marked Gnd and +5V (these pins are on the lower left of the EasyDriver board) and cut the jumper on the printed circuit board marked APWR (this disconnects the on-board regulator and powers the motor and EasyDriver board from an external 5V supply).

You can reduce current consumption when the motor is not stepping by connecting the Enable pin to a spare digital output and setting this `HIGH` to disable output (a `LOW` value enables output).

Stepping options are selected by connecting the MS1 and MS2 pins to +5V (`HIGH`) or Gnd (`LOW`), as shown in Table 8-2. The default options with the board connected as shown in Figure 8-17 will use eighth-step resolution (MS1 and MS2 are `HIGH`, Reset is `HIGH`, and Enable is `LOW`).

Table 8-2. Microstep options

Resolution	MS1	MS2
Full step	LOW	LOW
Half step	HIGH	LOW
Quarter step	LOW	HIGH
Eighth step	HIGH	HIGH

You can modify the code so that the speed value determines the revolutions per second as follows:

```
// use the following for speed given in RPM
int speed = 100;     // desired speed in RPM
int stepsPerRevolution = 200;  // this line sets steps for one revolution
```

Change the step function so that the first line is as follows:

```
int stepDelay = 60L * 1000L / stepsPerRevolution / speed; // speed as RPM
```

Everything else can remain the same, but now the speed command you send will be the RPM of the motor when it steps.

8.14 Driving a Unipolar Stepper Motor (ULN2003A)

Problem

You have a unipolar (five- or six-wire) stepper motor and you want to control it using a ULN2003A Darlington driver chip.

Solution

Connect a unipolar stepper as shown in Figure 8-18. The +V connection goes to a power supply rated for the voltage and current needed by your motor.

The following sketch steps the motor using commands from the serial port. A numeric value followed by a + steps in one direction; a - steps in the other:

```
/*
 * Stepper sketch
 *
 * stepper is controlled from the serial port.
 * a numeric value followed by '+' or '-' steps the motor
 *
 *
 * http://www.arduino.cc/en/Reference/Stepper
 */

#include <Stepper.h>

// change this to the number of steps on your motor
#define STEPS 24

// create an instance of the stepper class, specifying
// the number of steps of the motor and the pins it's
// attached to
Stepper stepper(STEPS, 2, 3, 4, 5);

int steps = 0;


void setup()
```

```
{
  stepper.setSpeed(30);    // set the speed of the motor to 30 RPMs
  Serial.begin(9600);
}

void loop()
{
  if ( Serial.available()) {
    char ch = Serial.read();

    if(isDigit(ch)){                // is ch a number?
      steps = steps * 10 + ch - '0';        // yes, accumulate the value
    }
    else if(ch == '+'){
      stepper.step(steps);
      steps = 0;
    }
    else if(ch == '-'){
      stepper.step(steps * -1);
      steps = 0;
    }
    else if(ch == 's'){
      stepper.setSpeed(steps);
      Serial.print("Setting speed to ");
      Serial.println(steps);
      steps = 0;
    }
  }
}
```

Discussion

This type of motor has two pairs of coils, and each coil has a connection to the center. Motors with only five wires have both center connections brought out on a single wire. If the connections are not marked, you can identify the wiring using a multimeter. Measure the resistance across pairs of wires to find the two pairs of wires that have the maximum resistance. The center tap wire should have half the resistance of the full coil. A step-by-step procedure is available at *http://techref.massmind.org/techref/io/stepper/wires.asp*.

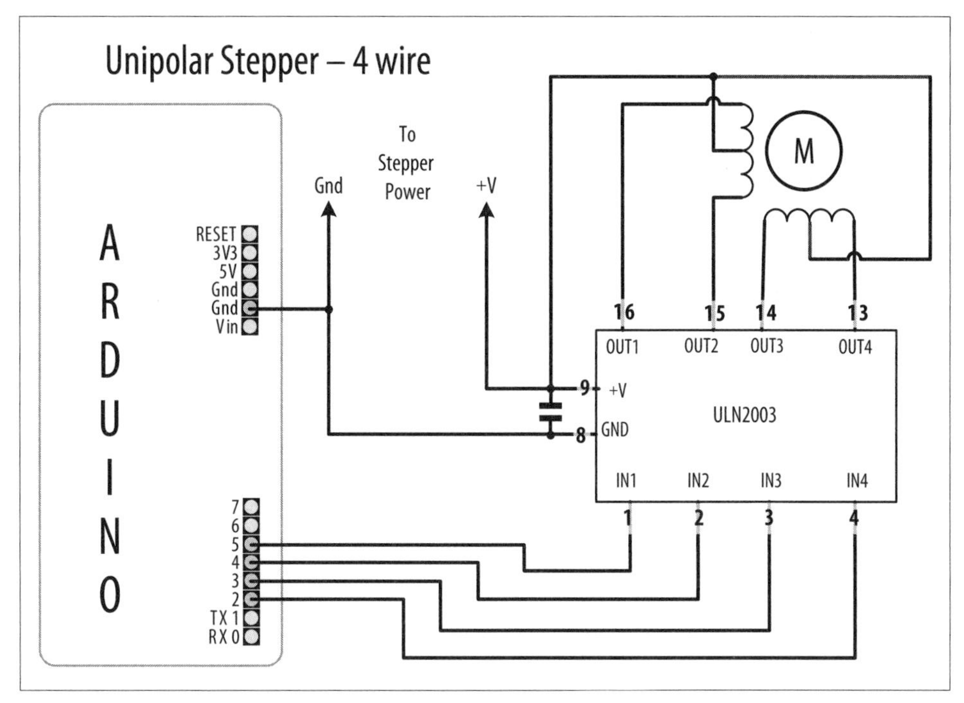

Figure 8-18. Unipolar stepper connected using ULN2003 driver

CHAPTER 9

Audio Output

9.0 Introduction

The Arduino isn't built to be a synthesizer, but it can certainly produce sound through an output device such as a speaker.

Sound is produced by vibrating air. A sound has a distinctive pitch if the vibration repeats regularly. The Arduino can create sound by driving a loudspeaker or *Piezo device* (a small ceramic transducer that produces sound when pulsed), converting electronic vibrations into speaker pulses that vibrate the air. The pitch (frequency) of the sound is determined by the time it takes to pulse the speaker in and out; the shorter the amount of time, the higher the frequency.

The unit of frequency is measured in hertz, and it refers to the number of times the signal goes through its repeating cycle in one second. The range of human hearing is from around 20 hertz (Hz) up to 20,000 hertz (although it varies by person and changes with age).

The Arduino software includes a `tone` function for producing sound. Recipes 9.1 and 9.2 show how to use this function to make sounds and tunes. The `tone` function uses hardware timers. On a standard Arduino board, only one tone can be produced at a time. Sketches where the timer (`timer2`) is needed for other functions, such as `analog Write` on pin 9 or 10, cannot use the `tone` function. To overcome this limitation, Recipe 9.3 shows how to use an enhanced tone library for multiple tones, and Recipe 9.4 shows how sound can be produced without using the `tone` function or hardware timers.

The sound that can be produced by pulsing a speaker is limited and does not sound very musical. The output is a square wave (see Figure 9-1), which sounds harsh and more like an antique computer game than a musical instrument.

It is difficult for Arduino to produce more musically complex sounds without external hardware. You can add a shield that extends Arduino's capabilities; Recipe 9.5 shows how to use the Adafruit Wave Shield to play back audio files from a memory card on the shield.

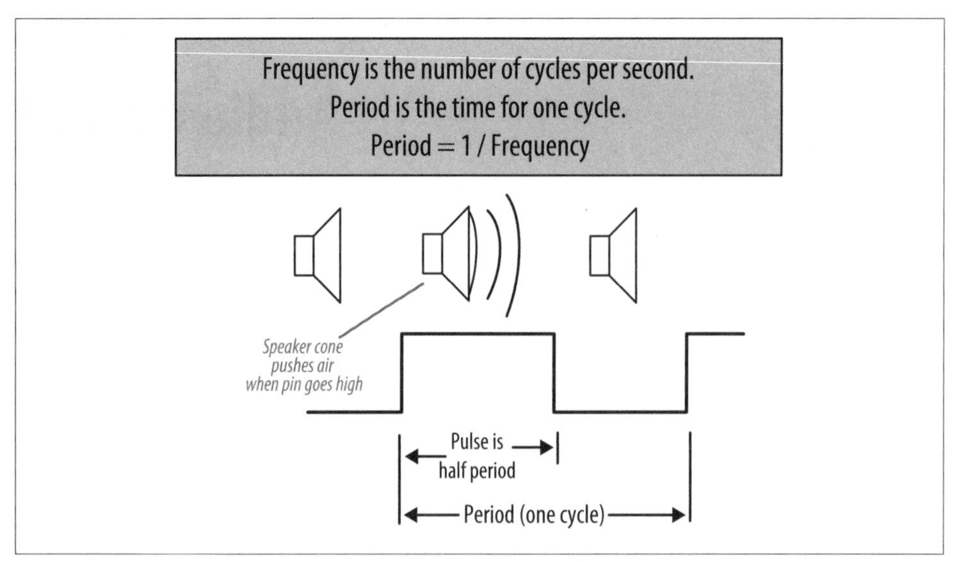

Figure 9-1. Generating sound using digital pulses

You can also use Arduino to control an external device that is built to make sound. Recipe 9.6 shows how to send Musical Instrument Digital Interface (MIDI) messages to a MIDI device. These devices produce high-quality sounds of a huge variety of instruments and can produce the sounds of many instruments simultaneously. The sketch in Recipe 9.6 shows how to generate MIDI messages to play a musical scale.

Recipe 9.7 provides an overview of an application called Auduino that uses complex software processing to synthesize sound.

This chapter covers the many ways you can generate sound electronically. If you want to make music by getting Arduino to play acoustic instruments (such as glockenspiels, drums, and acoustic pianos), you can employ actuators such as solenoids and servos that are covered in Chapter 8.

Many of the recipes in this chapter will drive a small speaker or Piezo device. The circuit for connecting one of these to an Arduino pin is shown in Figure 9-2.

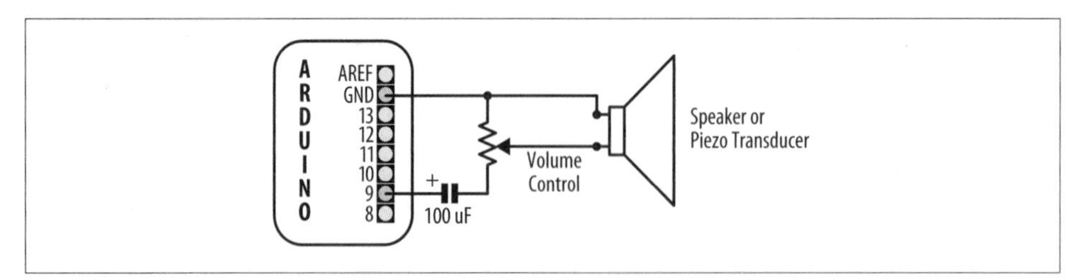

Figure 9-2. Connecting to an audio transducer

The volume control is a variable resistor, the value is not critical and anything from 200 to 500 ohms would work. The capacitor is a 100 microfarad electrolytic with the positive end connected to the Arduino pin. A speaker will work regardless of which wire is attached to ground, but a Piezo is polarized, so connect the negative wire (usually black) to the Gnd pin.

Alternatively, you can connect the output to an external audio amplifier. Recipe 9.7 shows how an output pin can be connected to an audio jack.

The voltage level (5 volts) is higher than audio amplifiers expect, so you may need to use a 4.7K variable resistor to reduce the voltage (connect one end to pin 9 and the other end to ground; then connect the slider to the tip of the jack plug. The barrel of the jack plug is connected to ground).

9.1 Playing Tones

Problem

You want to produce audio tones through a speaker or other audio transducer. You want to specify the frequency and duration of the tone.

Solution

Use the Arduino tone function. This sketch plays a tone with the frequency set by a variable resistor (or other sensor) connected to analog input 0 (see Figure 9-3):

```
/*
 * Tone sketch
 *
 * Plays tones through a speaker on digital pin 9
 * frequency determined by values read from analog port
 */

const int speakerPin = 9;      // connect speaker to pin 9
const int pitchPin = 0;        // pot that will determine the frequency of the tone

void setup()
{
}

void loop()
{
   int sensor0Reading = analogRead(pitchPin);     // read input to set frequency
   // map the analog readings to a meaningful range
   int frequency  = map(sensor0Reading, 0, 1023, 100,5000); // 100Hz to 5kHz
   int duration = 250;   // how long the tone lasts
   tone(speakerPin, frequency, duration); // play the tone
   delay(1000); // pause one second
}
```

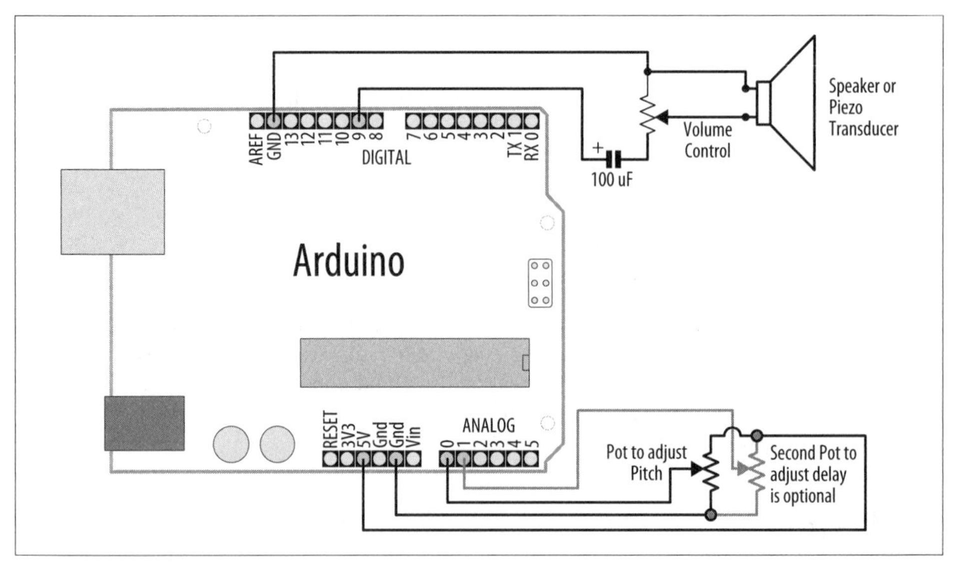

Figure 9-3. Connections for the Tone sketch

The `tone` function can take up to three parameters: the pin attached to the speaker, the frequency to play (in hertz), and the length of time (in milliseconds) to play the note. The third parameter is optional. If it is omitted, the note will continue until there is another call to `tone`, or a call to `noTone`. The value for the frequency is mapped to sensible values for audio frequencies in the following line:

```
int frequency  = map(sensor0Reading, 0, 1023, 100,5000); //100Hz to 5kHz
```

This variation uses a second variable resistor (the bottom right pot in Figure 9-3) to set the duration of the tone:

```
const int speakerPin = 9;      // connect speaker to pin 9
const int pitchPin = 0;        // input that determines frequency of the tone
const int durationPin = 1;     // input that will determine the duration of the tone

void setup()
{
}

void loop()
{
  int sensor0Reading = analogRead(pitchPin);    // read input for frequency
  int sensor1Reading = analogRead(durationPin); // read input for duration

  // map the analog readings to a meaningful range
  int frequency  = map(sensor0Reading, 0, 1023, 100,5000); // 100Hz to 5kHz
  int duration = map(sensor1Reading, 0, 1023, 100,1000);   // dur 0.1-1 second
  tone(speakerPin, frequency, duration); // play the tone
  delay(duration); //wait for the tone to finish
}
```

Another variation is to add a switch so that tones are generated only when the switch is pressed.

Enable pull-up resistors in `setup` with this line (see Recipe 5.2 for a connection diagram and explanation):

```
digitalWrite(inputPin,HIGH);  // turn on internal pull-up on the inputPin
```

Modify the `loop` code so that the `tone` and `delay` functions are only called when the switch is pressed:

```
if( digitalRead(inputPin) = LOW)  // read input value
{
   tone(speakerPin, frequency, duration); // play the tone
   delay(duration); //wait for the tone to finish
}
```

You can use almost any audio transducer to produce sounds with Arduino. Small speakers work very well. Piezo transducers also work and are inexpensive, robust, and easily salvaged from old audio greeting cards. Piezos draw little current (they are high-resistance devices), so they can be connected directly to the pin. Speakers are usually of much lower resistance and need a resistor to limit the current flow. The components to build the circuit pictured in Figure 9-3 should be easy to find; see this book's website (*http://shop.oreilly.com/product/0636920022244.do*) for suggestions on getting parts.

See Also

You can achieve enhanced functionality using the Tone library by Brett Hagman that is described in Recipe 9.3.

9.2 Playing a Simple Melody

Problem

You want Arduino to play a simple melody.

Solution

You can use the `tone` function described in Recipe 9.1 to play sounds corresponding to notes on a musical instrument. This sketch uses `tone` to play a string of notes, the "Hello world" of learning the piano, "Twinkle, Twinkle Little Star":

```
/*
 * Twinkle sketch
 *
 * Plays "Twinkle, Twinkle Little Star"
 *
 * speaker on digital pin 9
 */
```

```
const int speakerPin = 9; // connect speaker to pin 9

char noteNames[] =     {'C','D','E','F','G','a','b'};
unsigned int frequencies[] = {262,294,330,349,392,440,494};
const byte noteCount = sizeof(noteNames); // number of notes (7 here)

//notes, a space represents a rest
char score[] = "CCGGaaGFFEEDDC GGFFEEDGGFFEED CCGGaaGFFEEDDC ";
const byte scoreLen = sizeof(score); // the number of notes in the score

void setup()
{
}

void loop()
{
  for (int i = 0; i < scoreLen; i++)
  {
    int duration = 333;  // each note lasts for a third of a second
    playNote(score[i], duration); // play the note
  }

  delay(4000); // wait four seconds before repeating the song
}

void playNote(char note, int duration)
{
  // play the tone corresponding to the note name
  for (int i = 0; i < noteCount; i++)
  {
    // try and find a match for the noteName to get the index to the note
    if (noteNames[i] == note) // find a matching note name in the array
      tone(speakerPin, frequencies[i], duration); //  play the note
  }
  // if there is no match then the note is a rest, so just do the delay
  delay(duration);
}
```

`noteNames` is an array of characters to identify notes in a score. Each entry in the array is associated with a frequency defined in the `notes` array. For example, note C (the first entry in the `noteNames` array) has a frequency of 262 Hz (the first entry in the `notes` array).

`score` is an array of notes representing the note names you want to play:

```
// a space represents a rest
char score[] = "CCGGaaGFFEEDDC GGFFEEDGGFFEED CCGGaaGFFEEDDC ";
```

Each character in the score that matches a character in the `noteNames` array will make the note play. The space character is used as a rest, but any character not defined in `noteNames` will also produce a rest (no note playing).

The sketch calls `playNote` with each character in the score and a duration for the notes of one-third of a second.

The playNote function does a lookup in the noteNames array to find a match and uses the corresponding entry in the frequencies array to get the frequency to sound.

Every note has the same duration. If you want to specify the length of each note, you can add the following code to the sketch:

```
byte beats[scoreLen] = {1,1,1,1,1,1,2,  1,1,1,1,1,1,2,1,
                        1,1,1,1,1,1,2,  1,1,1,1,1,1,2,1,
                        1,1,1,1,1,1,2,  1,1,1,1,1,1,2};
byte beat = 180; // beats per minute for eighth notes
unsigned int speed = 60000 / beat; // the time in ms for one beat
```

beats is an array showing the length of each note: 1 is an eighth note, 2 a quarter note, and so on.

beat is the number of beats per minute.

speed is the calculation to convert beats per minute into a duration in milliseconds.

The only change to the loop code is to set the duration to use the value in the beats array. Change:

```
int duration = 333;  // each note lasts for a third of a second
```

to:

```
int duration = beats[i] * speed; // use beats array to determine duration
```

9.3 Generating More Than One Simultaneous Tone

Problem

You want to play two tones at the same time. The Arduino Tone library only produces a single tone on a standard board, and you want two simultaneous tones. Note that the Mega board has more timers and can produce up to six tones.

Solution

The Arduino Tone library is limited to a single tone because a different timer is required for each tone, and although a standard Arduino board has three timers, one is used for the millis function and another for servos. This recipe uses a library written by Brett Hagman, the author of the Arduino tone function. The library enables you to generate multiple simultaneous tones. You can download it from *http://code.google.com/p/rogue-code/wiki/ToneLibraryDocumentation*.

This is an example sketch from the download that plays two tones selectable from the serial port:

```
/*
 * Dual Tones - Simultaneous tone generation.
 *  plays notes 'a' through 'g' sent over the Serial Monitor.
 * lowercase letters for the first tone and uppercase for the second.
```

```
 * 's' stops the current playing tone.
 */
#include <Tone.h>

int notes[] = { NOTE_A3,
                NOTE_B3,
                NOTE_C4,
                NOTE_D4,
                NOTE_E4,
                NOTE_F4,
                NOTE_G4 };

// You can declare the tones as an array
Tone notePlayer[2];

void setup(void)
{
  Serial.begin(9600);
  notePlayer[0].begin(11);
  notePlayer[1].begin(12);
}

void loop(void)
{
  char c;

  if(Serial.available())
  {
    c = Serial.read();

    switch(c)
    {
      case 'a'...'g':
        notePlayer[0].play(notes[c - 'a']);
        Serial.println(notes[c - 'a']);
        break;
      case 's':
        notePlayer[0].stop();
        break;

      case 'A'...'G':
        notePlayer[1].play(notes[c - 'A']);
        Serial.println(notes[c - 'A']);
        break;
      case 'S':
        notePlayer[1].stop();
        break;

      default:
        notePlayer[1].stop();
        notePlayer[0].play(NOTE_B2);
        delay(300);
        notePlayer[0].stop();
        delay(100);
        notePlayer[1].play(NOTE_B2);
```

```
        delay(300);
        notePlayer[1].stop();
        break;
    }
  }
}
```

Discussion

To mix the output of the two tones to a single speaker, use 500 ohm resistors from each output pin and tie them together at the speaker. The other speaker lead connects to Gnd, as shown in the previous sketches.

On a standard Arduino board, the first tone will use timer 2 (so PWM on pins 9 and 10 will not be available); the second tone uses timer 1 (preventing the Servo library and PWM on pins 11 and 12 from working). On a Mega board, each simultaneous tone will use timers in the following order: 2, 3, 4, 5, 1, 0.

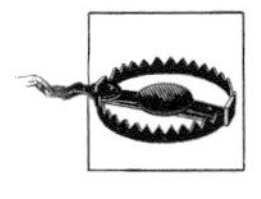

Playing three simultaneous notes on a standard Arduino board, or more than six on a Mega, is possible, but `millis` and `delay` will no longer work properly. It is safest to use only two simultaneous tones (or five on a Mega).

9.4 Generating Audio Tones and Fading an LED

Problem

You want to produce sounds through a speaker or other audio transducer, and you need to generate the tone in software instead of with a timer; for example, if you need to use `analogWrite` on pin 9 or 10.

Solution

The `tone` function discussed in earlier recipes is easier to use, but it requires a hardware timer, which may be needed for other tasks such as `analogWrite`. This code does not use a timer, but it will not do anything else while the note is played. Unlike the Arduino `tone` function, the `playTone` function described here will block—it will not return until the note has finished.

The sketch plays six notes, each one twice the frequency of (an octave higher than) the previous one. The `playTone` function generates a tone for a specified duration on a speaker or Piezo device connected to a digital output pin and ground; see Figure 9-4:

```
byte speakerPin = 9;
byte ledPin = 10;

void setup()
{
  pinMode(speakerPin, OUTPUT);
```

```
}

void playTone(int period, int duration)
{
// period is one cycle of tone
// duration is how long the pulsing should last in milliseconds
  int pulse = period / 2;
  for (long i = 0; i < duration * 1000L; i += period )
  {
    digitalWrite(speakerPin, HIGH);
    delayMicroseconds(pulse);
    digitalWrite(speakerPin, LOW);
    delayMicroseconds(pulse);
  }
}

void fadeLED(){
 for (int brightness = 0; brightness < 255; brightness++)
 {
   analogWrite(ledPin, brightness);
   delay(2);
  }
  for (int brightness = 255; brightness >= 0; brightness--) {
    analogWrite(ledPin, brightness);
    delay(2);
  }



}
void loop()
{
  // a note with period of 15289 is deep C (second lowest C note on piano)
  for(int period=15289; period >= 477;  period=period / 2)  // play 6 octaves
  {
     playTone( period, 200); // play tone for 200 milliseconds
  }
  fadeLED();
}
```

Figure 9-4. Connections for speaker and LED

Discussion

Two values are used by playTone: period and duration. The variable period represents the time for one cycle of the tone to play. The speaker is pulsed high and then low for the number of microseconds given by period. The for loop repeats the pulsing for the number of milliseconds given in the duration argument.

If you prefer to work in frequency rather than period, you can use the reciprocal relationship between frequency and period; period is equal to 1 divided by frequency. You need the period value in microseconds; because there are 1 million microseconds in one second, the period is calculated as 1000000L / frequency (the "L" at the end of that number tells the compiler that it should calculate using long integer math to prevent the calculation from exceeding the range of a normal integer—see the explanation of long integers in Recipe 2.2):

```
void playFrequency(int frequency, int duration)
{
  int period = 1000000L / frequency;
  int pulse = period / 2;
```

The rest of the code is the same as playTone:

```
  for (long i = 0; i < duration * 1000L; i += period )
  {
    digitalWrite(speakerPin, HIGH);
    delayMicroseconds(pulse);
    digitalWrite(speakerPin, LOW);
    delayMicroseconds(pulse);
  }
}
```

The code in this recipe stops and waits until a tone has completed before it can do any other processing. It is possible to produce the sound in the background (without waiting for the sound to finish) by putting the sound generation code in an interrupt handler. The source code for the tone function that comes with the Arduino distribution shows how this is done.

See Also

Recipe 9.7

Here are some examples of more complex audio synthesis that can be accomplished with the Arduino:

Pulse-Code Modulation
: PCM allows you to approximate analog audio using digital signaling. An Arduino wiki article that explains how to produce 8-bit PCM using a timer is available at *http://www.arduino.cc/playground/Code/PCMAudio*.

Pocket Piano shield
: Critter and Guitari's Pocket Piano shield gives you a piano-like keyboard, wave table synthesis, FM synthesis, and more; see *http://www.critterandguitari.com/home/store/arduino-piano.php*.

9.5 Playing a WAV File

Problem

Under program control, you want Arduino to trigger the playing of a WAV file.

Solution

This sketch uses the Adafruit Wave Shield and is based on one of the example sketches linked from the product page at *http://www.adafruit.com/index.php?main_page=product_info&products_id=94*.

This sketch will play one of nine files depending on readings taken from a variable resistor connected to analog input 0 when pressing a button connected to pin 15 (analog input 1):

```
/*
 * WaveShieldPlaySelection sketch
 *
 * play a selected WAV file
 *
 * Position of variable resistor slider when button pressed selects file to play
 *
 */

#include <FatReader.h>
#include <SdReader.h>

#include "WaveHC.h"
#include "WaveUtil.h"

SdReader card;    // This object holds the information for the card
FatVolume vol;    // This holds the information for the partition on the card
FatReader root;   // This holds the information for the volumes root directory
FatReader file;   // This object represents the WAV file
WaveHC wave;      // Only wave (audio) object - only one file played at a time


const int buttonPin = 15;
const int potPin = 0; // analog input pin 0

char * wavFiles[] = {
"1.WAV","2.WAV","3.WAV","4.WAV","5.WAV","6.WAV","7.WAV","8.WAV","9.WAV"};

void setup()
{
```

```
  Serial.begin(9600);
  pinMode(buttonPin, INPUT);
  digitalWrite(buttonPin, HIGH); // turn on pull-up resistor

  if (!card.init())
  {
    // Something went wrong, sdErrorCheck prints an error number
    putstring_nl("Card init. failed!");
    sdErrorCheck();
    while(1);                         // then 'halt' - do nothing!
  }


  // enable optimized read - some cards may time out
  card.partialBlockRead(true);

  // find a FAT partition!
  uint8_t part;
  for (part = 0; part < 5; part++)     // we have up to 5 slots to look in
  {
    if (vol.init(card, part))
      break;                          // found one so break out of this for loop
  }
  if (part == 5)                      // valid parts are 0 to 4, more not valid
  {
    putstring_nl("No valid FAT partition!");
    sdErrorCheck();                   // Something went wrong, print the error
    while(1);                         // then 'halt' - do nothing!
  }

  // tell the user about what we found
  putstring("Using partition ");
  Serial.print(part, DEC);
  putstring(", type is FAT");
  Serial.println(vol.fatType(),DEC);     // FAT16 or FAT32?

  // Try to open the root directory
  if (!root.openRoot(vol))
  {
    putstring_nl("Can't open root dir!"); // Something went wrong,
    while(1);                          // then 'halt' - do nothing!
  }

  // if here then all the file prep succeeded.
  putstring_nl("Ready!");
}

void loop()
{
  if(digitalRead(buttonPin) == LOW)
  {
    int value = analogRead(potPin);
    int index = map(value,0,1023,0,8); // index into one of the 9 files
    playcomplete(wavFiles[index]);
    Serial.println(value);
```

```
  }
}

// Plays a full file from beginning to end with no pause.
void playcomplete(char *name)
{
  // call playfile find and play this name
  playfile(name);
  while (wave.isplaying) {
    // do nothing while it's playing
  }
  // now it's done playing
}

void playfile(char *name) {
  // see if the wave object is currently doing something
  if (wave.isplaying) {
  // already playing something, so stop it!
    wave.stop(); // stop it
  }
  // look in the root directory and open the file
  if (!file.open(root, name)) {
    putstring("Couldn't open file ");
    Serial.print(name);
    return;
  }
  // read the file and turn it into a wave object
  if (!wave.create(file)) {
    putstring_nl("Not a valid WAV");
    return;
  }
  // start playback
  wave.play();
}

void sdErrorCheck(void)
{
  if (!card.errorCode()) return;
  putstring("\n\rSD I/O error: ");
  Serial.print(card.errorCode(), HEX);
  putstring(", ");
  Serial.println(card.errorData(), HEX);
  while(1)
    ; // stay here if there is an error
}
```

Discussion

The wave shield reads data stored on an SD card. It uses its own library that is available from the Ladyada website (*http://www.ladyada.net/make/waveshield/*). The WAV files to be played need to be put on the memory card using a computer. They must be 22 kHz, 12-bit uncompressed mono files, and the filenames must be in 8.3 format. The open source audio utility Audacity can be used to edit or convert audio files to the correct format. The wave shield accesses the audio file from the SD card, so the length of the audio is only limited by the size of the memory card.

See Also

The Ladyada wave shield library and documentation: *http://www.ladyada.net/make/waveshield/*

Audacity audio editing and conversion software: *http://audacity.sourceforge.net/*

SparkFun offers a range of audio modules, including an Audio-Sound Module (*http://www.sparkfun.com/products/9534*) and MP3 breakout board (*http://www.sparkfun.com/products/8954*).

9.6 Controlling MIDI

Problem

You want to get a MIDI synthesizer to play music using Arduino.

Solution

To connect to a MIDI device, you need a five-pin DIN plug or socket. If you use a socket, you will also need a lead to connect to the device. Connect the MIDI connector to Arduino using a 220 ohm resistor, as shown in Figure 9-5.

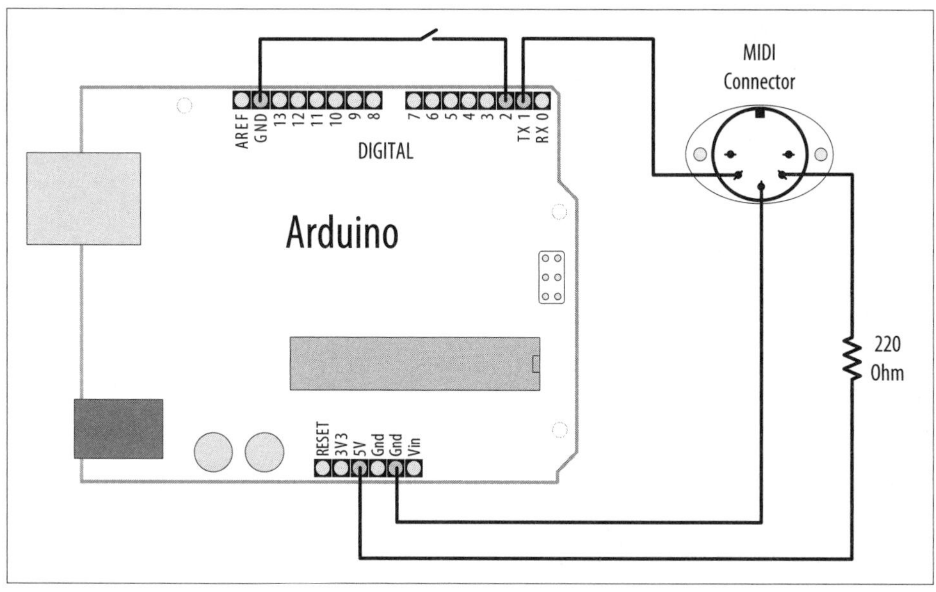

Figure 9-5. MIDI connections

To upload the code onto Arduino, you should disconnect the MIDI device, as it may interfere with the upload. After the sketch is uploaded, connect a MIDI sound device to the Arduino output. A musical scale will play each time you press the button connected to pin 2:

```
/*
midiOut sketch
sends MIDI messages to play a scale on a MIDI instrument
each time the switch on pin 2 is pressed
*/

//these numbers specify which note
const byte notes[8] = {60, 62, 64, 65, 67, 69, 71, 72};
//they are part of the MIDI specification
const int length = 8;
const int switchPin = 2;
const int ledPin = 13;

void setup() {
  Serial.begin(31250);
  pinMode(switchPin, INPUT);
  digitalWrite(switchPin, HIGH);
  pinMode(ledPin, OUTPUT);
}

void loop() {
  if (digitalRead(switchPin == LOW))
  {
    for (byte noteNumber = 0; noteNumber < 8; noteNumber++)
    {
      playMidiNote(1, notes[noteNumber], 127);
      digitalWrite(ledPin, HIGH);
      delay(70);
      playMidiNote(1, notes[noteNumber], 0);
      digitalWrite(ledPin, HIGH);
      delay(30);
    }
  }
}

void playMidiNote(byte channel, byte note, byte velocity)
{
  byte midiMessage= 0x90 + (channel - 1);
  Serial.write(midiMessage);
  Serial.write(note);
  Serial.write(velocity);
}
```

Discussion

This sketch uses the serial port to send MIDI information. The circuit connected to pin 1 may interfere with uploading code to the board. Remove the wire from pin 1 while you upload, and plug it back in afterward.

MIDI was originally used to connect digital musical instruments together so that one could control another. The MIDI specification describes the electrical connections and the messages you need to send.

MIDI is actually a serial connection (at a nonstandard serial speed, 31,250 baud), so Arduino can send MIDI messages using its serial port hardware from pins 0 and 1. Because the serial port is occupied by MIDI messages, you can't print messages to the Serial Monitor, so the sketch flashes the LED on pin 13 each time it sends a note.

Each MIDI message consists of at least one byte. This byte specifies what is to be done. Some commands need no other information, but other commands need data to make sense. The message in this sketch is *note on*, which needs two pieces of information: which note and how loud. Both of these bits of data are in the range of zero to 127.

The sketch initializes the serial port to a speed of 31,250 baud; the other MIDI-specific code is in the function `playMidiNote`:

```
void playMidiNote(byte channel, byte note, byte velocity)
{
  byte midiMessage= 0x90 + (channel - 1);
  Serial.write(midiMessage);
  Serial.write(note);
  Serial.write(velocity);
}
```

This function takes three parameters and calculates the first byte to send using the channel information.

MIDI information is sent on different channels between 1 and 16. Each channel can be set to be a different instrument, so multichannel music can be played. The command for *note on* (to play a sound) is a combination of 0x90 (the top four bits at b1001), with the bottom four bits set to the numbers between b0000 and b1111 to represent the MIDI channels. The byte represents channels using 0 to 15 for channels 1 to 16, so 1 is subtracted first.

Then the note value and the volume (referred to as *velocity* in MIDI, as it originally related to how fast the key was moving on a keyboard) are sent.

The serial `write` statements specify that the values must be sent as bytes (rather than as the ASCII value). `println` is not used because a line return character would insert additional bytes into the signal that are not wanted.

The sound is turned off by sending a similar message, but with velocity set to 0.

This recipe works with MIDI devices having five-pin DIN MIDI in connectors. If your MIDI device only has a USB connector, this will not work. It will not enable the Arduino to control MIDI music programs running on your computer without additional hardware (a MIDI-to-USB adapter). Although Arduino has a USB connector, your computer recognizes it as a serial device, not a MIDI device.

See Also

To send and receive MIDI, have a look at the MIDI library available at *http://www.arduino.cc/playground/Main/MIDILibrary*.

MIDI messages are described in detail at *http://www.midi.org/techspecs/midimessages.php*.

For more information on the SparkFun MIDI breakout shield (BOB-09598), see *http://www.sparkfun.com/products/9598*.

To set an Arduino Uno up as a native USB MIDI device, see Recipe 18.14.

9.7 Making an Audio Synthesizer

Problem

You want to generate complex sounds similar to those used to produce electronic music.

Solution

The simulation of audio oscillators used in a sound synthesizer is complex, but Peter Knight has created a sketch called Auduino that enables Arduino to produce more complex and interesting sounds.

Download the sketch by following the link on *http://code.google.com/p/tinkerit/wiki/Auduino*.

Connect five 4.7K ohm linear potentiometers to analog pins 0 through 4, as shown in Figure 9-6. Potentiometers with full-size shafts are better than small presets because you can easily twiddle the settings. Pin 5 is used for audio output and is connected to an amplifier using a jack plug.

Discussion

The sketch code is complex because it is directly manipulating hardware timers to generate the desired frequencies, which are transformed in software to produce the audio effects. It is not included in the text because you do not need to understand the code to use Auduino.

Auduino uses a technique called *granular synthesis* to generate the sound. It uses two electronically produced sound sources (called *grains*). The variable resistors control the frequency and decay of each grain (inputs 0 and 2 for one grain and inputs 3 and 1 for the other). Input 4 controls the synchronization between the grains.

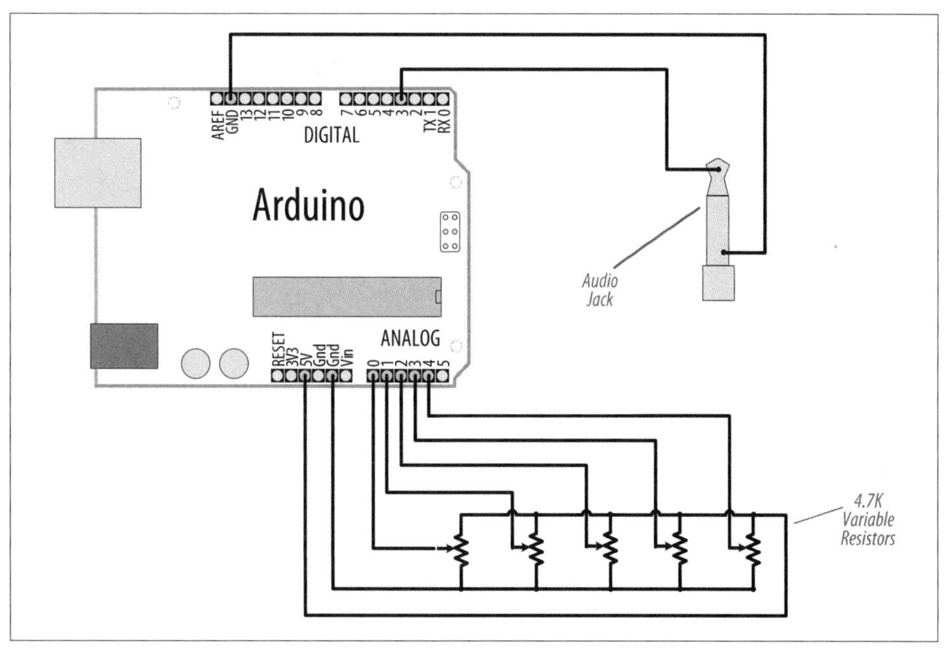

Figure 9-6. Auduino

If you want to tweak the code, you can change the scale used to calculate the frequency. The default setting is pentatonic, but you can comment that out and uncomment another option to use a different scale.

Be careful when adding code to the main loop, because the sketch is highly optimized and additional code could slow things down too much, causing the audio synthesis to not work well.

You can replace any of the pots with sensors that can produce an analog voltage signal (see Chapter 6). For example, a light-dependent resistor (see Recipe 6.2) or a distance sensor (the analog output described toward the end of Recipe 6.4) connected to one of the frequency inputs (pin 0 or 3) would enable you to control the pitch by moving your hand closer to or farther from the sensor (look up "theremin" in Wikipedia or on Google to read more about this musical instrument that is played by sensing hand movement).

See Also

Video demonstration of Auduino: *http://www.vimeo.com/2266458*

Wikipedia article explaining granular synthesis: *http://en.wikipedia.org/wiki/Granular_synthesis*

Wikipedia article on the theremin: *http://en.wikipedia.org/wiki/Theremin*

CHAPTER 10

Remotely Controlling External Devices

10.0 Introduction

The Arduino can interact with almost any device that uses some form of remote control, including TVs, audio equipment, cameras, garage doors, appliances, and toys. Most remote controls work by sending digital data from a transmitter to a receiver using infrared light (IR) or wireless radio technology. Different protocols (signal patterns) are used to translate key presses into a digital signal, and the recipes in this chapter show you how to use commonly found remote controls and protocols.

An IR remote works by turning an LED on and off in patterns to produce unique codes. The codes are typically 12 to 32 bits (pieces of data). Each key on the remote is associated with a specific code that is transmitted when the key is pressed. If the key is held down, the remote usually sends the same code repeatedly, although some remotes (e.g., NEC) send a special repeat code when a key is held down. For Philips RC-5 or RC-6 remotes, a bit in the code is toggled each time a key is pressed; the receiver uses this toggle bit to determine when a key is pressed a second time. You can read more about the technologies used in IR remote controls at *http://www.sbprojects.com/knowledge/ir/ir.htm*.

The recipes here use a low-cost IR receiver module to detect the signal and provide a digital output that the Arduino can read. The digital output is then decoded by a library called IRremote, which was written by Ken Shirriff and can be downloaded from *http://www.arcfn.com/2009/08/multi-protocol-infrared-remote-library.html*.

The same library is used in the recipes in which Arduino sends commands to act like a remote control.

To install the library, place it in the folder named *libraries* in your Arduino sketch folder. If you need help installing libraries, see Chapter 16.

Remote controls using wireless radio technology are more difficult to emulate than IR controls. However, the button contacts on these controls can be activated by Arduino. The recipes using wireless remotes simulate button presses by closing the button contacts circuit inside the remote control. With wireless remotes, you may need to take apart the remote control and connect wires from the contacts to Arduino to be able to use these devices. Components called *optocouplers* are used to provide electrical separation between Arduino and the remote control. This isolation prevents voltages from Arduino from harming the remote control, and vice versa.

Optocouplers (also called *optoisolators*) enable you to safely control another circuit that may be operating at different voltage levels from Arduino. As the "isolator" part of the name implies, optoisolators provide a way to keep things electrically separated. These devices contain an LED, which can be controlled by an Arduino digital pin. The light from the LED in the optocoupler shines onto a light-sensitive transistor. Turning on the LED causes the transistor to conduct, closing the circuit between its two connections—the equivalent of pressing a switch.

10.1 Responding to an Infrared Remote Control

Problem

You want to respond to any key pressed on a TV or other remote control.

Solution

Arduino responds to IR remote signals using a device called an *IR receiver module*. Common devices are the TSOP4838, PNA4602, and TSOP2438. The first two have the same connections, so the circuit is the same; the TSOP2438 has the +5V and Gnd pins reversed. Check the data sheet for your device to ensure that you connect it correctly.

This recipe uses the IRremote library from *http://www.arcfn.com/2009/08/multi-protocol-infrared-remote-library.html*. Connect the IR receiver module according to your data sheet. The Arduino wiring in Figure 10-1 is for the TSOP4838/PNA4602 devices.

This sketch will toggle an LED when any button on an infrared remote control is pressed:

```
/*
  IR_remote_detector sketch
  An IR remote receiver is connected to pin 2.
  The LED on pin 13 toggles each time a button on the remote is pressed.
 */

#include <IRremote.h>                    //adds the library code to the sketch

const int irReceiverPin = 2;             //pin the receiver is connected to
const int ledPin = 13;

IRrecv irrecv(irReceiverPin);            //create an IRrecv object
decode_results decodedSignal;            //stores results from IR detector
```

```
void setup()
{
  pinMode(ledPin, OUTPUT);
  irrecv.enableIRIn();                    // Start the receiver object
}

boolean lightState = false;     //keep track of whether the LED is on
unsigned long last = millis();  //remember when we last received an IR message

void loop()
{
  if (irrecv.decode(&decodedSignal) == true) //this is true if a message has
                                             //been received
  {
    if (millis() - last > 250) {       //has it been 1/4 sec since last message?
      lightState = !lightState;              //Yes: toggle the LED
      digitalWrite(ledPin, lightState);
    }
    last = millis();
    irrecv.resume();                         // watch out for another message
  }
}
```

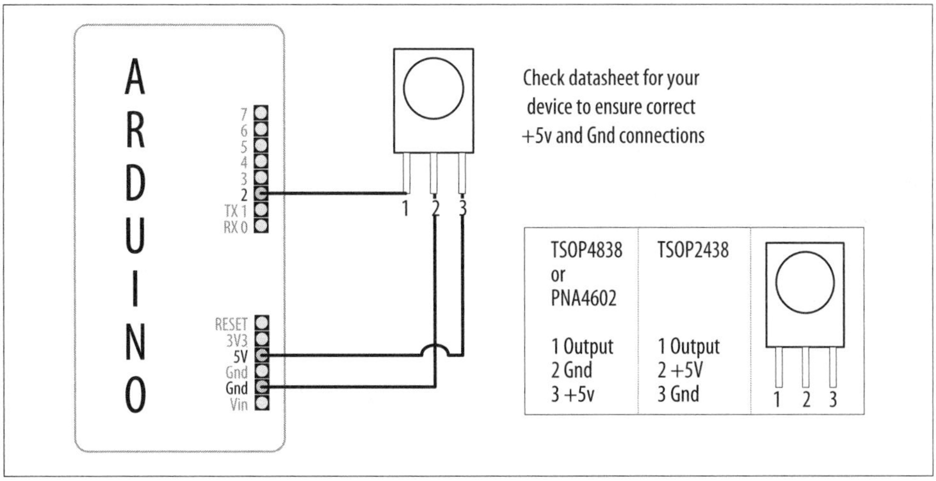

Figure 10-1. Connecting an infrared receiver module

Discussion

The IR receiver converts the IR signal to digital pulses. These are a sequence of ones and zeros that correspond to buttons on the remote. The IRremote library decodes these pulses and provides a numeric value for each key (the actual values that your sketch will receive are dependent on the specific remote control you use).

`#include <IRremote.h>` at the top of the sketch makes the library code available to your sketch, and the line `IRrecv irrecv(irReceiverPin);` creates an `IRrecv` object named irrecv to receive signals from an IR receiver module connected to irReceiverPin (pin 2 in the sketch). Chapter 16 has more on using libraries.

You use the irrecv object to access the signal from the IR receiver. You can give it commands to look for and decode signals. The decoded responses provided by the library are stored in a variable named decode_results. The receiver object is started in setup with the line irrecv.enableIRIn();. The results are checked in loop by calling the function irrecv.decode(&decodedSignal).

The decode function returns true if there is data, which will be placed in the decoded Signal variable. Recipe 2.11 explains how the ampersand symbol is used in function calls where parameters are modified so that information can be passed back.

If a remote message has been received, the code *toggles* the LED (flips its state) if it is more than one-quarter of a second since the last time it was toggled (otherwise, the LED will get turned on and off quickly by remotes that send codes more than once when you press the button, and may appear to be flashing randomly).

The decodedSignal variable will contain a value associated with a key. This value is ignored in this recipe (although it is used in the next recipe)—you can print the value by adding to the sketch the Serial.println line highlighted in the following code:

```
if (irrecv.decode(&decodedSignal) == true) //this is true if a message has
                                            // been received
{
   Serial.println(results.value);  // add this line to see decoded results
```

The library needs to be told to continue monitoring for signals, and this is achieved with the line irrecv.resume();.

This sketch flashes an LED when any button on the remote control is pressed, but you can control other things—for example, you can use a servo motor to dim a lamp (for more on controlling physical devices, see Chapter 8).

10.2 Decoding Infrared Remote Control Signals

Problem

You want to detect a specific key pressed on a TV or other remote control.

Solution

This sketch uses remote control key presses to adjust the brightness of an LED. The code prompts for remote control keys 0 through 4 when the sketch starts. These codes are stored in Arduino memory (RAM), and the sketch then responds to these keys by setting the brightness of an LED to correspond with the button pressed, with 0 turning the LED off and 1 through 4 providing increased brightness:

```
/*
  RemoteDecode sketch
 Infrared remote control signals are decoded to control LED brightness
 The values for keys 0 through 4 are detected and stored when the sketch starts
 key 0 turns the LED off, the brightness increases in steps with keys 1 through 4
 */

#include <IRremote.h>           // IR remote control library

const int irReceivePin = 2;     // pin connected to IR detector output
const int ledPin       = 9;     // LED is connected to a PWM pin

const int numberOfKeys = 5;     //  5 keys are learned (0 through 4)
long irKeyCodes[numberOfKeys]; // holds the codes for each key

IRrecv irrecv(irReceivePin);    // create the IR library
decode_results results;         // IR data goes here

void setup()
{
  Serial.begin(9600);
  pinMode(irReceivePin, INPUT);
  pinMode(ledPin, OUTPUT);
  irrecv.enableIRIn();              // Start the IR receiver
  learnKeycodes();                  // learn remote control key  codes
  Serial.println("Press a remote key");
}

void loop()
{
  long key;
  int  brightness;

  if (irrecv.decode(&results))
  {
    // here if data is received
    irrecv.resume();
    key = convertCodeToKey(results.value);
    if(key >= 0)
    {
      Serial.print("Got key ");
      Serial.println(key);
      brightness = map(key, 0,numberOfKeys-1, 0, 255);
      analogWrite(ledPin, brightness);
    }
  }
}

/*
 * get remote control codes
 */
void learnKeycodes()
{
  while(irrecv.decode(&results))   // empty the buffer
    irrecv.resume();
```

```
  Serial.println("Ready to learn remote codes");
  long prevValue = -1;
  int i=0;
  while( i < numberOfKeys)
  {
    Serial.print("press remote key ");
    Serial.print(i);
    while(true)
    {
      if( irrecv.decode(&results) )
      {
          if(results.value != -1 && results.value != prevValue)
          {
            showReceivedData();
            irKeyCodes[i] = results.value;
            i = i + 1;
            prevValue = results.value;
            irrecv.resume(); // Receive the next value
            break;
          }
        irrecv.resume(); // Receive the next value
      }
    }
  }
  Serial.println("Learning complete");
}

/*
 * converts a remote protocol code to a logical key code
 * (or -1 if no digit received)
 */
int convertCodeToKey(long code)
{
  for( int i=0; i < numberOfKeys; i++)
  {
    if( code == irKeyCodes[i])
    {
      return i; // found the key so return it
    }
  }
  return -1;
}

/*
 * display the protocol type and value
 */
void showReceivedData()
{
  if (results.decode_type == UNKNOWN)
  {
    Serial.println("-Could not decode message");
  }
  else
  {
```

```
    if (results.decode_type == NEC) {
      Serial.print("- decoded NEC: ");
    }
    else if (results.decode_type == SONY) {
      Serial.print("- decoded SONY: ");
    }
    else if (results.decode_type == RC5) {
      Serial.print("- decoded RC5: ");
    }
    else if (results.decode_type == RC6) {
      Serial.print("- decoded RC6: ");
    }
    Serial.print("hex value = ");
    Serial.println( results.value, HEX);
  }
}
```

Discussion

This solution is based on the IRremote library; see this chapter's introduction for details.

The sketch starts the remote control library with the following code:

```
irrecv.enableIRIn(); // Start the IR receiver
```

It then calls the `learnKeyCodes` function to prompt the user to press keys 0 through 4. The code for each key is stored in an array named `irKeyCodes`. After all the keys are detected and stored, the `loop` code waits for a key press and checks if this was one of the digits stored in the `irKeyCodes` array. If so, the value is used to control the brightness of an LED using `analogWrite`.

See Recipe 5.7 for more on using the `map` function and `analogWrite` to control the brightness of an LED.

The library should be capable of working with most any IR remote control; it can discover and remember the timings and repeat the signal on command.

You can permanently store the key code values so that you don't need to learn them each time you start the sketch. Replace the declaration of `irKeyCodes` with the following lines to initialize the values for each key. Change the values to coincide with the ones for your remote (these will be displayed in the Serial Monitor when you press keys in the `learnKeyCodes` function):

```
long irKeyCodes[numberOfKeys] = {
  0x18E758A7,  //0 key
  0x18E708F7,  //1 key
  0x18E78877,  //2 key
  0x18E748B7,  //3 key
```

```
  0x18E7C837,  //4 key
  };
```

See Also

Recipe 18.1 explains how you can store learned data in EEPROM (nonvolatile memory).

10.3 Imitating Remote Control Signals

Problem

You want to use Arduino to control a TV or other remotely controlled appliance by emulating the infrared signal. This is the inverse of Recipe 10.2—it sends commands instead of receiving them.

Solution

This sketch uses the remote control codes from Recipe 10.2 to control a device. Five buttons select and send one of five codes. Connect an infrared LED to send the signal as shown in Figure 10-2:

```
/*
  irSend sketch
  this code needs an IR LED connected to pin 3
  and 5 switches connected to pins 4 - 8
*/

#include <IRremote.h>        // IR remote control library

const int numberOfKeys = 5;
const int firstKey = 4;   // the first pin of the 5 sequential pins connected
                          // to buttons
boolean buttonState[numberOfKeys];
boolean lastButtonState[numberOfKeys];
long irKeyCodes[numberOfKeys] = {
  0x18E758A7,  //0 key
  0x18E708F7,  //1 key
  0x18E78877,  //2 key
  0x18E748B7,  //3 key
  0x18E7C837,  //4 key
};


IRsend irsend;

void setup()
{
  for (int i = 0; i < numberOfKeys; i++){
    buttonState[i]=true;
    lastButtonState[i]=true;
    int physicalPin=i + firstKey;
```

```
    pinMode(physicalPin, INPUT);
    digitalWrite(physicalPin, HIGH);  // turn on pull-ups
  }
  Serial.begin(9600);
}

void loop() {
  for (int keyNumber=0; keyNumber<numberOfKeys; keyNumber++)
  {
    int physicalPinToRead=keyNumber+4;
    buttonState[keyNumber] = digitalRead(physicalPinToRead);
    if (buttonState[keyNumber] != lastButtonState[keyNumber])
    {
      if (buttonState[keyNumber] == LOW)
      {
        irsend.sendSony(irKeyCodes[keyNumber], 32);
        Serial.println("Sending");
      }
      lastButtonState[keyNumber] = buttonState[keyNumber];
    }
  }
}
```

You won't see anything when the codes are sent because the light from the infrared LED isn't visible to the naked eye.

However, you can verify that an infrared LED is working with a digital camera—you should be able to see it flashing in the camera's LCD viewfinder.

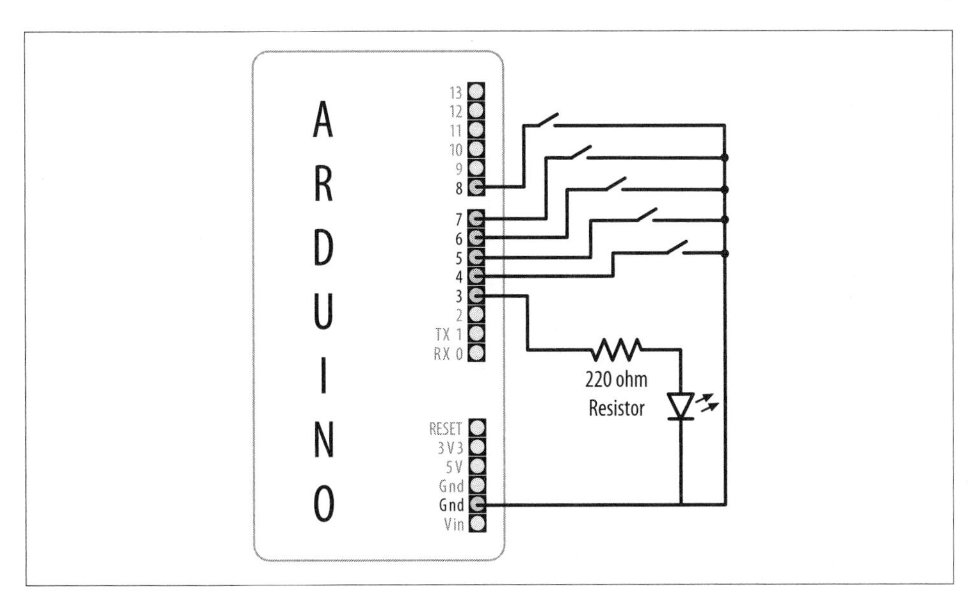

Figure 10-2. Buttons and LED for IR sender

Discussion

Here Arduino controls the device by flashing an IR LED to duplicate the signal that would be sent from your remote control. This requires an IR LED. The specifications are not critical; see Appendix A for suitable components.

The IR library handles the translation from numeric code to IR LED flashes. You need to create an object for sending IR messages. The following line creates an `IRsend` object that will control the LED on pin 3 (you are not able to specify which pin to use; this is hardcoded within the library):

```
IRsend irsend;
```

The code uses an array (see Recipe 2.4) called `irKeyCodes` to hold the range of values that can be sent. It monitors five switches to see which one has been pressed and sends the relevant code in the following line:

```
irsend.sendSony(irKeyCodes[keyNumber], 32);
```

The `irSend` object has different functions for various popular infrared code formats, so check the library documentation if you are using one of the other remote control formats. You can use Recipe 10.2 if you want to display the format used in your remote control.

The sketch passes the code from the array, and the number after it tells the function how many bits long that number is. The 0x at the beginning of the numbers in the definition of `irKeyCodes` at the top of the sketch means the codes are written in hex (see Chapter 2 for details about hex numbers). Each character in hex represents a 4-bit value. The codes here use eight characters, so they are 32 bits long.

The LED is connected with a current-limiting resistor (see the introduction to Chapter 7).

If you need to increase the sending range, you can use multiple LEDs or select one with greater output.

See Also

Chapter 7 provides more information on controlling LEDs.

Mitch Altman's TV-B-Gone is a clever remote control application; see *http://www.ladyada.net/make/tvbgone/* for construction details.

10.4 Controlling a Digital Camera

Problem

You want Arduino to control a digital camera to take pictures under program control. You may want to do time lapse photography or take pictures triggered by an event detected by the Arduino.

Solution

There are a few ways to do this. If your camera has an infrared remote, use Recipe 10.2 to learn the relevant remote codes and Recipe 10.3 to get Arduino to send those codes to the camera.

If your camera doesn't have an infrared remote but does have a socket for a wired remote, you can use this recipe to control the camera.

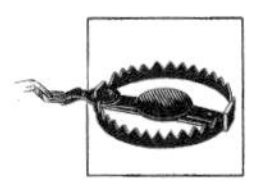

A camera shutter connector, usually called a *TRS* (tip, ring, sleeve) connector, typically comes in 2.5 mm or 3.5 mm sizes, but the length and shape of the tip may be nonstandard. The safest way to get the correct plug is to buy a cheap wired remote switch for your model of camera and modify that or buy an adapter cable from a specialist supplier (Google "TRS camera shutter").

You connect the Arduino to a suitable cable for your camera using optocouplers, as shown in Figure 10-3.

This sketch takes a picture every 20 seconds:

```
/*
  camera sketch
  takes 20 pictures with a digital camera
  using pin 4 to trigger focus
  pin 3 to trigger the shutter
 */

int focus = 4;              //optocoupler attached to focus
int shutter = 3;            //optocoupler attached to shutter
long exposure = 250;        //exposure time in milliseconds
long interval = 10000;      //time between shots, in milliseconds

void setup()
{
  pinMode(focus, OUTPUT);
  pinMode(shutter, OUTPUT);
  for (int i=0; i<20; i++)  //camera will take 20 pictures
  {
    takePicture(exposure);       //takes picture
    delay(interval);             //wait to take the next picture
  }
}
```

```
void loop()
{
                  //once it's taken 20 pictures it is done,
                  //so loop is empty
                  //but loop still needs to be here or the
                  //sketch won't compile
}

void takePicture(long exposureTime)
{
  int wakeup = 10;       //camera will take some time to wake up and focus
                         //adjust this to suit your camera

  digitalWrite(focus, HIGH);                //wake the camera and focus
  delay(wakeup);                            //wait for it to wake up and focus
  digitalWrite(shutter, HIGH);              //open the shutter
  delay(exposureTime);                      //wait for the exposure time
  digitalWrite(shutter, LOW);               //release shutter
  digitalWrite(focus, LOW);                 //release the focus
}
```

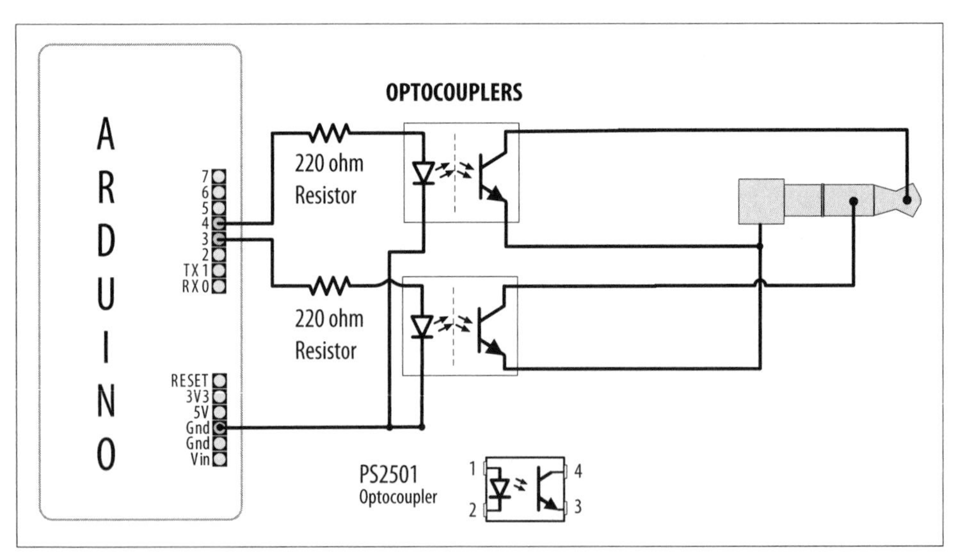

Figure 10-3. Using optocouplers with a TRS camera connector

Discussion

It's not advisable to connect Arduino pins directly to a camera—the voltages may not be compatible and you risk damaging your Arduino or your camera. Optocouplers are used to isolate Arduino from your camera; see the introduction of this chapter for more about these devices.

You will need to check the user manual for your camera to identify the correct TRS connector to use.

You may need to change the order of the pins turning on and off in the `takePicture` function to get the behavior you want. For a Canon camera to do bulb exposures, you need to turn on the focus, then open the shutter without releasing the focus, then release the shutter, and then release the focus (as in the sketch). To take a picture and have the camera calculate the exposure, press the focus button, release it, and then press the shutter.

See Also

If you want to control aspects of a camera's operation, have a look at the Canon Hack Development Kit at *http://chdk.wikia.com/wiki/CHDK*.

Also see *The Canon Camera Hackers Manual: Teach Your Camera New Tricks* by Berthold Daum (Rocky Nook).

It is also possible to control video cameras in a similar fashion using LANC. You can find details on this by searching for "LANC" in the Arduino Playground.

10.5 Controlling AC Devices by Hacking a Remote-Controlled Switch

Problem

You want to safely switch AC line currents on and off to control lights and appliances using a remote controlled switch.

Solution

Arduino can trigger the buttons of a remote controlled switch using an optocoupler. This may be necessary for remotes that use wireless instead of infrared technology. This technique can be used for almost any remote control. Hacking a remote is particularly useful to isolate potentially dangerous AC voltages from you and Arduino because only the battery-operated controller is modified.

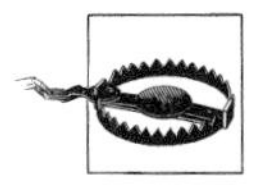

Opening the remote control will void the warranty and can potentially damage the device. The infrared recipes in this chapter are preferable because they avoid modifying the remote control.

If you want to use this recipe to control a switch, but you want to keep using the remote control, consider purchasing a spare remote control for hacking. Most manufacturers will be happy to sell you a spare (but make sure you choose the right frequency for the variant of appliance, light, or outlet you want to control). After you receive the spare, you may need to configure the channel that it uses.

Open the remote control and connect the optocoupler so that the photo-emitter (pins 1 and 2 in Figure 10-4) is connected to Arduino and the photo-transistor (pins 3 and 4) is connected across the remote control contacts.

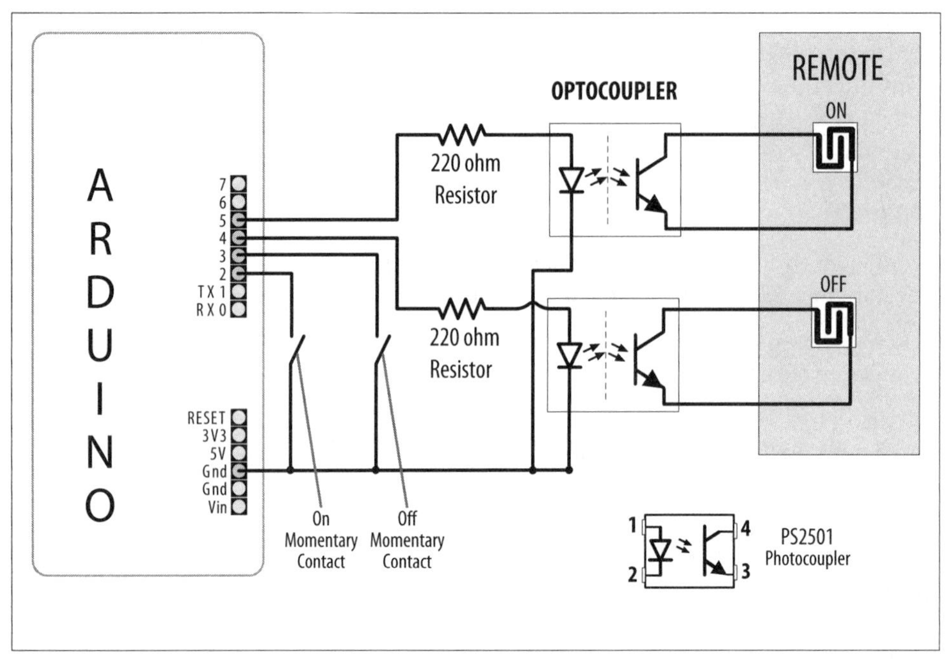

Figure 10-4. Optocouplers connected to remote control contacts

This sketch uses momentary contact switches (push and release) to turn the remote ON and OFF buttons:

```
/*
 OptoRemote sketch
 Switches connected to pins 2 and 3 turns a remote device on and off
 using optocouplers.

 The outputs are pulsed for at least half a second when a switch is pressed
 */
const int onSwitchPin  = 2;         // input pin for the On switch
const int offSwitchPin = 3;         // input pin for the Off switch
const int remoteOnPin  = 4;         // output pin to turn the remote on
const int remoteOffPin = 5;         // output pin to turn the remote off
const int PUSHED       = LOW;       // value when button is pressed


void setup() {
  Serial.begin(9600);
  pinMode(remoteOnPin, OUTPUT);
  pinMode(remoteOffPin, OUTPUT);
  pinMode(onSwitchPin, INPUT);
  pinMode(offSwitchPin, INPUT);
```

```
    digitalWrite(onSwitchPin,HIGH);  // turn on internal pull-up on the inputPins
    digitalWrite(offSwitchPin,HIGH);
}

void loop(){
  int val = digitalRead(onSwitchPin);  // read input value
  // if the switch is pushed then switch on if not already on
  if( val == PUSHED)
  {
    pulseRemote(remoteOnPin);
  }
  val = digitalRead(offSwitchPin);  // read input value
  // if the switch is pushed then switch on if not already on
  if( val == PUSHED)
  {
    pulseRemote(remoteOffPin);
  }
}

// turn the optocoupler on for half a second to blip the remote control button
void pulseRemote(int pin )
{
  digitalWrite(pin, HIGH);     // turn the optocoupler on
  delay(500);                  // wait half a second
  digitalWrite(pin, LOW);      // turn the optocoupler off
}
```

Discussion

The switches in most remote controls consist of interleaved bare copper traces with a conductive button that closes a connection across the traces when pressed. Less common are controls that contain conventional push switches; these are easier to use as the legs of the switches provide a convenient connection point.

Although the original remote button and the optocoupler can be used together—the switching action will be performed if either method is activated (pressing the button or turning on the optocoupler), the wires tethered to Arduino can make this inconvenient.

The transistor in the optocoupler will only allow electricity to flow in one direction, so if it doesn't work the first time, try switching the transistor side connections over and see if that fixes it.

Some remotes have one side of all of the switches connected together (usually to the ground of that circuit). You can trace the connections on the board to check for this or use a multimeter to see what the resistance is between the traces on different switches. If traces have common connections, it is only necessary to connect one wire to each common group. Fewer traces are easier because connecting the wires can be fiddly if the remote is small.

Optocouplers are explained in Recipe 10.4, so check that out if you are unfamiliar with optocouplers.

The remote control may have multiple contacts corresponding to each button. You may need more than one optocoupler for each button position to connect the contacts. Figure 10-5 shows three optocouplers that are controlled from a single Arduino pin.

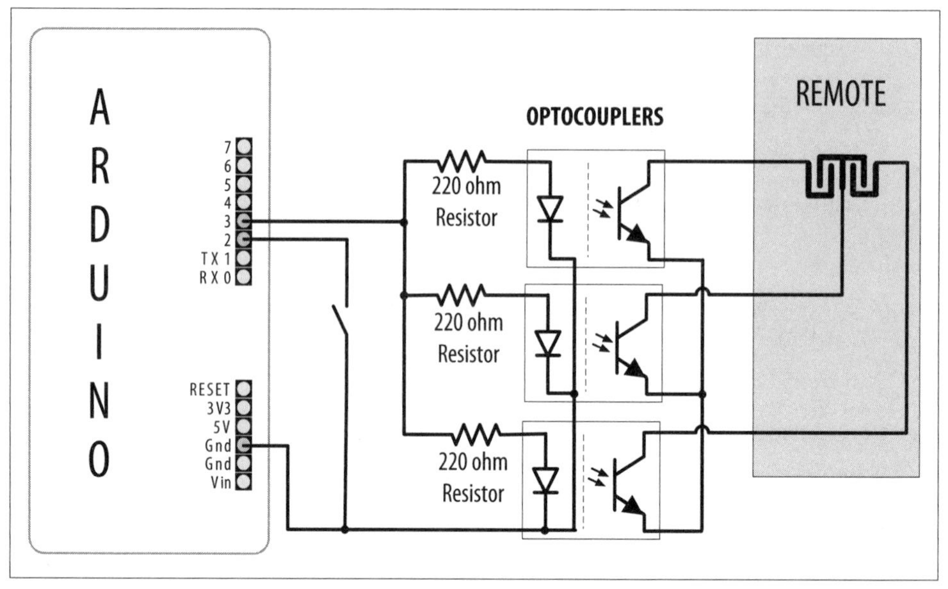

Figure 10-5. Multiple optocouplers connected to a single remote control button

See Also

Another approach to controlling AC line currents is to use an isolated relay such as the PowerTailSwitch that can be switched on and off directly from Arduino pins. See *http://powerswitchtail.com/default.aspx*.

CHAPTER 11
Using Displays

11.0 Introduction

Liquid crystal displays (LCDs) offer a convenient and inexpensive way to provide a user interface for a project. This chapter explains how to connect and use common text and graphical LCD panels with Arduino. By far the most popular LCD is the text panel based on the Hitachi HD44780 chip. This displays two or four lines of text, with 16 or 20 characters per line (32- and 40-character versions are available, but usually at much higher prices). A library for driving text LCD displays is provided with Arduino, and you can print text on your LCD as easily as on the Serial Monitor (see Chapter 4), because LCD and serial share the same underlying `print` functions.

LCDs can do more than display simple text: words can be scrolled or highlighted and you can display a selection of special symbols and non-English characters.

You can create your own symbols and block graphics with a text LCD, but if you want fine graphical detail, you need a graphical display. Graphical LCD (GLCD) displays are available at a small price premium over text displays, and many popular GLCD panels can display up to eight lines of 20 text characters in addition to graphics.

LCD displays have more wires connecting to Arduino than most other recipes in this book. Incorrect connections are the major cause of problems with LCDs, so take your time wiring things up and triple-check that things are connected correctly. An inexpensive multimeter capable of measuring voltage and resistance is a big help for verifying that your wiring is correct. It can save you a lot of head scratching if nothing is being displayed. You don't need anything fancy, as even the cheapest multimeter will help you verify that the correct pins are connected and that the voltages are correct.

You can even find a video tutorial and PDF explaining how to use a multimeter at *http://blog.makezine.com/archive/2007/01/multimeter_tutorial_make_1.html*.

For projects that require a bigger display than available in inexpensive LCD panels, Recipe 11.11 shows how you can use a television as an output device for Arduino.

11.1 Connecting and Using a Text LCD Display

Problem

You have a text LCD based on the industry-standard HD44780 or a compatible controller chip, and you want to display text and numeric values.

Solution

The Arduino software includes the LiquidCrystal library for driving LCD displays based on the HD44780 chip.

Most text LCDs supplied for use with Arduino will be compatible with the Hitachi HD44780 controller. If you are not sure about your controller, check the data sheet to see if it is a 44780 or compatible.

To get the display working, you need to wire the power, data, and control pins. Connect the data and status lines to digital output pins, and wire up a contrast potentiometer and connect the power lines. If your display has a backlight, this needs connecting, usually through a resistor.

Figure 11-1 shows the most common LCD connections. It's important to check the data sheet for your LCD to verify the pin connections. Table 11-1 shows the most common pin connections, but if your LCD uses different pins, make sure it is compatible with the Hitachi HD44780—this recipe will only work on LCD displays that are compatible with that chip. The LCD will have 16 pins (or 14 pins if there is no backlight)—make sure you identify pin 1 on your panel; it may be in a different position than shown in the figure.

You may wonder why LCD pins 7 through 10 are not connected. The LCD display can be connected using either four pins or eight pins for data transfer. This recipe uses the four-pin mode because this frees up the other four Arduino pins for other uses. There is a theoretical performance improvement using eight pins, but it's insignificant and not worth the loss of four Arduino pins.

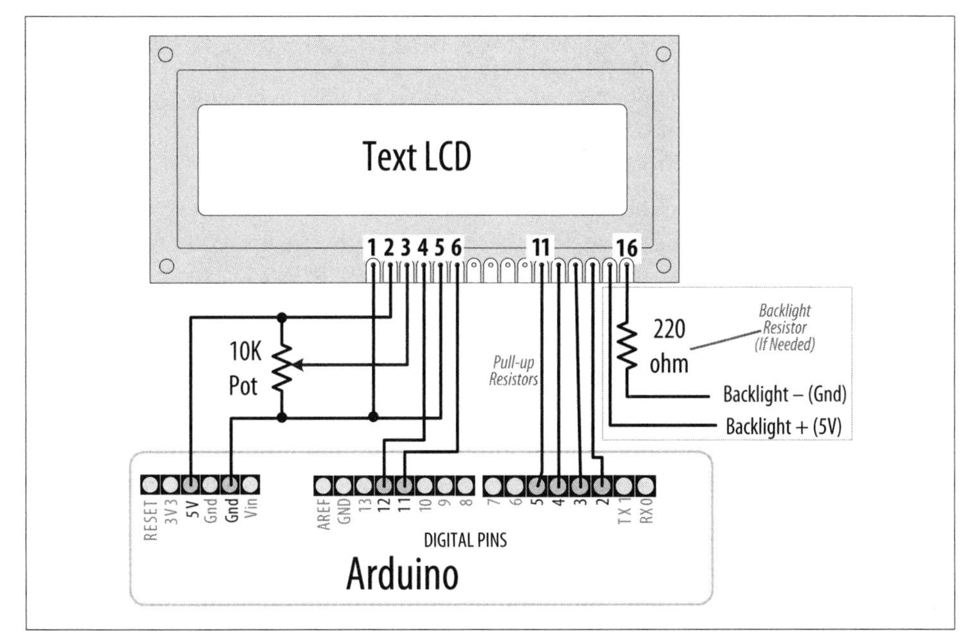

Figure 11-1. Connections for a text LCD

Table 11-1. LCD pin connections

LCD pin	Function	Arduino pin
1	Gnd or 0V or Vss	Gnd
2	+5V or Vdd	5V
3	Vo or contrast	
4	RS	12
5	R/W	Gnd
6	E	11
7	D0	
8	D1	
9	D2	
10	D3	
11	D4	5
12	D5	4
13	D6	3
14	D7	2
15	A or anode	
16	K or cathode	

You will need to connect a 10K potentiometer to provide the contrast voltage to LCD pin 3. Without the correct voltage on this pin, you may not see anything displayed. In Figure 11-1, one side of the pot connects to Gnd (ground), the other side connects to Arduino +5V, and the center of the pot goes to LCD pin 3. The LCD is powered by connecting Gnd and +5V from Arduino to LCD pins 1 and 2.

Many LCD panels have an internal lamp called a *backlight* to illuminate the display. Your data sheet should indicate whether there is a backlight and if it requires an external resistor—many do need this to prevent burning out the backlight LED assembly (if you are not sure, you can be safe by using a 220 ohm resistor). The backlight is polarized, so make sure pin 15 is connected to +5V and pin 16 to Gnd. (The resistor is shown connected between pin 16 and Gnd, but it can also be connected between pin 15 and +5V.)

Double-check the wiring before you apply power, as you can damage the LCD if you connect the power pins incorrectly. To run the HelloWorld sketch provided with Arduino, click the IDE Files menu item and navigate to Examples→Library→LiquidCrystal→HelloWorld.

The following code is modified slightly to print numbers in addition to "hello world." Change `numRows` and `numCols` to match the rows and columns in your LCD:

```
/*
  LiquidCrystal Library - Hello World

  Demonstrates the use of a 16 × 2 LCD display.
  http://www.arduino.cc/en/Tutorial/LiquidCrystal
 */

#include <LiquidCrystal.h> // include the library code

//constants for the number of rows and columns in the LCD
const int numRows = 2;
const int numCols = 16;


// initialize the library with the numbers of the interface pins
LiquidCrystal lcd(12, 11, 5, 4, 3, 2);

void setup()
{
  lcd.begin(numCols, numRows);
  lcd.print("hello, world!");  // Print a message to the LCD.
}

void loop()
{
  // set the cursor to column 0, line 1
  // (note: line 1 is the second row, since counting begins with 0):
  lcd.setCursor(0, 1);
  // print the number of seconds since reset:
```

```
    lcd.print(millis()/1000);
}
```

Run the sketch; you should see "hello world" displayed on the first line of your LCD. The second line will display a number that increases by one every second.

Discussion

If you don't see any text and you have double-checked that all wires are connected correctly, you may need to adjust the contrast pot. With the pot shaft rotated to one side (usually the side connected to Gnd), you will have maximum contrast and should see blocks appear in all the character positions. With the pot rotated to the other extreme, you probably won't see anything at all. The correct setting will depend on many factors, including viewing angle and temperature—turn the pot until you get the best-looking display.

If you can't see blocks of pixels appear at any setting of the pot, check that the LCD is being driven on the correct pins.

Once you can see text on the screen, using the LCD in a sketch is easy. You use similar print commands to those for serial printing, covered in Chapter 4. The next recipe reviews the print commands and explains how to control text position.

See Also

See the LiquidCrystal reference: *http://arduino.cc/en/Reference/LiquidCrystalPrint*.

See Chapter 4 for details on `print` commands.

The data sheet for the Hitachi HD44780 LCD controller is the definitive reference for detailed, low-level functionality. The Arduino library insulates you from most of the complexity, but if you want to read about the raw capabilities of the chip, you can download the data sheet from *http://www.sparkfun.com/datasheets/LCD/HD44780.pdf*.

The LCD page in the Arduino Playground contains software and hardware tips and links: *http://www.arduino.cc/playground/Code/LCD*.

11.2 Formatting Text

Problem

You want to control the position of text displayed on the LCD screen; for example, to display values in specific positions.

Solution

This sketch displays a countdown from 9 to 0. It then displays a sequence of digits in three columns of four characters. Change numRows and numCols to match the rows and columns in your LCD:

```
/*
  LiquidCrystal Library - FormatText
 */

#include <LiquidCrystal.h> // include the library code:


//constants for the number of rows and columns in the LCD
const int numRows = 2;
const int numCols = 16;

int count;

// initialize the library with the numbers of the interface pins
LiquidCrystal lcd(12, 11, 5, 4, 3, 2);

void setup()
{
  lcd.begin(numCols, numRows);
  lcd.print("Starting in ");  // this string is 12 characters long
  for(int i=9; i > 0; i--)    // count down from 9
  {
    // the top line is row 0
    lcd.setCursor(12,0); // move the cursor to the end of the string
    lcd.print(i);
    delay(1000);
  }
}

void loop()
{
  int columnWidth = 4;          //spacing for the columns
  int displayColumns = 3;       //how many columns of numbers

  lcd.clear();
  for( int col=0; col < displayColumns; col++)
  {
    lcd.setCursor(col * columnWidth, 0);
    count = count+ 1;
    lcd.print(count);
  }
  delay(1000);
}
```

Discussion

The `lcd.print` functions are similar to `Serial.print`. In addition, the LCD library has commands that control the cursor location (the row and column where text will be printed).

The `lcd.print` statement displays each new character after the previous one. Text printed beyond the end of a line may not be displayed or may be displayed on another line. The `lcd.setCursor()` command enables you to specify where the next `lcd.print` will start. You specify the column and row position (the top-left corner is 0,0). Once the cursor is positioned, the next `lcd.print` will start from that point, and it will overwrite existing text. The sketch in this recipe's Solution uses this to print numbers in fixed locations.

For example, in `setup`:

```
lcd.setCursor(12,0); // move the cursor to the 13th position
lcd.print(i);
```

`lcd.setCursor(12,0)` ensures that each number is printed in the same position, the thirteenth column, first row, producing the digit shown at a fixed position, rather than each number being displayed after the previous number.

Rows and columns start from zero, so `setCursor(4,0)` would set the cursor to the fifth column on the first row. This is because there are five characters located in positions 0 through 4. If that is not clear, it may help you if you count this out on your fingers starting from zero.

The following lines use `setCursor` to space out the start of each column to provide `columnwidth` spaces from the start of the previous column:

```
lcd.setCursor(col * columnWidth, 0);
count = count+ 1;
lcd.print(count);
lcd.clear();
```

`lcd.clear` clears the screen and moves the cursor back to the top-left corner.

Here is a variation on `loop` that displays numbers using all the rows of your LCD. Replace your `loop` code with the following (make sure you set `numRows` and `numCols` at the top of the sketch to match the rows and columns in your LCD):

```
void loop()
{
int columnWidth = 4;
int displayColumns = 3;

  lcd.clear();
  for(int row=0; row < numRows; row++)
  {
    for( int col=0; col < displayColumns; col++)
```

```
    {
      lcd.setCursor(col * columnWidth, row);
      count = count+ 1;
      lcd.print(count);
    }
  }
  delay(1000);
}
```

The first `for` loop steps through the available rows, and the second `for` loop steps through the columns.

To adjust how many numbers are displayed in a row to fit the LCD, calculate the `displayColumns` value rather than setting it. Change:

```
int displayColumns = 3;
```

to:

```
int displayColumns = numCols / columnWidth;
```

See Also

The LiquidCrystal library tutorial: *http://arduino.cc/en/Reference/LiquidCrystal?from=Tutorial.LCDLibrary*

11.3 Turning the Cursor and Display On or Off

Problem

You want to blink the cursor and turn the display on or off. You may also want to draw attention to a specific area of the display.

Solution

This sketch shows how you can cause the cursor (a flashing block at the position where the next character will be displayed) to blink. It also illustrates how to turn the display on and off; for example, to draw attention by blinking the entire display:

```
/*
  blink
 */

// include the library code:
#include <LiquidCrystal.h>

// initialize the library with the numbers of the interface pins
LiquidCrystal lcd(12, 11, 5, 4, 3, 2);

void setup()
{
  // set up the LCD's number of columns and rows and:
  lcd.begin(16, 2);
```

```
  // Print a message to the LCD.
  lcd.print("hello, world!");
}

void loop()
{
  lcd.setCursor(0, 1);

  lcd.print("cursor blink");
  lcd.blink();
  delay(2000);

  lcd.noBlink();
  lcd.print(" noBlink");
  delay(2000);

  lcd.clear();

  lcd.print("Display off ...");
  delay(1000);
  lcd.noDisplay();
  delay(2000);

  lcd.display();  // turn the display back on

  lcd.setCursor(0, 0);
  lcd.print(" display flash !");
  displayBlink(2, 250);  // blink twice
  displayBlink(2, 500);  // and again for twice as long

  lcd.clear();
}

void displayBlink(int blinks, int duration)
{
  while(blinks--)
  {
    lcd.noDisplay();
    delay(duration);
    lcd.display();
    delay(duration);
  }
}
```

Discussion

The sketch calls `blink` and `noBlink` functions to toggle cursor blinking on and off.

The code to blink the entire display is in a function named `displayBlink` that makes the display flash a specified number of times. The function uses `lcd.display()` and `lcd.noDisplay()` to turn the display text on and off (without clearing it from the screen's internal memory).

11.4 Scrolling Text

Problem

You want to scroll text; for example, to create a marquee that displays more characters than can fit on one line of the LCD display.

Solution

This sketch demonstrates both `lcd.ScrollDisplayLeft` and `lcd.ScrollDisplayRight`.

It scrolls a line of text to the left when tilted and to the right when not tilted. Connect one side of a tilt sensor to pin 7 and the other pin to Gnd (see Recipe 6.1 if you are not familiar with tilt sensors):

```
/*
   Scroll
  * this sketch scrolls text left when tilted
  * text scrolls right when not tilted.
 */

#include <LiquidCrystal.h>

// initialize the library with the numbers of the interface pins
LiquidCrystal lcd(12, 11, 5, 4, 3, 2);
const int numRows = 2;
const int numCols = 16;

const int tiltPin = 7; // pin connected to tilt sensor

const char textString[] = "tilt to scroll";
const int textLen = sizeof(textString) -1; // the number of characters
boolean isTilted = false;

void setup()
{
  // set up the LCD's number of columns and rows:
  lcd.begin(numCols, numRows);
  digitalWrite(tiltPin, HIGH); // turn on pull-ups for the tilt sensor
  lcd.print(textString);
}

void loop()
{
  if(digitalRead(tiltPin) == LOW && isTilted == false )
  {
    // here if tilted left so scroll text left
    isTilted = true;
    for (int position = 0; position  < textLen; position++)
    {
      lcd.scrollDisplayLeft();
      delay(150);
    }
```

```
  }
  if(digitalRead(tiltPin) == HIGH && isTilted == true )
  {
    // here if previously tilted but now flat, so scroll text right
    isTilted = false;
    for (int position = 0; position  < textLen; position++)
    {
      lcd.scrollDisplayRight();
      delay(150);
    }
  }
}
```

Discussion

The first half of the `loop` code handles the change from not tilted to tilted. The code checks to see if the tilt switch is closed (`LOW`) or open (`HIGH`). If it's `LOW` and the current state (stored in the `isTilted` variable) is not tilted, the text is scrolled left. The delay in the `for` loop controls the speed of the scroll; adjust the delay if the text moves too fast or too slow.

The second half of the code uses similar logic to handle the change from tilted to not tilted.

A scrolling capability is particularly useful when you need to display more text than can fit on an LCD line.

This sketch has a `marquee` function that will scroll text up to 32 characters in length:

```
/*
  Marquee
  * this sketch can scroll a very long line of text
 */

#include <LiquidCrystal.h>

// initialize the library with the numbers of the interface pins
LiquidCrystal lcd(12, 11, 5, 4, 3, 2);
const int numRows = 2;
const int numCols = 16;


void setup()
{
  // set up the LCD's number of columns and rows:
  lcd.begin(numCols, numRows);
}

void loop()
{
  marquee("A message too long to fit !");
  delay(1000);
  lcd.clear();
}
```

```
// this version of marquee uses manual scrolling for very long messages
void marquee( char *text)
{
  int length = strlen(text); // the number of characters in the text
  if(length < numCols)
    lcd.print(text);
  else
  {
    int pos;
    for( pos = 0; pos < numCols; pos++)
       lcd.print(text[pos]);
    delay(1000); // allow time to read the first line before scrolling
    pos=1;
    while(pos <= length - numCols)
    {
      lcd.setCursor(0,0);
      for( int i=0; i < numCols; i++)
        lcd.print(text[pos+i]);
      delay(300);
      pos = pos + 1;
    }
  }
}
```

The sketch uses the `lcd.scrollDisplayLeft` function to scroll the display when the text is longer than the width of the screen.

The LCD chip has internal memory that stores the text. This memory is limited (32 bytes on most four-line displays). If you try to use longer messages, they may start to wrap over themselves. If you want to scroll longer messages (e.g., a tweet), or control scrolling more precisely, you need a different technique. The following function stores the text in RAM on Arduino and sends sections to the screen to create the scrolling effect. These messages can be any length that can fit into Arduino memory:

```
// this version of marquee uses manual scrolling for very long messages
void marquee( char *text)
{
  int length = strlen(text); // the number of characters in the text
  if(length < numCols)
    lcd.print(text);
  else
  {
    int pos;
    for( pos = 0; pos < numCols; pos++)
       lcd.print(text[pos]);
    delay(1000); // allow time to read the first line before scrolling
    pos=1;
    while(pos <= length - numCols)
    {
      lcd.setCursor(0,0);
      for( int i=0; i < numCols; i++)
        lcd.print(text[pos+i]);
      delay(300);
```

```
      pos = pos + 1;
    }
  }
}
```

11.5 Displaying Special Symbols

Problem

You want to display special symbols: ° (degrees), ¢, ÷, π (pi), or any other symbol stored in the LCD character memory.

Solution

Identify the character code you want to display by locating the symbol in the character pattern table in the LCD data sheet. This sketch prints some common symbols in `setup`. It then shows all displayable symbols in `loop`:

```
/*
 LiquidCrystal Library - Special Chars
*/

#include <LiquidCrystal.h>

//set constants for number of rows and columns to match your LCD
const int numRows = 2;
const int numCols = 16;

// defines for some useful symbols
const byte degreeSymbol = B11011111;
const byte piSymbol     = B11110111;
const byte centsSymbol  = B11101100;
const byte sqrtSymbol   = B11101000;
const byte omegaSymbol  = B11110100;  // the symbol used for ohms

byte charCode = 32; // the first printable ascii character
int col;
int row;

// initialize the library with the numbers of the interface pins
LiquidCrystal lcd(12, 11, 5, 4, 3, 2);

void setup()
{
  lcd.begin(numRows, numCols);

  showSymbol(degreeSymbol, "degrees");
  showSymbol  (piSymbol, "pi");
  showSymbol(centsSymbol, "cents");
  showSymbol(sqrtSymbol, "sqrt");
  showSymbol(omegaSymbol, "ohms");
  lcd.clear();
```

```
}

void loop()
{
   lcd.print(charCode);
   calculatePosition();
    if(charCode == 255)
    {
        // finished all characters so wait another few seconds and start over
        delay(2000);
        lcd.clear();
        row = col = 0;
        charCode = 32;
    }
    charCode = charCode + 1;
}

void calculatePosition()
{
   col = col + 1;
   if( col == numCols)
   {
      col = 0;
      row = row + 1;
      if( row == numRows)
      {
         row = 0;
         delay(2000); // pause
         lcd.clear();
      }
      lcd.setCursor(col, row);
    }
}

// function to display a symbol and its description
void showSymbol( byte symbol, char * description)
{
  lcd.clear();
  lcd.write(symbol);
  lcd.print(' '); // add a space before the description
  lcd.print(description);
  delay(3000);
}
```

Discussion

A table showing the available character patterns is in the data sheet for the LCD controller chip (you can find it on page 17 of the data sheet at *http://www.sparkfun.com/datasheets/LCD/HD44780.pdf*).

To use the table, locate the symbol you want to display. The code for that character is determined by combining the binary values for the column and row for the desired symbol (see Figure 11-2).

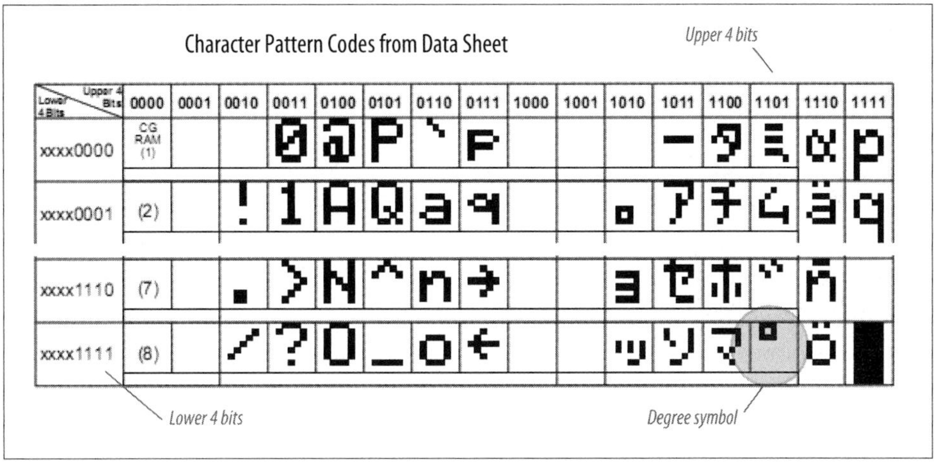

Figure 11-2. Using data sheet to derive character codes

For example, the degree symbol (°) is the third-from-last entry at the bottom row of the table shown in Figure 11-2. Its column indicates the upper four bits are 1101 and its row indicates the lower four bits are 1111. Combining these gives the code for this symbol: B11011111. You can use this binary value or convert this to its hex value (0xDF) or decimal value (223). Note that Figure 11-2 shows only 4 of the 16 actual rows in the data sheet.

The LCD screen can also show any of the displayable ASCII characters by using the ASCII value in `lcd.print`.

The sketch uses a function named `showSymbol` to print the symbol and its description:

```
void showSymbol( byte symbol, char * description)
```

(See Recipe 2.6 if you need a refresher on using character strings and passing them to functions.)

See Also

Data sheet for Hitachi HD44780 display: *http://www.sparkfun.com/datasheets/LCD/HD44780.pdf*

11.6 Creating Custom Characters

Problem

You want to define and display characters or symbols (glyphs) that you have created. The symbols you want are not predefined in the LCD character memory.

Solution

Uploading the following code will create an animation of a face, switching between smiling and frowning:

```
/*
  custom_char sketch
  creates an animated face using custom characters
*/

#include <LiquidCrystal.h>
LiquidCrystal lcd(12, 11, 5, 4, 3, 2);

byte happy[8] =
{
  B00000,
  B10001,
  B00000,
  B00000,
  B10001,
  B01110,
  B00000,
  B00000
};

byte saddy[8] =
{
  B00000,
  B10001,
  B00000,
  B00000,
  B01110,
  B10001,
  B00000,
  B00000
};

void setup() {
  lcd.createChar(0, happy);
  lcd.createChar(1, saddy);
  lcd.begin(16, 2);

}

void loop() {
  for (int i=0; i<2; i++)
  {
    lcd.setCursor(0,0);
    lcd.write(i);
    delay(500);
  }
}
```

Discussion

The LiquidCrystal library enables you to create up to eight custom characters, which can be printed as character codes 0 through 8. Each character on the screen is drawn on a grid of 5×8 pixels. To define a character, you need to create an array of eight bytes. Each byte defines one of the rows in the character. When written as a binary number, the 1 indicates a pixel is on, 0 is off (any values after the fifth bit are ignored). The sketch example creates two characters, named happy and saddy (see Figure 11-3).

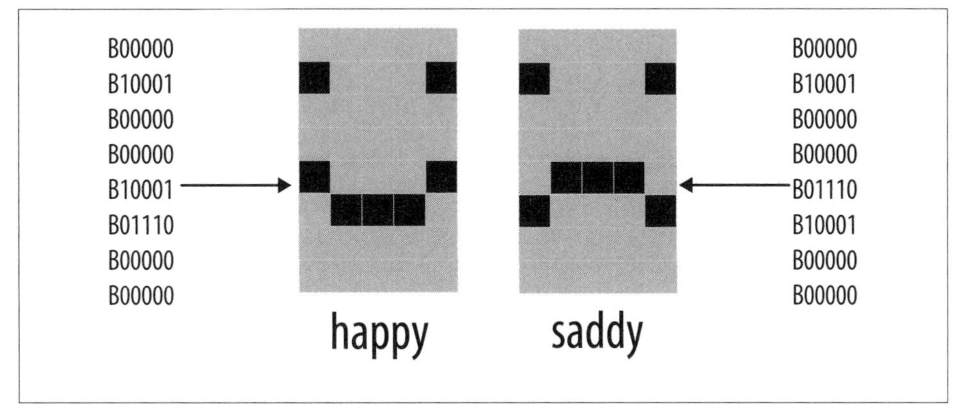

Figure 11-3. Defining custom characters

The following line in setup creates the character using data defined in the happy array that is assigned to character 0:

```
lcd.createChar(0, happy);
```

To print the custom character to the screen you would use this line:

```
lcd.write(0);
```

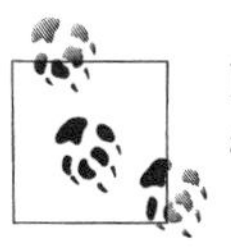

Note the difference between writing a character with or without an apostrophe. The following will print a zero, not the happy symbol:

```
lcd.write('0'); // this prints a zero
```

Code in the for loop switches between character 0 and character 1 to produce an animation.

11.7 Displaying Symbols Larger Than a Single Character

Problem

You want to combine two or more custom characters to print symbols larger than a single character; for example, double-height numbers on the screen.

Solution

The following sketch writes double-height numbers using custom characters:

```
/*
 * customChars
 *
 * This sketch displays double-height digits
 * the bigDigit arrays were inspired by Arduino forum member dcb
 */

#include <LiquidCrystal.h>

LiquidCrystal lcd(12, 11, 5, 4, 3, 2);

byte glyphs[5][8] = {
  { B11111,B11111,B00000,B00000,B00000,B00000,B00000,B00000 },
  { B00000,B00000,B00000,B00000,B00000,B00000,B11111,B11111 },
  { B11111,B11111,B00000,B00000,B00000,B00000,B11111,B11111 },
  { B11111,B11111,B11111,B11111,B11111,B11111,B11111,B11111 } ,
  { B00000,B00000,B00000,B00000,B00000,B01110,B01110,B01110 } };

const int digitWidth = 3; // the width in characters of a big digit
                          // (excludes space between characters)

//arrays  to index into custom characters that will comprise the big numbers
// digits 0 - 4                                0      1       2      3      4
const char bigDigitsTop[10][digitWidth]={ 3,0,3, 0,3,32, 2,2,3, 0,2,3, 3,1,3,
  // digits 5-9                                 5      6      7        8      9
                                         3,2,2, 3,2,2, 0,0,3,   3,2,3, 3,2,3};

const char bigDigitsBot[10][ digitWidth]={ 3,1,3, 1,3,1,  3,1,1, 1,1,3, 32,32,3,
                                          1,1,3, 3,1,3, 32,32,3, 3,1,3, 1,1,3};

char buffer[12]; // used to convert a number into a string
void setup ()
{
  lcd.begin(20,4);
  // create the custom glyphs
  for(int i=0; i < 5; i++)
    lcd.createChar(i, glyphs[i]);    // create the 5 custom glyphs
  // show a countdown timer
  for(int digit = 9; digit >= 0; digit--)
  {
    showDigit(digit, 2);  // show the digit
    delay(1000);
  }
  lcd.clear();
}

void loop ()
{
  // now show the number of seconds since the sketch started
  int number = millis() / 1000;
  showNumber( number, 0);
```

```
    delay(1000);
  }
  void showDigit(int digit, int position)
  {
    lcd.setCursor(position * (digitWidth + 1), 0);
    for(int i=0; i < digitWidth; i++)
      lcd.write(bigDigitsTop[digit][i]);
    lcd.setCursor(position * (digitWidth + 1), 1);
    for(int i=0; i < digitWidth; i++)
      lcd.write(bigDigitsBot[digit][i]);
  }
  void showNumber(int value, int position)
  {
    int index; // index to the digit being printed, 0 is the leftmost digit
    itoa(value, buffer, 10); // see Recipe 2.8 for more on using itoa
    // display each digit in sequence
    for(index = 0; index < 10; index++) // display up to ten digits
    {
      char c = buffer[index];
      if( c == 0)  // check for null (not the same as '0')
        return; // the end of string character is a null, see Chapter 2
      c = c - 48; // convert ascii value to a numeric value  (see Recipe 2.9)
      showDigit(c, position + index);
    }
  }
```

Discussion

The LCD display has fixed-size characters, but you can create larger symbols by combining characters. This recipe creates five custom characters using the technique described in Recipe 11.6. These symbols (see Figure 11-4) can be combined to create double-sized digits (see Figure 11-5). The sketch displays a countdown from 9 to 0 on the LCD using the big digits. It then displays the number of seconds since the sketch started.

The `glyphs` array defines pixels for the five custom characters. The array has two dimensions given in the square brackets:

```
byte glyphs[5][8] = {
```

`[5]` is the number of glyphs and `[8]` is the number of rows in each glyph. Each element contains 1s and 0s to indicate whether a pixel is on or off in that row. If you compare the values in `glyph[0]` (the first glyph) with Figure 11-2, you can see that the 1s correspond to dark pixels:

```
  { B11111,B11111,B00000,B00000,B00000,B00000,B00000,B00000  } ,
```

Each big number is built from six of these glyphs, three forming the upper half of the big digit and three forming the lower half. `bigDigitsTop` and `bigDigitsBot` are arrays defining which custom glyph is used for the top and bottom rows on the LCD screen.

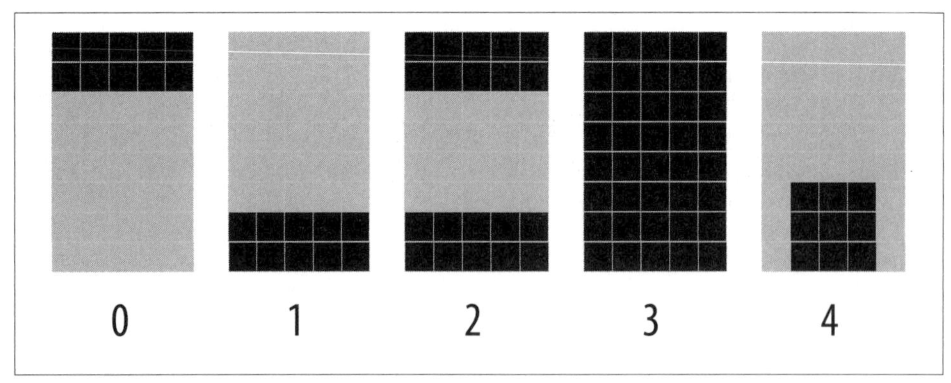

Figure 11-4. Custom characters used to form big digits

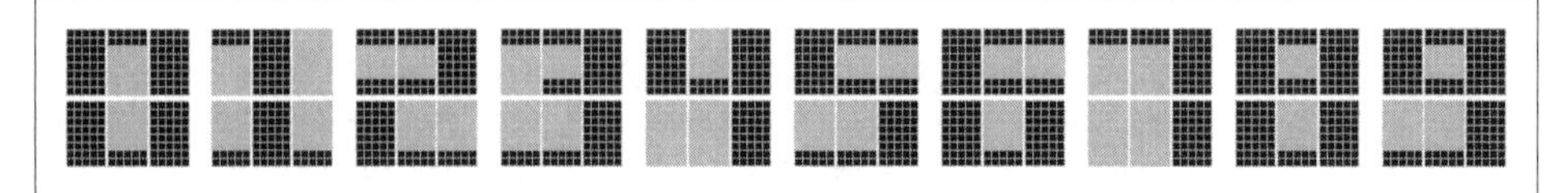

Figure 11-5. Ten big digits composed of custom glyphs

See Also

See Chapter 7 for information on 7-segment LED displays if you need really big numerals. Note that 7-segment displays can give you digit sizes from one-half inch to two inches or more. They can use much more power than LCD displays and don't present letters and symbols very well, but they are a good choice if you need something big.

11.8 Displaying Pixels Smaller Than a Single Character

Problem

You want to display information with finer resolution than an individual character; for example, to display a bar chart.

Solution

Recipe 11.7 describes how to build big symbols composed of more than one character. This recipe uses custom characters to do the opposite; it creates eight small symbols, each a single pixel higher than the previous one (see Figure 11-6).

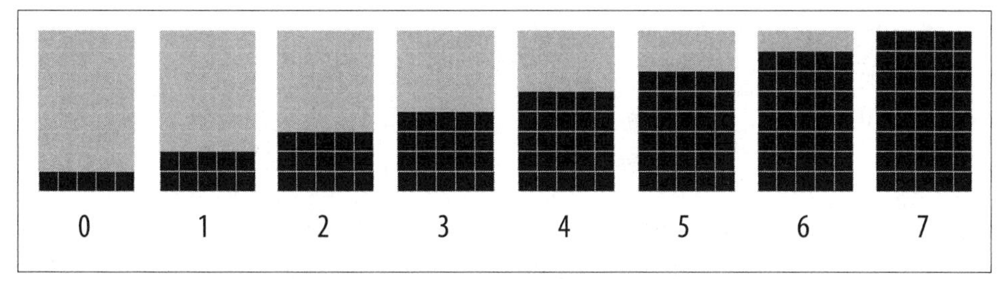

Figure 11-6. Eight custom characters used to form vertical bars

These symbols are used to draw bar charts, as shown in the sketch that follows:

```
/*
 * customCharPixels
 */

#include <LiquidCrystal.h>

LiquidCrystal lcd(12, 11, 5, 4, 3, 2);

//set constants for number of rows and columns to match your LCD
const int numRows = 2;
const int numCols = 16;


// array of bits defining pixels for 8 custom characters
// ones and zeros indicate if a pixel is on or off

  byte glyphs[8][8]  = {
      {B00000,B00000,B00000,B00000,B00000,B00000,B00000,B11111},   // 0
      {B00000,B00000,B00000,B00000,B00000,B00000,B11111,B11111},   // 1
      {B00000,B00000,B00000,B00000,B00000,B11111,B11111,B11111},   // 2
      {B00000,B00000,B00000,B00000,B11111,B11111,B11111,B11111},   // 3
      {B00000,B00000,B00000,B11111,B11111,B11111,B11111,B11111},   // 4
      {B00000,B00000,B11111,B11111,B11111,B11111,B11111,B11111},   // 5
      {B00000,B11111,B11111,B11111,B11111,B11111,B11111,B11111},   // 6
      {B11111,B11111,B11111,B11111,B11111,B11111,B11111,B11111}};  // 7

void setup ()
{
  lcd.begin(numCols, numRows);
  for(int i=0; i < 8; i++)
    lcd.createChar(i, glyphs[i]);     // create the custom glyphs
  lcd.clear();
}


void loop ()
{
  for( byte i=0; i < 8; i++)
    lcd.write(i);    // show all eight single height bars
  delay(2000);
```

```
  lcd.clear();
}
```

Discussion

The sketch creates eight characters, each a single pixel higher than the previous one; see Figure 11-6. These are displayed in sequence on the top row of the LCD. These "bar chart" characters can be used to display values in your sketch that can be mapped to a range from 0 to 7. For example, the following will display a value read from analog input 0:

```
int value = analogRead(0);
byte glyph = map(value, 0, 1023,0,8);// returns a proportional value
from 0 through 7
lcd.print(glyph);
```

You can stack the bars for greater resolution. The `doubleHeightBars` function shown in the following code displays a value from 0 to 15 with a resolution of 16 pixels, using two lines of the display:

```
void  doubleHeightBars(int value, int column)
{
char upperGlyph;
char lowerGlyph;


  if(value < 8)
  {
     upperGlyph = ' ';  // no pixels lit
     lowerGlyph = value;
  }
  else
  {
     upperGlyph = value - 8;
     lowerGlyph = 7; // all pixels lit
  }


  lcd.setCursor(column, 0); // do the upper half
  lcd.write(upperGlyph);
  lcd.setCursor(column, 1); // now to the lower half
  lcd.write(lowerGlyph);
}
```

The `doubleHeightBars` function can be used as follows to display the value of an analog input:

```
for( int i=0; i < 16; i++)
{
  int value = analogRead(0);
  value = map(value, 0, 1023,0,16);
  doubleHeightBars(value, i); // show a value from 0 to 15
  delay(1000);  // one second interval between readings
}
```

If you want horizontal bars, you can define five characters, each a single pixel wider than the previous one, and use similar logic to the vertical bars to calculate the character to show.

A more complex example of this technique can be found in a sketch implementing the well-known computer simulation known as John Conway's Game of Life. The sketch can be downloaded from this book's website (*http://shop.oreilly.com/product/0636920022244.do*).

11.9 Connecting and Using a Graphical LCD Display

Problem

You want to display graphics and text on an LCD that uses the KS0108 or compatible LCD driver chip.

Solution

This Solution uses the Arduino GLCD library to control the display. You can download it from *http://code.google.com/p/glcd-arduino/downloads/list* (see Chapter 16 if you need help installing libraries).

There are many different types of GLCD controllers, so check that yours is a KS0108 or compatible.

The pin connections for GLCD displays are not standardized, and it is important to check the data sheet for your panel to confirm how it should be wired. Incorrect connections of the signal lines are the most common cause of problems, and particular care should be taken with the power leads, as wiring these incorrectly can destroy a panel.

Most GLCD panels require an external variable resistor to set the LCD working voltage (contrast) and may require a fixed resistor to limit the current in the backlight. The data sheet for your panel should provide specific information on the wiring and choice of components for this.

Table 11-2 indicates the default connections from a KS0108 panel to an Arduino (or Mega). You will need to check the documentation for your particular panel to find where each function is connected on your display. The table shows the three most common panel layouts: the first, labeled "Panel A" in the table, is the one illustrated in Figure 11-7. The documentation with the GLCD library download includes color wiring diagrams for the more common displays.

Table 11-2. Default connections from a KS0108 panel to an Arduino or Mega

Arduino pins	Mega pins	GLCD function	Panel A	Panel B	Panel C	Comments
5V	5V	+5 volts	1	2	13	
Gnd	Gnd	Gnd	2	1	14	
N/A	N/A	Contrast in	3	3	12	Wiper of contrast pot
8	22	D0	4	7	1	
9	23	D1	5	8	2	
10	24	D2	6	9	3	
11	25	D3	7	10	4	
4	26	D4	8	11	5	
5	27	D5	9	12	6	
6	28	D6	10	13	7	
7	29	D7	11	14	8	
14 (analog 0)	33	CSEL1	12	15	15	Chip 1 select
15 (analog 1)	34	CSEL2	13	16	16	Chip 2 select
Reset		Reset	14	17	18	Connect to reset
16 (analog 2)	35	R_W	15	5	10	Read/write
17 (analog 3)	36	D_I	16	4	11	Data/instruction (RS)
18 (analog 4)	37	EN	17	6	9	Enable
N/A	N/A	Contrast out	18	18	17	10K or 20K preset
N/A	N/A	Backlight +5	19	19	19	See data sheet
Gnd	Gnd	Backlight Gnd	20	20	20	See data sheet

The numbers under the Arduino and Mega columns are the Arduino (or Mega) pins used in the configuration file provided in the library. It is possible to use other pins if these pins conflict with something else you want to connect. If you do change the connections, you will also need to change the pin assignments in the configuration file and should study the library documentation to learn how to edit the configuration file.

Wiring the panel using the default configuration and running the sketch in this recipe enables you to test that everything is working before you modify the configuration. A configuration that does not match the wiring is the most common source of problems, so testing with minimal changes makes it more likely that things will work the first time.

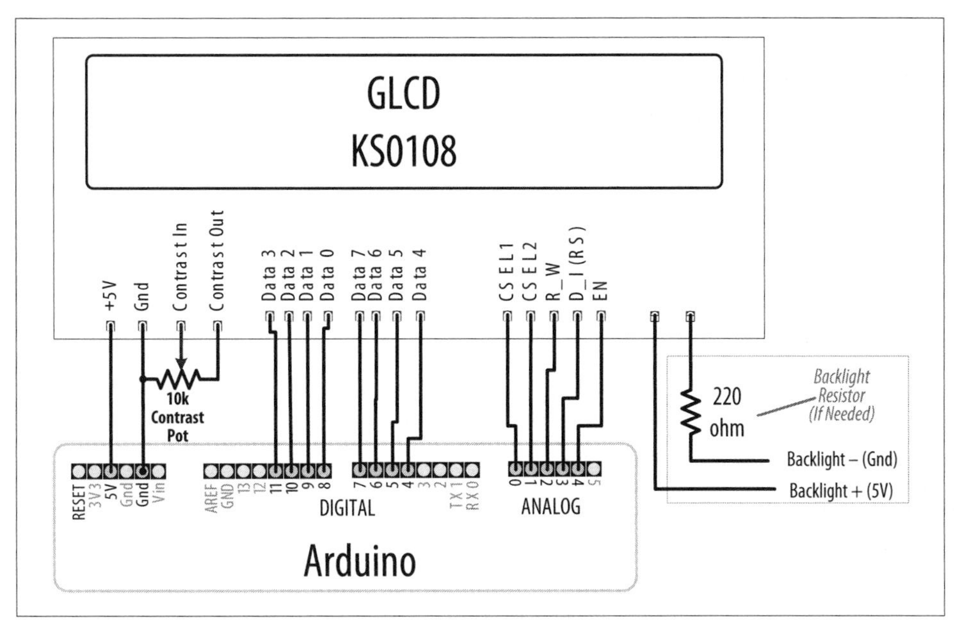

Figure 11-7. GLCD wiring for type A panels; check your data sheet for pinout

The following sketch prints some text and then draws some graphical objects:

```
/*
   glcd
 */

#include <glcd.h>

#include "fonts/allFonts.h"         // for access to all distributed fonts

int count = 0;

void setup()
{
  GLCD.Init(NON_INVERTED);          // initialize the library
  GLCD.ClearScreen();
  GLCD.SelectFont(System5x7);       // select fixed width system font
  GLCD.print("hello world");        // print a message
  delay(3000);
}


void loop()
{

    GLCD.ClearScreen();
    GLCD.DrawRect(0, 0, 64, 61, BLACK); // rectangle in left side of screen
    // rounded rectangle around text area
    GLCD.DrawRoundRect(68, 0, 58, 61, 5, BLACK);
    for(int i=0; i < 62; i += 4)
```

```
    {
      // draw lines from upper left down right side of rectangle
      GLCD.DrawLine(1,1,63,i, BLACK);
    }
    GLCD.DrawCircle(32,31,30,BLACK);   // circle centered on left side of screen
    GLCD.FillRect(92,40,16,16, WHITE); // clear previous spinner position
    GLCD.CursorTo(5,5);                // locate cursor for printing text
    GLCD.PrintNumber(count);           // print a number at current cursor position
    count = count + 1;
    delay(1000);
}
```

Discussion

The library provides a wide range of basic high-level graphical drawing functions, some of which are demonstrated in this sketch. All the functions are described in the documentation provided with the library.

Graphic and text screen coordinates start at 0,0 in the top-lefthand corner. Most popular GLCD panels are 128 × 64 pixels, and the library uses this resolution by default. If your screen is a different resolution, you will need to edit the configuration file in the library to match your panel (up to 255 × 255 pixel panels are currently supported).

GLCD enables printing text to the screen using statements similar to Arduino print commands used for printing to the serial port. In addition, you can specify the type and size of font. You can also specify an area of the screen that can be used as a text window. This enables you to define an area on the screen and then send text to that area, providing you with a "virtual terminal" that will contain and scroll text within the bounds you define. For instance, the following code creates an area 32 pixels square in the center of the screen:

```
gText  myTextArea = gText(GLCD.CenterX-16, GLCD.CenterY -16, GLCD.CenterX +16,
GLCD.CenterY+16);
```

You can select a font and print to the text area using code such as the following:

```
myTextArea.SelectFont(System5x7); // select the system font for the text area
name textTop
myTextArea.println("Go");         // print a line of text to the text area.
```

The example sketch supplied with the library download has a demo that shows how multiple text areas can be used along with graphical drawings.

These graphical displays have many more connections than the text LCD displays, and care should be taken to ensure that your panel is connected correctly.

If there are no pixels visible on the display, or the pixels are garbled, do the following:

- Check +5V and Gnd connections between Arduino and the GLCD panel.
- Check that all data and command pins are wired according to the data sheet and match the configuration settings. This is the most common cure for this problem.

- Check the data sheet for your panel to verify that appropriate timing values are set in the configuration file.
- Check the contrast voltage (typically between –3 and –4 volts) on the contrast-in pin of the LCD panel. While the sketch is operating, try gradually adjusting the pot through its range. Some displays are very sensitive to this setting.
- Check that the sketch has compiled correctly and has downloaded to Arduino.
- Run the GLCDdiags test sketch. The test sketch is available from the menu Examples→GLCD→GLCDdiags.

If the left and right sides of the image are reversed, swap the CSEL1 and CSEL2 wires (you can also swap pin assignments in the configuration file).

11.10 Creating Bitmaps for Use with a Graphical Display

Problem

You want to create and use your own graphical images (bitmaps) with the GLCD display discussed in Recipe 11.9. You want the font definition and text stored in program memory to minimize RAM usage.

Solution

You can use bitmaps distributed with the library or create your own. Bitmaps are defined in header files with an extension of *.h*; for example, an Arduino icon image named *ArduinoIcon.h* is stored in the *bitmap* folder of the GLCD library directory. This folder also contains a file named *allBitmaps.h* that has details of all the distributed bitmaps, so you can include this to make all the supplied (or newly created) bitmaps available:

```
#include "bitmaps/allBitmaps.h"  // this line includes all distributed bitmaps
```

Note that including all the bitmaps will not consume any memory if they are not explicitly referenced in your sketch with the `DrawBitmap` function.

To enable you to add your own bitmaps, the GLCD library includes a utility called glcdMakeBitmap which converts a *.gif*, *.jpg*, *.bmp*, *.tga*, or *.png* file to a header file that can be used by the GLCD library. The file *glcdMakeBitmap.pde* is a Processing sketch that can be run using the Processing environment. The sketch is located in the *bitmaps/utils/glcdMakeBitmap* directory. For more information on Processing, see *http://processing.org/*.

There is also a *.java* (Java) runtime file (*glcdMakeBitmap.jar*) and a *.java* (Java) source (*glcdMakeBitmap.java*) in the *bitmaps/utils/Java* directory.

Run the utility by loading the sketch into Processing (or click on the *.jar* file) and drag and drop the image file to be converted into the window. The utility will create a header file with the same name as the image file dropped into the window. The file is saved in the *bitmaps* directory and an entry is automatically added to the *allBitMaps.h* file so that the new image can be used in your sketch.

To demonstrate this, rename an image on your computer as *me.jpg*. Then run glcdMakeBitmap and drop the image into the window that appears.

Compile and upload the following sketch to show the supplied Arduino icon followed by the image you created:

```
/*
 * GLCDImage
 * Display an image defined in me.h
 */

#include <glcd.h>

#include "bitmaps/allBitmaps.h"       // all images in the bitmap folder

void setup()
{
  GLCD.Init();   // initialize the library
  GLCD.ClearScreen();
  GLCD.DrawBitmap(ArduinoIcon, 0,0); // draw the supplied bitmap
  delay(5000);
  GLCD.ClearScreen();
  GLCD.DrawBitmap(me, 0,0); // draw your bitmap
}

void  loop()
{

}
```

The following line draws the image defined in the file *ArduinoIcon.h* that is supplied with the library:

```
GLCD.DrawBitmap(ArduinoIcon, 0,0); // draw the supplied bitmap
```

After a delay, the following line draws the image you created that is stored in the file *me.h*:

```
GLCD.DrawBitmap(me, 0,0);
```

See Also

See the documentation supplied with the library for more on creating and using graphical images.

The documentation also describes how you can create your own fonts.

11.11 Displaying Text on a TV

Problem

You want to display text on a television or monitor with a video input.

Solution

This recipe uses a shield called TellyMate to print text or block graphics to a television. The shield plugs in to Arduino and has an output jack that connects to the video input of a television.

The sketch prints all the characters the TellyMate can display on a TV screen:

```
/*
  TellyMate
  Simple demo for TellyMate Shield
*/

const byte ESC = 0x1B;  // ASCII escape character used in TellyMate commands

void setup()
{
  Serial.begin(57600); // 57k6 baud is default TellyMate speed
  clear();  // clear the screen
  Serial.print("   TellyMate Character Set"); // write some text
  delay(2000);
}

void loop()
{

  byte charCode = 32;     // characters 0 through 31 are control codes
  for(int row=0; row < 7; row++) // show 7 rows
  {
    setCursor(2, row + 8); // center the display
    for(int col= 0; col < 32; col++) // 32 characters per row
    {
       Serial.print(charCode);
       charCode = charCode + 1;
       delay(20);
    }
  }
  delay(5000);
  clear();
}

// TellyMate helper functions

void clear( )  // clear the screen
{ // <ESC>E
  Serial.print(ESC);
  Serial.print('E');
}
```

```
void setCursor( int col, int row) // set the cursor
{ // <ESC>Yrc
  Serial.print(ESC);
  Serial.print('Y' ) ;
  Serial.print((unsigned char)(32 + row)) ;
  Serial.print((unsigned char)(32 + col)) ;
}
```

Discussion

Arduino controls the TellyMate display by sending commands to the serial port.

TellyMate communicates with the Arduino through the serial port, so you may need to unplug the shield to upload sketches.

Figure 11-8 shows the characters that can be displayed. You can find a table of values for each character at *http://en.wikipedia.org/wiki/Code_page_437*.

Figure 11-8. TellyMate character set (code page 437)

Characters 0 through 31 are interpreted as screen control commands, so only characters 32 to 255 can be displayed.

The sketch uses nonprintable codes, called *escape codes*, to differentiate printable characters from commands to control the screen. Control codes consist of the ESC (short for *escape*) character (hex value 0x1b) followed by one or more characters indicating the nature of the control function. Details of all the control codes are covered in the TellyMate documentation.

The sketch has a number of helper functions that send the appropriate sequence of characters to achieve the desired results, enabling you to concentrate on the higher level activity of the sketch—what you want it to do, rather than the details of how it will do it.

The screen will show a flashing cursor; you can turn this off using a control code. Adding the `cursorHide` function will turn off the cursor when the function is called:

```
void cursorHide()
{ // <ESC>f
  Serial.write(ESC) ;     // the escape character
  Serial.print('f' ) ;    // ... followed by the letter f will turn off the cursor.
}
```

To add a box around the edge of the screen, add the `drawBox` and `showXY` functions at the bottom of the previous sketch. To get the sketch to use them, add this line just inside the opening bracket of the loop:

```
drawBox(1,0, 38, 24); // the screen is 38 characters wide and 25 high
```

The `drawBox` function prints characters for the four corners and the top, bottom, and side edges using the line drawing character codes:

```
// characters that form the box outline
// see http://en.wikipedia.org/wiki/Code_page_437
const  byte boxUL = 201;
const  byte boxUR = 187;
const  byte boxLL = 200;
const  byte boxLR = 188;
const  byte HLINE = 205;  // horizontal line
const  byte VLINE = 186;  // vertical line

void drawBox( int startRow, int startCol, int width, int height)
{
  // draw top line
  showXY(boxUL, startCol,startRow); // the upper-left corner
  for(int col = startCol + 1; col < startCol + width-1; col++)
     Serial.print(HLINE); // the line characters
  Serial.print(boxUR);  // upper-right character

  // draw left and right edges
  for(int row = startRow + 1; row < startRow + height -1; row++)
  {
     showXY(VLINE, startCol,row);  // left edge
     showXY(VLINE, startCol + width-1,row);  // right edge
  }
  // draw bottom line
  showXY(boxLL, 0, startRow+height-1);  // the lower-left corner character
  for(int col = startCol + 1; col < startCol + width-1; col++)
     Serial.write(HLINE);
  Serial.write(boxLR);

}
```

A convenience function used by `drawBox`, named `showXY`, combines cursor positioning and printing:

```
void showXY( char ch, int x, int y){
  // display the given character at the screen x and y location
  setCursor(x,y);
```

```
  Serial.write(ch);
}
```

Here is an additional sketch that uses the cursor control commands to animate a ball bouncing around the screen:

```
 /*
  TellyBounce
*/

// define the edges of the screen:
const int HEIGHT = 25;    // the number of text rows
const int WIDTH  = 38;    // the number of characters in a row
const int LEFT   =  0;    // useful constants derived from the above
const int RIGHT  = WIDTH -1;
const int TOP    = 0;
const int BOTTOM = HEIGHT-1;

const byte BALL  = 'o';       // character code for ball
const byte ESC = 0x1B;        // ASCII escape character used in TellyMate commands

int ballX = WIDTH/2;           // X position of the ball
int ballY = HEIGHT/2;          // Y position of the ball
int ballDirectionY = 1;        // X direction of the ball
int ballDirectionX = 1;        // Y direction of the ball

// this delay moves ball across the 38-character screen in just under 4 seconds
long interval = 100;

void setup()
{
  Serial.begin(57600);    // 57k6 baud is default TellyMate speed
  clear();                // clear the screen
  cursorHide();           // turn cursor off
}

void loop()
{
  moveBall();
  delay(interval);
}

void moveBall() {
  // if the ball goes off the top or bottom, reverse its Y direction
  if (ballY == BOTTOM || ballY == TOP)
    ballDirectionY = -ballDirectionY;

  // if the ball goes off the left or right, reverse its X direction
  if ((ballX == LEFT) || (ballX == RIGHT))
    ballDirectionX = -ballDirectionX;

  // clear the ball's previous position
  showXY(' ', ballX, ballY);

  // increment the ball's position in both directions
  ballX = ballX + ballDirectionX;
```

```
  ballY = ballY + ballDirectionY;

  // show the new position
  showXY(BALL, ballX, ballY);
}

// TellyMate helper functions

void clear( )  // clear the screen
{ // <ESC>E
  Serial.write(ESC);
  Serial.write('E');
}

void setCursor( int col, int row) // set the cursor
{ // <ESC>Yrc
  Serial.write(ESC);
  Serial.write('Y' ) ;
  Serial.write((unsigned char)(32 + row)) ;
  Serial.write((unsigned char)(32 + col)) ;
}

void cursorShow( )
{ // <ESC>e
  Serial.write(ESC) ;
  Serial.write('e') ;
}

void cursorHide()
{ // <ESC>f
  Serial.write(ESC) ;
  Serial.write('f' ) ;
}

void showXY( char ch, int x, int y){
  // display the given character at the screen x and y location
  setCursor(x,y);
  Serial.write(ch);
}
```

See Also

Detailed information on the TellyMate shield is available at *http://www.batsocks.co.uk/products/Shields/index_Shields.htm*.

Much more information on code page 437, including a table of characters, is available at *http://en.wikipedia.org/wiki/Code_page_437*.

CHAPTER 12

Using Time and Dates

12.0 Introduction

Managing time is a fundamental element of interactive computing. This chapter covers built-in Arduino functions and introduces many additional techniques for handling time delays, time measurement, and real-world times and dates.

12.1 Creating Delays

Problem

You want your sketch to pause for some period of time. This may be some number of milliseconds, or a time given in seconds, minutes, hours, or days.

Solution

The Arduino `delay` function is used in many sketches throughout this book. `delay` pauses a sketch for the number of milliseconds specified as a parameter. (There are 1,000 milliseconds in one second.) The sketch that follows shows how you can use `delay` to get almost any interval:

```
/*
 * delay sketch
 */

const long oneSecond = 1000;  // a second is a thousand milliseconds
const long oneMinute = oneSecond * 60;
const long oneHour   = oneMinute * 60;
const long oneDay    = oneHour * 24;


void setup()
{
  Serial.begin(9600);
}
```

```
void loop()
{
  Serial.println("delay for 1 millisecond");
  delay(1);
  Serial.println("delay for 1 second");
  delay(oneSecond);
  Serial.println("delay for 1 minute");
  delay(oneMinute);
  Serial.println("delay for 1 hour");
  delay(oneHour);
  Serial.println("delay for 1 day");
  delay(oneDay);
  Serial.println("Ready to start over");
}
```

Discussion

The `delay` function has a range from one one-thousandth of a second to around 25 days (just less than 50 days if using an unsigned long variable type; see Chapter 2 for more on variable types).

The `delay` function pauses the execution of your sketch for the duration of the delay. If you need to perform other tasks within the delay period, using `millis`, as explained in Recipe 12.2, is more suitable.

You can use `delayMicroseconds` to delay short periods. There are 1,000 microseconds in one millisecond, and 1 million microseconds in one second. `delayMicroseconds` will pause from one microsecond to around 16 milliseconds, but for delays longer than a few thousand microseconds you should use `delay` instead:

```
delayMicroseconds(10);  // delay for 10 microseconds
```

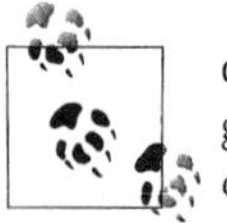

`delay` and `delayMicroseconds` will delay for at least the amount of time given as the parameter, but they could delay a little longer if interrupts occur within the delay time.

See Also

The Arduino reference for `delay`: *http://www.arduino.cc/en/Reference/Delay*

12.2 Using millis to Determine Duration

Problem

You want to know how much time has elapsed since an event happened; for example, how long a switch has been held down.

Solution

Arduino has a function named `millis` (short for milliseconds) that is used in the following sketch to print how long a button was pressed (see Recipe 5.2 for details on how to connect the switch):

```
/*
  millisDuration sketch
  returns the number of milliseconds that a button has been pressed
 */

const int switchPin = 2;                     // the number of the input pin

long startTime; // the value returned from millis when the switch is pressed
long duration;  // variable to store the duration

void setup()
{
  pinMode(switchPin, INPUT);
  digitalWrite(switchPin, HIGH); // turn on pull-up resistor
  Serial.begin(9600);
}


void loop()
{
  if(digitalRead(switchPin) == LOW)
  {
     // here if the switch is pressed
     startTime = millis();
     while(digitalRead(switchPin) == LOW)
        ; // wait while the switch is still pressed
     long duration = millis() - startTime;
     Serial.println(duration);
  }
}
```

Discussion

The `millis` function returns the number of milliseconds since the current sketch started running.

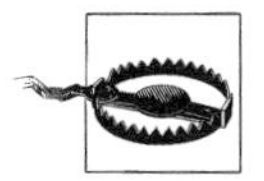

The `millis` function will *overflow* (go back to zero) after approximately 50 days. See Recipes 12.4 and 12.5 for information about using the Time library for handling intervals from seconds to years.

By storing the start time for an event, you can determine the duration of the event by subtracting the start time from the current time, as shown here:

```
long duration = millis() - startTime;
```

You can create your own delay function using `millis` that can continue to do other things while checking repeatedly to see if the delay period has passed. One example of this can be found in the BlinkWithoutDelay example sketch provided with the Arduino distribution. The following fragments from that sketch explain the loop code:

```
void loop()
{
  // here is where you'd put code that needs to be running all the time...
```

The next line checks to see if the desired interval has passed:

```
  if (millis() - previousMillis > interval)
  {
    // save the last time you blinked the LED
```

If the interval has passed, the current `millis` value is saved in the variable `previousMillis`:

```
    previousMillis = millis();

    // if the LED is off turn it on and vice versa:
    if (ledState == LOW)
      ledState = HIGH;
    else
      ledState = LOW;

    // set the LED with the ledState of the variable:
    digitalWrite(ledPin, ledState);
  }
}
```

Here is a way to package this logic into a function named `myDelay` that will delay the code in `loop` but can perform some action during the delay period. You can customize the functionality for your application, but in this example, an LED is flashed five times per second even while the print statement in `loop` is delayed for four-second intervals:

```
// blink an LED for a set amount of time
const int ledPin =  13;         // the number of the LED pin

int ledState = LOW;             // ledState used to set the LED
long previousMillis = 0;        // will store last time LED was updated

void setup()
{
  pinMode(ledPin, OUTPUT);
  Serial.begin(9600);
}

void loop()
{
  Serial.println(millis() / 1000); // print elapsed seconds every four seconds
  // wait four seconds (but at the same time, quickly blink an LED)
  myDelay(4000);
}
```

```
// duration is delay time in milliseconds
void myDelay(unsigned long duration)
{
  unsigned long start = millis();
  while (millis() - start <= duration)
  {
    blink(100);  // blink the LED inside the while loop
  }
}


// interval is the time that the LED is on and off
void blink(long interval)
{
  if (millis() - previousMillis > interval)
  {
    // save the last time you blinked the LED
    previousMillis = millis();
    // if the LED is off turn it on and vice versa:
    if (ledState == LOW)
      ledState = HIGH;
    else
      ledState = LOW;
    digitalWrite(ledPin, ledState);
  }
}
```

You can put code in the `myDelay` function for an action that you want to happen repeatedly while the function waits for the specified time to elapse.

Another approach is to use a third-party library available from the Arduino Playground, called TimedAction (*http://www.arduino.cc/playground/Code/TimedAction*):

```
#include <TimedAction.h>

//initialize a TimedAction class to change LED state every second.
TimedAction timedAction = TimedAction(NO_PREDELAY,1000,blink);


const int ledPin =  13;      // the number of the LED pin
boolean ledState = LOW;


void setup()
{
  pinMode(ledPin,OUTPUT);
  digitalWrite(ledPin,ledState);
}

void loop()
{
  timedAction.check();
}

void blink()
{
```

```
  if (ledState == LOW)
    ledState = HIGH;
  else
    ledState = LOW;

  digitalWrite(ledPin,ledState);
}
```

See Also

The Arduino reference for `millis`: *http://www.arduino.cc/en/Reference/Millis*

See Recipes 12.4 and 12.5 for information about using the Time library to handle intervals from seconds to years.

12.3 More Precisely Measuring the Duration of a Pulse

Problem

You want to determine the duration of a pulse with microsecond accuracy; for example, to measure the exact duration of `HIGH` or `LOW` pulses on a pin.

Solution

The `pulseIn` function returns the duration in microseconds for a changing signal on a digital pin. This sketch prints the time in microseconds of the `HIGH` and `LOW` pulses generated by `analogWrite` (see the section on "Analog Output" on page 241 in Chapter 7). Because the `analogWrite` pulses are generated internally by Arduino, no external wiring is required:

```
/*
  PulseIn sketch
  displays duration of high and low pulses from analogWrite
 */


const int inputPin = 3;    // analog output pin to monitor
unsigned long val;  // this will hold the value from pulseIn

void setup()
{
  Serial.begin(9600);

  analogWrite(inputPin, 128);
  Serial.print("Writing 128 to pin ");
  Serial.print(inputPin);
  printPulseWidth(inputPin);

  analogWrite(inputPin, 254);
  Serial.print("Writing 254 to pin ");
  Serial.print(inputPin);
```

```
  printPulseWidth(inputPin);

}

void loop()
{
}

void printPulseWidth(int pin)
{
   val = pulseIn(pin, HIGH);
   Serial.print(": High Pulse width = ");
   Serial.print(val);
   val = pulseIn(pin, LOW);
   Serial.print(", Low Pulse width = ");
   Serial.println(val);
}
```

Discussion

The Serial monitor will display :

```
Writing 128 to pin 3: High Pulse width = 989, Low Pulse width = 997
Writing 254 to pin 3: High Pulse width = 1977, Low Pulse width = 8
```

`pulseIn` can measure how long a pulse is either `HIGH` or `LOW`:

```
pulseIn(pin, HIGH); // returns microseconds that pulse is HIGH
pulseIn(pin, LOW)   // returns microseconds that pulse is LOW
```

The `pulseIn` function waits for the pulse to start (or for a timeout if there is no pulse). By default, it will stop waiting after one second, but you can change that by specifying the time to wait in microseconds as a third parameter (note that 1,000 microseconds equals 1 millisecond):

```
pulseIn(pin, HIGH, 5000); // wait 5 milliseconds for the pulse to start
```

The timeout value only matters if the pulse does not start within the given period. Once the start of a pulse is detected, the function will start timing and will not return until the pulse ends.

`pulseIn` can measure values between around 10 microseconds to three minutes in duration, but the value of long pulses may not be very accurate.

See Also

The Arduino reference for `pulseIn`: *http://www.arduino.cc/en/Reference/PulseIn*

Recipe 6.4 shows `pulseIn` used to measure the pulse width of an ultrasonic distance sensor.

Recipe 18.2 provides more information on using hardware interrupts.

12.4 Using Arduino as a Clock

Problem

You want to use the time of day (hours, minutes, and seconds) in a sketch, and you don't want to connect external hardware.

Solution

This sketch uses the Time library to display the time of day. The Time library can be downloaded from: *http://www.arduino.cc/playground/Code/Time*.

```
/*
 * Time sketch
 *
 */

#include <Time.h>


void setup()
{
  Serial.begin(9600);
  setTime(12,0,0,1,1,11); // set time to noon Jan 1 2011
}

void loop()
{
  digitalClockDisplay();
  delay(1000);
}

void digitalClockDisplay(){
  // digital clock display of the time
  Serial.print(hour());
  printDigits(minute());
  printDigits(second());
  Serial.print(" ");
  Serial.print(day());
  Serial.print(" ");
  Serial.print(month());
  Serial.print(" ");
  Serial.print(year());
  Serial.println();
}

void printDigits(int digits){
  // utility function for clock display: prints preceding colon and leading 0
  Serial.print(":");
  if(digits < 10)
    Serial.print('0');
  Serial.print(digits);
}
```

Discussion

The Time library enables you to keep track of the date and time. Many Arduino boards use a quartz crystal for timing, and this is accurate to a couple of seconds per day, but it does not have a battery to remember the time when power is switched off. Therefore, time will restart from 0 each time a sketch starts, so you need to set the time using the `setTime` function. The sketch sets the time to noon on January 1 each time it starts.

The Time library uses a standard known as Unix (also called POSIX) time. The values represent the number of elapsed seconds since January 1, 1970. Experienced C programmers may recognize that this is the same as the `time_t` used in the ISO standard C library for storing time values.

Of course, it's more useful to set the time to your current local time instead of a fixed value. The following sketch gets the numerical time value (the number of elapsed seconds since January 1, 1970) from the serial port to set the time. You can enter a value using the Serial Monitor (the current Unix time can be found on a number of websites using the Google search terms "Unix time convert"):

```
/*
 * TimeSerial sketch
 * example code illustrating Time library set through serial port messages.
 *
 * Messages consist of the letter T followed by ten digit time
 * (as seconds since Jan 1 1970)
 * You can send the text on the next line using Serial Monitor to set the
 * clock to noon Jan 1 2011:
 * T1293883200
 *
 * A Processing example sketch to automatically send the messages is
 * included in the Time library download
 */

#include <Time.h>

#define TIME_MSG_LEN  11   // time sync consists of a HEADER followed by ten
                           // ascii digits
#define TIME_HEADER  'T'   // Header tag for serial time sync message

void setup()  {
  Serial.begin(9600);
  Serial.println("Waiting for time sync message");
}

void loop(){
  if(Serial.available() )
  {
    processSyncMessage();
  }
  if(timeStatus()!= timeNotSet)
  {
```

```
    // here if the time has been set
    digitalClockDisplay();
  }
  delay(1000);
}

void digitalClockDisplay(){
  // digital clock display of the time
  Serial.print(hour());
  printDigits(minute());
  printDigits(second());
  Serial.print(" ");
  Serial.print(day());
  Serial.print(" ");
  Serial.print(month());
  Serial.print(" ");
  Serial.print(year());
  Serial.println();
}

void printDigits(int digits){
  // utility function for digital clock display: prints preceding colon
  // and leading 0
  Serial.print(":");
  if(digits < 10)
    Serial.print('0');
  Serial.print(digits);
}

void processSyncMessage() {
  // if time sync available from serial port, update time and return true
  // time message consists of a header and ten ascii digits
  while(Serial.available() >=  TIME_MSG_LEN ){
    char c = Serial.read() ;
    Serial.print(c);
    if( c == TIME_HEADER ) {
      time_t pctime = 0;
      for(int i=0; i < TIME_MSG_LEN -1; i++){
        c = Serial.read();
        if( isDigit(c)) {
          pctime = (10 * pctime) + (c - '0') ; // convert digits to a number
        }
      }
      setTime(pctime);   // Sync clock to the time received on serial port
    }
  }
}
```

The code to display the time and date is the same as before, but now the sketch waits to receive the time from the serial port. See the Discussion in Recipe 4.3 if you are not familiar with how to receive numeric data using the serial port.

A processing sketch named SyncArduinoClock is included with the Time library examples (it's in the *Time/Examples/Processing/SyncArduinoClock* folder). This Processing sketch will send the current time from your computer to Arduino at the click of a mouse. Run SyncArduinoClock in Processing, ensuring that the serial port is the one connected to Arduino (Chapter 4 describes how to run a Processing sketch that talks to Arduino). You should see the message `Waiting for time sync message` sent by Arduino and displayed in the Processing text area (the black area for text messages at the bottom of the Processing IDE). Click the Processing application window (it's a 200-pixel gray square) and you should see the text area display the time as printed by the Arduino sketch.

You can also set the clock from the Serial Monitor if you can get the current Unix time; *http://www.epochconverter.com/* is one of many websites that provide the time in this format. Copy the 10-digit number indicated as the current Unix time and paste this into the Serial Monitor Send window. Precede the number with the letter *T* and click Send. For example, if you send this:

```
T1282041639
```

Arduino should respond by displaying the time every second:

```
10:40:49 17 8 2010
10:40:50 17 8 2010
10:40:51 17 8 2010
10:40:52 17 8 2010
10:40:53 17 8 2010
10:40:54 17 8 2010
. . .
```

You can also set the time using buttons or other input devices such as tilt sensors, a joystick, or a rotary encoder.

The following sketch uses two buttons to move the clock "hands" forward or backward. Figure 12-1 shows the connections (see Recipe 5.2 if you need help using switches):

```
/*
   AdjustClockTime sketch
   buttons on pins 2 and 3 adjust the time
 */

#include <Time.h>

const int  btnForward = 2;  // button to move time forward
const int  btnBack = 3;     // button to move time back

unsigned long  prevtime;   // when the clock was last displayed

void setup()
{
  digitalWrite(btnForward, HIGH);  // enable internal pull-up resistors
  digitalWrite(btnBack, HIGH);
  setTime(12,0,0,1,1,11); // start with the time set to noon Jan 1 2011
  Serial.begin(9600);
  Serial.println("ready");
}

void loop()
```

```
{
  prevtime = now();   // note the time
  while( prevtime == now() )    // stay in this loop till the second changes
  {
      // check if the set button pressed while waiting for second to roll over
     if(checkSetTime())
         prevtime = now();   // time changed so reset start time
  }
  digitalClockDisplay();
}

// functions checks to see if the time should be adjusted
// returns true if time was changed
boolean checkSetTime()
{
int step;   // the number of seconds to move (backwards if negative)
boolean isTimeAdjusted = false;  // set to true if the time is adjusted
  step = 1;   // ready to step forwards
  while(digitalRead(btnForward)== LOW)
  {
     adjustTime(step);
     isTimeAdjusted = true; // to tell the user that the time has changed
     step = step + 1; // next step will be bigger
     digitalClockDisplay(); // update clock
     delay(100);
  }
  step = -1;   // negative numbers step backwards
  while(digitalRead(btnBack)== LOW)
  {
     adjustTime(step);
     isTimeAdjusted = true; // to tell the user that the time has changed
     step = step - 1; // next step will be a bigger negative number
     digitalClockDisplay(); // update clock
     delay(100);
  }
  return isTimeAdjusted;  // tell the user if the time was adjusted
}

void digitalClockDisplay(){
  // digital clock display of the time
  Serial.print(hour());
  printDigits(minute());
  printDigits(second());
  Serial.print(" ");
  Serial.print(day());
  Serial.print(" ");
  Serial.print(month());
  Serial.print(" ");
  Serial.print(year());
  Serial.println();
}

void printDigits(int digits){
  // utility function for clock display: prints preceding colon and leading 0
  Serial.print(":");
```

```
  if(digits < 10)
    Serial.print('0');
  Serial.print(digits);
}
```

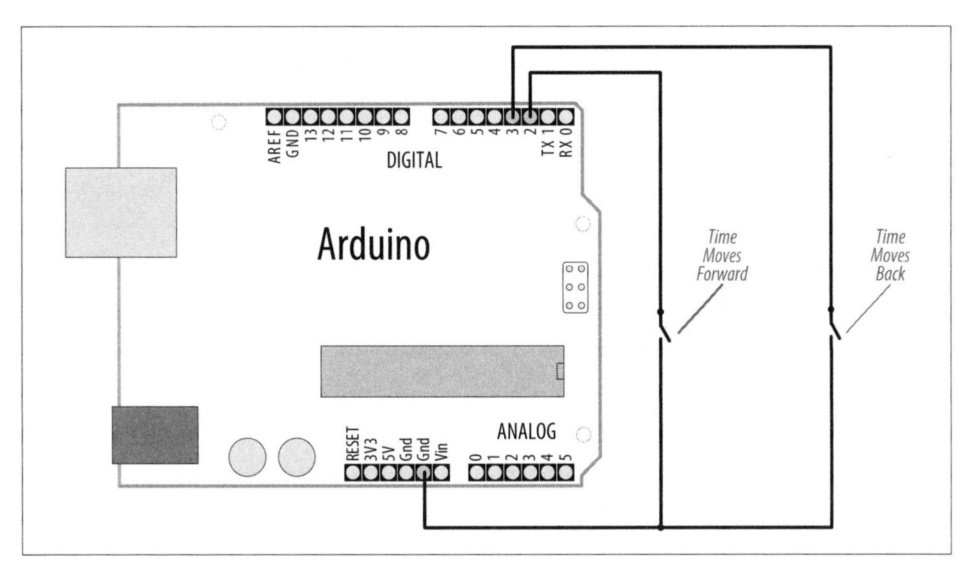

Figure 12-1. Two buttons used to adjust the time

The sketch uses the same `digitalClockDisplay` and `printDigits` functions from Recipe 12.3, so copy those prior to running the sketch.

Here is a variation on this sketch that uses the position of a variable resistor to determine the direction and rate of adjustment when a switch is pressed:

```
#include <Time.h>

const int  potPin = 0;      // pot to determine direction and speed
const int  buttonPin = 2;  // button enables time adjustment

unsigned long  prevtime;   // when the clock was last displayed

void setup()
{
  digitalWrite(buttonPin, HIGH);  // enable internal pull-up resistors
  setTime(12,0,0,1,1,11); // start with the time set to noon Jan 1 2011
  Serial.begin(9600);
}

void loop()
{
  prevtime = now();   // note the time
  while( prevtime == now() )    // stay in this loop till the second changes
  {
```

```
        // check if the set button pressed while waiting for second to roll over
       if(checkSetTime())
          prevtime = now();    //  time changed so reset start time
    }
    digitalClockDisplay();
  }

  // functions checks to see if the time should be adjusted
  // returns true if time was changed
  boolean checkSetTime()
  {
  int value;  // a value read from the pot
  int step;   // the number of seconds to move (backwards if negative)
  boolean isTimeAdjusted = false;  // set to true if the time is adjusted

    while(digitalRead(buttonPin)== LOW)
    {
      // here while button is pressed
       value = analogRead(potPin);  // read the pot value
       step = map(value, 0,1023, 10, -10);  // map value to the desired range
       if( step != 0)
       {
          adjustTime(step);
          isTimeAdjusted = true; // to tell the user that the time has changed
          digitalClockDisplay(); // update clock
          delay(100);
       }
    }
    return isTimeAdjusted;
  }
```

The preceding sketch uses the same `digitalClockDisplay` and `printDigits` functions from Recipe 12.3, so copy those prior to running the sketch. Figure 12-2 shows how the variable resistor and switch are connected.

All these examples print to the serial port, but you can print the output to LEDs or LCDs. The download for the Graphical LCD covered in Recipe 11.9 contains example sketches for displaying and setting time using an analog clock display drawn on the LCD.

The Time library includes convenience functions for converting to and from various time formats. For example, you can find out how much time has elapsed since the start of the day and how much time remains until the day's end.

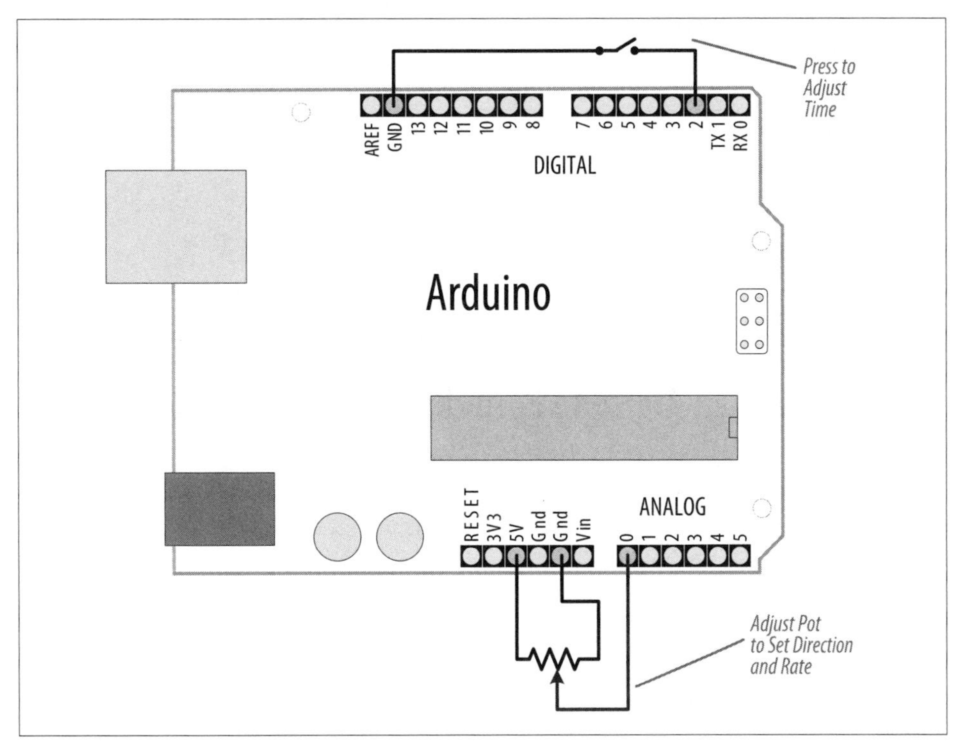

Figure 12-2. A variable resistor used to adjust the time

You can look in *Time.h* in the *libraries* folder for the complete list. More details are available in Chapter 16:

```
dayOfWeek( now() );           // the day of the week (Sunday is day 1)
elapsedSecsToday( now() );    // returns the number of seconds since the start
                              // of today
nextMidnight( now() );        // how much time to the end of the day
elapsedSecsThisWeek( now() ); // how much time has elapsed since the start of
                              // the week
```

You can also print text strings for the days and months; here is a variation on the digital clock display code that prints the names of the day and month:

```
void digitalClockDisplay(){
  // digital clock display of the time
  Serial.print(hour());
  printDigits(minute());
  printDigits(second());
  Serial.print(" ");
  Serial.print(dayStr(weekday())); // print the day of the week
  Serial.print(" ");
  Serial.print(day());
  Serial.print(" ");
  Serial.print(monthShortStr(month())); // print the month (abbreviated)
  Serial.print(" ");
```

```
  Serial.print(year());
  Serial.println();
}
```

See Also

Arduino Time library reference: *http://www.arduino.cc/playground/Code/Time*

Wikipedia article on Unix time: *http://en.wikipedia.org/wiki/Unix_time*

http://www.epochconverter.com/ and *http://www.onlineconversion.com/unix_time.htm* are two popular Unix time conversion tools.

12.5 Creating an Alarm to Periodically Call a Function

Problem

You want to perform some action on specific days and at specific times of the day.

Solution

TimeAlarms is a companion library included in the Time library download discussed in Recipe 12.4 (installing the Time library will also install the TimeAlarms library). TimeAlarms makes it easy to create time and date alarms:

```
/*
 * TimeAlarmsExample sketch
 *
 * This example calls alarm functions at 8:30 am and at 5:45 pm (17:45)
 * and simulates turning lights on at night and off in the morning
 *
 * A timer is called every 15 seconds
 * Another timer is called once only after 10 seconds
 *
 * At startup the time is set to Jan 1 2010  8:29 am
 */

#include <Time.h>
#include <TimeAlarms.h>

void setup()
{
  Serial.begin(9600);
  Serial.println("TimeAlarms Example");
  Serial.println("Alarms are triggered daily at 8:30 am and 17:45 pm");
  Serial.println("One timer is triggered every 15 seconds");
  Serial.println("Another timer is set to trigger only once after 10 seconds");
  Serial.println();

  setTime(8,29,40,1,1,10); // set time to 8:29:40am Jan 1 2010

  Alarm.alarmRepeat(8,30,0, MorningAlarm);  // 8:30am every day
```

```
  Alarm.alarmRepeat(17,45,0,EveningAlarm);  // 5:45pm every day

  Alarm.timerRepeat(15, RepeatTask);            // timer for every 15 seconds
  Alarm.timerOnce(10, OnceOnlyTask);            // called once after 10 seconds
}

void MorningAlarm()
{
  Serial.println("Alarm: - turn lights off");
}

void EveningAlarm()
{
  Serial.println("Alarm: - turn lights on");
}

void RepeatTask()
{
  Serial.println("15 second timer");
}

void OnceOnlyTask()
{
  Serial.println("This timer only triggers once");
}

void  loop()
{
  digitalClockDisplay();
  Alarm.delay(1000); // wait one second between clock display
}

void digitalClockDisplay()
{
  // digital clock display of the time
  Serial.print(hour());
  printDigits(minute());
  printDigits(second());
  Serial.println();
}

// utility function for digital clock display: prints preceding colon and
// leading 0.
//
void printDigits(int digits)
{
  Serial.print(":");
  if(digits < 10)
    Serial.print('0');
  Serial.print(digits);
}
```

Discussion

You can schedule tasks to trigger at a particular time of day (these are called *alarms*) or schedule tasks to occur after an interval of time has elapsed (called *timers*). Each of these tasks can be created to continuously repeat or to occur only once.

To specify an alarm to trigger a task repeatedly at a particular time of day use:

```
Alarm.alarmRepeat(8,30,0, MorningAlarm);
```

This calls the function `MorningAlarm` at 8:30 a.m. every day.

If you want the alarm to trigger only once, you can use the `alarmOnce` method:

```
Alarm.alarmOnce(8,30,0, MorningAlarm);
```

This calls the function `MorningAlarm` a single time only (the next time it is 8:30 a.m.) and will not trigger again.

Timers trigger tasks that occur after a specified interval of time has passed rather than at a specific time of day. The timer interval can be specified in any number of seconds, or in hour, minutes, and seconds:

```
Alarm.timerRepeat(15, Repeats);            // timer task every 15 seconds
```

This calls the `Repeats` function in your sketch every 15 seconds.

If you want a timer to trigger once only, use the `timerOnce` method:

```
Alarm.timerOnce(10, OnceOnly);             // called once after 10 seconds
```

This calls the `onceOnly` function in a sketch 10 seconds after the timer is created.

Your code needs to call `Alarm.delay` regularly because this function checks the state of all the scheduled events. Failing to regularly call `Alarm.delay` will result in the alarms not being triggered. You can call `Alarm.delay(0)` if you need to service the scheduler without a delay. Always use `Alarm.delay` instead of `delay` when using TimeAlarms in a sketch.

The TimeAlarms library requires the Time library to be installed—see Recipe 12.4. No internal or external hardware is required to use the TimeAlarms library. The scheduler does not use interrupts, so the task-handling function is the same as any other functions you create in your sketch (code in an interrupt handler has restrictions that are discussed in Chapter 18, but these do not apply to TimeAlarms functions).

Timer intervals can range from one second to several years. (If you need timer intervals shorter than one second, the TimedAction library by Alexander Brevig may be more suitable; see *http://www.arduino.cc/playground/Code/TimedAction*.)

Tasks are scheduled for specific times designated by the system clock in the Time library (see Recipe 12.4 for more details). If you change the system time (e.g., by calling setTime), the trigger times are not adjusted. For example, if you use setTime to move one hour ahead, all alarms and timers will occur one hour sooner. In other words, if it's 1:00 and a task is set to trigger in two hours (at 3:00), and then you change the current time to 2:00, the task will trigger in one hour. If the system time is set backward—for example, to 12:00—the task will trigger in three hours (i.e., when the system time indicates 3:00). If the time is reset to earlier than the time at which a task was scheduled, the task will be triggered immediately (actually, on the next call to Alarm.delay).

This is the expected behavior for alarms—tasks are scheduled for a specific time of day and will trigger at that time—but the effect on timers may be less clear. If a timer is scheduled to trigger in five minutes' time and then the clock is set back by one hour, that timer will not trigger until one hour and five minutes have elapsed (even if it is a repeating timer—a repeat does not get rescheduled until after it triggers).

Up to six alarms and timers can be scheduled to run at the same time. You can modify the library to enable more tasks to be scheduled; Recipe 16.3 shows you how to do this.

onceOnly alarms and timers are freed when they are triggered, and you can reschedule these as often as you want so long as there are no more than six pending at one time. The following code gives one example of how a timerOnce task can be rescheduled:

```
Alarm.timerOnce(random(10), randomTimer);  // trigger after random
                                           // number of seconds

void randomTimer(){
  int period = random(2,10);              // get a new random period
  Alarm.timerOnce(period, randomTimer);  // trigger for another random period
}
```

12.6 Using a Real-Time Clock

Problem

You want to use the time of day provided by a real-time clock (RTC). External boards usually have battery backup, so the time will be correct even when Arduino is reset or turned off.

Solution

The simplest way to use an RTC is with a companion library for the Time library, named *DS1307RTC.h*. This recipe is for the widely used DS1307 and DS1337 RTC chips:

```
/*
 * TimeRTC sketch
 * example code illustrating Time library with real-time clock.
 *
 */

#include <Time.h>
#include <Wire.h>
```

```
#include <DS1307RTC.h>  // a basic DS1307 library that returns time as a time_t

void setup()  {
  Serial.begin(9600);
  setSyncProvider(RTC.get);   // the function to get the time from the RTC
  if(timeStatus()!= timeSet)
     Serial.println("Unable to sync with the RTC");
  else
     Serial.println("RTC has set the system time");
}

void loop()
{
   digitalClockDisplay();
   delay(1000);
}

void digitalClockDisplay(){
  // digital clock display of the time
  Serial.print(hour());
  printDigits(minute());
  printDigits(second());
  Serial.print(" ");
  Serial.print(day());
  Serial.print(" ");
  Serial.print(month());
  Serial.print(" ");
  Serial.print(year());
  Serial.println();
}

// utility function for digital clock display: prints preceding colon and
// leading 0.
//
void printDigits(int digits){
  Serial.print(":");
  if(digits < 10)
    Serial.print('0');
  Serial.print(digits);
}
```

Most RTC boards for Arduino use the I2C protocol for communicating (see Chapter 13 for more on I2C). Connect the line marked "SCL" (or "Clock") to Arduino analog pin 5 and "SDA" (or "Data") to analog pin 4, as shown in Figure 12-3. (Analog pins 4 and 5 are used for I2C; see Chapter 13). Take care to ensure that you connect the +5V power line and Gnd pins correctly.

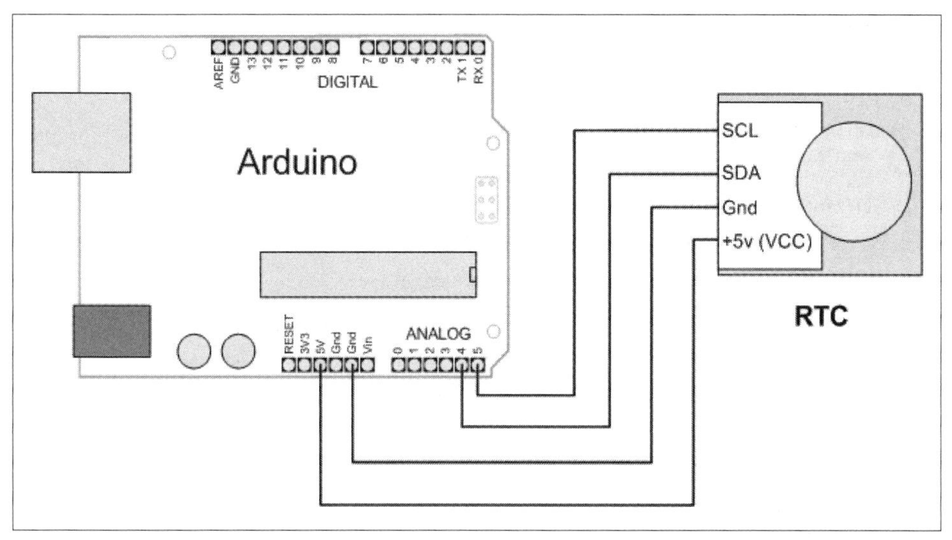

Figure 12-3. Connecting a real-time clock

Discussion

The code is similar to other recipes using the Time library, but it gets its value from the RTC rather than from the serial port or hardcoded value. The only additional line needed is this:

```
setSyncProvider(RTC.get);   // the function to get the time from the RTC
```

The `setSyncProvider` function tells the Time library how it should get information for setting (and updating) the time. `RTC.get` is a method within the RTC library that returns the current time in the format used by the Time library (Unix time).

Each time Arduino starts, the `setup` function will call `RTC.get` to set the time from the RTC hardware.

Before you can get the correct time from the module, you need to set its time. Here is a sketch that enables you to set the time on the RTC hardware—you only need to do this when you first attach the battery to the RTC, when replacing the battery, or if the time needs to be changed:

```
/*
 * TimeRTCSet sketch
 * example code illustrating Time library with real-time clock.
 *
 * RTC is set in response to serial port time message
 * A Processing example sketch to set the time is included in the download
 */
```

```
#include <Time.h>
#include <Wire.h>
#include <DS1307RTC.h>  // a basic DS1307 library that returns time as a time_t

void setup()  {
  Serial.begin(9600);
  setSyncProvider(RTC.get);   // the function to get the time from the RTC
  if(timeStatus()!= timeSet)
     Serial.println("Unable to sync with the RTC");
  else
     Serial.println("RTC has set the system time");
}

void loop()
{
  if(Serial.available())
  {
     time_t t = processSyncMessage();
     if(t >0)
     {
        RTC.set(t);   // set the RTC and the system time to the received value
        setTime(t);
     }
  }
   digitalClockDisplay();
   delay(1000);
}

void digitalClockDisplay(){
  // digital clock display of the time
  Serial.print(hour());
  printDigits(minute());
  printDigits(second());
  Serial.print(" ");
  Serial.print(day());
  Serial.print(" ");
  Serial.print(month());
  Serial.print(" ");
  Serial.print(year());
  Serial.println();
}

// utility function for digital clock display: prints preceding colon and
// leading 0.
//
void printDigits(int digits){
  Serial.print(":");
  if(digits < 10)
    Serial.print('0');
  Serial.print(digits);
}

/*  code to process time sync messages from the serial port   */
#define TIME_MSG_LEN  11   // time sync to PC is HEADER followed by Unix time_t
                           // as ten ascii digits
```

```
#define TIME_HEADER  'T'   // Header tag for serial time sync message

time_t processSyncMessage() {
  // return the time if a valid sync message is received on the serial port.
  // time message consists of a header and ten ascii digits
  while(Serial.available() >=  TIME_MSG_LEN ){
    char c = Serial.read() ;
    Serial.print(c);
    if( c == TIME_HEADER ) {
      time_t pctime = 0;
      for(int i=0; i < TIME_MSG_LEN -1; i++){
        c = Serial.read();
        if( c >= '0' && c <= '9'){
          pctime = (10 * pctime) + (c - '0') ; // convert digits to a number
        }
      }
      return pctime;
    }
  }
  return 0;
}
```

This sketch is almost the same as the TimeSerial sketch in Recipe 12.4 for setting the time from the serial port, but here the following function is called when a time message is received from the computer to set the RTC:

```
RTC.set(t);   // set the RTC and the system time to the received value

setTime(t);
```

The RTC chip uses I2C to communicate with Arduino. I2C is explained in Chapter 13; see Recipe 13.3 if you are interested in more details on I2C communication with the RTC chip.

See Also

The SparkFun BOB-00099 data sheet: *http://store.gravitech.us/i2crecl.html*

CHAPTER 13

Communicating Using I2C and SPI

13.0 Introduction

The I2C (Inter-Integrated Circuit) and SPI (Serial Peripheral Interface) standards were created to provide simple ways for digital information to be transferred between sensors and microcontrollers such as Arduino. Arduino libraries for both I2C and SPI make it easy for you to use both of these protocols.

The choice between I2C and SPI is usually determined by the devices you want to connect. Some devices provide both standards, but usually a device or chip supports one or the other.

I2C has the advantage that it only needs two signal connections to Arduino—using multiple devices on the two connections is fairly easy, and you get acknowledgment that signals have been correctly received. The disadvantages are that the data rate is slower than SPI and data can only be traveling in one direction at a time, lowering the data rate even more if two-way communication is needed. It is also necessary to connect pull-up resistors to the connections to ensure reliable transmission of signals (see the introduction to Chapter 5 for more on pull-ups).

The advantages of SPI are that it runs at a higher data rate, and it has separate input and output connections, so it can send and receive at the same time. It uses one additional line per device to select the active device, so more connections are required if you have many devices to connect.

Most Arduino projects use SPI devices for high data rate applications such as Ethernet and memory cards, with just a single device attached. I2C is more typically used with sensors that don't need to send a lot of data.

This chapter shows how to use I2C and SPI to connect to common devices. It also shows how to connect two or more Arduino boards together using I2C for multiboard applications.

I2C

The two connections for the I2C bus are called *SCL* and *SDA*. These are available on a standard Arduino board using analog pin 5 for SCL, which provides a clock signal, and analog pin 4 for SDL, which is for transfer of data (on the Mega, use digital pin 20 for SDA and pin 21 for SCL). Uno rev 3 boards have extra pins (shown back in Recipe 1.2) that duplicate pins 4 and 5. If you have such a board, you can use either set of pins. One device on the I2C bus is considered the *master* device. Its job is to coordinate the transfer of information between the other devices (*slaves*) that are attached. There must be only one master, and in most cases the Arduino is the master, controlling the other chips attached to it. Figure 13-1 depicts an I2C master with multiple I2C slaves.

Boards introduced with Arduino 1.0 such as the Leonardo board have the SCL and SDA lines duplicated on pins next to the AREF pin. This new location for these pins enables future boards to always have the I2C connections in the same physical position.

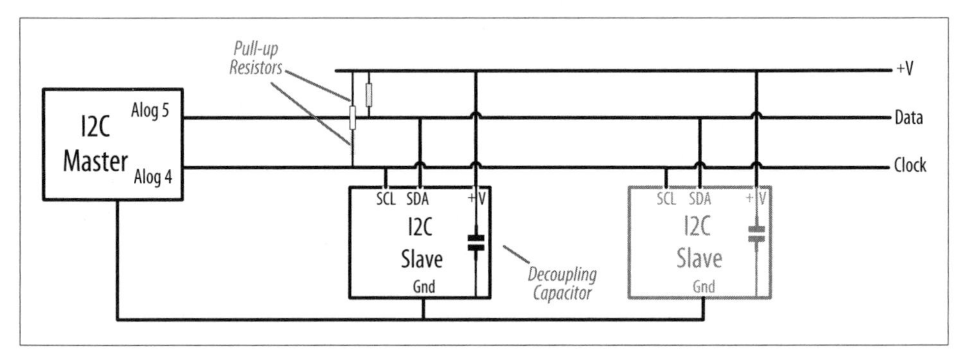

Figure 13-1. An I2C master with one or more I2C slaves

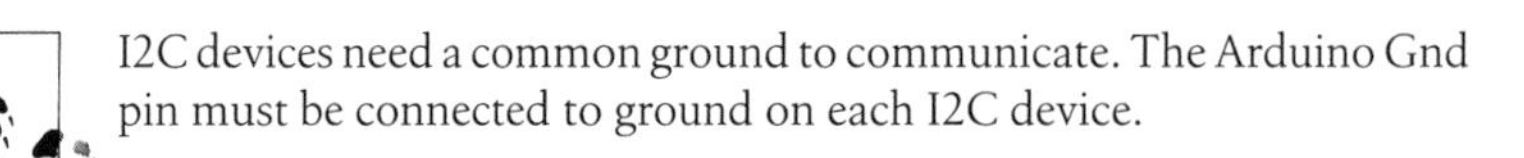

I2C devices need a common ground to communicate. The Arduino Gnd pin must be connected to ground on each I2C device.

Slave devices are identified by their address number. Each slave must have a unique address. Some I2C devices have a fixed address (an example is the nunchuck in Recipe 13.2) while others allow you to configure their address by setting pins high or low (see Recipe 13.7) or by sending initialization commands.

Arduino uses 7-bit values to specify I2C addresses. Some device data sheets use 8-bit address values. If yours does, divide that value by 2 to get the correct 7-bit value.

I2C and SPI only define how communication takes place between devices—the messages that need to be sent depend on each individual device and what it does. You will need to consult the data sheet for your device to determine what commands are required to get it to function, and what data is required, or returned.

The Arduino Wire library hides all the low-level functionality for I2C and enables simple commands to be used to initialize and communicate with devices. Recipe 13.1 provides a basic introduction to the library and its use.

Migrating Wire code to Arduino 1.0

The Arduino Wire library has been changed in release 1.0 and you will need to modify sketches written for previous releases to compile them in 1.0. The send and receive methods have been renamed for consistency with other libraries:

Change `Wire.send()` to `Wire.write()`.
Change `Wire.receive()` to `Wire.read()`.

You now need to specify the variable type for literal constant arguments to write, for example:

Change `Wire.write(0x10)` to `Wire.write((byte)0x10)`.

Using 3.3 Volt Devices with 5 Volt Boards

Many I2C devices are intended for 3.3 volt operation and can be damaged when connected to a 5 volt Arduino board. You can use a logic-level translator such as the BOB-08745 breakout board from SparkFun to enable connection by converting the voltage levels (see Figure 13-2). The level translator board has a low-voltage (LV) side for 3.3 volts and a high-voltage (HV) side for 5 volts.

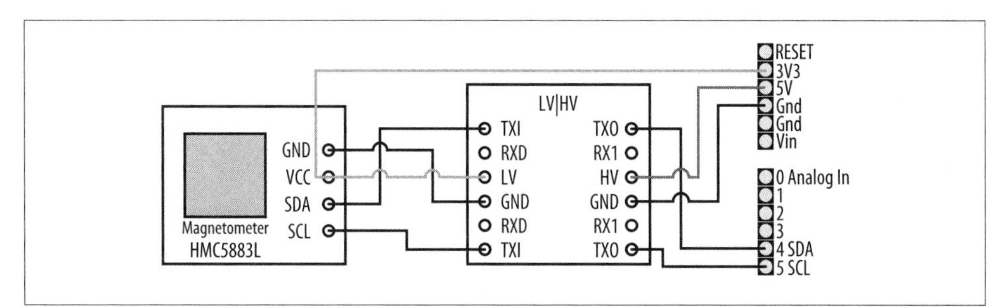

Figure 13-2. Using a 3.3V device with a logic-level translator

For a 3.3V I2C device, connect the LV side as follows:

- Upper TXI pin to I2C SDA pin
- Lower TXI pin to I2C SCL pin
- LV pin to I2C VCC (power) and 3.3 volt power source
- GND pin to I2C Gnd

Connect the HV side as follows:

- Upper TXO pin to I2C SDA pin
- Lower TXO pin to I2C SCL pin
- HV pin to Arduino 5 volt power source
- GND pin to Arduino Gnd

You can connect multiple I2C devices using a single logic-level translator, as in Figure 13-3.

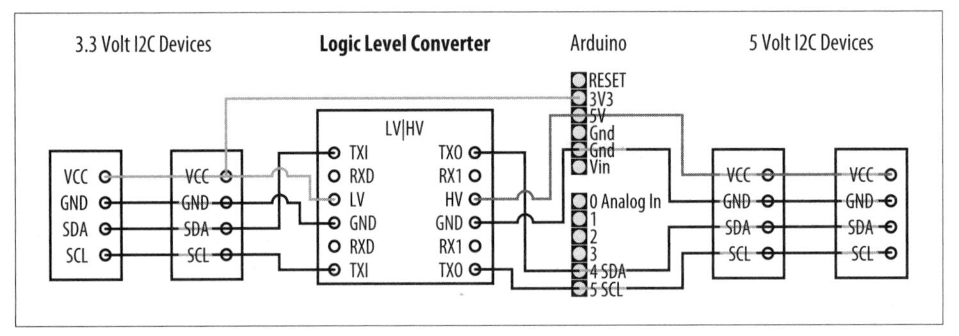

Figure 13-3. Connecting multiple 3.3V and 5V I2C devices

For examples that use a logic-level translator, see the discussion on the ITG-3200 in Recipe 6.15 and the HMC5883 in Recipe 6.16.

SPI

Recent Arduino releases (from release 0019) include a library that allows communication with SPI devices. SPI has separate input (labeled "MOSI") and output (labeled "MISO") lines and a clock line. These three lines are connected to the respective lines on one or more slaves. Slaves are identified by signaling with the Slave Select (SS) line. Figure 13-4 shows the SPI connections.

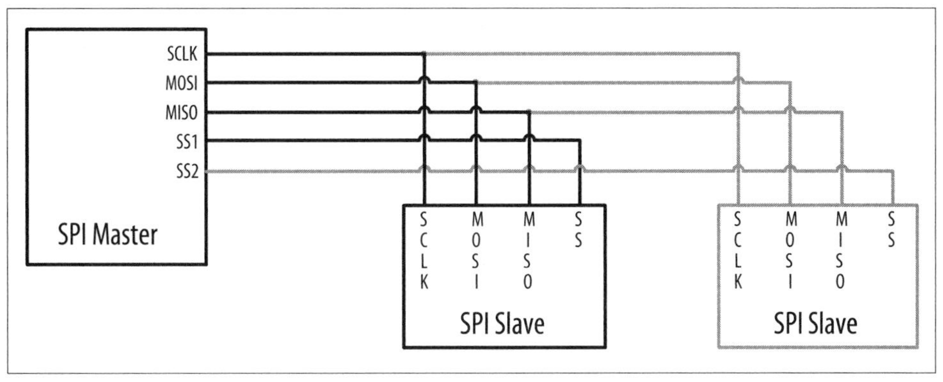

Figure 13-4. Signal connections for SPI master and slaves

The pin numbers to use for the SPI pins are shown in Table 13-1.

Table 13-1. Arduino digital pins used for SPI

SPI signal	Standard Arduino board	Arduino Mega
SCLK (clock)	13	52
MISO (data out)	12	50
MOSI (data in)	11	51
SS (slave select)	10	53

See Also

Applications note comparing I2C to SPI: *http://www.maxim-ic.com/app-notes/index.mvp/id/4024*

Arduino Wire library reference: *http://www.arduino.cc/en/Reference/Wire*

Arduino SPI library reference: *http://www.arduino.cc/playground/Code/Spi*

13.1 Controlling an RGB LED Using the BlinkM Module

Problem

You want to control I2C-enabled LEDs such as the BlinkM module.

Solution

BlinkM is a preassembled color LED module that gets you started with I2C with minimal fuss.

Insert the BlinkM pins onto analog pins 2 through 5, as shown in Figure 13-5.

The following sketch is based on Recipe 7.4, but instead of directly controlling the voltage on the red, green, and blue LED elements, I2C commands are sent to the BlinkM module with instructions to produce a color based on the red, green, and blue levels. The hueToRGB function is the same as what we used in Recipe 7.4 and is not repeated here, so copy the function into the bottom of your sketch before compiling (this book's website has the complete sketch):

```
 /*
 * BlinkM sketch
 * This sketch continuously fades through the color wheel
 */

#include <Wire.h>

const int address = 0;  // Default I2C address for BlinkM

int color = 0; // a value from 0 to 255 representing the hue
byte R, G, B;  // the Red, Green, and Blue color components

void setup()
{
  Wire.begin(); // set up Arduino I2C support

  // turn on power pins for BlinkM
  pinMode(17, OUTPUT);    // pin 17 (analog out 3) provides +5V to BlinkM
  digitalWrite(17, HIGH);
  pinMode(16, OUTPUT);    // pin 16 (analog out 2) provides Ground
  digitalWrite(16, LOW);
}


void loop()
{
  int brightness = 255; // 255 is maximum brightness
  hueToRGB(color, brightness);  // call function to convert hue to RGB
  // write the RGB values to BlinkM

  Wire.beginTransmission(address);// join I2C, talk to BlinkM
  Wire.write('c');             // 'c' == fade to color
  Wire.write(R);               // value for red channel
  Wire.write(B);               // value for blue channel
  Wire.write(G);               // value for green channel
  Wire.endTransmission();      // leave I2C bus

  color++;             // increment the color
  if (color > 255)
     color = 0;
     delay(10);
}
```

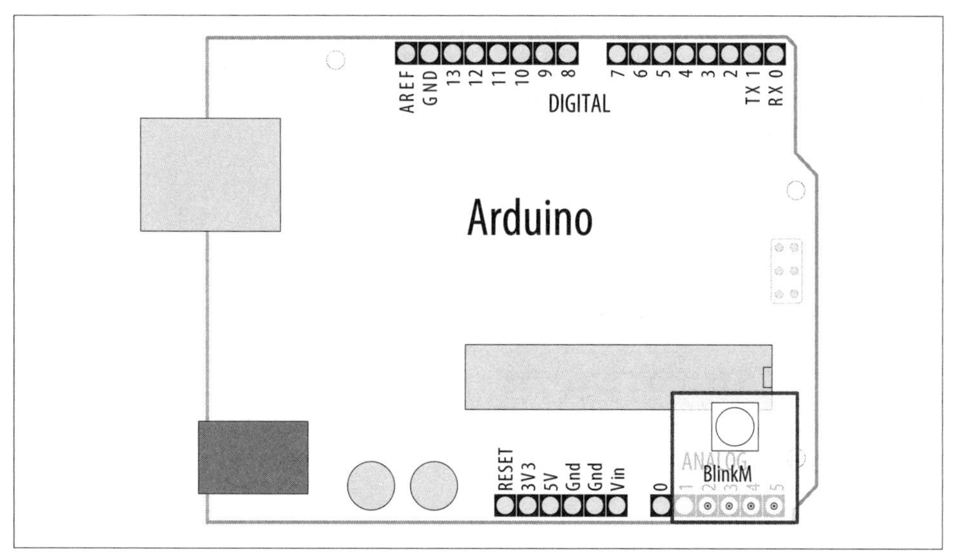

Figure 13-5. BlinkM module plugged in to analog pins

Discussion

The Wire library is added to the sketch using the following:

```
#include <Wire.h>
```

For more details about using libraries, see Chapter 16.

The code in `setup` initializes the Wire library and the hardware in the Arduino to drive SCA and SDL on analog pins 4 and 5 and turns on the pins used to power the BlinkM module.

The `loop` code calls the function `hueToRGB` to calculate the red, green, and blue values for the color.

The R, G, and B values are sent to BlinkM using this sequence:

```
Wire.beginTransmission(address); // start an I2C message to the BlinkM address
Wire.write('c');                 // 'c' is a command to fade to the color that follows
Wire.write(R);                   // value for red
Wire.write(B);                   // value for blue
Wire.write(G);                   // value for green
Wire.endTransmission();         // complete the I2C message
```

All data transmission to I2C devices follows this pattern: `beginTransmission`, a number of `write` messages, and `endTransmission`.

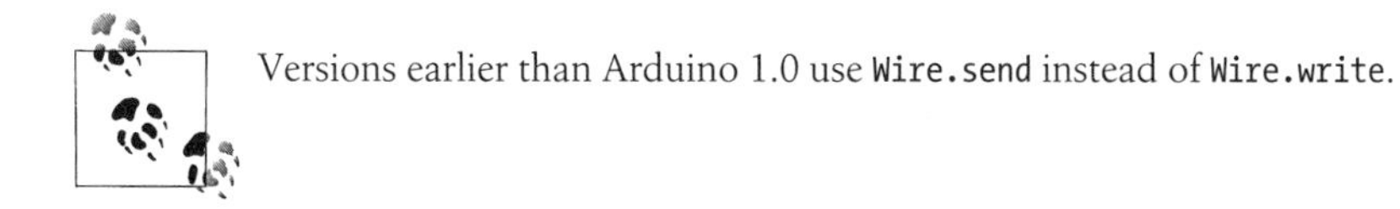

Versions earlier than Arduino 1.0 use `Wire.send` instead of `Wire.write`.

I2C supports up to 127 devices connected to the clock and data pins, and the address determines which device will respond. The default address for BlinkM is 0, but this can be altered by sending a command to change the address—see the BlinkM user manual for information on all commands.

To connect multiple BlinkMs, connect all the clock pins (marked "c" on BlinkM, analog pin 5 on Arduino) and all the data pins (marked "d" on BlinkM, analog pin 4 on Arduino), as shown in Figure 13-6. The power pins should be connected to +5V and Gnd on Arduino or an external power source, as the analog pins cannot provide enough current for more than a couple of modules.

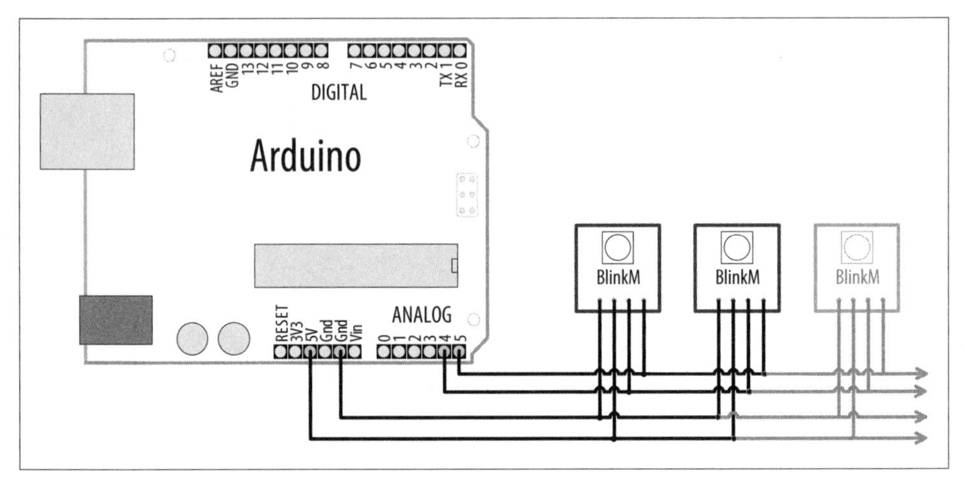

Figure 13-6. Multiple BlinkM modules connected together

Each BlinkM can draw up to 60 mA, so if you're using more than a handful, they should be powered using an external supply.

You need to set each BlinkM to a different I2C address, and you can use the BlinkM-Tester sketch that comes with the BlinkM examples downloadable from *http://code.google.com/p/blinkm-projects/*.

Compile and upload the BlinkMTester sketch. Plug each BlinkM module in to Arduino one at a time (switch off power when connecting and disconnecting the modules). Use the BlinkMTester scan command, `s`, to display the current address, and use the `A` command to set each module to a different address.

BlinkMTester communicates at 19,200 baud, so you may need to set the baud rate in the Serial Monitor to this speed to get a readable display.

After all the BlinkMs have a unique address, you can set the `address` variable in the preceding sketch to the address of the BlinkM you want to control. This example assumes addresses from 9 to 11:

```
#include <Wire.h>

int addressA = 9;  // I2C address for BlinkM
int addressB = 10;
int addressC = 11;

int color = 0; // a value from 0 to 255 representing the hue
byte R, G, B;  // the red, green, and blue color components

void setup()
{
  Wire.begin(); // set up Arduino I2C support

  // turn on power pins for BlinkM
  pinMode(17, OUTPUT);     // pin 17 (analog out 4) provides +5V to BlinkM
  digitalWrite(17, HIGH);
  pinMode(16, OUTPUT);  // pin 16 (analog out 3) provides Ground
  digitalWrite(16, LOW);
}


void loop()
{
  int brightness = 255; // 255 is maximum brightness
  hueToRGB( color, brightness);  // call function to convert hue to RGB
  // write the RGB values to each BlinkM
  setColor(addressA, R,G,B);
  setColor(addressB, G,B,R);
  setColor(addressA, B,R,G);

  color++;            // increment the color
  if(color > 255)     // ensure valid value
     color = 0;
       delay(10);
}

void setColor(int address, byte R, byte G, byte B)
{
  Wire.beginTransmission(address);// join I2C, talk to BlinkM
  Wire.write('c');                 // 'c' == fade to color
  Wire.write(R);                   // value for red channel
  Wire.write(B);                   // value for blue channel
  Wire.write(G);                   // value for green channel
  Wire.endTransmission();         // leave I2C bus
}
```

```
// Use hueToRGB function from previous sketch
```

The `setColor` function writes the given RGB values to the BlinkM at the given address.

The code uses the `hueToRGB` function from earlier in this recipe to convert an integer value into its red, green, and blue components.

See Also

The BlinkM User Manual: *http://thingm.com/fileadmin/thingm/downloads/BlinkM_datasheet.pdf*

Example Arduino sketches: *http://code.google.com/p/blinkm-projects/*

13.2 Using the Wii Nunchuck Accelerometer

Problem

You want to connect a Wii nunchuck to your Arduino as a convenient and inexpensive way to use accelerometer input. The nunchuck is a popular low-cost game device that can be used to indicate the orientation of the device by measuring the effects of gravity.

Solution

The nunchuck uses a proprietary plug. If you don't want to use your nunchuck with your Wii again, you can cut the lead to connect it. Alternatively, it is possible to use a small piece of matrix board to make the connections in the plug if you are careful (the pinouts are shown in Figure 13-7) or you can buy an adapter made by Todbot (*http://todbot.com/blog/2008/02/18/wiichuck-wii-nunchuck-adapter-available/*).

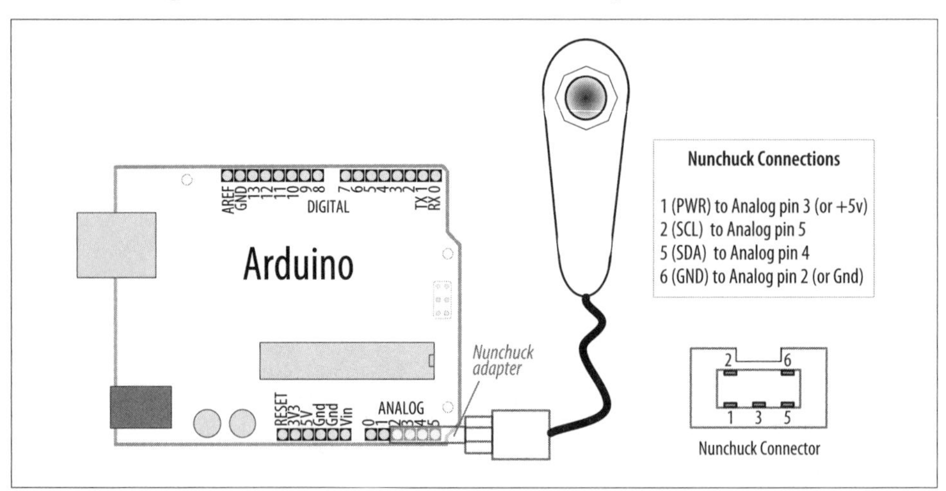

Figure 13-7. Connecting a nunchuck to Arduino

```
/*
 * nunchuck_lines sketch
 * sends data to Processing to draw line that follows nunchuck movement
 */

#include <Wire.h> // initialize wire

const int vccPin = A3;   // +v provided by pin 17
const int gndPin = A2;   // gnd provided by pin 16

const int dataLength = 6;          // number of bytes to request
static byte rawData[dataLength];   // array to store nunchuck data

enum nunchuckItems { joyX, joyY, accelX, accelY, accelZ, btnZ, btnC };

void setup() {
  pinMode(gndPin, OUTPUT); // set power pins to the correct state
  pinMode(vccPin, OUTPUT);
  digitalWrite(gndPin, LOW);
  digitalWrite(vccPin, HIGH);
  delay(100);  // wait for things to stabilize

  Serial.begin(9600);
  nunchuckInit();
}

void loop(){
  nunchuckRead();
  int acceleration = getValue(accelX);
  if((acceleration >= 75) && (acceleration <= 185))
  {
    //map returns a value from 0 to 63 for values from 75 to 185
    byte  x = map(acceleration, 75, 185, 0, 63);
    Serial.write(x);
  }
  delay(20); // the time in milliseconds between redraws
}

void nunchuckInit(){
  Wire.begin();                   // join i2c bus as master
  Wire.beginTransmission(0x52);// transmit to device 0x52
  Wire.write((byte)0x40);         // sends memory address
  Wire.write((byte)0x00);         // sends sent a zero.
  Wire.endTransmission();         // stop transmitting
}

// Send a request for data to the nunchuck
static void nunchuckRequest(){
  Wire.beginTransmission(0x52);// transmit to device 0x52
  Wire.write((byte)0x00);         // sends one byte
  Wire.endTransmission();         // stop transmitting
}

// Receive data back from the nunchuck,
// returns true if read successful, else false
```

```
boolean nunchuckRead(){
  int cnt=0;
  Wire.requestFrom (0x52, dataLength); // request data from nunchuck
  while (Wire.available ()) {
    rawData[cnt] = nunchuckDecode(Wire.read());
    cnt++;
  }
  nunchuckRequest();  // send request for next data payload
  if (cnt >= dataLength)
    return true;     // success if all 6 bytes received
  else
    return false;    //failure
}

// Encode data to format that most wiimote drivers accept
static char nunchuckDecode (byte x) {
  return (x ^ 0x17) + 0x17;
}

int getValue(int item){
  if (item <= accelZ)
    return (int)rawData[item];
  else if (item  == btnZ)
    return bitRead(rawData[5], 0) ? 0: 1;
  else if (item  == btnC)
    return bitRead(rawData[5], 1) ? 0: 1;
}
```

Discussion

I2C is often used in commercial products such as the nunchuck for communication between devices. There are no official data sheets for this device, but the nunchuck signaling was analyzed (reverse engineered) to determine the commands needed to communicate with it.

You can use the following Processing sketch to display a line that follows the nunchuck movement, as shown in Figure 13-8 (see Chapter 4 for more on using Processing to receive Arduino serial data; also see Chapter 4 for advice on setting up and using Processing with Arduino):

```
 // Processing sketch to draw line that follows nunchuck data

import processing.serial.*;

Serial myPort;  // Create object from Serial class
public static final short portIndex = 1;

void setup()
{
  size(200, 200);
  // Open whatever port is the one you're using - See Chapter 4
  myPort = new Serial(this,Serial.list()[portIndex], 9600);
}
```

```
void draw()
{
  if ( myPort.available() > 0) {  // If data is available,
    int y = myPort.read();        // read it and store it in val
    background(255);              // Set background to white
    line(0,63-y,127,y);           // draw the line
  }
}
```

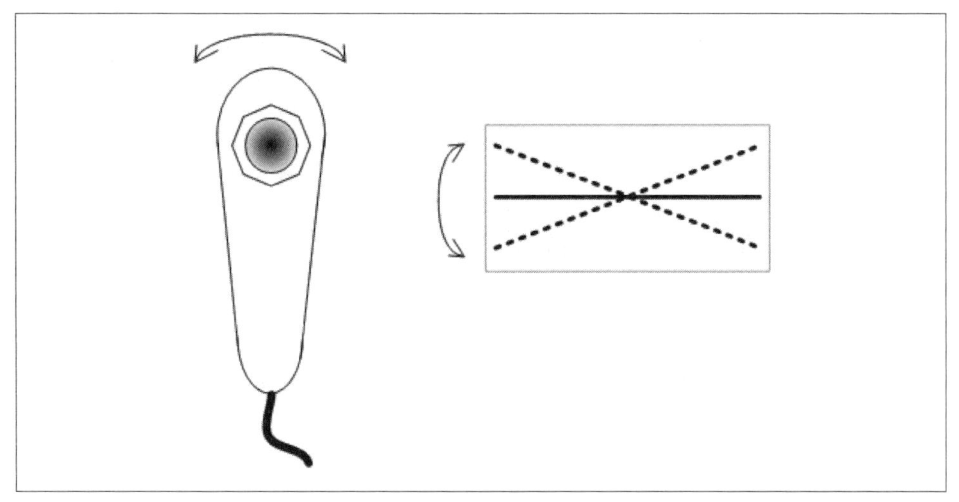

Figure 13-8. Nunchuck movement represented by tilted line in Processing

The sketch includes the Wire library for I2C communication and defines the pins used to power the nunchuck:

```
#include <Wire.h> // initialize wire

const int vccPin = A3;    // +v (vcc) provided by pin 17
const int gndPin = A2;    // gnd provided by pin 16
```

Wire.h is the I2C library that is included with the Arduino release. `A3` is analog pin 3 (digital pin 17), `A2` is analog pin 2 (digital pin 16); these pins provide power to the nunchuck.

```
enum nunchuckItems { joyX, joyY, accelX, accelY, accelZ, btnZ, btnC };
```

`enum` is the construct to create an enumerated list of constants, in this case a list of the sensor values returned from the nunchuck. These constants are used to identify requests for one of the nunchuck sensor values.

`setup` initializes the pins used to power the nunchuck by setting the `vccPin` `HIGH` and `gndPin` `LOW`. This is only needed if the nunchuck adapter is providing the power source. Using digital pins as a power source is not usually recommended, unless you are certain, as with the nunchuck, that the device being powered will not exceed a pin's maximum current capability (40 mA; see Chapter 5).

The function nunchuckInit establishes I2C communication with the nunchuck.

I2C communication starts with Wire.begin(). In this example, Arduino as the master is responsible for initializing the desired slave device, the nunchuck, on address 0x52.

The following line tells the Wire library to prepare to send a message to the device at hexadecimal address 52 (0x52):

```
beginTransmission(0x52);
```

I2C documentation typically shows addresses with hexadecimal values, so it's convenient to use this notation in your sketch.

Wire.send puts the given values into a buffer within the Wire library where data is stored until Wire.endTransmission is called to actually do the sending.

nunchuckRequest and nunchuckRead are used to request and read data from the nunchuck:

This Wire library requestFrom function is used to get six bytes of data from device 0x52 (the nunchuck).

The nunchuck returns its data using six bytes as follows:

Byte number	Description
Byte 1	x-axis analog joystick value
Byte 2	y-axis analog joystick value
Byte 3	x-axis acceleration value
Byte 4	y-axis acceleration value
Byte 5	z-axis acceleration value
Byte 6	Button states and least significant bits of acceleration

Wire.available works like Serial.available (see Chapter 4) to indicate how many bytes have been received, but over the I2C interface rather than the serial interface. If data is available, it is read using Wire.read and then decoded using nunchuckDecode. Decoding is required to convert the values sent into numbers that are usable by your sketch, and these are stored in a buffer (named rawData). A request is sent for the next six bytes of data so that it will be ready and waiting for the next call to get data:

```
int acceleration  = getValue(accelX);
```

The function getValue is passed one of the constants from the enumerated list of sensors, in this case the item accelX for acceleration in the x-axis.

You can send additional fields by separating them using commas (see Recipe 4.4); here is the revised loop function to achieve this:

```
void loop(){
  nunchuckRead();
  Serial.print("H,"); // header
  for(int i=0; i < 3; i++)
  {
     Serial.print(getValue(accelX+ i), DEC);
     if( i > 2)
        Serial.write(',');
     else
        Serial.write('\n') ;
  }
  delay(20); // the time in milliseconds between redraws
}
```

See Also

See Recipe 16.5 for a library for interfacing with the nunchuck, and the Discussion of Recipe 4.4 for a Processing sketch that displays a real-time bar chart showing each of the nunchuck values.

13.3 Interfacing to an External Real-Time Clock

Problem

You want to use the time of day provided by an external real-time clock (RTC).

Solution

This solution uses the Wire library to access an RTC. It uses the same hardware as in Recipe 12.6. Figure 12-3 shows the connections:

```
/*
 * I2C_RTC sketch
 * example code for using Wire library to access real-time clock
 */

#include <Wire.h>

const byte DS1307_CTRL_ID = 0x68; // address of the DS1307 real-time clock
const byte NumberOfFields = 7; // the number of fields (bytes) to
                               // request from the RTC
int Second ;
int Minute;
int Hour;
int Day;
int Wday;
int Month;
int Year;

void setup()  {
  Serial.begin(9600);
  Wire.begin();
```

```
}

void loop()
{
  Wire.beginTransmission(DS1307_CTRL_ID);
  Wire.write((byte)0x00);
  Wire.endTransmission();

  // request the 7 data fields (secs, min, hr, dow, date, mth, yr)
  Wire.requestFrom(DS1307_CTRL_ID, NumberOfFields);

  Second = bcd2dec(Wire.read() & 0x7f);
  Minute = bcd2dec(Wire.read() );
  Hour   = bcd2dec(Wire.read() & 0x3f);  // mask assumes 24hr clock
  Wday   = bcd2dec(Wire.read() );
  Day    = bcd2dec(Wire.read() );
  Month  = bcd2dec(Wire.read() );
  Year   = bcd2dec(Wire.read() );
  Year   = Year + 2000; // RTC year 0 is year 2000

  digitalClockDisplay(); // display the time
  delay(1000);
}

// Convert Binary Coded Decimal (BCD) to Decimal
byte bcd2dec(byte num)
{
  return ((num/16 * 10) + (num % 16));
}

void digitalClockDisplay(){
  // digital clock display of the time
  Serial.print(Hour);
  printDigits(Minute);
  printDigits(Second);
  Serial.print(" ");
  Serial.print(Day);
  Serial.print(" ");
  Serial.print(Month);
  Serial.print(" ");
  Serial.print(Year);
  Serial.println();
}

// utility function for clock display: prints preceding colon and leading 0
void printDigits(int digits){
  Serial.print(":");
  if(digits < 10)
    Serial.print('0');
  Serial.print(digits);
}
```

The `requestFrom` method of the Wire library is used to request seven time fields from the clock (`DS1307_CTRL_ID` is the address identifier of the clock):

```
Wire.requestFrom(DS1307_CTRL_ID, NumberOfFields);
```

The date and time values are obtained by making seven calls to the `Wire.receive` method:

The values returned by the module are binary coded decimal (BCD) values, so the function `bcd2dec` is used to convert each value as it is received. (BCD is a method for storing decimal values in four bits of data.)

See Also

Recipe 12.6 provides details on how to set the time on the clock.

13.4 Adding External EEPROM Memory

Problem

You need more permanent data storage than Arduino has onboard, and you want to use an external memory chip to increase the capacity.

Solution

This recipe uses the 24LC128 I2C-enabled serial EEPROM from Microchip Technology. Figure 13-9 shows the connections.

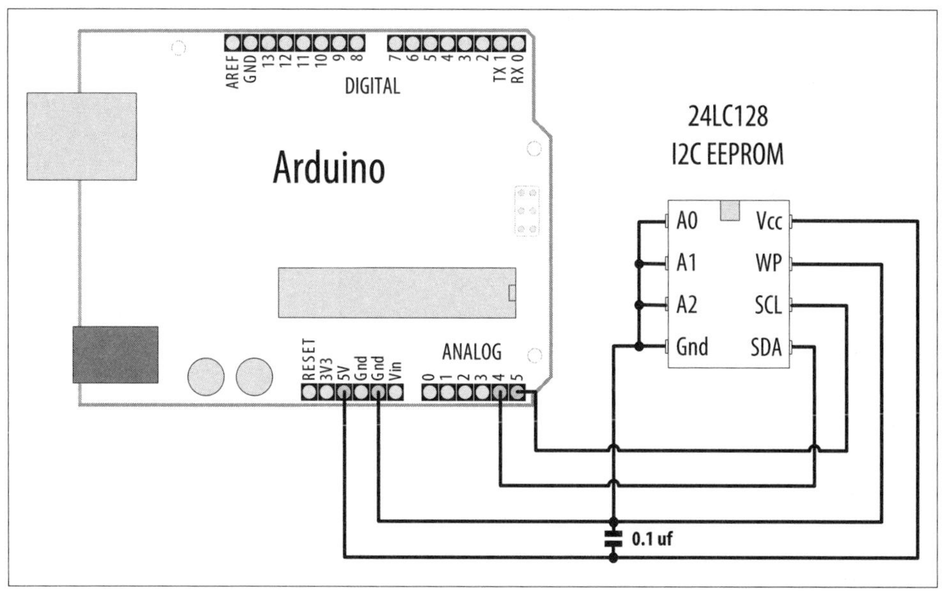

Figure 13-9. I2C EEPROM connections

This recipe provides functionality similar to the Arduino EEPROM library (see Recipe 18.1), but it uses an external EEPROM connected using I2C to provide greatly increased storage capacity:

```
/*
 * I2C EEPROM sketch
 * this version for 24LC128
 */
#include <Wire.h>

const byte EEPROM_ID = 0x50;      // I2C address for 24LC128 EEPROM

// first visible ASCII character '!' is number 33:
int thisByte = 33;

void setup()
{
  Serial.begin(9600);
  Wire.begin();

  Serial.println("Writing 1024 bytes to EEPROM");
  for (int i=0; i < 1024; i++)
  {
    I2CEEPROM_Write(i, thisByte);
    // go on to the next character
    thisByte++;
    if (thisByte == 126)   // you could also use if (thisByte == '~')
      thisByte = 33;     // start over
  }

  Serial.println("Reading 1024 bytes from EEPROM");
  int thisByte = 33;
  for (int i=0; i < 1024; i++)
  {
    char c = I2CEEPROM_Read(i);
    if( c != thisByte)
    {
      Serial.println("read error");
      break;
    }
    else
    {
      Serial.print(c);
    }
    thisByte++;
    if(thisByte == 126)
    {
      Serial.println();
      thisByte = 33;     // start over on a new line
    }
  }
  Serial.println();
}

void loop()
```

```
{

}

// This function is similar to EEPROM.write()
void I2CEEPROM_Write( unsigned int address, byte data )
{
  Wire.beginTransmission(EEPROM_ID);
  Wire.write((int)highByte(address) );
  Wire.write((int)lowByte(address) );
  Wire.write(data);
  Wire.endTransmission();
  delay(5); // wait for the I2C EEPROM to complete the write cycle
}

// This function is similar to EEPROM.read()
byte I2CEEPROM_Read(unsigned int address )
{
  byte data;
  Wire.beginTransmission(EEPROM_ID);
  Wire.write((int)highByte(address) );
  Wire.write((int)lowByte(address) );
  Wire.endTransmission();
  Wire.requestFrom(EEPROM_ID,(byte)1);
  while(Wire.available() == 0) // wait for data
    ;
  data = Wire.read();
  return data;
}
```

Discussion

This recipe shows the 24LC128, which has 128K of memory; although there are similar chips with higher and lower capacities (the Microchip link in this recipe's See Also section has a cross-reference). The chip's address is set using the three pins marked A0 through A2 and is in the range 0x50 to 0x57, as shown in Table 13-2.

Table 13-2. Address values for 24LC128

A0	A1	A2	Address
Gnd	Gnd	Gnd	0x50
+5V	Gnd	Gnd	0x51
Gnd	+5V	Gnd	0x52
+5V	+5V	Gnd	0x53
Gnd	Gnd	+5V	0x54
+5V	Gnd	+5V	0x55
+5V	+5V	Gnd	0x56
+5V	+5V	+5V	0x57

Use of the Wire library in this recipe is similar to its use in Recipes 13.1 and 13.2, so read through those for an explanation of the code that initializes and requests data from an I2C device.

The write and read operations that are specific to the EEPROM are contained in the functions `i2cEEPROM_Write` and `i2cEEPROM_Read`. These operations start with a `Wire.beginTransmission` to the device's I2C address. This is followed by a 2-byte value indicating the memory location for the read or write operation. In the write function, the address is followed by the data to be written—in this example, one byte is written to the memory location.

The read operation sends a memory location to the EEPROM, which is followed by `Wire.requestFrom(EEPROM_ID,(byte)1);`. This returns one byte of data from the memory at the address just set.

If you need to speed up writes, you can replace the 5 ms delay with a status check to determine if the EEPROM is ready to write a new byte. See the "Acknowledge Polling" technique described in Section 7 of the data sheet. You can also write data in pages of 64 bytes rather than individually; details are in Section 6 of the data sheet.

The chip remembers the address it is given and will move to the next sequential address each time a read or write is performed. If you are reading more than a single byte, you can set the start address and then perform multiple requests and receives.

The Wire library can read or write up to 32 bytes in a single request. Attempting to read or write more than this can result in bytes being discarded.

The pin marked WP is for setting write protection. It is connected to ground in the circuit here to enable the Arduino to write to memory. Connecting it to 5V prevents any writes from taking place. This could be used to write persistent data to memory and then prevent it from being overwritten accidentally.

See Also

The 24LC128 data sheet: *http://ww1.microchip.com/downloads/en/devicedoc/21191n.pdf*

If you need to speed up writes, you can replace the 5 ms delay with a status check to determine if the EEPROM is ready to write a new byte. See the "Acknowledge Polling" technique described in Section 7 of the data sheet.

A cross-reference of similar I2C EEPROMs with a wide range of capacities is available at *http://ww1.microchip.com/downloads/en/DeviceDoc/21621d.pdf*.

A shield is available that combines reading temperature, storing in EEPROM, and 7-segment display: *http://store.gravitech.us/7segmentshield.html*.

13.5 Reading Temperature with a Digital Thermometer

Problem

You want to measure temperature, perhaps using more than one device, so you can take readings in different locations.

Solution

This recipe uses the TMP75 temperature sensor from Texas Instruments. You connect a single TMP75 as shown in Figure 13-10:

```
/*
 * I2C_Temperature sketch
 * I2C access the TMP75 digital Thermometer
 */

#include <Wire.h>

const byte TMP75_ID = 0x49; // address of the TMP75
const byte NumberOfFields = 2; // the number of fields (bytes) to request

// high byte of temperature (this is the signed integer value in degrees c)
char tempHighByte;
// low byte of temperature  (this is the fractional temperature)
char tempLowByte;

float temperature;  // this will hold the floating-point temperature


void setup()  {
  Serial.begin(9600);
  Wire.begin();

  Wire.beginTransmission(TMP75_ID);
  Wire.write(1);           // 1 is the configuration register
  // set default configuration, see data sheet for significance of config bits
  Wire.write((byte)0);
  Wire.endTransmission();

  Wire.beginTransmission(TMP75_ID);
  Wire.write((byte)0);   // set pointer register to 0 (the 12-bit temperature)
  Wire.endTransmission();

}

void loop()
{
    Wire.requestFrom(TMP75_ID, NumberOfFields);
    tempHighByte = Wire.read();
    tempLowByte = Wire.read();
    Serial.print("Integer temperature is ");
```

```
    Serial.print(tempHighByte, DEC);
    Serial.print(",");

    // least significant 4 bits of LowByte is the fractional temperature
    int t = word( tempHighByte, tempLowByte) / 16 ;
    temperature = t / 16.0; // convert the value to a float
    Serial.println(temperature);
    delay(1000);
}
```

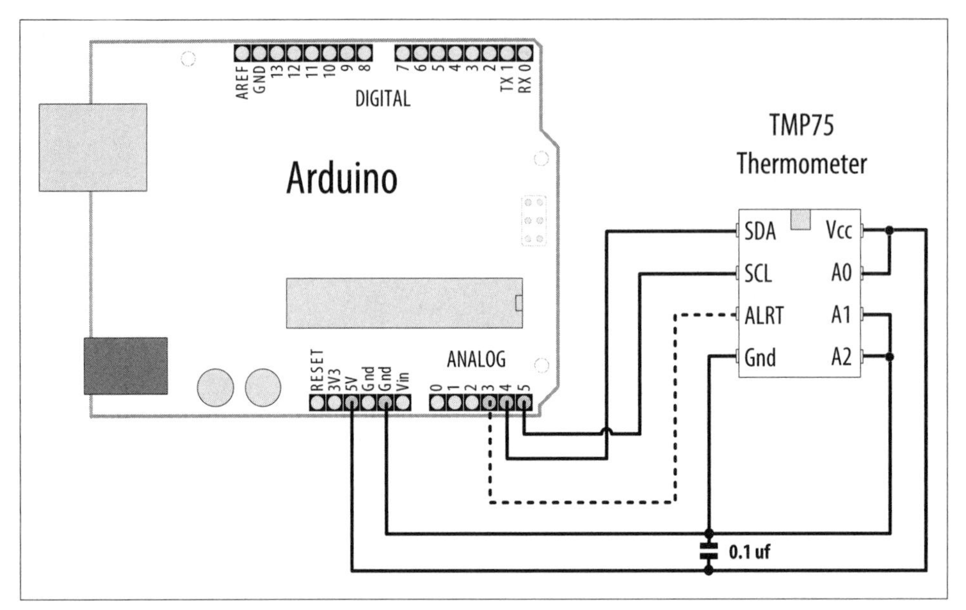

Figure 13-10. TMP75 I2C thermometer

Discussion

As with all the I2C devices in this chapter, signaling is through the two-wire SCL and SDA pins. Power and ground need to be connected to the device, as well, to power it.

`Setup` sends data to configure the device for normal operation—there are a number of options for specialized applications (interrupts, power down, etc.), but the value used here is for normal mode with a precision of .5°C.

To get the device to send the temperature, with the Arduino (as the master), the code in `loop` tells the slave (at the address given by the constant `TMP75_ID`) that it wants two bytes of data:

```
Wire.requestFrom(TMP75_ID, NumberOfFields);
```

`Wire.read` gets the two bytes of information (the data sheet has more detail on how data is requested from this device):

```
tempHighByte = Wire.read();
tempLowByte = Wire.read();
```

The first byte is the integer value of the temperature in degrees Celsius. The second byte contains four significant bits indicating the fractional temperature.

The two bytes are converted to a 16-bit word (see Chapter 3) and then shifted by four to form the 12-bit number. As the first four bits are the fractional temperature, the value is again shifted by four to get the floating-point value.

The TMP75 can be configured for eight different addresses, allowing eight devices to be connected to the same bus (see Figure 13-11). This sketch uses I2C address 0x48 (the TMP75 address pins labeled A connected to +5V, and A1 and A2 connected to Gnd). Table 13-3 shows the connections for the eight addresses.

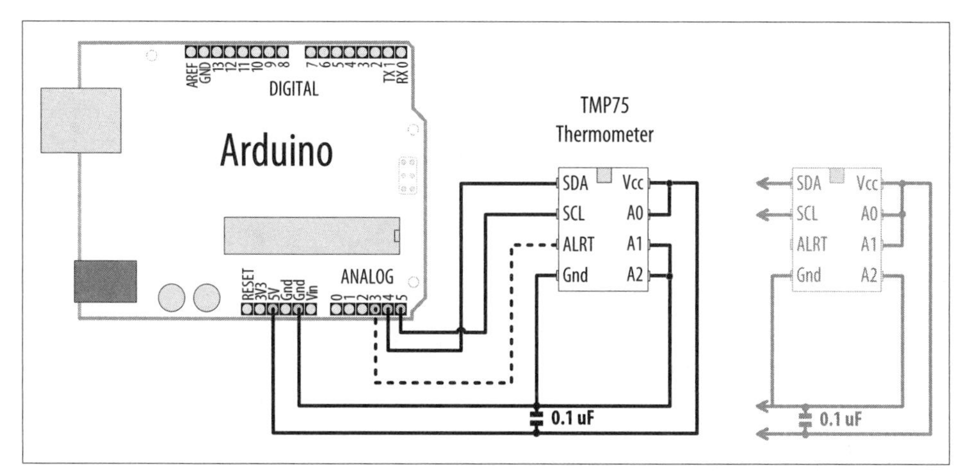

Figure 13-11. Multiple devices with SDA and SCL connected in parallel with different addresses

Table 13-3. Address values for TMP75

A0	A1	A2	Address
Gnd	Gnd	Gnd	0x48
+5V	Gnd	Gnd	0x49
Gnd	+5V	Gnd	0x4A
+5V	+5V	Gnd	0x4B
Gnd	Gnd	+5V	0x4C
+5V	Gnd	+5V	0x4D
+5V	+5V	Gnd	0x4E
+5V	+5V	+5V	0x4F

When connecting more than one I2C device, you wire all the SDA lines together and all the SCL lines together. Each device connects to power and should have 0.1uF bypass capacitors. The Gnd lines must be connected together, even if the devices use separate power supplies (e.g., batteries).

This sketch prints the temperature of two devices with consecutive addresses starting from 0x49:

```
#include <Wire.h>

const byte TMP75_ID = 0x49; // address of the first TMP75

const byte NumberOfFields  = 2; // the number of fields (bytes) to request
const byte NumberOfDevices = 2; // nbr TMP75 devices with consecutive addresses


char tempHighByte;  // high byte of temperature (this is
                    // the signed integer value in degrees c)
char tempLowByte;   // low byte of temperature  (this is
                    // the fractional temperature)

float temperature;  // this will hold the floating-point temperature

void setup()  {
  Serial.begin(9600);
  Wire.begin();

  for (int i=0; i < NumberOfDevices; i++)
  {
    Wire.beginTransmission(TMP75_ID+i);
    Wire.write(1);
    // set default configuration, see data sheet for significance of config bits
    Wire.write((byte)0);
    Wire.endTransmission();

    Wire.beginTransmission(TMP75_ID+i);
    Wire.write((byte)0);   // set pointer register to 0 (the 12-bit temperature)
    Wire.endTransmission();
  }
}

void loop()
{
  for (int i=0; i < NumberOfDevices; i++)
  {
    byte id = TMP75_ID + i;   // address IDs are consecutive
    Wire.requestFrom(id, NumberOfFields);
    tempHighByte = Wire.read();
    tempLowByte = Wire.read();
    Serial.print(id,HEX); // print the device address
    Serial.print(": integer temperature is ");
    Serial.print(tempHighByte, DEC);
    Serial.print(",");

    // least significant 4 bits of LowByte is the fractional temperature
```

```
    int t = word( tempHighByte, tempLowByte) / 16 ;
    temperature = t / 16.0; // convert the value to a float
    Serial.println(temperature);
  }
  delay(1000);
}
```

You can add more devices by changing the constant `NumberOfDevices` and wiring the devices to use addresses that are consecutive, in this example starting from 0x49.

The Alert line (pin 3) can be programmed to provide a signal when the temperature reaches a threshold. See the data sheet for details if you want to use this feature.

See Also

The TMP75 data sheet: *http://focus.ti.com/docs/prod/folders/print/tmp75.html*

See Recipe 3.15 for more on the `word` function.

13.6 Driving Four 7-Segment LEDs Using Only Two Wires

Problem

You want to use a multidigit, 7-segment display, and you need to minimize the number of Arduino pins required for the connections.

Solution

This recipe uses the Gravitech 7-segment display shield that has the SAA1064 I2C to 7-segment driver from Philips (see Figure 13-12).

This simple sketch lights each segment in sequence on all the digits:

```
/*
 * I2C_7Segment sketch
 */

#include <Wire.h>

const byte LedDrive = 0x38;   // I2C address for 7-Segment

int segment,decade;

void setup()  {
  Serial.begin(9600);
  Wire.begin();            // Join I2C bus

  Wire.beginTransmission(LedDrive);
  Wire.write((byte)0);
```

```
  Wire.write(B01000111); // show digits 1 through 4, use maximum drive current
  Wire.endTransmission();
}

void loop()
{
  for (segment = 0; segment < 8; segment++)
  {
    Wire.beginTransmission(LedDrive);
    Wire.write(1);
    for (decade = 0 ; decade < 4; decade++)
    {
      byte bitValue = bit(segment);
      Wire.write(bitValue);
    }
    Wire.endTransmission();
    delay (250);
  }
}
```

Figure 13-12. Gravitech I2C shield

Discussion

The SAA1064 chip (using address 0x38) is initialized in `setup`. The value used configures the chip to display four digits using maximum drive current (see the data sheet section on control bits for configuration details).

The `loop` code lights each segment in sequence on all the digits. The `Wire.send(1);` command tells the chip that the next received byte will drive the first digit and subsequent bytes will drive sequential digits.

Initially, a value of `1` is sent four times and the chip lights the A (top) segment on all four digits. (See Chapter 2 for more on using the `bit` function.)

The value of segment is incremented in the for loop, and this shifts the bitValue to light the next LED segment in sequence.

Each bit position corresponds to a segment of the digit. These bit position values can be combined to create a number that will turn on more than one segment.

The following sketch will display a count from 0 to 9999. The array called lookup[10] contains the values needed to create the numerals from 0 to 9 in a segment:

```
#include <Wire.h>

const byte LedDrive = 0x38;    // I2C address for 7-Segment

// lookup array containing segments to light for each digit
const int lookup[10] = {0x3F,0x06,0x5B,0x4F,0x66,0x6D,0x7D,0x07,0x7F,0x6F};

int count;

void setup()
{
  Wire.begin();          // join I2C bus (address optional for master)
}

void loop()
{
  Wire.beginTransmission(LedDrive);
  Wire.write((byte)0);
  Wire.write(B01000111); // init the 7-segment driver - see data sheet
  Wire.endTransmission();

  // show numbers from 0 to 9999
  for (count = 0; count <= 9999; count++)
  {
    displayNumber(count);
    delay(10);
  }
}

// function to display up to four digits on a 7-segment I2C display
void displayNumber( int number)
{
  number = constrain(number, 0, 9999);
  Wire.beginTransmission(LedDrive);
  Wire.write(1);
  for(int i =0; i < 4; i++)
  {
    byte digit = number % 10;
    {
       Wire.write(lookup[digit]);
    }
    number = number / 10;
  }
  Wire.endTransmission();
}
```

The function `displayNumber` is given a number to be displayed. The value to be sent for each segment in the `for` loop is handled in two steps. First, the digit is determined by taking the remainder after dividing the number by 10. This value (a digit from 0 through 9) is used to get the bit pattern from the `lookup[]` array to light the segments needed to display the digit.

Each successive digit is obtained by looking at the remainder after dividing the number by 10. When the remainder becomes 0, all digits have been sent.

You can suppress *leading zeros* (unnecessary zeros in front of digits) by changing the `displayNumber` function as follows:

```
// function to display up to four digits on a 7-segment I2C display
void displayNumber( int number)
{
  number = constrain(number, 0, 9999);
  Wire.beginTransmission(LedDrive);
  Wire.write(1);
  for(int i =0; i < 4; i++)
  {
    byte digit = number % 10;
    // this check will suppress leading zeros
    if ((number == 0) && (i > 0)) {
       Wire.write((byte)0);  // turn off all segments to suppress leading zeros
    }
    else {
       Wire.write(lookup[digit]);
    }
    number = number / 10;
  }
  Wire.endTransmission();
}
```

The following statement checks if the value is `0` and it's not the first (least significant) digit:

```
if ((number == 0) && (i > 0))
   Wire.write((byte)0);  // turn off all segments to suppress leading zeros
```

If so, it sends a value of `0`, which turns off all segments for that digit. This suppresses leading zeros, but it displays a single zero if the number passed to the function was `0`.

The expression `(byte)0` is needed in the `Wire.write` statement to clarify to the compiler that the constant 0 is a byte. Without this you will get an error message saying that the "call of overloaded 'write(int)' is ambiguous."

See Also

SAA1064 data sheet: *http://www.nxp.com/documents/data_sheet/SAA1064_CNV.pdf*

A shield is available that combines reading temperature, storing in EEPROM, and 7-segment display: *http://store.gravitech.us/7segmentshield.html*.

13.7 Integrating an I2C Port Expander

Problem

You want to use more input/output ports than your board provides.

Solution

You can use an external port expander, such as the PCF8574A, which has eight input/output pins that can be controlled using I2C. The sketch creates a bar graph with eight LEDs. Figure 13-13 shows the connections.

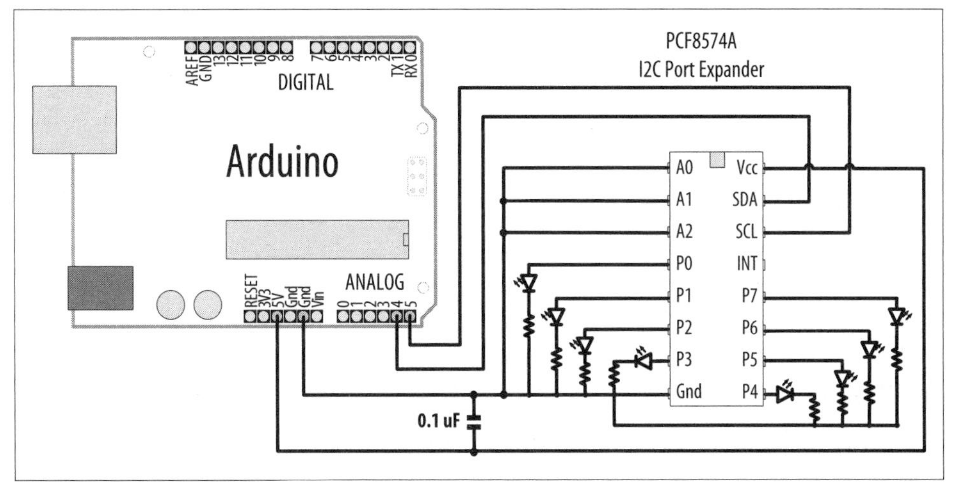

Figure 13-13. PCF8574A port expander driving eight LEDs

The sketch has the same functionality as described in Recipe 7.5, but it uses the I2C port expander to drive the LEDs so that only two pins are required:

```
/*
 * I2C_7segment
 * Uses I2C port to drive a bar graph
 * Turns on a series of LEDs proportional to a value of an analog sensor.
 * see Recipe 7.5
 */

#include <Wire.h>

//address for PCF8574 with pins connected as shown in Figure 13-12
const int address = 0x38;
const int NbrLEDs = 8;
```

```
const int analogInPin = 0; // Analog input pin connected
                           // to the variable resistor

int sensorValue = 0;       // value read from the sensor
int ledLevel = 0;          // sensor value converted into LED 'bars'
int ledBits = 0;           //  bits for each LED will be set to 1 to turn on LED

void setup()
{
  Wire.begin(); // set up Arduino I2C support
  Serial.begin(9600);
}

void loop() {
  sensorValue = analogRead(analogInPin);             // read the analog in value
  ledLevel = map(sensorValue, 0, 1023, 0, NbrLEDs);  // map to number of LEDs
  for (int led = 0; led < NbrLEDs; led++)
  {
    if (led < ledLevel ) {
      bitWrite(ledBits,led, HIGH);    // turn on LED if less than the level
    }
    else {
      bitWrite(ledBits,led, LOW);    // turn off LED if higher than the level
    }
    // send the value to I2C
    Wire.beginTransmission(address);
    Wire.write(ledBits);
    Wire.endTransmission();
  }
  delay(100);
}
```

Discussion

The resistors should be 220 ohms or more (see Chapter 7 for information on selecting resistors).

The PCF8574A has a lower capacity for driving LEDs than Arduino. If you need more capacity (refer to the data sheet for details) see Recipe 13.8 for a more appropriate device.

You can change the address by changing the connections of the pins marked A0, A1, and A2, as shown in Table 13-4.

Table 13-4. Address values for PCF8574A

A0	A1	A2	Address
Gnd	Gnd	Gnd	0x38
+5V	Gnd	Gnd	0x39
Gnd	+5V	Gnd	0x3A
+5V	+5V	Gnd	0x3B
Gnd	Gnd	+5V	0x3C
+5V	Gnd	+5V	0x3D
+5V	+5V	Gnd	0x3E
+5V	+5V	+5V	0x3F

To use the port expander for input, read a byte from the expander as follows:

```
Wire.requestFrom(address, 1);
if(Wire.available())
{
  data = Wire.receive();
  Serial.println(data,BIN);
}
```

See Also

PCF8574 data sheet: *http://www.nxp.com/documents/data_sheet/PCF8574.pdf*

13.8 Driving Multidigit, 7-Segment Displays Using SPI

Problem

You want to control 7-segment displays, but you don't want to use many pins.

Solution

This recipe provides similar functionality to Recipe 7.12, but it only requires three output pins. The text here explains the SPI commands used to communicate with the MAX7221 device (Figure 13-14 shows the connections):

```
/*
 * SPI_Max7221_0019
 */

#include <SPI.h>

const int slaveSelect = 10; // pin used to enable the active slave

const int numberOfDigits = 2; // change to match the number of digits wired up
const int maxCount = 99;
```

```
int count = 0;

void setup()
{
  SPI.begin();   // initialize SPI
  pinMode(slaveSelect, OUTPUT);
  digitalWrite(slaveSelect,LOW);  // select slave
  // prepare the 7221 to display 7-segment data - see data sheet
  sendCommand(12,1);  // normal mode (default is shutdown mode);
  sendCommand(15,0);  // Display test off
  sendCommand(10,8);  // set medium intensity (range is 0-15)
  sendCommand(11,numberOfDigits);  // 7221 digit scan limit command
  sendCommand(9,255); // decode command, use standard 7-segment digits
  digitalWrite(slaveSelect,HIGH);  // deselect slave
}

void loop()
{
  displayNumber(count);
  count = count + 1;
  if (count > maxCount)
    count = 0;
  delay(100);
}

// function to display up to four digits on a 7-segment display
void displayNumber( int number)
{
  for (int i = 0; i < numberOfDigits; i++)
  {
    byte character = number % 10;  // get the value of the rightmost decade
    // send digit number as command, first digit is command 1
    sendCommand(numberOfDigits-i, character);
    number = number / 10;
  }
}

void sendCommand( int command, int value)
{
  digitalWrite(slaveSelect,LOW); // chip select is active low
  // 2 byte data transfer to the 7221
  SPI.transfer(command);
  SPI.transfer(value);
  digitalWrite(slaveSelect,HIGH); // release chip, signal end transfer
}
```

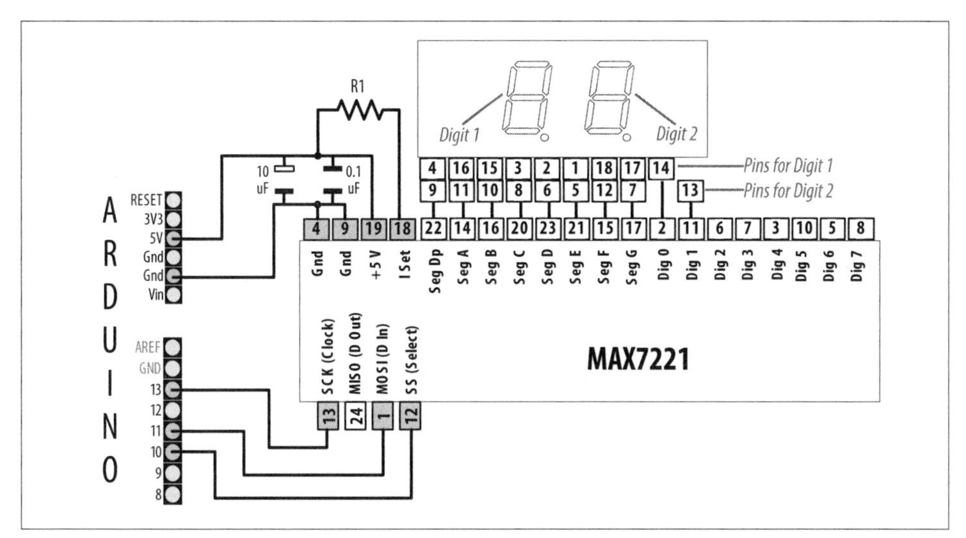

Figure 13-14. Connections for MAX7221 with Lite-On LTD-6440G

Discussion

The MAX7221 needs a common cathode LED. The pinouts in Figure 13-14 are for a Lite-On LTD-6440G (Jameco part #2005366). This is a two-digit, 7-segment LED and the corresponding segments for each digit must be connected together. For example, the decimal point is on pin 4 for digit 1 and pin 9 for digit 2. The figure indicates that pins 4 and 9 are connected together and wired to the MAX7221 pin 22.

The MAX7221 can display up to eight digits (or an 8 × 8 matrix) and is controlled by sending commands that determine which LED segment is lit.

After initializing the library, the SPI code is contained within the `sendCommand` function. Because SPI uses the select slave wire connected to the chip, the chip is selected by setting that pin `LOW`. All SPI commands are then received by that chip until it is set `HIGH`. `SPI.transfer` is the library function for sending an SPI message. This consists of two parts: a numerical code to specify which register should receive the message, followed by the actual data. The details for each SPI device can be found in the data sheet.

`Setup` initializes the 7221 by sending commands to wake up the chip (it starts up in a low-power mode), adjust the display intensity, set the number of digits, and enable decoding for 7-segment displays. Each command consists of a command identifier (referred to as a *register* in the data sheet) and a value for that command.

For example, command (register) 10 is for intensity, so it sets medium intensity (the intensity range is from 0 to 15):

```
sendCommand(10,8);  // set medium intensity (range is 0-15)
```

Command numbers 1 through 8 are used to control the digits. The following code would light the segments to display the number 5 in the first (leftmost) digit. Note that digit numbers shown in the data sheet (and Figure 13-14) start from 0, so you must remember that you control digit 0 with command 1, digit 1 with command 2, and so on:

```
sendCommand(1, 5); // display 5 on the first digit
```

You can suppress leading zeros by adding two lines of code in `displayNumber` that send 0xf to the 7221 to blank the segments if the residual value is 0:

```
void displayNumber( int number)
{
  for (int i = 0; i < numberOfDigits; i++)
  {
    byte character = number % 10;
```

The next two lines are added to suppress leading zeros:

```
    if ((number == 0) && (i > 0))
       character = 0xf;  // value to blank the 7221 segments
    sendCommand(numberOfDigits-i, character);
    number = number / 10;
  }
}
```

13.9 Communicating Between Two or More Arduino Boards

Problem

You want to have two or more Arduino boards working together. You may want to increase the I/O capability or perform more processing than can be achieved on a single board. You can use I2C to pass data between boards so that they can share the workload.

Solution

The two sketches in this recipe show how I2C can be used as a communications link between two or more Arduino boards. Figure 13-15 shows the connections.

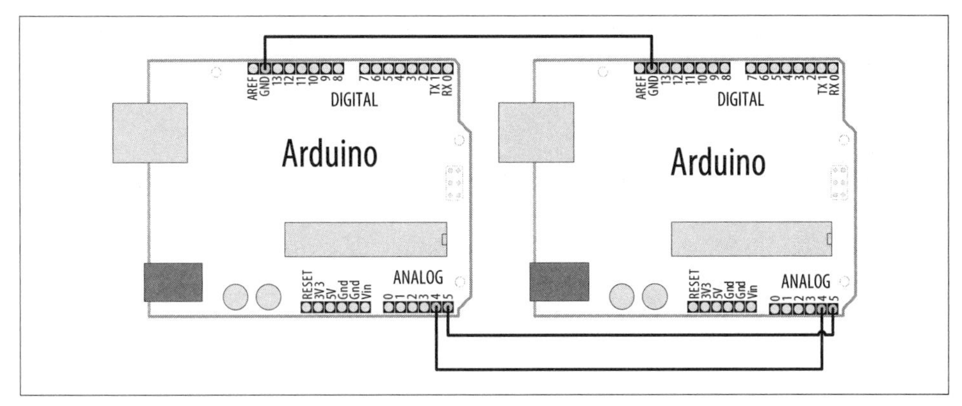

Figure 13-15. Arduino as I2C master and slave

The master sends characters received on the serial port to an Arduino slave using I2C:

```
/*
 * I2C_Master
 * Echo Serial data to an I2C slave
 */

#include <Wire.h>

const int address = 4;  // the address to be used by the communicating devices

void setup()
{
   Wire.begin();
}


void loop()
{
  char c;
  if(Serial.available() > 0 )
  {
     // send the data
     Wire.beginTransmission(address); // transmit to device
     Wire.write(c);
     Wire.endTransmission();
  }
}
```

The slave prints characters received over I2C to its serial port:

```
/*
 * I2C_Slave
 * monitors I2C requests and echoes these to the serial port
 */

#include <Wire.h>
```

```
const int address = 4;  // the address to be used by the communicating devices

void setup()
{
  Serial.begin(9600);
  Wire.begin(address); // join I2C bus using this address
  Wire.onReceive(receiveEvent); // register event to handle requests
}

void loop()
{
  // nothing here, all the work is done in receiveEvent
}

void receiveEvent(int howMany)
{
  while(Wire.available() > 0)
  {
    char c = Wire.read(); // receive byte as a character
    Serial.write(c); // echo
  }
}
```

Discussion

This chapter focused on Arduino as the I2C master accessing various I2C slaves. Here a second Arduino acts as an I2C slave that responds to requests from another Arduino. Techniques covered in Chapter 4 for sending bytes of data can be applied here. Arduino 1.0 added a print capability to the wire library method so you can now send data using the print method.

The following sketch sends its output over I2C using `Wire.println`. Using this with the I2C slave sketch shown previously enables you to print data on the master without using the serial port (the slave's serial port is used to display the output):

```
/*
 * I2C_Master
 * Sends sensor data to an I2C slave using print
 */

#include <Wire.h>

const int address = 4;     // the address to be used by the communicating devices
const int sensorPin = 0;   // select the analog input pin for the sensor
int val;                   // variable to store the sensor value

void setup()
{
   Wire.begin();
}

void loop()
{
   val = analogRead(sensorPin);        // read the voltage on the pot
```

```
                                    // (val ranges from 0 to 1023)
    Wire.beginTransmission(address); // transmit to device
    Wire.println(val);
    Wire.endTransmission();
    delay(1000);
}
```

See Also

Chapter 4 has more information on using the Arduino print functionality.

CHAPTER 14

Wireless Communication

14.0 Introduction

Arduino's ability to interact with the world is wonderful, but sometimes you might want to communicate with your Arduino from a distance, without wires, and without the overhead of a full TCP/IP network connection. This chapter covers various simple wireless modules for applications where low cost is the primary requirement but most of the recipes focus on the versatile XBee wireless modules.

XBee provides flexible wireless capability to the Arduino, but that very flexibility can be confusing. This chapter provides examples ranging from simple "wireless serial port replacements" through to mesh networks connecting multiple boards to multiple sensors.

A number of different XBee modules are available. The most popular are the XBee 802.15.4 (also known as XBee Series 1) and XBee ZB Series 2. Series 1 is easier to use than Series 2, but it does not support mesh networks. See *http://www.digi.com/support/kbase/kbaseresultdetl.jsp?id=2213*.

14.1 Sending Messages Using Low-Cost Wireless Modules

Problem

You want to transmit data between two Arduino boards using simple, low-cost wireless modules.

Solution

This recipe uses simple transmit and receive modules such as the SparkFun 315 MHz: WRL-10535 and WRL-10533, or the 434 MHz: WRL-10534 and WRL-10532.

Wire the transmitter as shown in Figure 14-1 and the receiver as in Figure 14-2. Some modules have the power line labeled VDD instead of Vcc.

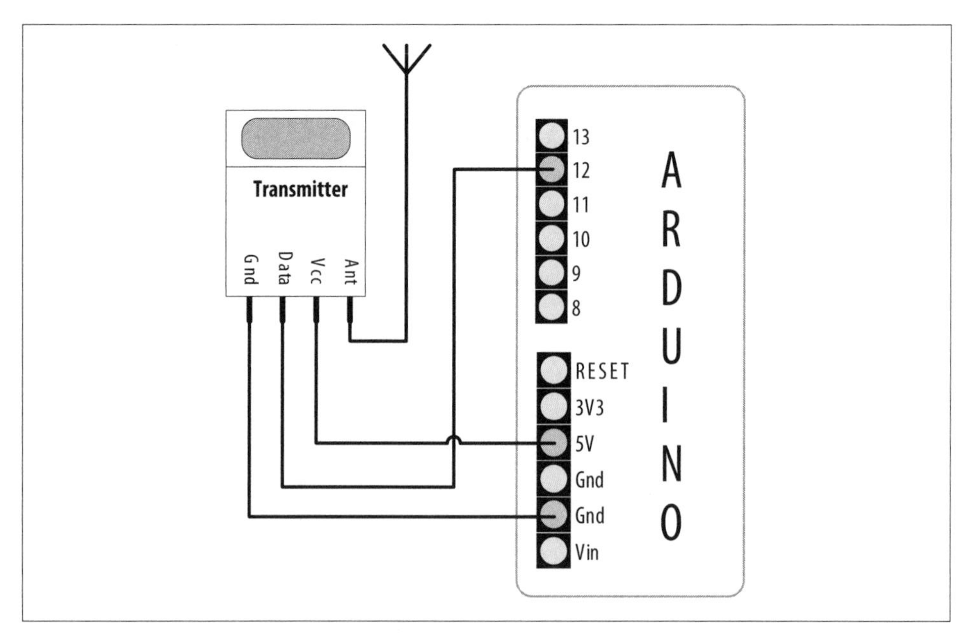

Figure 14-1. Simple wireless transmitter using VirtualWire

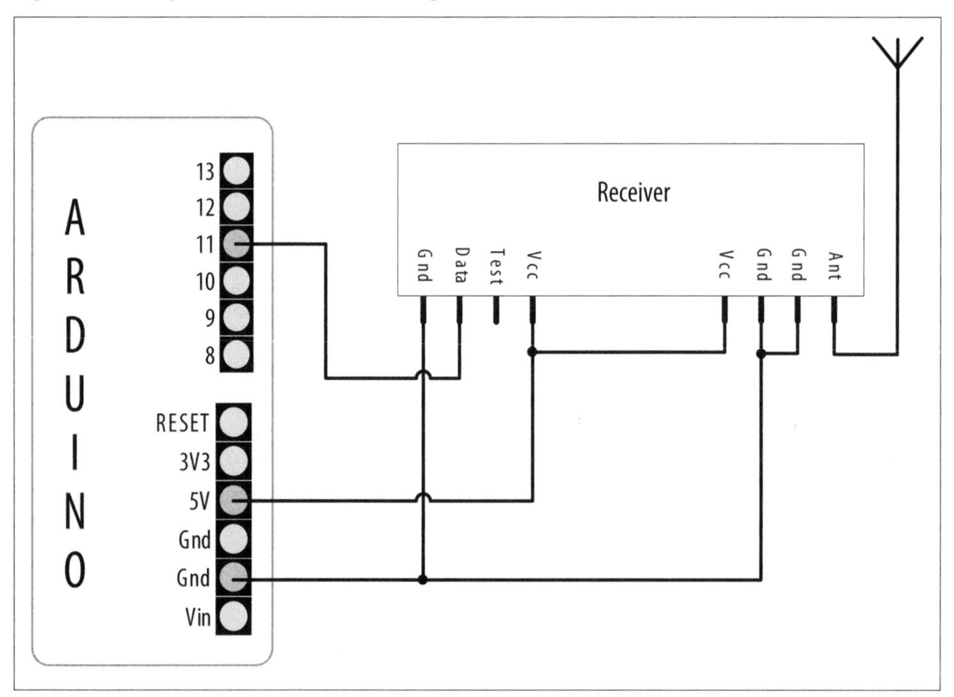

Figure 14-2. Simple wireless receiver using VirtualWire

The transmit sketch sends a simple text message to the receive sketch, which echoes the text to the Serial Monitor. The transmit and receive sketches use the VirtualWire library written by Mike McCauley to provide the interface to the wireless hardware. The library can be downloaded from *http://www.open.com.au/mikem/arduino/VirtualWire-1.5.zip*:

```
/*
  SimpleSend
  This sketch transmits a short text message using the VirtualWire library
  connect the Transmitter data pin to Arduino pin 12
*/

#include <VirtualWire.h>

void setup()
{
    // Initialize the IO and ISR
    vw_setup(2000);           // Bits per sec
}

void loop()
{
    send("hello");
    delay(1000);
}

void send (char *message)
{
  vw_send((uint8_t *)message, strlen(message));
  vw_wait_tx(); // Wait until the whole message is gone
}
```

The receive sketch also uses the VirtualWire library:

```
/*
  SimpleReceive
  This sketch  displays text strings received using VirtualWire
  Connect the Receiver data pin to Arduino pin 11
*/
#include <VirtualWire.h>

byte message[VW_MAX_MESSAGE_LEN];    // a buffer to hold the incoming messages
byte msgLength = VW_MAX_MESSAGE_LEN; // the size of the message


void setup()
{
    Serial.begin(9600);
    Serial.println("Ready");

    // Initialize the IO and ISR
    vw_setup(2000);             // Bits per sec
    vw_rx_start();              // Start the receiver
}
```

```
void loop()
{
    if (vw_get_message(message, &msgLength)) // Non-blocking
    {
        Serial.print("Got: ");
    for (int i = 0; i < msgLength; i++)
    {
        Serial.write(message[i]);
    }
    Serial.println();
    }
}
```

Discussion

The VirtualWire library defaults to pin 12 for transmit and pin 11 for receive, but see the documentation link at the end of this recipe if you want to use different pins. `Setup` initializes the library. The `loop` code simply calls a `send` function that calls the library `vw_send` and waits for the message to be transmitted.

The receive side initializes the library receive logic and then waits in `loop` for the message. `vw_get_message` will return `true` if a message is available, and if so, each character in the message is printed to the Serial Monitor.

The VirtualWire library handles the assembly of multiple bytes into packets, so sending binary data consists of passing the address of the data and the number of bytes to send.

The sending sketch that follows is similar to the transmit sketch in this recipe's Solution, but it fills the message buffer with binary values from reading the analog input ports using `analogRead`. The size of the buffer is the number of integers to be sent multiplied by the number of bytes in an integer (the six analog integer values take 12 bytes because each `int` is two bytes):

```
/*
  SendBinary
  Sends digital and analog pin values as binary data using VirtualWire library
  See SendBinary in Chapter 4
*/

#include <VirtualWire.h>

const int numberOfAnalogPins = 6; // how many analog pins to read

int data[numberOfAnalogPins];  // the data buffer

const int dataBytes = numberOfAnalogPins * sizeof(int); // the number of bytes
                                                         // in the data buffer

void setup()
{
    // Initialize the IO and ISR
    vw_setup(2000);           // Bits per sec
}
```

```
void loop()
{
  int values = 0;
  for(int i=0; i <= numberOfAnalogPins; i++)
  {
    // read the analog ports
    data[i] = analogRead(i); // store the values into the data buffer
  }
  send((byte*)data, dataBytes);
  delay(1000); //send every second
}


void send (byte *data, int nbrOfBytes)
{
  vw_send(data, nbrOfBytes);
  vw_wait_tx(); // Wait until the whole message is gone
}
```

The `sizeof` operator is used to determine the number of bytes in an `int`.

The receive side waits for messages, checks that they are the expected length, and converts the buffer back into the six integer values for display on the Serial Monitor:

```
/*
 SendBinary
 Sends digital and analog pin values as binary data using VirtualWire library
 See SendBinary in Chapter 4
 */

#include <VirtualWire.h>

const int numberOfAnalogPins = 6; // how many analog integer values to receive
int data[numberOfAnalogPins];  // the data buffer

// the number of bytes in the data buffer
const int dataBytes = numberOfAnalogPins * sizeof(int);

byte msgLength = dataBytes;


void setup()
{
  Serial.begin(9600);
  Serial.println("Ready");

  // Initialize the IO and ISR
  vw_set_ptt_inverted(true); // Required for DR3100
  vw_setup(2000);            // Bits per sec
```

```
  vw_rx_start();              // Start the receiver
}

void loop()
{
  if (vw_get_message((byte*)data, &msgLength)) // Non-blocking
  {
    Serial.println("Got: ");
    if(msgLength == dataBytes)
    {
      for (int i = 0; i <  numberOfAnalogPins; i++)
      {
        Serial.print("pin ");
        Serial.print(i);
        Serial.print("=");
        Serial.println(data[i]);
      }
    }
    else
    {
       Serial.print("unexpected msg length of ");
       Serial.println(msgLength);
    }
    Serial.println();
  }
}
```

The Serial Monitor will display the analog values on the sending Arduino:

```
Got:
pin 0=1023
pin 1=100
pin 2=227
pin 3=303
pin 4=331
pin 5=358
```

Bear in mind that the maximum buffer size for VirtualWire is 30 bytes long (the constant `VW_MAX_MESSAGE_LEN` is defined in the library header file).

Wireless range can be up to 100 meters or so depending on supply voltage and antenna and is reduced if there are obstacles between the transmitter and the receiver.

Also note that the messages are not guaranteed to be delivered, and if you get out of range or there is excessive radio interference some messages could get lost. If you need a guaranteed wireless delivery mechanism, the ZigBee API used in recipes that follow is a better choice, but these inexpensive modules work well for tasks such as displaying the status of Arduino sensors—each message contains the current sensor value to display and any lost messages get replaced by messages that follow.

See Also

A technical document on the VirtualWire Library can be downloaded from *http://www.open.com.au/mikem/arduino/VirtualWire.pdf*.

Data sheets for the transmitter and receiver modules can be found at *http://www.spark fun.com/datasheets/Wireless/General/MO-SAWR.pdf* and *http://www.sparkfun.com/da tasheets/Wireless/General/MO-RX3400.pdf*.

14.2 Connecting Arduino to a ZigBee or 802.15.4 Network

Problem

You'd like your Arduino to participate in a ZigBee or 802.15.4 network.

802.15.4 is an IEEE standard for low-power digital radios that are implemented in products such as the inexpensive XBee modules from Digi International. *ZigBee* is an alliance of companies and also the name of a standard maintained by that alliance. ZigBee is based on IEEE 802.15.4 and is a superset of it. ZigBee is implemented in many products, including certain XBee modules from Digi.

Only XBee modules that are listed as ZigBee-compatible, such as the XBee ZB modules, are guaranteed to be ZigBee-compliant. That being said, you can use a subset of the features (IEEE 802.15.4) of ZigBee even with the older XBee Series 1 modules. In fact, all the recipes here will work with the Series 1 modules.

Troubleshooting XBee

If you have trouble getting your XBees to talk, make sure they both have the same type of firmware (e.g., XB24-ZB under the Modem: XBEE setting shown in Figure 14-5), and that they are both running the most current version of the firmware (the Version setting shown in Figure 14-5). For a comprehensive set of XBee troubleshooting tips, see Robert Faludi's "Common XBee Mistakes" at *http://www.faludi.com/projects/com mon-xbee-mistakes/*. For extensive details on working with XBees, see his book, *Building Wireless Sensor Networks*, published by O'Reilly (search for it on oreilly.com).

Solution

Obtain two or more XBee modules, configure them (as described in "Discussion" on page 467) to communicate with one another, and hook them up to at least one Arduino. You can connect the other XBee modules to another Arduino, a computer, or an analog sensor (see Recipe 14.4).

If you connect the Arduino to the XBee and run this simple sketch, the Arduino will reply to any message it receives by simply echoing what the other XBee sends it:

```
/*
  XBeeEcho
  Reply with whatever you receive over the serial port
```

```
 */

void setup()
{
  Serial.begin(9600);
}

void loop()
{
  while (Serial.available() ) {
    Serial.write(Serial.read()); // reply with whatever you receive
  }
}
```

Figure 14-3 shows the connection between an Adafruit XBee Adapter and Arduino. Notice that the Arduino's RX is connected to the XBee's TX and vice versa.

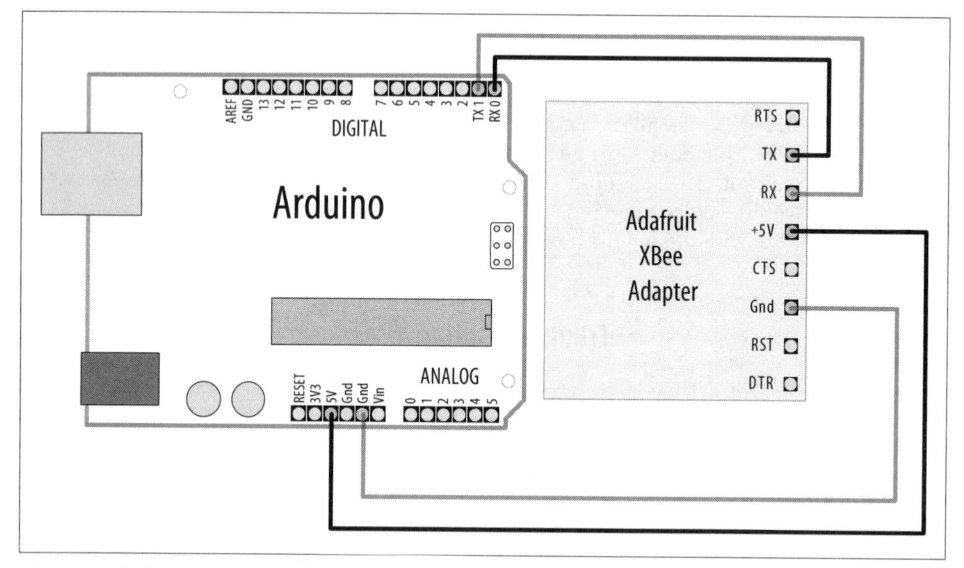

Figure 14-3. Connecting an Arduino to an XBee using the Adafruit XBee Adapter

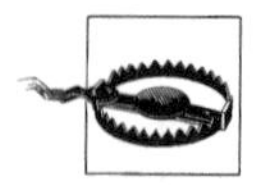

If you are using a different adapter that does not have an on-board voltage regulator, it will be sending voltage directly into the XBee. If this is the case, you must connect the 3V3 pin from the Arduino to the adapter's power supply, or you risk burning out your XBee.

With the XBees configured and connected to a computer and/or Arduino, you can send messages back and forth.

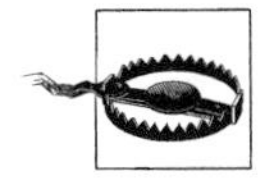

You must disconnect the Arduino from the XBee before you attempt to program the Arduino. This is because Arduino uses pins 0 and 1 for programming, and the signals will get crossed if anything else, such as an XBee, is connected to those pins.

Discussion

To configure your XBees, plug them in to an XBee adapter such as the Adafruit XBee Adapter kit ($10; Maker Shed part number MKAD13, Adafruit 126) and use a USB-to-TTL serial adapter such as the TTL-232R ($20; Maker Shed TTL232R, Adafruit 70) to connect the adapter to a computer.

You should purchase at least two adapters (and if needed, two cables), which will allow you to have two XBees connected to your computer at the same time. These same adapters can be used to connect an XBee to an Arduino.

You could also use an all-in-one XBee USB adapter, such as the Parallax XBee USB Adapter ($20; Adafruit 247, Parallax 32400) or the SparkFun XBee Explorer USB ($25; SparkFun WRL-08687).

Figure 14-4 shows the Adafruit XBee Adapter and the SparkFun XBee Explorer USB with Series 2 XBee modules connected.

Series 2 configuration

For the initial configuration of Series 2 XBees, you will need to plug your XBees in to a Windows computer (the configuration utility is not available for Mac or Linux). Plug only one in to a USB port for now. The TTL-232R and Parallax XBee USB Adapter both use the same USB-to-serial driver as the Arduino itself, so you should not need to install an additional driver.

1. Open Device Manager (press Windows-R, type **`devmgmt.msc`**, and press Enter), expand the Ports (COM & LPT) section, and take note of the number of the USB Serial Port the XBee you just plugged in is connected to (unplug it and plug it back in if it's not obvious which port is correct). Exit Device Manager.
2. Run the X-CTU application (*http://www.digi.com/support/productdetl.jsp?pid=3352&osvid=0&tp=5&tp2=0*), then select the serial port you identified in the previous step, and press Test/Query to ensure that X-CTU recognizes your XBee. (If not, see the support document at *http://www.digi.com/support/kbase/kbaseresultdetl.jsp?id=2103*.)
3. Switch to the Modem Configuration tab, and click Read. X-CTU will determine which model of XBee you are using as well as the current configuration.
4. Under Function Set, choose ZIGBEE COORDINATOR AT (*not* API).

Figure 14-4. Two XBees, one connected to an Adafruit adapter and the other connected to a SparkFun adapter

5. Click the Version menu and pick the highest numbered version of the firmware available.
6. Click Show Defaults.
7. Change the PAN ID setting from 0 to 1234 (or any hexadecimal number you want, as long as you use the same PAN ID for all devices on the same network), as shown in Figure 14-5.
8. Click Write.
9. Click the Terminal tab.
10. Next, leave X-CTU running and leave that XBee plugged in. Plug your second XBee in to a different USB port. Repeat the preceding steps (in step 2, you will be starting up a second copy of X-CTU), but instead of choosing ZIGBEE COORDINATOR AT in step 4, choose ZIGBEE ROUTER AT. On this XBee, you should also set Channel Verification (JV) to 1 to make sure it will confirm that it's on the right channel, which makes its connection to the coordinator more reliable.

If you have two computers running Windows, you can connect each XBee into a separate computer.

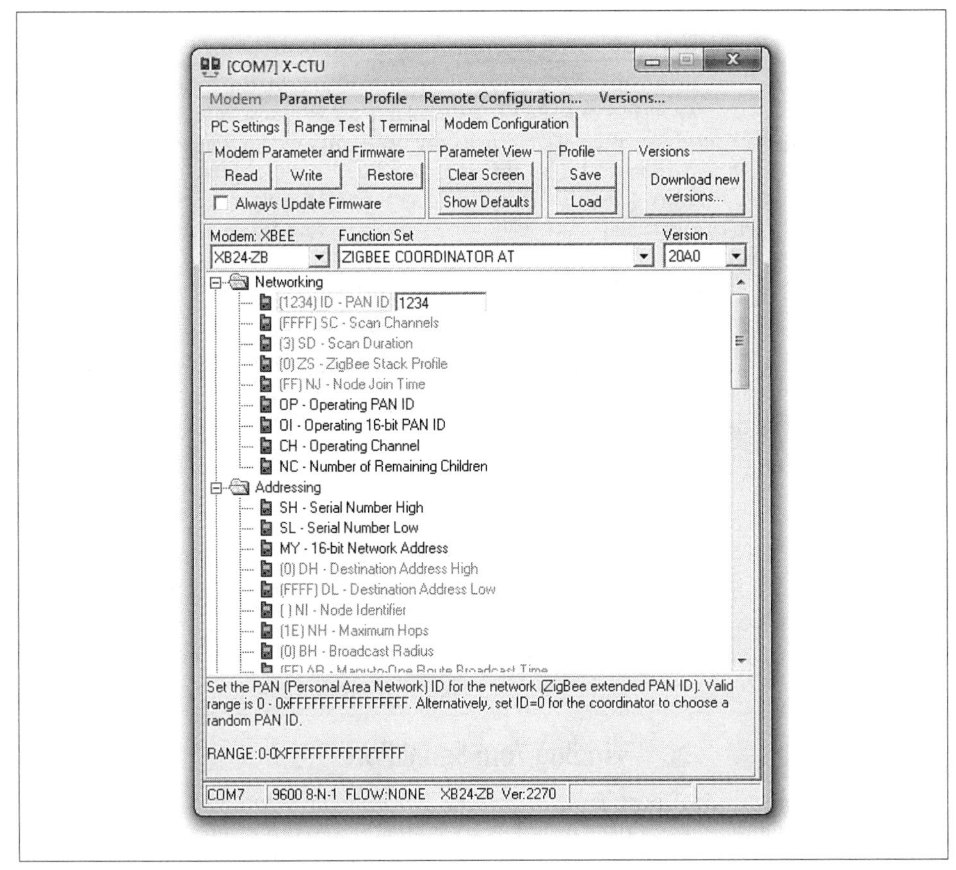

Figure 14-5. Configuring the XBee

With both XBees connected and two copies of X-CTU showing their Terminal tab, type into either Terminal window. You'll see whatever you type into one XBee appear on the Terminal of the other one. You've set up your first simple XBee Personal Area Network (PAN). Now you can connect XBees to two Arduino boards and run the sketch as described in "Talking to the Arduino" on page 471.

Series 1 configuration

For Series 1 XBees, you can use a Mac or a PC running Linux or Windows. However, if you wish to update the firmware on the XBees, you will need to use the X-CTU utility described in "Series 2 configuration" on page 467.

Determine which serial port your XBee is using, as described in "Finding Your Serial Port" on page 470. Connect to this port in your serial terminal program. To connect to your XBee using CoolTerm (Windows or Mac), follow these steps:

1. Run CoolTerm.

You can download CoolTerm for Windows and Mac at *http://freeware.the-meiers.org/*. PuTTY is available for Windows and Linux at *http://www.chiark.greenend.org.uk/~sgtatham/putty/download.html*. You may also be able to install PuTTY under Linux using your Linux system's package manager. For example, on Ubuntu, PuTTY is available in the Universe repository with `apt-get install putty`.

2. Click the Options button in the toolbar.
3. Select the USB serial port (such as usbserial-A700eYw1 on a Mac or COM8 on a PC). Make sure it is set to a baud rate of 9,600, 8 data bits, no parity, 1 stop bit (these are the defaults).
4. Check the box labeled Local Echo.
5. Click OK.
6. Click the Save button in the toolbar and save your session settings.
7. In future sessions, you can skip steps 2 through 6 by clicking Open and selecting the settings file you saved.
8. Click the Connect button in the toolbar.

Finding Your Serial Port

To determine the serial port assigned to your XBee under Windows, see step 1 in "Series 2 configuration" on page 467. To determine the serial port under Mac OS X, open the Mac OS X Terminal window (located in */Applications/Utilities*) and type this command: **`ls /dev/tty.usbserial-*`**. On Linux, open an xterm or similar console terminal and type **`ls /dev/ttyUSB*`**.

If you see more than one result here, unplug all USB serial devices except the XBee you wish to configure and type the command again. You should only see one result.

You'll see output like this on the Mac:

```
/dev/tty.usbserial-A700eYw1
```

And like this on Linux:

```
/dev/ttyUSB0
```

The result you see is the filename that corresponds to your XBee's USB serial port.

To connect to your XBee using PuTTY (Windows or Linux), follow these steps:

1. Run PuTTY.
2. Click Serial under Connection Type.

3. Type the name of your serial port in the Serial Line field (such as /dev/ttyUSB0 on Linux or COM7 on Windows). Make sure Speed is set to 9600 (the default).
4. On the left side of the window, under Category, click Terminal.
5. Under Local Echo, choose Force On.
6. Under "Set various terminal options," choose Implicit LF in Every CR.
7. On the left side of the window, under Category, click Session.
8. Type a name for the session, such as "XBee 1," then click Save.
9. In future sessions, you can skip steps 2 through 8 by double-clicking the saved session name. This will open the serial connection.

Now that you're connected, configure the first XBee with the following AT commands. You will need to type `+++` and wait a second to get the XBee's attention (it will respond with "OK"):

```
ATMY1234
ATDL5678
ATDH0
ATID0
ATWR
```

Keep your Serial Terminal up and running so that you can continue to type commands into it. Next, plug in the second XBee, and follow the earlier instructions to connect to it with PuTTY or CoolTerm (to open a new PuTTY window, you can simply launch the program again; you can start a new CoolTerm window with File→New). Then, configure the second XBee with these commands:

```
ATMY5678
ATDL1234
ATDH0
ATID0
ATWR
```

Now you can type commands into the Serial Terminal window for one XBee and they will appear in the Serial Terminal window for the other XBee (and vice versa).

The `ATMY` command sets the identifier for an XBee. `ATDL` and `ATDH` set the low byte and the high byte of the destination XBee. `ATID` sets the network ID (it needs to be the same for XBees to talk to one another) and `ATWR` saves the settings into the XBee so that it remembers the settings even if you power it down and back up.

Talking to the Arduino

Now that you've got your XBee modules configured, pick one of the XBees and close the Serial Terminal that was connected to it, and disconnect it from your computer. Next, program your Arduino with the code shown in this recipe's Solution, and connect the XBee to your Arduino as shown in Figure 14-3. When you type characters into the Serial Terminal program connected to your other XBee, you'll see the characters echoed back (if you type `a`, you'll see `aa`).

If you see each character echoed back to you twice, it's because you enabled local echo in the terminal program earlier in this recipe. You can disconnect and reconnect with Local Echo turned off (follow the earlier instructions for CoolTerm or PuTTY and make sure Local Echo is off) if you'd like.

See Also

Recipes 14.3, 14.4, and 14.5

14.3 Sending a Message to a Particular XBee

Problem

You want to configure which node your message goes to from your Arduino sketch.

Solution

Send the AT commands directly from your Arduino sketch:

```
/*
  XBeeMessage
 Send a message to an XBee using its address
 */

boolean configured;

boolean configureRadio() {

  // put the radio in command mode:
  Serial.print("+++");

  String ok_response = "OK\r"; // the response we expect.

  // Read the text of the response into the response variable
  String response = String("");
  while (response.length() < ok_response.length()) {
    if (Serial.available() > 0) {
      response += (char) Serial.read();
    }
  }

  // If we got the right response, configure the radio and return true.
  if (response.equals(ok_response)) {
    Serial.print("ATDH0013A200\r"); // destination high-REPLACE THIS
    Serial.print("ATDL403B9E1E\r"); // destination low-REPLACE THIS
    Serial.print("ATCN\r");     // back to data mode
    return true;
  } else {
    return false; // This indicates the response was incorrect.
  }
```

```
}

void setup () {
  Serial.begin(9600); // Begin serial
  configured = configureRadio();
}

void loop () {
  if (configured) {
    Serial.print("Hello!");
    delay(3000);
  }
  else {
    delay(30000);      // Wait 30 seconds
    configured = configureRadio(); // try again
  }
}
```

Discussion

Although the configurations in Recipe 14.2 work for two XBees, they are not as flexible when used with more than two.

For example, consider a three-node network of Series 2 XBees, with one XBee configured with the COORDINATOR AT firmware and the other two with the ROUTER AT firmware. Messages you send from the coordinator will be broadcast to the two routers. Messages you send from each router are sent to the coordinator.

The Series 1 configuration in that recipe is a bit more flexible, in that it specifies explicit destinations: by configuring the devices with AT commands and then writing the configuration, you effectively hardcode the destination addresses in the firmware.

This solution instead lets the Arduino code send the AT commands to configure the XBees on the fly. The heart of the solution is the `configureRadio()` function. It sends the `+++` escape sequence to put the XBee in command mode, just as the Series 1 configuration did at the end of Recipe 14.2. After sending this escape sequence, the Arduino sketch waits for the OK response before sending these AT commands:

```
ATDH0013A200
ATDL403B9E1E
ATCN
```

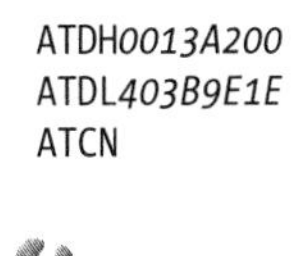

In your code, you must replace *0013A200* and *403B9E1E* with the high and low addresses of the destination radio.

The first two commands are similar to what is shown in the Series 1 configuration at the end of Recipe 14.2, but the numbers are longer. That's because the example shown in that recipe's Solution uses Series 2–style addresses. As you saw in Recipe 14.2, you can specify the address of a Series 1 XBee with the `ATMY` command, but in a Series 2

XBee, each module has a unique address that is embedded in each chip. You can look up the high (ATDH) and low (ATDL) portions of the serial number using X-CTU, as shown in Figure 14-6. The numbers are also printed on the label underneath the XBee.

The ATCN command exits command mode; think of it as the reverse of what the +++ sequence accomplishes.

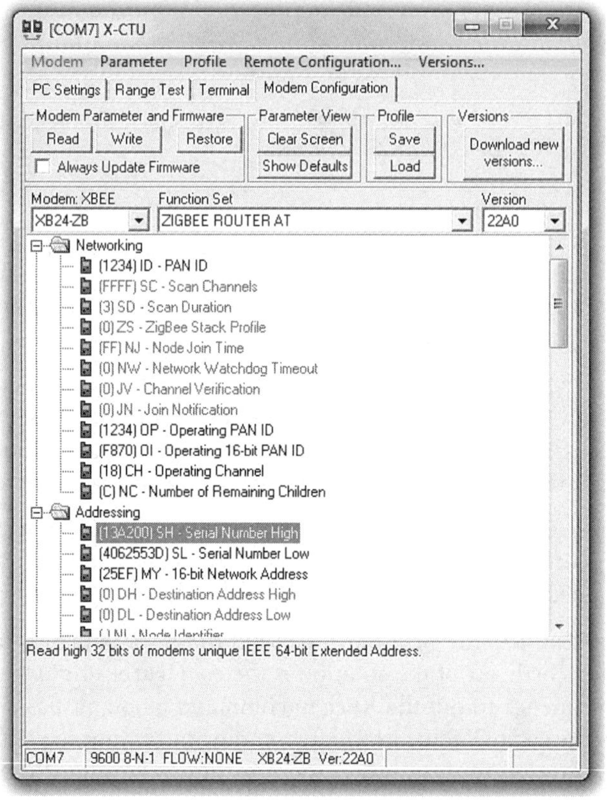

Figure 14-6. Looking up the high and low serial numbers in X-CTU

See Also

Recipe 14.2

14.4 Sending Sensor Data Between XBees

Problem

You want to send the status of digital and analog pins or control pins based on commands received from XBee.

Solution

Hook one of the XBees (the transmitting XBee) up to an analog sensor and configure it to read the sensor and transmit the value periodically. Connect the Arduino to an XBee (the receiving XBee) configured in API mode and read the value of the API frames that it receives from the other XBee.

Discussion

XBees have a built-in analog-to-digital converter (ADC) that can be polled on a regular basis. The XBee can be configured to transmit the values (between 0 and 1023) to other XBees in the network. The configuration and code differ quite a bit between Series 2 and Series 1 XBees.

Series 2 XBees

Using X-CTU (see "Series 2 configuration" on page 467 in Recipe 14.2), configure the transmitting XBee with the ZIGBEE ROUTER AT (*not* API) function set and the following settings (click Write when you are done):

> PAN ID: `1234` (or a number you pick, as long as you use the same one for both XBees)
> Channel Verification (JV): `1` (this makes sure the router will confirm that it's on the right channel when talking to the coordinator)
> Destination Address High (DH): the high address (SH) of the other XBee, usually 13A200
> Destination Address Low (DL): the low address (SL) of the other XBee
> Under I/O Settings, AD0/DIO0 Configuration (D0): `2`
> Under I/O Settings→Sampling Rate (IR): `64` (100 milliseconds in hex)

You can look up the high (`ATDH`) and low (`ATDL`) portions of the serial number using X-CTU, as shown earlier in Figure 14-6. The numbers are also printed on the label underneath the XBee.

Configure the receiving XBee with the ZIGBEE COORDINATOR API (*not* AT) function set with the following settings:

PAN ID: **1234** (or a number you pick, as long as you use the same one for both XBees)
Destination Address High (DH): the high address (SH) of the other XBee, usually 13A200
Destination Address Low (DL): the low address (SL) of the other XBee

Wire up the transmitting XBee to the sensor, as shown in Figure 14-7. The value of R1 should be double whatever your potentiometer is (if you are using a 10K pot, use a 20K resistor). This is because the Series 2 XBees' analog-to-digital converters read a range of 0 to 1.2 volts, and R1 reduces the 3.3 volts to stay below 1.2 volts.

Check the pinout of your XBee breakout board carefully, as the pins on the breakout board don't always match up exactly to the pins on the XBee itself. For example, on some breakout boards, the upper left pin is GND, and the pin below it is 3.3V.

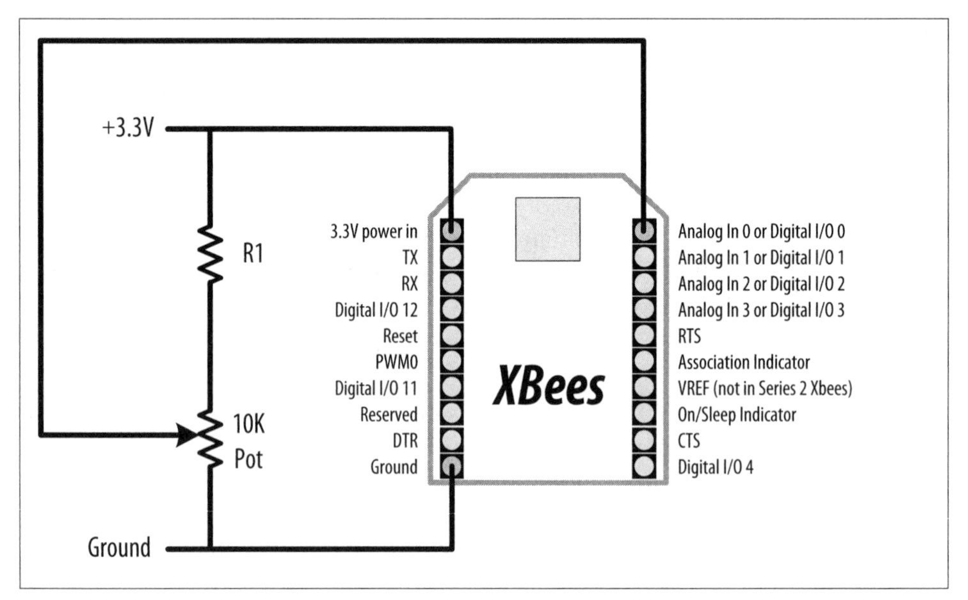

Figure 14-7. Connecting the receiving Series 2 XBee to an analog sensor

Next, load the following sketch onto the Arduino, and wire the transmitting XBee to the Arduino as shown in Recipe 14.2. If you need to reprogram the Arduino, remember to disconnect it from the XBee first:

```
/*
  XBeeAnalogReceive
 Read an analog value from an XBee API frame and set the brightness
 of an LED accordingly.
 */
```

```
#define LEDPIN 9

void setup() {
  Serial.begin(9600);
  pinMode(LEDPIN, OUTPUT);
}

void loop() {

  if (Serial.available() >= 21) { // Wait until we have a mouthful of data

     if (Serial.read() == 0x7E) { // Start delimiter of a frame

      // Skip over the bytes in the API frame we don't care about
      for (int i = 0; i < 18; i++) {
        Serial.read();
      }

      // The next two bytes are the high and low bytes of the sensor reading
      int analogHigh = Serial.read();
      int analogLow = Serial.read();
      int analogValue = analogLow + (analogHigh * 256);

      // Scale the brightness to the Arduino PWM range
      int brightness = map(analogValue, 0, 1023, 0, 255);

      // Light the LED
      analogWrite(LEDPIN, brightness);
    }
  }
}
```

Series 1 XBees

Using a terminal program as described in "Series 1 configuration" on page 469 in Recipe 14.2, send the following configuration commands to the transmitting XBee:

```
ATRE
ATMY1234
ATDL5678
ATDH0
ATID0
ATD02
ATIR64
ATWR
```

Next, send the following configuration commands to the receiving XBee:

```
ATRE
ATMY5678
ATDL1234
ATDH0
ATID0
ATWR
```

Both XBees

`ATRE` resets the XBee to factory defaults. The `ATMY` command sets the identifier for an XBee. `ATDL` and `ATDH` set the low byte and the high byte of the destination XBee. `ATID` sets the network ID (it needs to be the same for XBees to talk to one another). `ATWR` saves the settings into the XBee so that it remembers the settings even if you power it down and back up.

Transmitting XBee

`ATD02` configures pin 20 (analog or digital input 0) as an analog input; `ATIR64` tells the XBee to sample every 100 (64 hex) milliseconds and send the value to the XBee specified by `ATDL` and `ATDH`.

Wire up the transmitting XBee to the sensor, as shown in Figure 14-8.

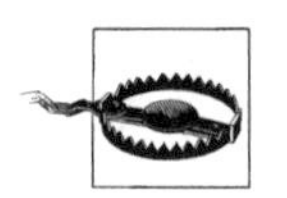

Check the pinout of your XBee breakout board carefully, as the pins on the breakout board don't always match up exactly to the pins on the XBee itself. For example, on some breakout boards, the upper left pin is GND, and the pin below it is 3.3V. Similarly, you might find that the VREF pin (labeled RES on the SparkFun XBee Explorer USB) is fifth from the bottom on the right, while it is fourth from the bottom on the XBee itself.

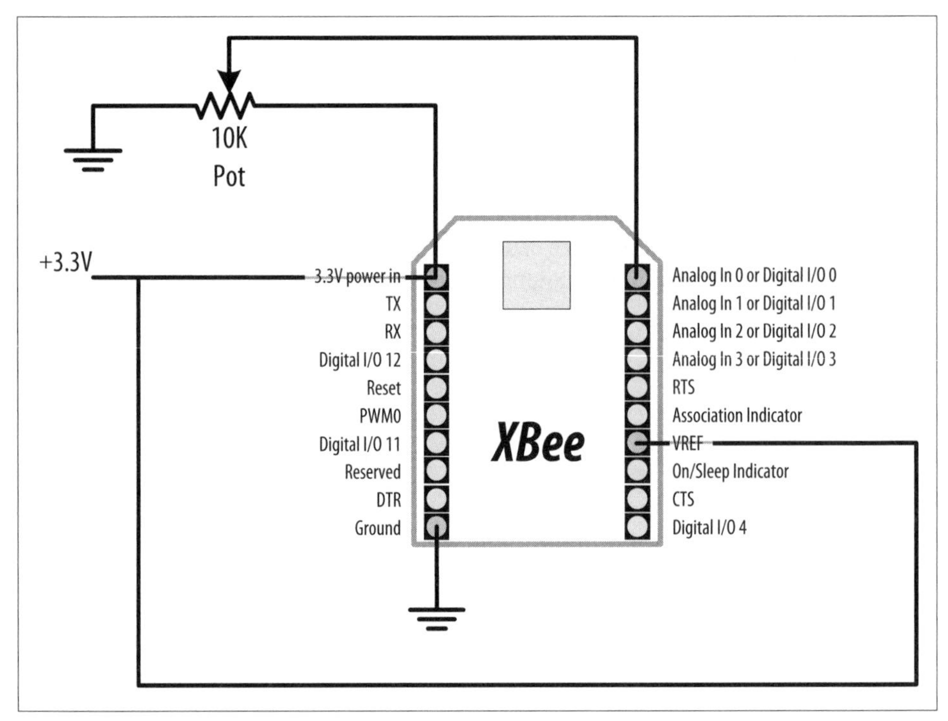

Figure 14-8. The receiving Series 1 XBee connected to an analog sensor

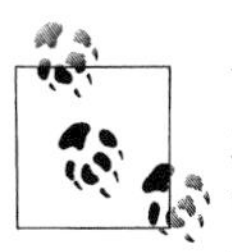

Unlike Series 2, Series 1 XBee uses an external reference connected to 3.3V. Because the voltage on the slider of the pot can never be greater than the reference voltage, the resistor shown in Figure 14-7 is not needed.

Next, load the following sketch onto the Arduino, and wire the transmitting XBee to the Arduino as shown in Recipe 14.2. If you need to reprogram the Arduino, disconnect it from the XBee first:

```
/*
   XBeeAnalogReceiveSeries1
  Read an analog value from an XBee API frame and set the brightness
  of an LED accordingly.
 */

const int ledPin = 9;

void setup() {
  Serial.begin(9600);
  pinMode(ledPin, OUTPUT);
  configureRadio(); // check the return value if you need error handling
}

boolean configureRadio() {

  // put the radio in command mode:
  Serial.flush();
  Serial.print("+++");
  delay(100);

  String ok_response = "OK\r"; // the response we expect.

  // Read the text of the response into the response variable
  String response = String("");
  while (response.length() < ok_response.length()) {
    if (Serial.available() > 0) {
      response += (char) Serial.read();
    }
  }

  // If we got the right response, configure the radio and return true.
  if (response.equals(ok_response)) {
    Serial.print("ATAP1\r"); // Enter API mode
    delay(100);
    Serial.print("ATCN\r");  // back to data mode
    return true;
  } else {
    return false; // This indicates the response was incorrect.
  }
}

void loop() {
```

```
    if (Serial.available() >= 14) { // Wait until we have a mouthful of data

      if (Serial.read() == 0x7E) { // Start delimiter of a frame

        // Skip over the bytes in the API frame we don't care about
        for (int i = 0; i < 10; i++) {
          Serial.read();
        }

        // The next two bytes are the high and low bytes of the sensor reading
        int analogHigh = Serial.read();
        int analogLow = Serial.read();
        int analogValue = analogLow + (analogHigh * 256);

        // Scale the brightness to the Arduino PWM range
        int brightness = map(analogValue, 0, 1023, 0, 255);

        // Light the LED
        analogWrite(ledPin, brightness);
      }
    }
  }
```

On the Series 1 XBees, the Arduino code needed to configure the radio for API mode with an AT command (ATAP1). On Series 2 XBees, this is accomplished by flashing the XBee with a different firmware version. The reason for the return to data mode (ATCN) is because command mode was entered earlier with +++ and a return to data mode to receive data is required.

See Also

Recipe 14.2

14.5 Activating an Actuator Connected to an XBee

Problem

You want to tell an XBee to activate a pin, which could be used to turn on an actuator connected to it, such as a relay or LED.

Solution

Configure the XBee connected to the actuator so that it will accept instructions from another XBee. Connect the other XBee to an Arduino to send the commands needed to activate the digital I/O pin that the actuator is connected to.

Discussion

The XBee digital/analog I/O pins can be configured for digital output. Additionally, XBees can be configured to accept instructions from other XBees to take those pins high or low. In Series 2 XBees, you'll be using the Remote AT Command feature. In Series 1 XBees, you can use the direct I/O, which creates a virtual wire between XBees.

Series 2 XBees

Using X-CTU (see "Series 2 configuration" on page 467), configure the *receiving* XBee with the ZIGBEE ROUTER AT (*not* API) function set and the following settings:

PAN ID: `1234` (or a number you pick, as long as you use the same one for both XBees)
Channel Verification (JV): `1` (this makes sure the router will confirm that it's on the right channel when talking to the coordinator)
Destination Address High (DH): the high address (SH) of the other XBee, usually 13A200
Destination Address Low (DL): the low address (SL) of the other XBee
Under I/O Settings, AD1/DIO1 Configuration (D1): `4` (digital output, low)

You can look up the high (`ATDH`) and low (`ATDL`) portions of the serial number using X-CTU, as shown earlier in Figure 14-6. The numbers are also printed on the label underneath the XBee.

Configure the *transmitting* XBee with the ZIGBEE COORDINATOR API (*not* AT) function set with the following settings:

PAN ID: `1234` (or a number you pick, as long as you use the same one for both XBees)
Destination Address High (DH): the high address (SH) of the other XBee, usually 13A200
Destination Address Low (DL): the low address (SL) of the other XBee

Wire up the receiving XBee to an LED, as shown in Figure 14-9.

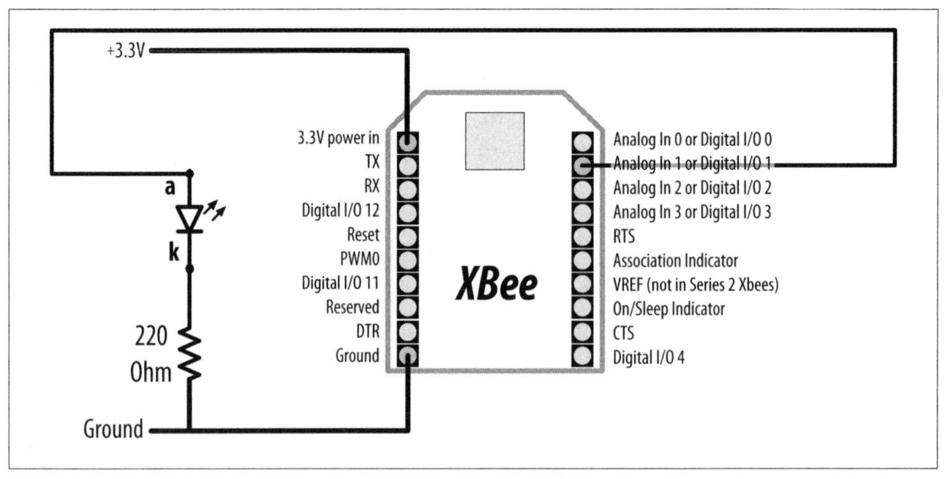

Figure 14-9. Connecting an LED to an XBee's digital I/O pin 1 (both Series 1 and Series 2)

Next, load the following sketch onto the Arduino, and wire the transmitting XBee to the Arduino as shown in Recipe 14.2. If you need to reprogram the Arduino, remember to disconnect it from the XBee first. This sketch sends a Remote AT command (`ATD14` or `ATD15`) that sets the state of pin 1 (`ATD1`) alternatingly on (digital out high, 5) and off (digital out low, 4):

```
/*
  XBeeActuate
  Send a Remote AT command to activate a digital pin on another XBee.
 */

const byte frameStartByte = 0x7E;
const byte frameTypeRemoteAT  = 0x17;
const byte remoteATOptionApplyChanges = 0x02;

void setup() {
  Serial.begin(9600);
}

void loop()
{

  toggleRemotePin(1);
  delay(3000);
  toggleRemotePin(0);
  delay(2000);
}


byte sendByte(byte value) {
  Serial.write(value);
  return value;
}
```

```
void toggleRemotePin(int value) {  // 0 = off, nonzero = on

  byte pin_state;
  if (value) {
    pin_state = 0x5;
  } else {
    pin_state = 0x4;
  }

  sendByte(frameStartByte); // Begin the API frame

  // High and low parts of the frame length (not counting checksum)
  sendByte(0x0);
  sendByte(0x10);

  long sum = 0; // Accumulate the checksum

  sum += sendByte(frameTypeRemoteAT); // Indicate this frame contains a
                                      // Remote AT command

  sum += sendByte(0x0);  // frame ID set to zero for no reply

  // The following 8 bytes indicate the ID of the recipient.
  // Use 0xFFFF to broadcast to all nodes.
  sum += sendByte(0x0);
  sum += sendByte(0x0);
  sum += sendByte(0x0);
  sum += sendByte(0x0);
  sum += sendByte(0x0);
  sum += sendByte(0x0);
  sum += sendByte(0xFF);
  sum += sendByte(0xFF);

  // The following 2 bytes indicate the 16-bit address of the recipient.
  // Use 0xFFFE to broadcast to all nodes.
  sum += sendByte(0xFF);
  sum += sendByte(0xFF);

  sum += sendByte(remoteATOptionApplyChanges); // send Remote AT options

  // The text of the AT command
  sum += sendByte('D');
  sum += sendByte('1');

  // The value (0x4 for off, 0x5 for on)
  sum += sendByte(pin_state);

  // Send the checksum
  sendByte( 0xFF - ( sum & 0xFF));

  delay(10); // Pause to let the microcontroller settle down if needed
}
```

Series 1 XBees

Using a terminal program as described in "Series 1 configuration" on page 469, send the following configuration commands to the *transmitting* XBee (the one you'll connect to the Arduino):

```
ATRE
ATMY1234
ATDL5678
ATDHO
ATIDO
ATD13
ATICFF
ATWR
```

Next, send the following configuration commands to the *receiving* XBee:

```
ATRE
ATMY5678
ATDL1234
ATDHO
ATIDO
ATD14
ATIUO
ATIA1234
ATWR
```

Both XBees

`ATRE` resets the XBee to factory defaults. The `ATMY` command sets the identifier for an XBee. `ATDL` and `ATDH` set the low byte and the high byte of the destination XBee. `ATID` sets the network ID (it needs to be the same for XBees to talk to one another). `ATWR` saves the settings into the XBee so that it remembers the settings even if you power it down and back up.

Transmitting XBee

`ATICFF` tells the XBee to check every digital input pin and send their values to the XBee specified by `ATDL` and `ATDH`. `ATD13` configures pin 19 (analog or digital input 1) to be in digital input mode. The state of this pin will be relayed from the transmitting XBee to the receiving XBee.

Receiving XBee

`ATIU1` tells the XBee to not send the frames it receives to the serial port. `ATIA1234` tells it to accept commands from the other XBee (whose MY address is 1234). `ATD14` configures pin 19 (analog or digital input 1) to be in low digital output mode (off by default).

Wire up the transmitting XBee to the Arduino, as shown in Figure 14-10.

Next, wire the receiving XBee to an Arduino, as shown in Recipe 14.2. Note that instead of sending AT commands over the serial port, we're using an electrical connection to take the XBee's pin high. The two 10K resistors form a voltage divider that drops the

Arduino's 5V logic to about 2.5 volts (high enough for the XBee to recognize, but low enough to avoid damaging the XBee's 3.3V logic pins).

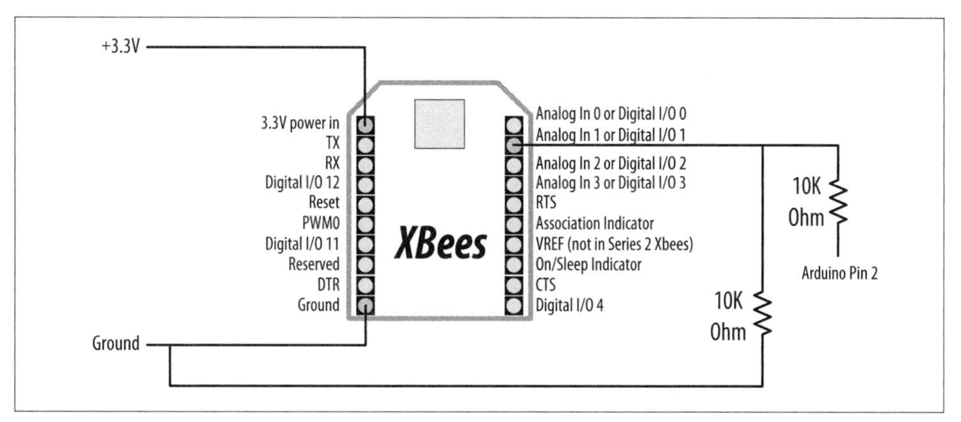

Figure 14-10. Connecting the Arduino to the Series 1 transmitting XBee's digital I/O pin 1

Next, load the following sketch onto the transmitting Arduino. This sketch takes the XBee's digital I/O pin 1 alternatingly on (digital out high, 5) and off (digital out low, 4). Because the transmitting XBee is configured to relay its pin states to the receiving XBee, when its pin 1 changes state the receiving XBee's pin 1 changes as well:

```
/*
  XBeeActuateSeries1
  Activate a digital pin on another XBee.
 */

const int xbeePin = 2;

void setup() {
  pinMode(xbeePin, OUTPUT);
}

void loop()
{

  digitalWrite(xbeePin, HIGH);
  delay(3000);
  digitalWrite(xbeePin, LOW);
  delay(3000);
}
```

See Also

Recipe 14.2

14.6 Sending Messages Using Low-Cost Transceivers

Problem

You want a low-cost wireless solution with more capability than the simple modules in Recipe 14.1.

Solution

Use the increasingly popular Hope RFM12B modules to send and receive data. This Recipe uses two Arduino boards and wireless modules. One pair reads and sends values and the other displays the received value—both pairs are wired the same way.

Connect the modules as shown in Figure 14-11. The Antenna is just a piece of wire cut to the correct length for the frequency of your modules; use 78 mm for 915 MHz, 82 mm for 868 MHz and 165 mm for 433 MHz.

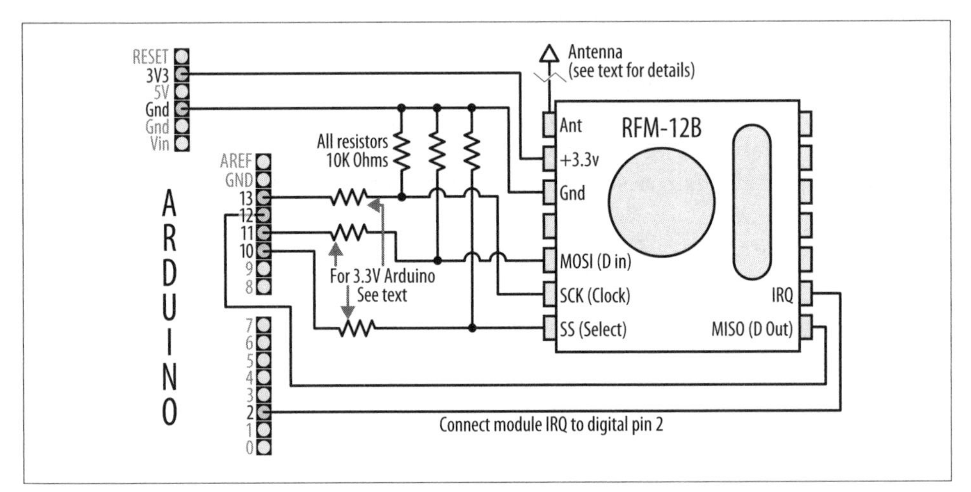

Figure 14-11. RFM12B Transceiver connections

If you are using a 3.3 volt Arduino such as the Fio or 3.3V Arduino Pro, eliminate the resistors and wire Arduino pins 10, 11, and 13 directly to the respective RFM12B pins.

The transmit sketch sends values from the six analog pins every second:

```
/*
 * SimpleSend
 * RFM12B wireless demo - transmitter - no ack
 * Sends values of analog inputs 0 through 6
 *
 */

#include <RF12.h>  //from jeelabs.org
#include <Ports.h>
```

```
// RF12B constants:
const byte network  = 100;   // network group (can be in the range 1-255).
const byte myNodeID = 1;     // unique node ID of receiver (1 through 30)

//Frequency of RF12B can be RF12_433MHZ, RF12_868MHZ or RF12_915MHZ.
const byte freq = RF12_868MHZ; // Match freq to module

const byte RF12_NORMAL_SENDWAIT = 0;

void setup()
{
  rf12_initialize(myNodeID, freq, network);   // Initialize RFM12
}

const int payloadCount = 6; // the number of integers in the payload message
int payload[payloadCount];

void loop()
{
  for( int i= 0; i < payloadCount; i++)
  {
    payload[i] = analogRead(i);
  }
  while (!rf12_canSend())  // is the driver ready to send?
    rf12_recvDone();       // no, so service the driver

  rf12_sendStart(rf12_hdr, payload, payloadCount*sizeof(int));
  rf12_sendWait(RF12_NORMAL_SENDWAIT); // wait for send completion

  delay(1000);  // send every second
}
```

The receive sketch displays the six analog values on the Serial Monitor:

```
/*
 * SimpleReceive
 * RFM12B wireless demo - receiver - no ack
 *
 */

#include <RF12.h>     //from jeelabs.org
#include <Ports.h>

// RFM12B constants:
const byte network  = 100;   // network group (can be in the range 1-255).
const byte myNodeID = 2;     // unique node ID of receiver (1 through 30)

// Frequency of RFM12B can be RF12_433MHZ, RF12_868MHZ or RF12_915MHZ.
const byte freq = RF12_868MHZ; // Match freq to module

void setup()
{
  rf12_initialize(myNodeID,freq,network);   // Initialize RFM12 with settings above
  Serial.begin(9600);
  Serial.println("RFM12B Receiver ready");
```

```
    Serial.println(network,DEC);    // print the network
    Serial.println(myNodeID,DEC);   // and node ID
  }

  const int payloadCount = 6; // the number of integers in the payload message

  void loop()
  {
    if (rf12_recvDone() && rf12_crc == 0 && (rf12_hdr & RF12_HDR_CTL) == 0)
    {
      int *payload = (int*)rf12_data;  // access rf12 data buffer as an arrya of ints
      for( int i= 0; i < payloadCount; i++)
      {
        Serial.print(payload[i]);
        Serial.print(" ");
      }
      Serial.println();
    }
  }
```

Discussion

The RFM12B modules are designed for 3.3 volts and the resistors shown in Figure 14-11 are needed to drop the voltage to the correct level. The JeeLabs website *http://jeelabs.com/products/rfm12b-board* has details on breakout boards and modules for the RFM12B.

The RF12 library provides for different groups of modules to be used in the same vicinity where each group is identified by a network ID. Your send and receive sketches must use the same network ID to communicate with each other. Each node must have a unique ID within a network. In this example, the network is set for 100 with the sender using ID 1 and the receiver using ID 2.

The loop code fills an array (see Recipe 2.4) named `payload` with the six integer values read from analog input ports 0 through 5.

The sending is achieved by calling `rf12_sendStart`; the `rf12-hdr` argument determines the target node, which by default will be 0 (sending to node 0 will broadcast to all nodes on the network); `&payload` is the address of the payload buffer; `payloadCount * sizeof(int)` is the number of bytes in the buffer. `rf12_sendWait` waits for completion of the send (see the RF12 documentation for information about power down options).

This code does not check to see if messages are acknowledged. In applications like this, that repeatedly send information, this is not a problem because the occasional lost message will be updated with the next send. See the example code in the library download for sketches that show other techniques for sending and receiving data.

Any kind of data that fits within a 66-byte buffer can be sent. For example, the following sketch sends a binary data structure consisting of an integer and floating point value:

```
/*
 * RFM12B wireless demo - struct sender - no ack
```

```
 * Sends a floating point value using a C structure
 */

#include <RF12.h> //from jeelabs.org
#include <Ports.h>

// RF12B constants:
const byte network  = 100;   // network group (can be in the range 1-255)
const byte myNodeID = 1;     // unique node ID of receiver (1 through 30)

// Frequency of RF12B can be RF12_433MHZ, RF12_868MHZ or RF12_915MHZ.
const byte freq = RF12_868MHZ; // Match freq to module

const byte RF12_NORMAL_SENDWAIT = 0;

void setup()
{
  rf12_initialize(myNodeID, freq, network);   // Initialize RFM12
}

typedef struct {  // Message data Structure, this must match Tx
  int pin;  // pin number used for this measurement
  float value;  // floating point measurement value
}
Payload;

Payload sample;  // declare an instance of type Payload named sample

void loop()
{
  int inputPin = 0; // the input pin
  float value = analogRead(inputPin) * 0.01; // a floating point value
  sample.pin = inputPin; // send demontx.ct1=emontx.ct1+1;
  sample.value = value;

  while (!rf12_canSend())  // is the driver ready to send?
    rf12_recvDone();       // no, so service the driver

  rf12_sendStart(rf12_hdr, &sample, sizeof sample);
  rf12_sendWait(RF12_NORMAL_SENDWAIT);  // wait for send completion

  Serial.print(sample.pin);
  Serial.print(" = ");
  Serial.println(sample.value);
  delay(1000);
}
```

Here is the sketch that receives and displays the struct data:

```
/*
 * RFM12B wireless demo - struct receiver - no ack
 *
 */


#include <RF12.h>  // from jeelabs.org
```

```
#include <Ports.h>

// RF12B constants:
const byte network  = 100;   // network group (can be in the range 1-255)
const byte myNodeID = 2;     // unique node ID of receiver (1 through 30)

// Frequency of RF12B can be RF12_433MHZ, RF12_868MHZ or RF12_915MHZ.
const byte freq = RF12_868MHZ; // Match freq to module

void setup()
{
  rf12_initialize(myNodeID,freq,network);   // Initialize RFM12 with settings above
  Serial.begin(9600);
  Serial.print("RFM12B Receiver ready");
}

typedef struct {  // Message data Structure, this must match Tx
  int   pin;      // pin number used for this measurement
  float value;      // floating point measurement value
}
Payload;

Payload sample;         // declare an instance of type Payload named sample

void loop() {

  if (rf12_recvDone() && rf12_crc == 0 && (rf12_hdr & RF12_HDR_CTL) == 0)
  {
    sample = *(Payload*)rf12_data;            // Access the payload
    Serial.print("AnalogInput ");
    Serial.print(sample.pin);
    Serial.print(" = ");
    Serial.println(sample.value);
  }
}
```

This code is similar to the previous pair of sketches with the `payload` buffer replaced by a pointer named `sample` that points to the `Payload` structure.

See Also

The libraries used in this recipe were developed by Jean-Claude Wippler. A wealth of information is available on his site: *http://www.jeelabs.com*.

Each function of the RF12 library is documented here: *http://jeelabs.net/projects/cafe/wiki/RF12*.

An example sketch for sending strings with the RFM12 can be found here: *http://jeelabs.org/2010/09/29/sending-strings-in-packets*.

An example using sleep mode to save power between sends can be found here: *https://github.com/openenergymonitor/emonTxFirmware*.

A breakout board for the RFM12B is available here: *http://jeelabs.com/products/rfm12b-board*.

JeeNode is a board that combines the RFM12B and an Arduino-compatible chip: *http://http://jeelabs.com/products/jeenode*.

RFM12B 915 MHz versions of the module for use in the USA are available from Modern Device: *http://shop.moderndevice.com/collections/jeelabs*.

A 433 MHz version of RFM12B that should work anywhere in the world is available from SparkFun: *http://www.sparkfun.com/products/9582*.

14.7 Communicating with Bluetooth Devices

Problem

You want to send and receive information to another device using Bluetooth; for example, a laptop or cellphone.

Solution

Connect Arduino to a Bluetooth module such as the BlueSMiRF, Bluetooth Mate, or Bluetooth Bee, as shown in Figure 14-12.

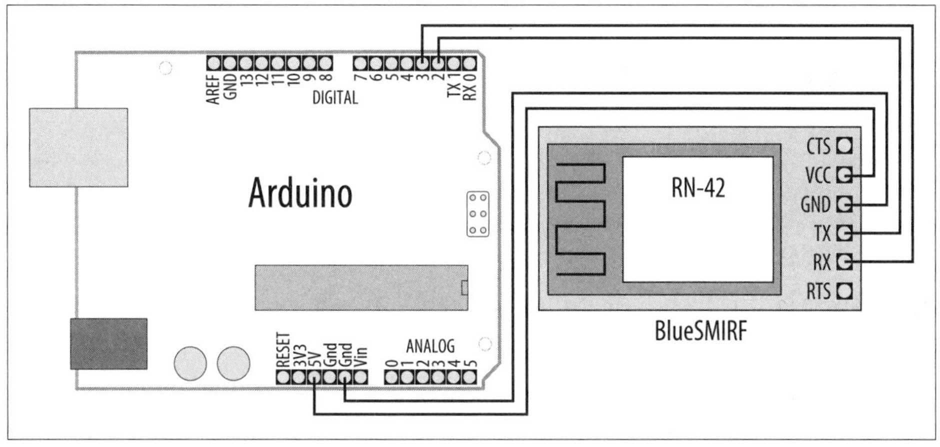

Figure 14-12. BlueSMiRF Bluetooth module wired to SoftwareSerial pins

This sketch is similar to the one in Recipe 4.13; it monitors characters received on the hardware serial port and a software serial port (connected to Bluetooth), so anything received on one is sent to the other:

```
/*
 * Use SoftwareSerial to talk to BlueSMiRF module
 * note pairing code is 1234
```

```
*/

#include <SoftwareSerial.h>

const int rxpin = 2;           // pin used to receive
const int txpin = 3;           // pin used to send to
SoftwareSerial bluetooth(rxpin, txpin); // new serial port on given pins

void setup()
{
  Serial.begin(9600);
  bluetooth.begin(9600); // initialize the software serial port
  Serial.println("Serial ready");
  bluetooth.println("Bluetooth ready");
}

void loop()
{
  if (bluetooth.available())
  {
    char c = (char)bluetooth.read();
    Serial.write(c);
  }
  if (Serial.available())
  {
    char c = (char)Serial.read();
    bluetooth.write(c);
  }
}
```

Discussion

You will need Bluetooth capability on your computer (or phone) to communicate with this sketch. Both sides participating in a Bluetooth conversation need to be paired—the ID of the module connected to Arduino needs to be known to the other end. The default ID for the BlueSMiRF is 1234. See the documentation for your computer/phone Bluetooth to set the pairing ID and accept the connection.

If you have a board that plugs in to an FTDI cable, you can directly plug in a Bluetooth Mate module. (See Figure 14-13.)

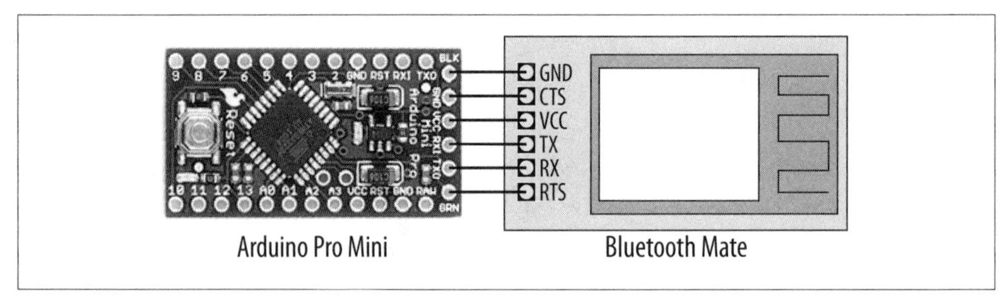

Figure 14-13. Bluetooth Mate uses similar connections as FTDI

The Bluetooth Mate can also be wired to use with a standard board, as shown in Figure 14-14.

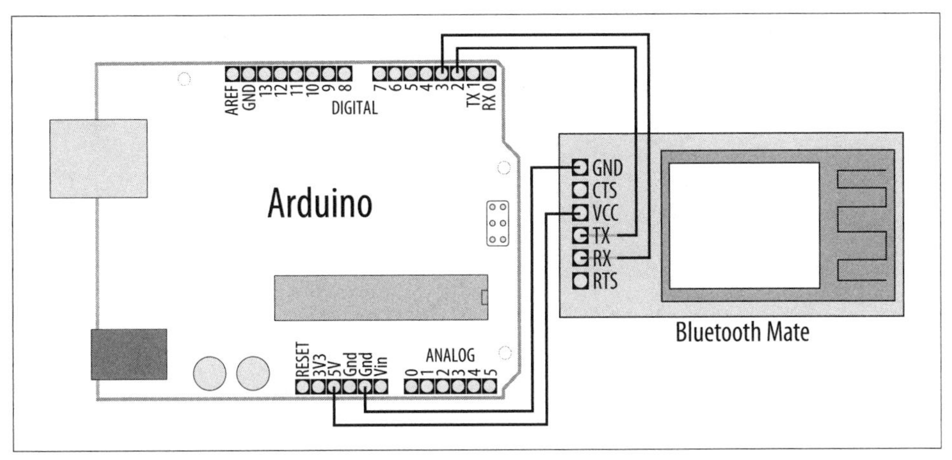

Figure 14-14. Bluetooth Mate wired for SoftwareSerial

All the common Bluetooth modules used with Arduino implement the Bluetooth Serial Port Profile (SPP). Once the devices are paired, the computer or phone will see the module as a serial port. These modules are not capable of appearing as other types of Bluetooth service, such as a Bluetooth mouse or keyboard.

Bluetooth range is between 5 and 100 meters, depending on whether you have class 3, 2, or 1 devices.

See Also

A SparkFun tutorial covering the installation and use of Bluetooth: *http://www.sparkfun.com/tutorials/67*

Bluetooth Bee is a Bluetooth module that plugs in to an XBee socket so you can use shields and adapters designed for XBee: *http://www.seeedstudio.com/depot/bluetooth-bee-p-598.html*.

CHAPTER 15

Ethernet and Networking

15.0 Introduction

Want to share your sensor data? Let other people take control of your Arduino's actions? Your Arduino can communicate with a broader world over Ethernet and networks. This chapter describes the many ways you can use Arduino with the Internet. It has examples that demonstrate how to build and use web clients and servers, and it shows how to use the most common Internet communication protocols with Arduino.

The Internet allows a client (e.g., your web browser) to request information from a server (a web server or other Internet service provider). This chapter contains recipes showing how to make an Internet client that retrieves information from a service such as Google or Yahoo! Other recipes in this chapter show how Arduino can be an Internet server that provides information to clients using Internet protocols and can act as a web server that creates pages for viewing in web browsers.

The Arduino Ethernet library supports a range of methods (protocols) that enable your sketches to be an Internet client or a server. The Ethernet library uses the suite of standard Internet protocols, and most of the low-level plumbing is hidden. Getting your clients or servers up and running and doing useful tasks will require some understanding of the basics of network addressing and protocols, and you may want to consult one of the many references available online or one of these introductory books:

- *Head First Networking* by Al Anderson and Ryan Benedetti (O'Reilly)
- *Network Know-How: An Essential Guide for the Accidental Admin* by John Ross (No Starch Press)
- *Windows NT TCP/IP Network Administration* by Craig Hunt and Robert Bruce Thompson (O'Reilly)
- *Making Things Talk* by Tom Igoe (O'Reilly)

(Search for O'Reilly titles on oreilly.com.)

Here are some of the key concepts in this chapter. You may want to explore them in more depth than is possible here:

Ethernet
: This is the low-level signaling layer providing basic physical message-passing capability. Source and destination addresses for these messages are identified by a Media Access Control (MAC) address. Your Arduino sketch defines a MAC address value that must be unique on your network.

TCP and IP
: Transmission Control Protocol (TCP) and Internet Protocol (IP) are core Internet protocols built above Ethernet. They provide a message-passing capability that operates over the global Internet. TCP/IP messages are delivered through unique IP addresses for the sender and receiver. A server on the Internet uses a numeric label (address) that no other server will have so that it can be uniquely identified. This address consists of four bytes, usually represented with dots separating the bytes (e.g., 64.233.187.64 is an IP address used by Google). The Internet uses the Domain Name System (DNS) service to translate the common service name (*www.google.com*) to the numeric IP address. This capability was added in Arduino 1.0; Recipe 15.3 shows how to use this capability in your sketches.

Local IP addresses
: If you have more than one computer connected to the Internet on your home network using a broadband router or gateway, each computer probably uses a local IP address that is provided by your router. The local address is created using a Dynamic Host Configuration Protocol (DHCP) service in your router. The Arduino Ethernet library now (from release 1.0) includes a DHCP service. Most of the recipes in this chapter show a user-selected IP address that you may need to modify to suit your network. Recipe 15.2 shows how the IP address can be obtained automatically using DHCP.

Web requests from a web browser and the resultant responses use Hypertext Transfer Protocol (HTTP) messages. For a web client or server to respond correctly, it must understand and respond to HTTP requests and responses. Many of the recipes in this chapter use this protocol, and referring to one of the references listed earlier for more details will help with understanding how these recipes work in detail.

Web pages are usually formatted using Hypertext Markup Language (HTML). Although it's not essential to use HTML if you are making an Arduino web server, as Recipe 15.9 illustrates, the web pages you serve can use this capability.

Extracting data from a web server page intended to be viewed by people using a web browser can be a little like finding a needle in a haystack because of all the extraneous text, images, and formatting tags used on a typical page. This task is simplified by using the Stream parsing functionality in Arduino 1.0 to find particular sequences of characters and to get strings and numeric values from a stream of data. If you are using an earlier Arduino release, you can download a library called TextFinder, available from the Arduino Playground. TextFinder extracts information from a stream of data. Stream parsing and TextFinder have similar functionality (Stream parsing is based on the TextFinder code that was written for the first edition of this book). However, some of the methods have been renamed; see the TextFinder documentation in the Playground if you need help migrating sketches from TextFinder to Arduino 1.0.

Web interchange formats have been developed to enable reliable extraction of web data by computer software. XML and JSON are two of the most popular formats, and Recipe 15.5 shows an example of how to do this using Arduino.

Arduino 1.0 Enhancements

The Arduino Ethernet library has had a number of improvements in the 1.0 release that make it easier to use and added capabilities such as DHCP and DNS that previously required the download of third-party libraries. Some of the class and method names have changed so sketches written for previous releases will require modification to compile with Arduino 1.0, here is a summary of the required changes to sketches written for earlier Arduino releases:

- `SPI.h` must be included before the Ethernet include at the top of the sketch (as of Arduino 0018).
- `Client client(server, 80);` changed to `EthernetClient client;`.
- `if(client.connect())` changed to `if(client.connect(serverName, 80)>0)`.
- `Server server(80)` changed to `EthernetServer server(80)`.
- DHCP does not require an external library (see Recipe 15.2).
- DNS does not require an external library (see Recipe 15.3).
- Word and number searching simplified through new Stream parsing capability (see Recipe 15.4).
- `F(text)` construct added to simplify storing text in flash memory (Recipe 15.11).

The code in this chapter is for Arduino release 1.0. If you are running an earlier version, use the download code from the first edition at *http://oreilly.com/catalog/9780596802486*.

The code in this book was tested with the Arduino 1.0 release candidates. Any updates to sketches will be listed in the *changelog.txt* file in the code download file at *http://shop.oreilly.com/product/0636920022244.do*.

Alternative Hardware for Low Cost Networking

If you want a low-cost DIY-friendly Ethernet board that doesn't require surface-mount technology, you can use the open source design created for a project called Nanode. This uses the same ATmega328 controller as Arduino but replaces the Wiznet chip with the lower cost ENC28J60 device. This chip is capable of providing the functionality described in this chapter, but it uses a different set of libraries, so you would need to use sketches written specifically for the ENC28J60.

For more information, see the Nanode home page at: *http://www.nanode.eu/*.

15.1 Setting Up the Ethernet Shield

Problem

You want to set up the Ethernet shield to use a hardcoded IP address.

Solution

This sketch is based on the Ethernet client example sketch distributed with Arduino. Check the documentation for your network to ensure that the Arduino IP address (the value of the `ip` variable) is valid for your network:

```
/*
 * Simple Web Client
 * Arduino 1.0 version
 */

#include <SPI.h>
#include <Ethernet.h>

byte mac[] = { 0xDE, 0xAD, 0xBE, 0xEF, 0xFE, 0xED };
byte ip[] = { 192, 168, 1, 177 };     // change to a valid address for your network
byte server[] = { 209,85,229,104  }; // Google
                                      // see text for more on IP addressing

EthernetClient client;

void setup()
{
  Serial.begin(9600);       // start the serial library:
  Ethernet.begin(mac,ip);
  delay(1000);              // give the ethernet hardware a second to initialize

  Serial.println("connecting...");

  if (client.connect(server, 80)) {
    Serial.println("connected");
    client.println("GET /search?q=arduino HTTP/1.0"); // the HTTP request
    client.println();
  }
  else {
    Serial.println("connection failed");
  }
}

void loop()
```

```
{
  if (client.available()) {
    char c = client.read();
    Serial.print(c);  // echo all data received to the Serial Monitor
  }

  if (!client.connected()) {
    Serial.println();
    Serial.println("disconnecting.");
    client.stop();
    for(;;)
      ;
  }
}
```

Discussion

This sketch performs a Google search using the word "arduino." Its purpose is to provide working code that you can use to verify that your network configuration is suitable for the Arduino Ethernet shield.

There are up to four addresses that may need to be configured correctly for the sketch to successfully connect and display the results of the search on the Serial Monitor:

```
byte mac[] = { 0xDE, 0xAD, 0xBE, 0xEF, 0xFE, 0xED };
```

The MAC address uniquely identifies your Ethernet shield. Every network device must have a different MAC address, and if you use more than one Arduino shield on your network, each must use a different address. Recent Ethernet shields have a MAC address printed on a sticker on the underside of the board. If you have a single Ethernet shield, you don't need to change this:

```
byte ip[] = { 192, 168 1, 177 };       // change to a valid address for your network
```

The IP address is used to identify something that is communicating on the Internet and must also be unique on your network. The address consists of four bytes, and the range of valid values for each byte depends on how your network is configured. IP addresses are usually expressed with dots separating the bytes—for example, 192.168.1.177. In all the Arduino sketches, commas are used instead of dots because the bytes are stored in an array (see Recipe 2.4).

If your network is connected to the Internet using a router or gateway, you may need to provide the IP address of the gateway when you call the `ethernet.begin` function. You can find the address of the gateway in the documentation for your router/gateway. Add two lines after the IP and server addresses at the top of the sketch with the address of your DNS server and gateway:

```
// add if needed by your router or gateway
byte dns_server[] = { 192, 168, 1, 2 }; // The address of your DNS server
byte gateway[] = { 192, 168, 1, 254 }; // your gateway address
```

And change the first line in `setup` to include the gateway address in the startup values for Ethernet:

```
Ethernet.begin(mac, ip, dns_server, gateway);
```

The server address consists of the 4-byte IP address of the server you want to connect to—in this case, Google. Server IP addresses change from time to time, so you may need to use the ping utility of your operating system to find a current IP address for the server you wish to connect to:

```
byte server[] = { 64, 233, 187, 99 }; // Google
```

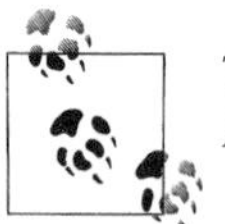

The line at the top of the sketch that includes `<SPI.h>` is required for Arduino releases starting at 0019.

See Also

The web reference for getting started with the Arduino Ethernet shield is at *http://arduino.cc/en/Guide/ArduinoEthernetShield*.

15.2 Obtaining Your IP Address Automatically

Problem

The IP address you use for the Ethernet shield must be unique on your network and you would like this to be allocated automatically. You want the Ethernet shield to obtain an IP address from a DHCP server.

Solution

This is similar to the sketch from Recipe 15.1 but without passing an IP address to the `Ethernet.begin` method:

```
/*
 * Simple Client to display IP address obtained from DHCP server
 * Arduino 1.0 version
 */

#include <SPI.h>
#include <Ethernet.h>

byte mac[] = { 0xDE, 0xAD, 0xBE, 0xEF, 0xFE, 0xED };
byte server[] = { 209,85,229,104  }; // Google

EthernetClient client;

void setup()
{
```

```
  Serial.begin(9600);
  if(Ethernet.begin(mac) == 0) { // start ethernet using mac & DHCP
    Serial.println("Failed to configure Ethernet using DHCP");
    while(true)   // no point in carrying on, so stay in endless loop:
      ;
  }
  delay(1000); // give the Ethernet shield a second to initialize

  Serial.print("This IP address: ");
  IPAddress myIPAddress = Ethernet.localIP();
  Serial.print(myIPAddress);
  if(client.connect(server, 80)>0) {
    Serial.println(" connected");
    client.println("GET /search?q=arduino HTTP/1.0");
    client.println();
  } else {
    Serial.println("connection failed");
  }
}

void loop()
{
  if (client.available()) {
    char c = client.read();
    // uncomment the next line to show all the received characters
    // Serial.print(c);
  }

  if (!client.connected()) {
    Serial.println();
    Serial.println("disconnecting.");
    client.stop();
    for(;;)
      ;
  }
}
```

Discussion

The library distributed with the Arduino 1.0 now supports DHCP (earlier releases required a third-party library from *http://blog.jordanterrell.com/post/Arduino-DHCP-Library-Version-04.aspx*.

The major difference from the sketch in Recipe 15.1 is that there is no IP (or gateway) address variable—these values are acquired from your DHCP server when the sketch starts. Also there is a check to confirm that the `ethernet.begin` statement was successful. This is needed to ensure that a valid IP address has been provided by the DHCP server (Internet access is not possible without a valid IP address).

This code prints the IP address to the Serial Monitor using a the `IPAddress.printTo` method introduced in Arduino 1.0:

```
  Serial.print("This IP address: ");
  IPAddress myIPAddress = Ethernet.localIP();
  Serial.print(myIPAddress);
```

The argument to `Serial.print` above may look odd but the new `IPAddress` class has the capability to output its value to objects such as `Serial` that derive from the `Print` class.

If you are not familiar with deriving functionality from classes, suffice it to say that the `IPAddress` object is smart enough to display its address when asked.

15.3 Resolving Hostnames to IP Addresses (DNS)

Problem

You want to use a server name—for example, yahoo.com—rather than a specific IP address. Web providers often have a range of IP addresses used for their servers and a specific address may not be in service when you need to connect.

Solution

You can use DNS to look up a valid IP address for the name you provide:

```
/*
 * Web Client DNS sketch
 * Arduino 1.0 version
 */

#include <SPI.h>
#include <Ethernet.h>

byte mac[] = {0xDE, 0xAD, 0xBE, 0xEF, 0xFE, 0xED };
char serverName[] = "www.google.com";

EthernetClient client;

void setup()
{
  Serial.begin(9600);
  if (Ethernet.begin(mac) == 0) { // start ethernet using mac & IP address
    Serial.println("Failed to configure Ethernet using DHCP");
    while(true)   // no point in carrying on, so stay in endless loop:
      ;
  }
  delay(1000); // give the Ethernet shield a second to initialize

  int ret = client.connect(serverName, 80);
  if (ret == 1) {
    Serial.println("connected"); //  report successful connection
    // Make an HTTP request:
```

```
    client.println("GET /search?q=arduino HTTP/1.0");
    client.println();
  }
  else {
    Serial.println("connection failed, err: ");
    Serial.print(ret,DEC);
  }
}

void loop()
{
  // Read and print incoming butes from the server:
  if (client.available()) {
    char c = client.read();
    Serial.print(c);
  }

  // stop the client if disconnected:
  if (!client.connected()) {
    Serial.println();
    Serial.println("disconnecting.");
    client.stop();

    while(true) ;  // endless loop
  }
}
```

Discussion

This code is similar to the code in Recipe 15.2; it does a Google search for "arduino." But in this version it is not necessary to provide the Google IP address—it is obtained through a request to the Internet DNS service.

The request is achieved by passing the "www.google.com" hostname instead of an IP address to the `client.connect` method:

```
char serverName[] = "www.google.com";

int ret = client.connect(serverName, 80);
if(ret == 1) {
    Serial.println("connected"); //  report successful connection
```

The function will return 1 if the hostname can be resolved to an IP address by the DNS server and the client can connect successfully. Here are the values that can be returned from `client.connect`:

```
 1 = success
 0 = connection failed
-1 = no DNS server given
-2 = No DNS  records found
-3 = timeout
```

If the error is –1, you will need to manually configure the DNS server to use it. The DNS server address is usually provided by the DHCP server, but if you're configuring the shield manually you'll have to provide it (otherwise connect will return –1).

15.4 Requesting Data from a Web Server

Problem

You want Arduino to get data from a web server. For example, you want to find and use specific values returned from a web server.

Solution

This sketch uses Yahoo! search to convert 50 kilometers to miles. It sends the query "what+is+50+km+in+mi" and prints the result to the Serial Monitor:

```
/*
  Simple Client Parsing sketch
  Arduino 1.0 version
 */
#include <SPI.h>
#include <Ethernet.h>

byte mac[] = { 0xDE, 0xAD, 0xBE, 0xEF, 0xFE, 0xED };
char serverName[] = "search.yahoo.com";

EthernetClient client;

int result; // the result of the calculation

void setup()
{
  Serial.begin(9600);
  if(Ethernet.begin(mac) == 0) { // start ethernet using mac & IP address
    Serial.println("Failed to configure Ethernet using DHCP");
    while(true)   // no point in carrying on, so stay in endless loop:
      ;
  }
  delay(1000); // give the Ethernet shield a second to initialize

  Serial.println("connecting...");
}

void loop()
{
  if (client.connect(serverName, 80)>0) {
    Serial.print("connected... ");
    client.println("GET /search?p=50+km+in+miles HTTP/1.0");
    client.println();
  } else {
    Serial.println("connection failed");
  }
```

```
    if (client.connected()) {
      if(client.find("<b>50 Kilometers")){
        if(client.find("=")){
           result = client.parseInt();
           Serial.print("50 km is " );
           Serial.print(result);
           Serial.println(" miles");
        }
      }
      else
         Serial.println("result not found");
      client.stop();
      delay(10000); // check again in 10 seconds
    }
    else {
      Serial.println();
      Serial.println("not connected");
      client.stop();
      delay(1000);
    }
  }
```

Discussion

The sketch assumes the results will be returned in bold (using the HTML `<b>` tag) followed by the value given in the query and the word *kilometers*.

The searching is done using the Stream parsing functionality described in this chapter's introduction. The `find` method searches through the received data and returns `true` if the target string is found. The code looks for text associated with the reply. In this example, it tries to find "<b>50 kilometers" using this line:

```
if (client.find("<b>50 kilometers")){
```

`client.find` is used again to find the equals sign (=) that precedes the numerical value of the result.

The result is obtained using the `parseInt` method and is printed to the Serial Monitor.

`parseInt` returns an integer value; if you want to get a floating-point value, use `parseFloat` instead:

```
float floatResult = client.parseInt();
    Serial.println(floatResult);
```

If you want your searches to be robust, you need to look for a unique tag that will only be found preceding the data you want. This is easier to achieve on pages that use unique tags for each field, such as this example that gets the Google stock price from Google Finance and writes the value to analog output pin 3 (see Chapter 7) and to the Serial Monitor:

```
/*
 * Web Client Google Finance sketch
 * get the stock value for google and write to analog pin 3.
 */

#include <SPI.h>       // needed for Arduino versions later than 0018
#include <Ethernet.h>

byte mac[] = { 0xDE, 0xAD, 0xBE, 0xEF, 0xFE, 0xED };
char serverName[] = "www.google.com";

EthernetClient client;
float value;

void setup()
{
  Serial.begin(9600);
  if(Ethernet.begin(mac) == 0) { // start ethernet using mac & IP address
    Serial.println("Failed to configure Ethernet using DHCP");
    while(true)   // no point in carrying on, so stay in endless loop:
      ;
  }
  delay(1000); // give the Ethernet shield a second to initialize
}

void loop()
{
  Serial.print("Connecting...");
  if (client.connect(serverName, 80)>0) {
    client.println("GET //finance?q=google HTTP/1.0");
    client.println("User-Agent: Arduino 1.0");
    client.println();
  }
  else
  {
    Serial.println("connection failed");
  }
  if (client.connected()) {
     if(client.find("<span class=\"pr\">"))
     {
       client.find(">");  // seek past the next '>'
       value = client.parseFloat();
       Serial.print("google stock is at ");
       Serial.println(value);  // value is printed
     }
    else
      Serial.print("Could not find field");
  }
  else {
    Serial.println("Disconnected");
  }
  client.stop();
  client.flush();
  delay(5000); // 5 seconds between each connect attempt
}
```

These examples use the GET command to request a specific page. Some web requests need data to be sent to the server within the body of the message, because there is more data to be sent than can be handled by the GET command. These requests are handled using the POST command. Here is an example of POST that uses the Babel Fish language translation service to translate text from Italian into English:

```
/*
 * Web Client Babel Fish sketch
 * Uses Post to get data from a web server
 */

#include <SPI.h>
#include <Ethernet.h>

byte mac[] = { 0xDE, 0xAD, 0xBE, 0xEF, 0xFE, 0xED };
char serverName[] = "babelfish.yahoo.com";

EthernetClient client;

// the text to translate
char * transText = "trtext=Ciao+mondo+da+Arduino.&lp=it_en";

const int MY_BUFFER_SIZE = 30;  // big enough to hold result
char buffer [MY_BUFFER_SIZE+1]; // allow for the terminating null

void setup()
{
  Serial.begin(9600);
  if(Ethernet.begin(mac) == 0) { // start ethernet using mac & IP address
    Serial.println("Failed to configure Ethernet using DHCP");
    while(true)   // no point in carrying on, so stay in endless loop:
      ;
  }
  delay(1000); // give the Ethernet shield a second to initialize
}

void loop()
{
  Serial.print("Connecting...");
  postPage( "/translate_txt", transText);
  delay(5000);
}

void postPage(char *webPage, char *parameter){
  if (client.connect(serverName,80)>0) {
    client.print("POST ");
    client.print(webPage);
    client.println("  HTTP/1.0");
    client.println("Content-Type: application/x-www-form-urlencoded");
    client.println("Host: babelfish.yahoo.com");
    client.print("Content-Length: ");
    client.println(strlen(parameter));
    client.println();
    client.println(parameter);
```

```
  }
  else {
    Serial.println(" connection failed");
  }
  if (client.connected()) {
     client.find("<div id=\"result\">");
     client.find( ">");
     memset(buffer,0, sizeof(buffer)); // clear the buffer
     client.readBytesUntil('<' ,buffer, MY_BUFFER_SIZE);
     Serial.println(buffer);
  }
  else {
    Serial.println("Disconnected");
  }
  client.stop();
  client.flush();
}
```

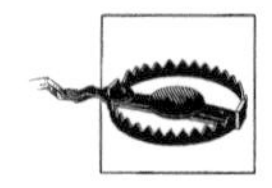

POST requires the content length to be sent to tell the server how much data to expect. Omitting or sending an incorrect value is a common cause of problems when using POST. See Recipe 15.12 for another example of a POST request.

Sites such as Google Weather and Google Finance generally keep the tags used to identify fields unchanged. But if some future update to a site does change the tags you are searching for, your sketch will not function correctly until you correct the search code. A more reliable way to extract data from a web server is to use a formal protocol, such as XML or JSON. The next recipe shows how to extract information from a site that uses XML.

15.5 Requesting Data from a Web Server Using XML

Problem

You want to retrieve data from a site that publishes information in XML format. For example, you want to use values from specific fields in one of the Google API services.

Solution

This sketch retrieves the weather in London from the Google Weather site. It uses the Google XML API:

```
/*
 * Simple Client Google Weather
 * gets xml data from http://www.google.com/ig/api?weather=london,uk
 * reads temperature from field:  <temp_f data="66" />
 * writes temperature  to analog output port.
 */

#include <SPI.h>         // needed for Arduino versions later than 0018
```

```
#include <Ethernet.h>

byte mac[] = { 0xDE, 0xAD, 0xBE, 0xEF, 0xFE, 0xED };
char serverName[] = "www.google.com";

const int temperatureOutPin = 3; // analog output for temperature
const int humidityOutPin    = 5; // analog output for humidity

EthernetClient client;

void setup()
{
  Serial.begin(9600);
  if(Ethernet.begin(mac) == 0) { // start ethernet using mac & IP address
    Serial.println("Failed to configure Ethernet using DHCP");
    while(true)   // no point in carrying on, so stay in endless loop:
      ;
  }
  delay(1000); // give the Ethernet shield a second to initialize

  Serial.println("connecting...");
}

void loop()
{
  if (client.connect(serverName,80)>0) {
  // get google weather for London
    client.println("GET /ig/api?weather=london HTTP/1.0");
    client.println();
  }
  else {
    Serial.println(" connection failed");
  }
  if (client.connected()) {
     // get temperature in fahrenheit (use field "<temp_c data=" for Celsius)
     if(client.find("<temp_f data=") )
     {
        int temperature = client.parseInt();
        analogWrite(temperatureOutPin, temperature); // write analog output
        Serial.print("Temperature is ");  // and echo it to the serial port
        Serial.println(temperature);
     }
    else
      Serial.print("Could not find temperature field");
     // get temperature in fahrenheit (use field "<temp_c data=" for Celsius)
     if(client.find("<humidity data=") )
     {
        int humidity = client.parseInt();
        analogWrite(humidityOutPin, humidity); // write value to analog port
        Serial.print("Humidity is ");  // and echo it to the serial port
        Serial.println(humidity);
     }
     else
       Serial.print("Could not find humidity field");
  }
```

```
  else {
    Serial.println("Disconnected");
  }
  client.stop();
  client.flush();
  delay(60000); // wait a minute before next update
}
```

Each field is preceded by a tag, and the one indicating the temperature in Fahrenheit on Google Weather is `"<temp_f data="`.

On this site, if you want the temperature in Celsius you would look for the tag `"<temp_c data="`.

You will need to consult the documentation for the page you are interested in to find the relevant tag for the data you want.

You select the page through the information sent in your `GET` statement. This also depends on the particular site, but in the preceding example, the city is selected by the text after the equals sign following the `GET` statement. For example, to change the location from London to Rome, change:

```
client.println("GET /ig/api?weather=london HTTP/1.0");  // weather for London
```

to:

```
client.println("GET /ig/api?weather=Rome HTTP/1.0"); // weather for Rome
```

You can use a variable if you want the location to be selected under program control:

```
char *cityString[4] = { "London", "New%20York", "Rome", "Tokyo"};
int   city;

void loop()
{
  city = random(4); // get a random city
  if (client.connect(serverName,80)>0) {
    Serial.print("Getting weather for ");
    Serial.println(cityString[city]);

    client.print("GET /ig/api?weather=");
    client.print(cityString[city]); // print one of 4 random cities
    client.println(" HTTP/1.0");
    client.println();
  }
  else {
    Serial.println(" connection failed");
  }
  if (client.connected()) {
     // get temperature in fahrenheit (use field "<temp_c data=\"" for Celsius)
     if(client.find("<temp_f data=") )
     {
        int temperature = client.parseInt();
        analogWrite(temperatureOutPin, temperature); // write analog output
        Serial.print(cityString[city]);
        Serial.print(" temperature is ");  // and echo it to the serial port
```

```
      Serial.println(temperature);
    }
   else
     Serial.println("Could not find temperature field");
    // get temperature in fahrenheit (use field "<temp_c data=\"" for Celsius)
    if(client.find("<humidity data=") )
    {
      int humidity = client.parseInt();
      analogWrite(humidityOutPin, humidity);    // write value to analog port
      Serial.print("Humidity is ");  // and echo it to the serial port
      Serial.println(humidity);
    }
    else
      Serial.println("Could not find humidity field");
  }
  else {
    Serial.println("Disconnected");
  }
  client.stop();
  client.flush();
  delay(60000); // wait a minute before next update
}

// the remainder of the code is the same as the previous sketch
```

Information sent in URLs cannot contain spaces, which is why New York is written as "New%20York". The encoding to indicate a space is %20. Your browser does the encoding before it sends a request, but on Arduino you must translate spaces to %20 yourself.

15.6 Setting Up an Arduino to Be a Web Server

Problem

You want Arduino to serve web pages. For example, you want to use your web browser to view the values of sensors connected to Arduino analog pins.

Solution

This is the standard Arduino Web Server example sketch distributed with Arduino that shows the value of the analog input pins. This recipe explains how this sketch works and how it can be extended:

```
/*
 * Web Server
 * A simple web server that shows the value of the analog input pins.
 */

#include <SPI.h>
#include <Ethernet.h>
```

```
byte mac[] = { 0xDE, 0xAD, 0xBE, 0xEF, 0xFE, 0xED };
byte ip[] = { 192, 168, 1, 177};  // IP address of this web server

EthernetServer server(80);

void setup()
{
  Ethernet.begin(mac, ip);
  server.begin();
}

void loop()
{
  EthernetClient client = server.available();
  if (client) {
    // an http request ends with a blank line
    boolean current_line_is_blank = true;
    while (client.connected()) {
      if (client.available()) {
        char c = client.read();
        // if we've gotten to the end of the line (received a newline
        // character) and the line is blank, the http request has ended,
        // so we can send a reply
        if (c == '\n' && current_line_is_blank) {
          // send a standard http response header
          client.println("HTTP/1.1 200 OK");
          client.println("Content-Type: text/html");
          client.println();

          // output the value of each analog input pin
          for (int i = 0; i < 6; i++) {
            client.print("analog input ");
            client.print(i);
            client.print(" is ");
            client.print(analogRead(i));
            client.println("<br />");
          }
          break;
        }
        if (c == '\n') {
          // we're starting a new line
          current_line_is_blank = true;
        } else if (c != '\r') {
          // we've gotten a character on the current line
          current_line_is_blank = false;
        }
      }
    }
    // give the web browser time to receive the data
    delay(1);
    client.stop();
  }
}
```

Discussion

As discussed in Recipe 15.1, all of the sketches using the Ethernet library need a unique MAC address and IP address. The IP address you assign in this sketch determines the address of the web server. In this example, typing 192.168.1.177 into your browser's address bar should display a page showing the values on analog input pins 0 through 6 (see Chapter 5 for more on the analog ports).

As described in this chapter's introduction, 192.168.1.177 is a local address that is only visible on your local network. If you want to expose your web server to the entire Internet, you will need to configure your router to forward incoming messages to Arduino. The technique is called *port forwarding* and you will need to consult the documentation for your router to see how to set this up. (For more on port forwarding in general, see *SSH, The Secure Shell: The Definitive Guide* by Daniel J. Barrett, Richard E. Silverman, and Robert G. Byrnes; search for it on oreilly.com.)

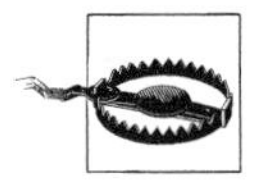

Configuring your Arduino Ethernet board to be visible on the Internet makes the board accessible to anyone with the IP address. The Arduino Ethernet library does not offer secure connections, so be careful about the information you expose.

The two lines in `setup` initialize the Ethernet library and configure your web server to the IP address you provide. The `loop` waits for and then processes each request received by the web server:

```
EthernetClient client = server.available();
```

The `client` object here is actually the web server—it processes messages for the IP address you gave the server.

`if (client)` tests that the client has been successfully started.

`while (client.connected())` tests if the web server is connected to a client making a request.

`client.available()` and `client.read()` check if data is available, and read a byte if it is. This is similar to `Serial.available()`, discussed in Chapter 4, except the data is coming from the Internet rather than the serial port. The code reads data until it finds the first line with no data, signifying the end of a request. An HTTP header is sent using the `client.println` commands followed by the printing of the values of the analog ports.

15.7 Handling Incoming Web Requests

Problem

You want to control digital and analog outputs with Arduino acting as a web server. For example, you want to control the values of specific pins through parameters sent from your web browser.

Solution

This sketch reads requests sent from a browser and changes the values of digital and analog output ports as requested.

The URL (text received from a browser request) contains one or more fields starting with the word *pin* followed by a *D* for digital or *A* for analog and the pin number. The value for the pin follows an equals sign.

For example, sending *http://192.168.1.177/?pinD2=1* from your browser's address bar turns digital pin 2 on; *http://192.168.1.177/?pinD2=0* turns pin 2 off. (See Chapter 7 if you need information on connecting LEDs to Arduino pins.)

Figure 15-1 shows what you will see on your web browser when connected to the web server code that follows.

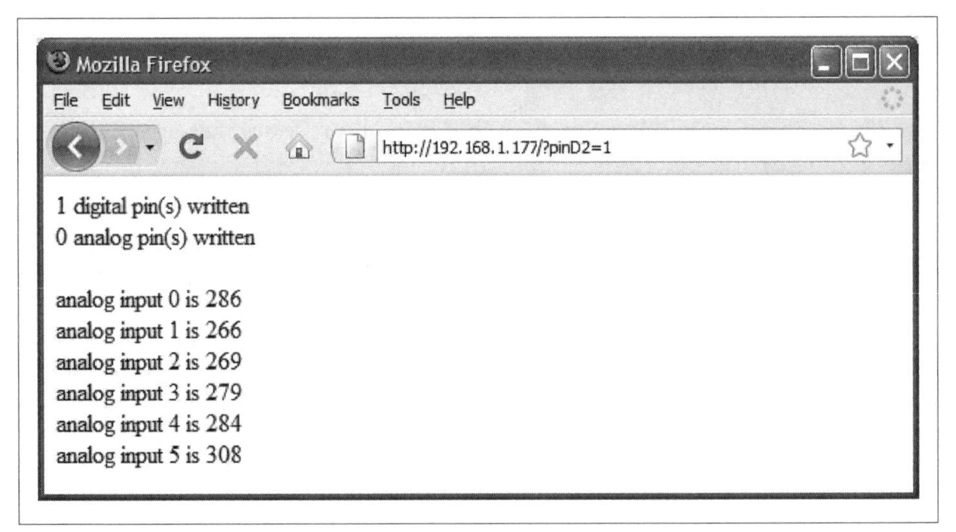

Figure 15-1. Browser page displaying output created by this recipe's Solution

```
/*
 * WebServerParsing
 * Respond to requests in the URL to change digital and analog output ports
 * show the number of ports changed and the value of the analog input pins.
 * for example:
```

```
 *   sending http://192.168.1.177/?pinD2=1 turns digital pin 2 on
 *   sending http://192.168.1.177/?pinD2=0 turns pin 2 off.
 * This sketch demonstrates text parsing using the 1.0 Stream class.
 */

#include <SPI.h>
#include <Ethernet.h>

byte mac[] = { 0xDE, 0xAD, 0xBE, 0xEF, 0xFE, 0xED };
byte ip[] = { 192,168,1,177 };

EthernetServer server(80);

void setup()
{
  Serial.begin(9600);
  Ethernet.begin(mac, ip);
  server.begin();
  Serial.println("ready");
}

void loop()
{
  EthernetClient client = server.available();
  if (client) {
    while (client.connected()) {
      if (client.available()) {
        // counters to show the number of pin change requests
        int digitalRequests = 0;
        int analogRequests = 0;
        if( client.find("GET /") ) {  // search for 'GET'
          // find tokens starting with "pin" and stop on the first blank line
          // search to the end of line for 'pin'
          while(client.findUntil("pin", "\n\r")){
            char type = client.read(); // D or A
            // the next ascii integer value in the stream is the pin
            int pin = client.parseInt();
            int val = client.parseInt(); // the integer after that is the value
            if( type == 'D') {
              Serial.print("Digital pin ");
              pinMode(pin, OUTPUT);
              digitalWrite(pin, val);
              digitalRequests++;
            }
            else if( type == 'A'){
              Serial.print("Analog pin ");
              analogWrite(pin, val);
              analogRequests++;
            }
            else {
              Serial.print("Unexpected type ");
              Serial.print(type);
            }
            Serial.print(pin);
            Serial.print("=");
```

```
        Serial.println(val);
      }
    }
    Serial.println();

    // the findUntil has detected the blank line (a lf followed by cr)
    // so the http request has ended and we can send a reply
    // send a standard http response header
    client.println("HTTP/1.1 200 OK");
    client.println("Content-Type: text/html");
    client.println();

    // output the number of pins handled by the request
    client.print(digitalRequests);
    client.print(" digital pin(s) written");
    client.println("<br />");
    client.print(analogRequests);
    client.print(" analog pin(s) written");
    client.println("<br />");
    client.println("<br />");


    // output the value of each analog input pin
    for (int i = 0; i < 6; i++) {
      client.print("analog input ");
      client.print(i);
      client.print(" is ");
      client.print(analogRead(i));
      client.println("<br />");
    }
    break;
  }
}
// give the web browser time to receive the data
delay(1);
client.stop();
}
}
```

Discussion

This is what was sent: *http://192.168.1.177/?pinD2=1*. Here is how the information is broken down: Everything before the question mark is treated as the address of the web server (192.168.1.177 in this example; this address is the IP address set at the top of the sketch for the Arduino board). The remaining data is a list of fields, each beginning with the word *pin* followed by a *D* indicating a digital pin or *A* indicating an analog pin. The numeric value following the *D* or *A* is the pin number. This is followed by an equals sign and finally the value you want to set the pin to. `pinD2=1` sets digital pin 2 `HIGH`. There is one field per pin, and subsequent fields are separated by an ampersand. You can have as many fields as there are Arduino pins you want to change.

The request can be extended to handle multiple parameters by using ampersands to separate multiple fields. For example:

http://192.168.1.177/?pinD2=1&pinD3=0&pinA9=128&pinA11=255

Each field within the ampersand is handled as described earlier. You can have as many fields as there are Arduino pins you want to change.

15.8 Handling Incoming Requests for Specific Pages

Problem

You want to have more than one page on your web server; for example, to show the status of different sensors on different pages.

Solution

This sketch looks for requests for pages named "analog" or "digital" and displays the pin values accordingly:

```
/*
 * WebServerMultiPage
 * Respond to requests in the URL to view digital and analog output ports
 * http://192.168.1.177/analog/   displays analog pin data
 * http://192.168.1.177/digital/  displays digital pin data
 */

#include <SPI.h>
#include <Ethernet.h>

byte mac[] = { 0xDE, 0xAD, 0xBE, 0xEF, 0xFE, 0xED };
byte ip[] =  { 192,168,1,177 };

const int MAX_PAGE_NAME_LEN = 8;  // max characters in a page name
char buffer[MAX_PAGE_NAME_LEN+1];  // page name + terminating null

EthernetServer server(80);
EthernetClient client;

void setup()
{
  Serial.begin(9600);
  Ethernet.begin(mac, ip);
  server.begin();
  Serial.println("Ready");
}

void loop()
{
  client = server.available();
  if (client) {
    while (client.connected()) {
      if (client.available()) {
        if( client.find("GET ") ) {
          // look for the page name
```

```
          memset(buffer,0, sizeof(buffer)); // clear the buffer
          if(client.find( "/"))
            if(client.readBytesUntil('/', buffer, MAX_PAGE_NAME_LEN ))
            {
              if(strcmp(buffer, "analog") == 0)
                showAnalog();
              else if(strcmp(buffer, "digital") == 0)
                showDigital();
              else
                unknownPage(buffer);
            }
        }
        Serial.println();
        break;
      }
    }
    // give the web browser time to receive the data
    delay(1);
    client.stop();
  }
}

void showAnalog()
{
  Serial.println("analog");
  sendHeader();
  client.println("<h1>Analog Pins</h1>");
  // output the value of each analog input pin

    for (int i = 0; i < 6; i++) {
    client.print("analog pin ");
    client.print(i);
    client.print(" = ");
    client.print(analogRead(i));
    client.println("<br />");
  }
}

void showDigital()
{
  Serial.println("digital");
  sendHeader();
  client.println("<h1>Digital Pins</h1>");
  // show the value of digital pins
  for (int i = 2; i < 8; i++) {
    pinMode(i, INPUT);
    client.print("digital pin ");
    client.print(i);
    client.print(" is ");
    if(digitalRead(i) == LOW)
      client.print("LOW");
    else
      client.print("HIGH");
    client.println("<br />");
  }
```

```
  client.println("</body></html>");
}

void unknownPage(char *page)
{
  sendHeader();
  client.println("<h1>Unknown Page</h1>");
  client.print(page);
  client.println("<br />");
  client.println("Recognized pages are:<br />");
  client.println("/analog/<br />");
  client.println("/digital/<br />");
  client.println("</body></html>");
}

void sendHeader()
{
  // send a standard http response header
  client.println("HTTP/1.1 200 OK");
  client.println("Content-Type: text/html");
  client.println();
  client.println("<html><head><title>Web server multi-page Example</title>");
  client.println("<body>");
}
```

Discussion

You can test this from your web browser by typing *http://192.168.1.177/analog/* or *http://192.168.1.177/digital/* (if you are using a different IP address for your web server, change the URL to match).

Figure 15-2 shows the expected output.

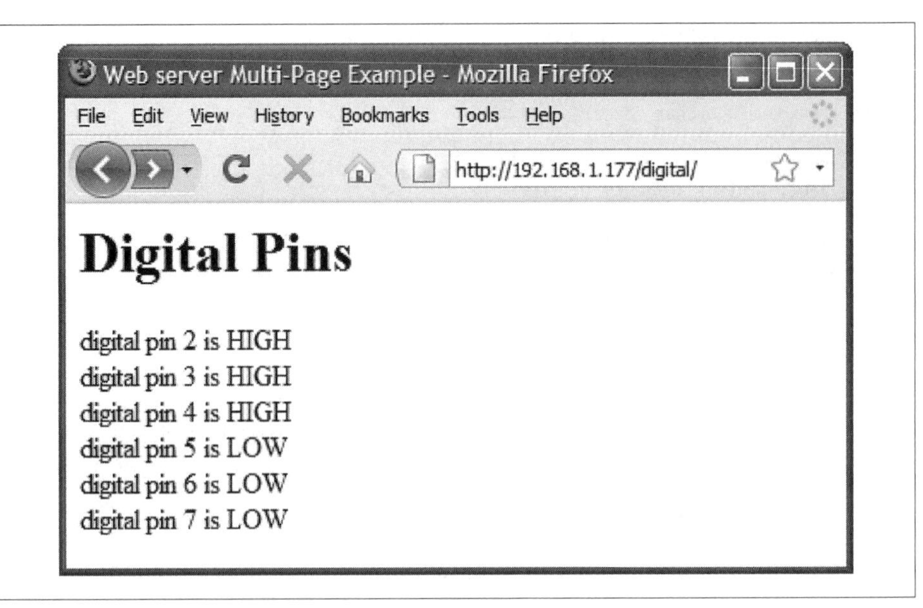

Figure 15-2. Browser output showing digital pin values

The sketch looks for the "/" character to determine the end of the page name. The server will report an unknown page if the "/" character does not terminate the page name.

You can easily enhance this with some code from Recipe 15.7 to allow control of Arduino pins from another page named `update`. Here is the new `loop` code:

```
void loop()
{
  client = server.available();
  if (client) {
    while (client.connected()) {
      if (client.available()) {
        if( client.find("GET ") ) {
          // look for the page name
          memset(buffer,0, sizeof(buffer)); // clear the buffer
          if(client.find( "/"))
            if(client.readBytesUntil('/', buffer, MAX_PAGE_NAME_LEN ))
            {
              if(strcmp(buffer, "analog") == 0)
                showAnalog();
              else if(strcmp(buffer, "digital") == 0)
                showDigital();
              // add this code for new page named: update
              else if(strcmp(buffer, "update") == 0)
                doUpdate();
              else
                unknownPage(buffer);
            }
        }
```

```
        Serial.println();
        break;
      }
    }
    // give the web browser time to receive the data
    delay(1);
    client.stop();
  }
}
```

Here is the doUpdate function:

```
void doUpdate()
{
  Serial.println("update");
  sendHeader();
  // find tokens starting with "pin" and stop on the first blank line
  while(client.findUntil("pin", "\n\r")){
    char type = client.read(); // D or A
    int pin = client.parseInt();
    int val = client.parseInt();
    if( type == 'D') {
      Serial.print("Digital pin ");
      pinMode(pin, OUTPUT);
      digitalWrite(pin, val);

    }
    else if( type == 'A'){
      Serial.print("Analog pin ");
      analogWrite(pin, val);

    }
    else {
      Serial.print("Unexpected type ");
      Serial.print(type);
    }
    Serial.print(pin);
    Serial.print("=");
    Serial.println(val);
  }
}
```

Sending *http://192.168.1.177/update/?pinA5=128* from your browser's address bar writes the value `128` to analog output pin 5.

15.9 Using HTML to Format Web Server Responses

Problem

You want to use HTML elements such as tables and images to improve the look of web pages served by Arduino. For example, you want the output from Recipe 15.8 to be rendered in an HTML table.

Solution

Figure 15-3 shows how the web server in this recipe's Solution formats the browser page to display pin values. (You can compare this to the unformatted values shown in Figure 15-2.)

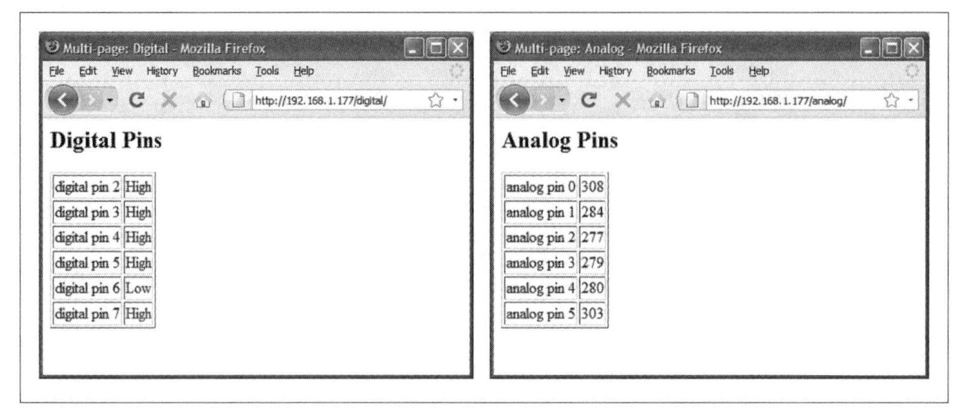

Figure 15-3. Browser pages using HTML formatting

This sketch shows the functionality from Recipe 15.8 with output formatted using HTML:

```
/*
 * WebServerMultiPageHTML
 * Arduino 1.0 version
 * Display analog and digital pin values using HTML formatting
 */

#include <SPI.h>         // needed for Arduino versions later than 0018
#include <Ethernet.h>

byte mac[] = { 0xDE, 0xAD, 0xBE, 0xEF, 0xFE, 0xED };
byte ip[] =  { 192,168,1,177 };

// Buffer must be big enough to hold requested page names and terminating null
const int MAX_PAGE_NAME_LEN = 8+1;  // max characters in a page name + null
char buffer[MAX_PAGE_NAME_LEN];

EthernetServer server(80);
EthernetClient client;

void setup()
{
  Serial.begin(9600);

  Ethernet.begin(mac, ip);
  server.begin();
  pinMode(13,OUTPUT);
  for(int i=0; i < 3; i++)
```

```
  {
     digitalWrite(13,HIGH);
     delay(500);
     digitalWrite(13,LOW);
     delay(500);
  }
}

void loop()
{
  client = server.available();
  if (client) {
    while (client.connected()) {
      if (client.available()) {
        if( client.find("GET ") ) {
          // look for the page name
          memset(buffer,0, sizeof(buffer)); // clear the buffer
          if(client.find( "/"))
            if(client.readBytesUntil('/', buffer, MAX_PAGE_NAME_LEN ))
            {
              if(strcasecmp(buffer, "analog") == 0)
                showAnalog();
              else if(strcasecmp(buffer, "digital") == 0)
                showDigital();
              else
                unknownPage(buffer);
            }
        }
        break;
      }
    }
    // give the web browser time to receive the data
    delay(1);
    client.stop();
  }
}

void showAnalog()
{
  sendHeader("Multi-page: Analog");
  client.println("<h2>Analog Pins</h2>");
  client.println("<table border='1' >");
  for (int i = 0; i < 6; i++) {
    // output the value of each analog input pin
    client.print("<tr><td>analog pin ");
    client.print(i);
    client.print(" </td><td>");
    client.print(analogRead(i));
    client.println("</td></tr>");
  }
  client.println("</table>");
  client.println("</body></html>");
}

void showDigital()
```

```
{
  sendHeader("Multi-page: Digital");
  client.println("<h2>Digital Pins</h2>");
  client.println("<table border='1'>");
  for (int i = 2; i < 8; i++) {
    // show the value of digital pins
    pinMode(i, INPUT);
    digitalWrite(i, HIGH); // turn on pull-ups
    client.print("<tr><td>digital pin ");
    client.print(i);
    client.print(" </td><td>");
    if(digitalRead(i) == LOW)
      client.print("Low");
    else
      client.print("High");
    client.println("</td></tr>");
  }
  client.println("</table>");
  client.println("</body></html>");
}

void unknownPage(char *page)
{
  sendHeader("Unknown Page");
  client.println("<h1>Unknown Page</h1>");
  client.print(page);
  client.println("<br />");
  client.println("Recognized pages are:<br />");
  client.println("/analog/<br />");
  client.println("/digital/<br />");
  client.println("</body></html>");
}

void sendHeader(char *title)
{
  // send a standard http response header
  client.println("HTTP/1.1 200 OK");
  client.println("Content-Type: text/html");
  client.println();
  client.print("<html><head><title>");
  client.println(title);
  client.println("</title><body>");
}
```

Discussion

The same information is provided as in Recipe 15.8, but here the data is formatted using an HTML table. The following code indicates that the web browser should create a table with a border width of 1:

```
client.println("<table border='1' >");
```

The `for` loop defines the table data cells with the `<td>` tag and the row entries with the `<tr>` tag. The following code places the string `"analog pin"` in a cell starting on a new row:

```
client.print("<tr><td>analog pin ");
```

This is followed by the value of the variable `i`:

```
client.print(i);
```

The next line contains the tag that closes the cell and begins a new cell:

```
client.print(" </td><td>");
```

This writes the value returned from `analogRead` into the cell:

```
client.print(analogRead(i));
```

The tags to end the cell and end the row are written as follows:

```
client.println("</td></tr>");
```

The `for` loop repeats this until all six analog values are written. Any of the books mentioned in "Series 1 configuration" on page 469 or one of the many HTML reference sites can provide more details on HTML tags.

See Also

Learning Web Design by Jennifer Niederst Robbins (O'Reilly)

Web Design in a Nutshell by Jennifer Niederst Robbins (O'Reilly)

HTML & XHTML: The Definitive Guide by Chuck Musciano and Bill Kennedy (O'Reilly)

(Search for O'Reilly titles on oreilly.com.)

15.10 Serving Web Pages Using Forms (POST)

Problem

You want to create web pages with forms that allow users to select an action to be performed by Arduino. Figure 15-4 shows the web page created by this recipe's Solution.

Figure 15-4. Web form with buttons

Solution

This sketch creates a web page that has a form with buttons. Users navigating to this page will see the buttons in the web browser and the Arduino web server will respond to the button clicks. In this example, the sketch turns a pin on and off depending on which button is pressed:

```
/*
 * WebServerPost sketch
 * Turns pin 8 on and off using HTML form
 */

#include <SPI.h>
#include <Ethernet.h>

byte mac[] = { 0xDE, 0xAD, 0xBE, 0xEF, 0xFE, 0xED };
byte ip[] =  { 192,168,1,177 };

const int MAX_PAGENAME_LEN = 8; // max characters in page name
char buffer[MAX_PAGENAME_LEN+1]; // additional character for terminating null

EthernetServer server(80);

void setup()
{
  Serial.begin(9600);
  Ethernet.begin(mac, ip);
  server.begin();
  delay(2000);
}

void loop()
{
  EthernetClient client = server.available();
  if (client) {
    int type = 0;
    while (client.connected()) {
          if (client.available()) {
```

```
        // GET, POST, or HEAD
        memset(buffer,0, sizeof(buffer)); // clear the buffer
        if(client.find("/"))
          if(client.readBytesUntil('/', buffer,sizeof(buffer))){
            Serial.println(buffer);
            if(strcmp(buffer,"POST ") == 0){
              client.find("\n\r"); // skip to the body
              // find string starting with "pin", stop on first blank line
              // the POST parameters expected in the form pinDx=Y
              // where x is the pin number and Y is 0 for LOW and 1 for HIGH
              while(client.findUntil("pinD", "\n\r")){
                int pin = client.parseInt();       // the pin number
                int val = client.parseInt();       // 0 or 1
                pinMode(pin, OUTPUT);
                digitalWrite(pin, val);
              }
            }
            sendHeader(client,"Post example");
            //create HTML button to control pin 8
            client.println("<h2>Click buttons to turn pin 8 on or off</h2>");
            client.print(
            "<form action='/' method='POST'><p><input type='hidden' name='pinD8'");
            client.println(" value='0'><input type='submit' value='Off'/></form>");
            //create HTML button to turn on pin 8
            client.print(
            "<form action='/' method='POST'><p><input type='hidden' name='pinD8'");
            client.print(" value='1'><input type='submit' value='On'/></form>");
            client.println("</body></html>");
            client.stop();
          }
        break;
      }
    }
    // give the web browser time to receive the data
    delay(1);
    client.stop();
  }
}
void sendHeader(EthernetClient client, char *title)
{
  // send a standard http response header
  client.println("HTTP/1.1 200 OK");
  client.println("Content-Type: text/html");
  client.println();
  client.print("<html><head><title>");
  client.print(title);
  client.println("</title><body>");
}
```

Discussion

A web page with a user interface form consists of HTML tags that identify the controls (buttons, checkboxes, labels, etc.) that comprise the user interface. This recipe uses buttons for user interaction.

These lines create a form with a button named `pinD8` that is labeled "OFF," which will send back a value of `0` (zero) when clicked:

```
client.print("<form action='/' method='POST'><p><input type='hidden' name='pinD8'");
client.println(" value='0'><input type='submit' value='Off'/></form>");
```

When the server receives a request from a browser, it looks for the `"POST "` string to identify the start of the posted form:

```
if (strcmp(buffer,"POST ") == 0) // find the start of the posted form

   client.find("\n\r"); // skip to the body
   // find parameters starting with "pin" and stop on the first blank line
   // the POST parameters expected in the form pinDx=Y
   // where x is the pin number and Y is 0 for LOW and 1 for HIGH
```

If the OFF button was pressed, the received page will contain the string `pinD8=0`, or `pinD8=1` for the ON button.

The sketch searches until it finds the button name (`pinD`):

```
while(client.findUntil("pinD", "\n\r"))
```

The `findUntil` method in the preceding code will search for "pinD" and stop searching at the end of a line (`\n\r` is the newline carriage return sent by the web browser at the end of a form).

The number following the name `pinD` is the pin number:

```
int pin = client.parseInt();        // the pin number
```

And the value following the pin number will be `0` if button OFF was pressed or `1` if button ON was pressed:

```
int val = client.parseInt();        // 0 or 1
```

The value received is written to the pin after setting the pin mode to output:

```
pinMode(pin, OUTPUT);
    digitalWrite(pin, val);
```

More buttons can be added by inserting tags for the additional controls. The following lines add another button to turn on digital pin 9:

```
//create HTML button to turn on pin 9
client.print("<form action='/' method='POST'><p><input type='hidden' name='pinD9'");
client.print(" value='1'><input type='submit' value='On'/></form>");
```

15.11 Serving Web Pages Containing Large Amounts of Data

Problem

Your web pages require more memory than you have available, so you want to use program memory (also known as *progmem* or *flash memory*) to store data (see Recipe 17.4).

Solution

This sketch combines the `POST` code from Recipe 15.10 with the HTML code from Recipe 15.9 and adds new code to access text stored in progmem. As in Recipe 15.9, the server can display analog and digital pin status and turn digital pins on and off (see Figure 15-5).

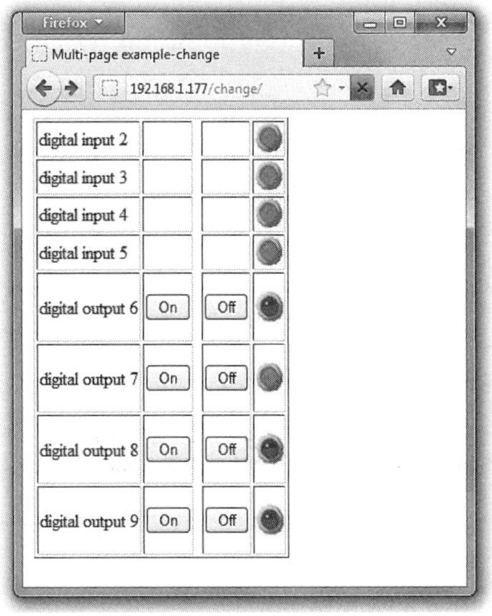

Figure 15-5. Web page with LED images

```
/*
 * WebServerMultiPageHTMLProgmem sketch
 *
 * Respond to requests in the URL to change digital and analog output ports
 * show the number of ports changed and the value of the analog input pins.
 *
 * http://192.168.1.177/analog/   displays analog pin data
 * http://192.168.1.177/digital/  displays digital pin data
 * http://192.168.1.177/change/  allows changing digital pin data
 *
 */

#include <SPI.h>         // needed for Arduino versions later than 0018
#include <Ethernet.h>

#include <avr/pgmspace.h> // for progmem
#define P(name)   static const prog_uchar name[] PROGMEM  // declare a static string

byte mac[] = { 0xDE, 0xAD, 0xBE, 0xEF, 0xFE, 0xED };
byte ip[] =  { 192,168,1,177 };
```

```
const int MAX_PAGENAME_LEN = 8;  // max characters in page name
char buffer[MAX_PAGENAME_LEN+1]; // additional character for terminating null

EthernetServer server(80);
EthernetClient client;

void setup()
{
  Serial.begin(9600);
  Ethernet.begin(mac, ip);
  server.begin();
  delay(1000);
  Serial.println(F("Ready"));
}

void loop()
{

  client = server.available();
  if (client) {
    int type = 0;
    while (client.connected()) {
      if (client.available()) {
        // GET, POST, or HEAD
        memset(buffer,0, sizeof(buffer)); // clear the buffer
        if(client.readBytesUntil('/', buffer,MAX_PAGENAME_LEN)){
          if(strcmp(buffer, "GET ") == 0 )
            type = 1;
          else if(strcmp(buffer,"POST ") == 0)
            type = 2;
          // look for the page name
          memset(buffer,0, sizeof(buffer)); // clear the buffer
          if(client.readBytesUntil( '/', buffer,MAX_PAGENAME_LEN ))
          {
            if(strcasecmp(buffer, "analog") == 0)
              showAnalog();
            else if(strcasecmp(buffer, "digital") == 0)
              showDigital();
            else if(strcmp(buffer, "change")== 0)
              showChange(type == 2);
            else
              unknownPage(buffer);
          }
        }
        break;
      }
    }
    // give the web browser time to receive the data
    delay(1);
    client.stop();
  }
}

void showAnalog()
{
```

```
  Serial.println(F("analog"));
  sendHeader("Multi-page example-Analog");
  client.println("<h1>Analog Pins</h1>");
  // output the value of each analog input pin

    client.println(F("<table border='1' >"));
  for (int i = 0; i < 6; i++) {
    client.print(F("<tr><td>analog pin "));
    client.print(i);
    client.print(F(" </td><td>"));
    client.print(analogRead(i));
    client.println(F("</td></tr>"));
  }
  client.println(F("</table>"));
  client.println(F("</body></html>"));
}

// mime encoded data for the led on and off images:
// see: http://www.motobit.com/util/base64-decoder-encoder.asp
P(led_on) =  "<img src=\"data:image/jpg;base64,"
"/9j/4AAQSkZJRgABAgAAZABkAAD/7AARRHVja3kAAQAEAAAAHgAA/+4ADkFkb2JlAGTAAAAAAf/b"
"AIQAEAsLCwwLEAwMEBcPDQ8XGxQQEBQbHxcXFxcXHx4XGhoaGhceHiMlJyUjHi8vMzMvL0BAQEBA"
"QEBAQEBAQEBAQAERDw8RExEVEhIVFBEUERQaFBYWFBomGhocGhomMCMeHh4eIzArLicnJy4rNTUw"
"MDU1QEA/QEBAQEBAQEBAQEBA/8AAEQgAGwAZAwEiAAIRAQMRAf/EAIIAAAICAwAAAAAAAAAAAAAA"
"AAUGAAcCAwQBAAMBAAAAAAAAAAAAAAAAAACBAUQAAECBAQBCgcAAAAAAAAAAAECAwARMRIhQQQF"
"UWFxkaHRMoITUwYiQnKSIxQ1EQAAAwYEBwAAAAAAAAAAAAAAARECEgMTBBQhQWEiMVGBMkJiJP/a"
"AAwDAQACEQMRAD8AcNz3BGibKieOnhCOv3A+teKJt8JmZEdHuZalOitgUoHnEpQEWtSyLqgACWFI"
"nixWiaQhsUFFBiQSbiMvvrmeCBp27eLnG7lFTDxs+Kra8oOyium3ltJUAcDIy4EUMN/7Dnq9cPMO"
"W9OE9kxeyF2d3HFOQ175olKudUm7TqlfKqDQEDOFR1sNqtC7k5ERYjndNPFSArtvnI/nV+ed9coI"
"ktd2BgozrSZO3J5jVEXRcwD2bbXNdqOzT+BohTyjgPp5SYdPJZ9NP2jsiIz7vhjLohtjnqJ/ouPK"
"co//2Q=="
"\"/>";

P(led_off) = "<img src=\"data:image/jpg;base64,"
"/9j/4AAQSkZJRgABAgAAZABkAAD/7AARRHVja3kAAQAEAAAAHgAA/+4ADkFkb2JlAGTAAAAAAf/b"
"AIQAEAsLCwwLEAwMEBcPDQ8XGxQQEBQbHxcXFxcXHx4XGhoaGhceHiMlJyUjHi8vMzMvL0BAQEBA"
"QEBAQEBAQEBAQAERDw8RExEVEhIVFBEUERQaFBYWFBomGhocGhomMCMeHh4eIzArLicnJy4rNTUw"
"MDU1QEA/QEBAQEBAQEBAQEBA/8AAEQgAHAAZAwEiAAIRAQMRAf/EAHgAAQEAAwAAAAAAAAAAAAAA"
"AAYFAgQHAQEBAQAAAAAAAAAAAAAAAAACAQQQAAECBQAHBQkAAAAAAAAAAAECAwAREhMEITFhoSIF"
"FUFROUIGgZHBMlIjM1MWEQABAwQDAQEAAAAAAAAAAAABABECIWESA1ETIyIE/9oADAMBAAIRAxEA"
"PwBvl5SWEkkylpJMGsj1XjXSE1kCQuJ8Iy9W5DoxradFa6VDf8IJZAQ6loNtBooTJaqp3DP5oBlV"
"nWrTpEouQS/Cf4POOuKbqWHGXTSlztSvuVFiZjmfLH3GUuMkzSoTMu8aiNsXet5/17hFyo6PR64V"
"ZnuqfqDDDySFpNpYH3E6aFjzGBr2DkMuFBSFDsWkilUdLftW13pWpcdWqnbBzI/l6hVXKZlROUSe"
"L1KX5zvAPXESjdHsTFWpxLKO0J54hIA1DZCj+Vx/3r96fCNrkvRaTO+V3zV/llplr9sVeHZui/ONk"
"H3dzt6cL/9k="
"\"/>";
;

void showDigital()
{
  Serial.println(F("digital"));
  sendHeader("Multi-page example-Digital");
  client.println(F("<h2>Digital Pins</h2>"));
  // show the value of digital pins
  client.println(F("<table border='1'>"));
```

```
  for (int i = 2; i < 8; i++) {
    pinMode(i, INPUT);
    digitalWrite(i, HIGH); // turn on pull-ups
    client.print(F("<tr><td>digital pin "));
    client.print(i);
    client.print(F(" </td><td>"));
    if(digitalRead(i) == LOW)
      printP(led_off);
    else
      printP(led_on);
    client.println(F("</td></tr>"));
  }
  client.println(F("</table>"));

  client.println(F("</body></html>"));
}


void showChange(boolean isPost)
{
  Serial.println(F("change"));
  if(isPost)
  {
    Serial.println("isPost");
    client.find("\n\r"); // skip to the body
    // find parameters starting with "pin" and stop on the first blank line
    Serial.println(F("searching for parms"));
    while(client.findUntil("pinD", "\n\r")){
      int pin = client.parseInt();       // the pin number
      int val = client.parseInt();       // 0 or 1
      Serial.print(pin);
      Serial.print("=");
      Serial.println(val);
      pinMode(pin, OUTPUT);
      digitalWrite(pin, val);
    }
  }
  sendHeader("Multi-page example-change");
  // table with buttons from 2 through 9
  // 2 to 5 are inputs, the other buttons are outputs

  client.println(F("<table border='1'>"));

  // show the input pins
  for (int i = 2; i < 6; i++) {  // pins 2-5 are inputs
    pinMode(i, INPUT);
    digitalWrite(i, HIGH); // turn on pull-ups
    client.print(F("<tr><td>digital input "));
    client.print(i);
    client.print(F(" </td><td>"));

    client.print(F("  </td><td>"));
    client.print(F(" </td><td>"));
    client.print(F("  </td><td>"));
```

```
    if(digitalRead(i) == LOW)
      //client.print("Low");
      printP(led_off);
    else
      //client.print("high");
      printP(led_on);
    client.println("</td></tr>");
  }

  // show output pins 6-9
  // note pins 10-13 are used by the ethernet shield
  for (int i = 6; i < 10; i++) {
    client.print(F("<tr><td>digital output "));
    client.print(i);
    client.print(F(" </td><td>"));
    htmlButton( "On", "pinD", i, "1");
    client.print(F(" </td><td>"));
    client.print(F(" </td><td>"));
    htmlButton("Off", "pinD", i, "0");
    client.print(F(" </td><td>"));

    if(digitalRead(i) == LOW)
      //client.print("Low");
      printP(led_off);
    else
      //client.print("high");
      printP(led_on);
    client.println(F("</td></tr>"));
  }
  client.println(F("</table>"));
}

// create an HTML button
void htmlButton( char * label, char *name, int nameId, char *value)
{
  client.print(F("<form action='/change/' method='POST'><p><input type='hidden' name='"));
  client.print(name);
  client.print(nameId);
  client.print(F("' value='"));
  client.print(value);
  client.print(F("'><input type='submit' value='"));
  client.print(label);
  client.print(F("'/></form>"));
}

void unknownPage(char *page)
{
  Serial.print(F("Unknown : "));
  Serial.println(F("page"));

  sendHeader("Unknown Page");
  client.println(F("<h1>Unknown Page</h1>"));
  client.println(page);
  client.println(F("</body></html>"));
```

```
}

void sendHeader(char *title)
{
  // send a standard http response header
  client.println(F("HTTP/1.1 200 OK"));
  client.println(F("Content-Type: text/html"));
  client.println();
  client.print(F("<html><head><title>"));
  client.println(title);
  client.println(F("</title><body>"));
}

void printP(const prog_uchar *str)
{
  // copy data out of program memory into local storage, write out in
  // chunks of 32 bytes to avoid extra short TCP/IP packets
  // from webduino library Copyright 2009 Ben Combee, Ran Talbott
  uint8_t buffer[32];
  size_t bufferEnd = 0;

  while (buffer[bufferEnd++] = pgm_read_byte(str++))
  {
    if (bufferEnd == 32)
    {
      client.write(buffer, 32);
      bufferEnd = 0;
    }
  }

  // write out everything left but trailing NUL
  if (bufferEnd > 1)
    client.write(buffer, bufferEnd - 1);
}
```

Discussion

The logic used to create the web page is similar to that used in the previous recipes. The form here is based on Recipe 15.10, but it has more elements in the table and uses embedded graphical objects to represent the state of the pins. If you have ever created a web page, you may be familiar with the use of JPEG images within the page. The Arduino Ethernet libraries do not have the capability to handle images in *.jpg* format.

Images need to be encoded using one of the Internet standards such as Multipurpose Internet Mail Extensions (MIME). This provides a way to represent graphical (or other) media using text. The sketch in this recipe's Solution shows what the LED images look like when they are MIME-encoded. Many web-based services will MIME-encode your images; the ones in this recipe were created using the service at *http://www.motobit.com/util/base64-decoder-encoder.asp*.

Even the small LED images used in this example are too large to fit into Arduino RAM. Program memory (flash) is used; see Recipe 17.3 for an explanation of the `P(name)` expression.

The images representing the LED on and off states are stored in a sequence of characters; the LED on array begins like this:

```
P(led_on) =  "<img src=\"data:image/jpg;base64,"
```

`P(led_on) =` defines `led_on` as the name of this array. The characters are the HTML tags identifying an image followed by the MIME-encoded data comprising the image.

This example is based on code produced for the Webduino web server. Webduino is highly recommended for building web pages if your application is more complicated than the examples shown in this chapter.

See Also

See Recipe 17.4 for more on using the `F("text")` construct for storing text in flash memory.

Webduino web page: *http://code.google.com/p/webduino/*

15.12 Sending Twitter Messages

Problem

You want Arduino to send messages to Twitter; for example, when a sensor detects some activity that you want to monitor via Twitter.

Solution

This sketch sends a Twitter message when a switch is closed. It uses a proxy at: *http://www.thingspeak.com* to provide authorization so you will need to register on that site to get a (free) API key. Click on the Sign Up button on the home page and fill in the form (your desired user ID, email, time zone, and password). Clicking the Create Account button will get you a ThingSpeak API key. To use the ThingSpeak service, you'll need to authorize your Twitter account to allow ThingTweet to post messages to your account. After that is set up, replace `"YourThingTweetAPIKey"` with the key string you are given and upload and run the following sketch:

```
/*
 * Send tweet when switch on pin 2 is pressed
 * uses api.thingspeak.com  as a Twitter proxy
 * see: http://community.thingspeak.com/documentation/apps/thingtweet/
 */

#include <SPI.h>
#include <Ethernet.h>
```

```
byte mac[] = { 0xDE, 0xAD, 0xBE, 0xEF, 0xFE, 0xED };
byte server[]  = { 184, 106, 153, 149 };  // IP Address for the ThingSpeak API

char *thingtweetAPIKey = "YourThingTweetAPIKey";  // your ThingTweet API key

EthernetClient client;

boolean MsgSent = false;
const int Sensor = 2;

void setup()
{
  Serial.begin(9600);
  if (Ethernet.begin(mac) == 0) { // start ethernet using mac & DHCP address
    Serial.println("Failed to configure Ethernet using DHCP");
    while(true)   // no point in carrying on, so stay in endless loop:
      ;
  }
  pinMode(Sensor, INPUT);
  digitalWrite(Sensor, HIGH);  //turn on  pull-up resistors
  delay(1000);
  Serial.println("Ready");
}

void loop()
{
  if(digitalRead(Sensor) == LOW)
  { // here if mailbox is open

    if(MsgSent == false){ // check if message already sent
       MsgSent = sendMessage("Mail has been delivered");
       if(MsgSent)
           Serial.println("tweeted successfully");
       else
           Serial.println("Unable tweet");
    }
  }
  else{
      MsgSent = false;  // door closed so reset the state
  }
  delay(100);
}

boolean sendMessage( char *message)
{
boolean result = false;

  const int tagLen = 16; // the number of tag character used to frame the message
  int msgLen = strlen(message) + tagLen + strlen(thingtweetAPIKey);
  Serial.println("connecting ...");
  if (client.connect(server, 80) ) {
    Serial.println("making POST request...");
    client.print("POST /apps/thingtweet/1/statuses/update HTTP/1.1\r\n");
    client.print("Host: api.thingspeak.com\r\n");
    client.print("Connection: close\r\n");
```

```
    client.print("Content-Type: application/x-www-form-urlencoded\r\n");
    client.print("Content-Length: ");
    client.print(msgLen);
    client.print("\r\n\r\n");
    client.print("api_key=");             // msg tag
    client.print(thingtweetAPIKey);       // api key
    client.print("&status=");             // msg tag
    client.print(message);                // the message
    client.println("\r\n");
  }
  else {
    Serial.println("Connection failed");
  }
  // response string
  if (client.connected()) {
    Serial.println("Connected");
    if(client.find("HTTP/1.1") && client.find("200 OK") ){
       result = true;
    }
    else
       Serial.println("Dropping connection - no 200 OK");
  }
  else {
    Serial.println("Disconnected");
  }
  client.stop();
  client.flush();

  return result;
}
```

Discussion

The sketch waits for a pin to go `LOW` and then posts your message to Twitter via the ThingTweet API.

The web interface is handled by the `sendMessage();` function, which will tweet the given message string. In this sketch it attempts to send the message string "Mail has been delivered" to Twitter and returns `true` if it is able to connect.

See the documentation on the ThingTweet web site for more details: *http://community.thingspeak.com/documentation/apps/thingtweet/*

The following version uses the same `sendMessage` function but can monitor an array of sensors:

```
/*
 * Send tweet selected by multiple sensors
 * uses api.thingspeak.com  as a Twitter proxy
 * see: http://community.thingspeak.com/documentation/apps/thingtweet/
 */

#include <SPI.h>
#include <Ethernet.h>
```

```
byte mac[] = { 0xDE, 0xAD, 0xBE, 0xEF, 0xFE, 0xED };
byte server[]  = { 184, 106, 153, 149 };  // IP Address for the ThingSpeak API

char *thingtweetAPIKey = "YourThingTweetAPIKey";  // your ThingTweet API key

EthernetClient client;
boolean MsgSent = false;

char frontOpen[] = "The front door was opened";
char backOpen[] = "The back door was opened";

const int frontSensor = 2;  // sensor pins
const int backSensor  = 3;

boolean frontMsgSent = false;
boolean backMsgSent = false;

void setup()
{
//  Ethernet.begin(mac,ip);
  Serial.begin(9600);
  if(Ethernet.begin(mac) == 0) { // start ethernet using mac & IP address
    Serial.println("Failed to configure Ethernet using DHCP");
    while(true)   // no point in carrying on, so stay in endless loop:
      ;
  }
  pinMode(frontSensor, INPUT);
  pinMode(backSensor, INPUT);
  digitalWrite(frontSensor, HIGH);  // pull-ups
  digitalWrite(backSensor, HIGH);
  delay(1000);
  Serial.println("ready");
}

void loop()
{
  if(digitalRead(frontSensor) == LOW)
  { // here if door is open
      if (frontMsgSent == false) { // check if message already sent
         frontMsgSent = sendMessage(frontOpen);
      }
  }
  else{
      frontMsgSent = false;  // door closed so reset the state
  }
  if(digitalRead(backSensor) == LOW)
  {
    if(frontMsgSent == false) {
       backMsgSent = sendMessage(backOpen);
    }
  }
  else {
      backMsgSent = false;
  }
  delay(100);
```

```
}

// add the sendMesage function from the sketch above
```

The code that communicates with Twitter is the same, but the message string here is constructed from the values read from sensors connected to two Arduino digital pins.

See Also

A ThingSpeak Arduino tutorial can be found here: *http://community.thingspeak.com/tutorials/arduino/using-an-arduino-ethernet-shield-to-update-a-thingspeak-channel/*

15.13 Sending and Receiving Simple Messages (UDP)

Problem

You want to send and receive simple messages over the Internet.

Solution

This sketch uses the Arduino UDP (User Datagram Protocol) library to send and receive strings. In this simple example, Arduino prints the received string to the Serial Monitor and a string is sent back to the sender saying "acknowledged":

```
/*
 * UDPSendReceiveStrings
 * This sketch receives UDP message strings, prints them to the serial port
 * and sends an "acknowledge" string back to the sender
 * Use with Arduino 1.0
 *
 */

#include <SPI.h>         // needed for Arduino versions later than 0018
#include <Ethernet.h>
#include <EthernetUdp.h> // Arduino 1.0 UDP library

byte mac[] = { 0xDE, 0xAD, 0xBE, 0xEF, 0xFE, 0xED }; // MAC address to use
byte ip[] = {192, 168, 1, 177 };     // Arduino's IP address

unsigned int localPort = 8888;      // local port to listen on

// buffers for receiving and sending data
char packetBuffer[UDP_TX_PACKET_MAX_SIZE]; //buffer to hold incoming packet,
char replyBuffer[] = "acknowledged";       // a string to send back

// A UDP instance to let us send and receive packets over UDP
EthernetUDP Udp;

void setup() {
    // start the Ethernet and UDP:
  Ethernet.begin(mac,ip);
  Udp.begin(localPort);
```

```
  Serial.begin(9600);
}

void loop() {
  // if there's data available, read a packet
  int packetSize =  Udp.parsePacket();
  if(packetSize)
  {
    Serial.print("Received packet of size ");
    Serial.println(packetSize);

    // read packet into packetBuffer and get sender's IP addr and port number
    Udp.read(packetBuffer,UDP_TX_PACKET_MAX_SIZE);
    Serial.println("Contents:");
    Serial.println(packetBuffer);

    // send a string back to the sender
    Udp.beginPacket(Udp.remoteIP(), Udp.remotePort());
    Udp.write(replyBuffer);
    Udp.endPacket();
  }
  delay(10);
}
```

You can test this by running the following Processing sketch on your computer (see Chapter 4 for guidance on installing and running Processing):

```
// Processing UDP example to send and receive string data from Arduino
// press any key to send the "Hello Arduino" message

import hypermedia.net.*;

UDP udp;  // define the UDP object

void setup() {
  udp = new UDP( this, 6000 );  // create datagram connection on port 6000
  //udp.log( true );            // <-- print out the connection activity
  udp.listen( true );           // and wait for incoming message
}

void draw()
{
}

void keyPressed() {
 String ip       = "192.168.1.177";     // the remote IP address
 int port        = 8888;      // the destination port

 udp.send("Hello World", ip, port );    // the message to send
}

void receive( byte[] data ) {           // <-- default handler
//void receive( byte[] data, String ip, int port ) { // extended handler

 for(int i=0; i < data.length; i++)
     print(char(data[i]));
```

```
  println();
}
```

Discussion

Plug the Ethernet shield into Arduino and connect the Ethernet cable to your computer. Upload the Arduino sketch and run the Processing sketch on your computer. Hit any key to send the "hello Arduino" message. Arduino sends back "acknowledged," which is displayed in the Processing text window. String length is limited by a constant set in the *EthernetUdp.h* library file; the default value is 24 bytes, but you can increase this by editing the following line in *Udp.h* if you want to send longer strings:

```
#define UDP_TX_PACKET_MAX_SIZE 24
```

UDP is a simple and fast way to send and receive messages over Ethernet. But it does have limitations—the messages are not guaranteed to be delivered, and on a very busy network some messages could get lost or get delivered in a different order than that in which they were sent. But UDP works well for things such as displaying the status of Arduino sensors—each message contains the current sensor value to display, and any lost messages get replaced by messages that follow.

This sketch demonstrates sending and receiving sensor messages. It receives messages containing values to be written to the analog output ports and replies back to the sender with the values on the analog input pins:

```
/*
 * UDPSendReceive sketch:
 */

#include <SPI.h>         // needed for Arduino versions later than 0018
#include <Ethernet.h>
#include <EthernetUDP.h>  // Arduino 1.0 UDP library

byte mac[] = { 0xDE, 0xAD, 0xBE, 0xEF, 0xFE, 0xED }; // MAC address to use
byte ip[] = {192, 168, 1, 177 };    // Arduino's IP address

unsigned int localPort = 8888;      // local port to listen on

char packetBuffer[UDP_TX_PACKET_MAX_SIZE]; //buffer to hold incoming packet,
int packetSize; // holds received packet size

const int analogOutPins[] = { 3,5,6,9};  // pins 10 and 11 used by ethernet shield

// A UDP instance to let us send and receive packets over UDP
EthernetUDP Udp;

void setup() {
  Ethernet.begin(mac,ip);
  Udp.begin(localPort);

  Serial.begin(9600);
  Serial.println("Ready");
}
```

```
void loop() {
  // if there's data available, read a packet
  packetSize = Udp.parsePacket();
  if(packetSize > 0 )
  {
    Serial.print("Received packet of size ");
    Serial.print(packetSize);
    Serial.println(" with contents:");
    // read packet into packetBuffer and get sender's IP addr and port number
    packetSize = min(packetSize,UDP_TX_PACKET_MAX_SIZE);
    Udp.read(packetBuffer,UDP_TX_PACKET_MAX_SIZE);

    for( int i=0; i < packetSize; i++)
    {
        byte value = packetBuffer[i];
        if( i < 4)
        {
           // only write to the first four analog out pins
           analogWrite(analogOutPins[i], value);
        }
        Serial.println(value, DEC);
    }
    Serial.println();
    // tell the sender the values of our analog ports
    sendAnalogValues(Udp.remoteIP(), Udp.remotePort());
  }
  //wait a bit
  delay(10);
}

void sendAnalogValues( IPAddress targetIp, unsigned int targetPort )
{
  int index = 0;
  for(int i=0; i < 6; i++)
  {
     int value = analogRead(i);

     packetBuffer[index++] = lowByte(value);  // the low byte);
     packetBuffer[index++] = highByte(value); // the high byte);    }
  }
  //send a packet back to the sender
  Udp.beginPacket(targetIp, targetPort);
  Udp.write(packetBuffer);
  Udp.endPacket();
}
```

The sketch sends and receives the values on analog ports 0 through 5 using binary data. If you are not familiar with messages containing binary data, see the introduction to Chapter 4, as well as Recipes 4.6 and 4.7, for a detailed discussion on how this is done on Arduino.

The difference here is that the data is sent using `Udp.write` instead of `Serial.write`.

Here is a Processing sketch you can use with the preceding sketch. It has six scroll bars that can be dragged with a mouse to set the six `analogWrite` levels; it prints the received sensor data to the Processing text window:

```
// Processing UDPTest
// Demo sketch sends & receives data to Arduino using UDP

import hypermedia.net.*;

UDP udp;  // define the UDP object

HScrollbar[] scroll = new HScrollbar[6];  //see: topics/gui/scrollbar

void setup() {
  size(256, 200);
  noStroke();
  for(int i=0; i < 6; i++) // create the scroll bars
    scroll[i] = new HScrollbar(0, 10 + (height / 6) * i, width, 10, 3*5+1);

  udp = new UDP( this, 6000 );  // create datagram connection on port 6000
  //udp.log( true );            // print out the connection activity
  udp.listen( true );           // and wait for incoming message
}

void draw()
{
  background(255);
  fill(255);
  for(int i=0; i < 6; i++) {
    scroll[i].update();
    scroll[i].display();
  }
}

void keyPressed() {
  String ip       = "192.168.1.177";    // the remote IP address
  int port        = 8888;               // the destination port
  byte[] message = new byte[6] ;

  for (int i=0; i < 6; i++){
    message[i] = byte(scroll[i].getPos());
    println(int(message[i]));
  }
  println();
  udp.send( message, ip, port );

}

void receive( byte[] data ) {                           // <-- default handler
//void receive( byte[] data, String ip, int port ) {    // <-- extended handler

  println("incoming data is:");
  for(int i=0; i < 6; i++){
     scroll[i].setPos(data[i]);
     println((int)data[i]);
```

```
  }
}

class HScrollbar
{
  int swidth, sheight;    // width and height of bar
  int xpos, ypos;         // x and y position of bar
  float spos, newspos;    // x position of slider
  int sposMin, sposMax;   // max and min values of slider
  int loose;              // how loose/heavy
  boolean over;           // is the mouse over the slider?
  boolean locked;
  float ratio;

  HScrollbar (int xp, int yp, int sw, int sh, int l) {
    swidth = sw;
    sheight = sh;
    int widthtoheight = sw - sh;
    ratio = (float)sw / (float)widthtoheight;
    xpos = xp;
    ypos = yp-sheight/2;
    spos = xpos + swidth/2 - sheight/2;
    newspos = spos;
    sposMin = xpos;
    sposMax = xpos + swidth - sheight;
    loose = l;
  }

  void update() {
    if (over()) {
      over = true;
    } else {
      over = false;
    }
    if (mousePressed && over) {
      locked = true;
    }
    if (!mousePressed) {
      locked = false;
    }
    if (locked) {
      newspos = constrain(mouseX-sheight/2, sposMin, sposMax);
    }
    if(abs(newspos - spos) > 1) {
      spos = spos + (newspos-spos)/loose;
    }
  }

  int constrain(int val, int minv, int maxv) {
    return min(max(val, minv), maxv);
  }

  boolean over() {
    if (mouseX > xpos && mouseX < xpos+swidth &&
    mouseY > ypos && mouseY < ypos+sheight) {
```

```
      return true;
    } else {
      return false;
    }
  }

  void display() {
    fill(255);
    rect(xpos, ypos, swidth, sheight);
    if (over || locked) {
      fill(153, 102, 0);
    } else {
      fill(102, 102, 102);
    }
    rect(spos, ypos, sheight, sheight);
  }

  float getPos() {
    return spos * ratio;
  }

  void setPos(int value) {
    spos = value / ratio;
  }
}
```

15.14 Getting the Time from an Internet Time Server

Problem

You want to get the current time from an Internet time server; for example, to synchronize clock software running on Arduino.

Solution

This sketch gets the time from a Network Time Protocol (NTP) server and prints the results as seconds since January 1, 1900 (NTP time) and seconds since January 1, 1970:

```
/*
 * UdpNtp sketch
 * Get the time from an NTP time server
 * Demonstrates use of UDP sendPacket and ReceivePacket
 */

#include <SPI.h>
#include <Ethernet.h>

#include <EthernetUDP.h>

byte mac[] = { 0xDE, 0xAD, 0xBE, 0xEF, 0xFE, 0xED }; // MAC address to use

unsigned int localPort = 8888;      // local port to listen for UDP packets
```

```
IPAddress timeServer(192, 43, 244, 18); // time.nist.gov NTP server
const int NTP_PACKET_SIZE= 48; // NTP time stamp is in the first 48
                               // bytes of the message
byte packetBuffer[ NTP_PACKET_SIZE]; // buffer to hold incoming/outgoing packets

// A UDP instance to let us send and receive packets over UDP
EthernetUDP Udp;

void setup()
{
  Serial.begin(9600);
  // start Ethernet and UDP
  if (Ethernet.begin(mac) == 0) {
    Serial.println("Failed to configure Ethernet using DHCP");
    // no point in carrying on, so do nothing forevermore:
    for(;;)
      ;
  }
  Udp.begin(localPort);
}

void loop()
{
  sendNTPpacket(timeServer); // send an NTP packet to a time server
  // wait to see if a reply is available
  delay(1000);
  if ( Udp.parsePacket() ) {
     Udp.read(packetBuffer,NTP_PACKET_SIZE);  // read packet into buffer

     //the timestamp starts at byte 40, convert four bytes into a long integer
     unsigned long hi = word(packetBuffer[40], packetBuffer[41]);
     unsigned long low = word(packetBuffer[42], packetBuffer[43]);
     unsigned long secsSince1900 = hi << 16 | low;  // this is NTP time
                                                    // (seconds since Jan 1 1900)

    Serial.print("Seconds since Jan 1 1900 = " );
    Serial.println(secsSince1900);

    Serial.print("Unix time = ");
    // Unix time starts on Jan 1 1970
    const unsigned long seventyYears = 2208988800UL;
    unsigned long epoch = secsSince1900 - seventyYears;  // subtract 70 years
    Serial.println(epoch);                               // print Unix time


    // print the hour, minute and second:
    // UTC is the time at Greenwich Meridian (GMT)
    Serial.print("The UTC time is ");
    // print the hour (86400 equals secs per day)
    Serial.print((epoch  % 86400L) / 3600);
    Serial.print(':');
    if ( ((epoch % 3600) / 60) < 10 ) {
     // Add leading zero for the first 10 minutes of each hour
      Serial.print('0');
    }
```

```
      // print the minute (3600 equals secs per minute)
      Serial.print((epoch  % 3600) / 60);
      Serial.print(':');
      if ( (epoch % 60) < 10 ) {
        // Add leading zero for the first 10 seconds of each minute
        Serial.print('0');
      }
      Serial.println(epoch %60); // print the second
    }
    // wait ten seconds before asking for the time again
    delay(10000);
  }

  // send an NTP request to the time server at the given address
  unsigned long sendNTPpacket(IPAddress& address)
  {
    memset(packetBuffer, 0, NTP_PACKET_SIZE);  // set all bytes in the buffer to 0

    // Initialize values needed to form NTP request
    packetBuffer[0] = B11100011;   // LI, Version, Mode
    packetBuffer[1] = 0;     // Stratum
    packetBuffer[2] = 6;     // Max Interval between messages in seconds
    packetBuffer[3] = 0xEC;  // Clock Precision
    // bytes 4 - 11 are for Root Delay and Dispersion and were set to 0 by memset
    packetBuffer[12]  = 49;  // four byte reference ID identifying
    packetBuffer[13]  = 0x4E;
    packetBuffer[14]  = 49;
    packetBuffer[15]  = 52;

    // all NTP fields have been given values, now
    // you can send a packet requesting a timestamp:
    Udp.beginPacket(address, 123); //NTP requests are to port 123
    Udp.write(packetBuffer,NTP_PACKET_SIZE);
    Udp.endPacket();
  }
```

Discussion

NTP is a protocol used to synchronize time through Internet messages. NTP servers provide time as a value of the number of seconds that have elapsed since January 1, 1900. NTP time is UTC (Coordinated Universal Time, similar to Greenwich Mean Time) and does not take time zones or daylight saving time into account.

NTP servers use UDP messages; see Recipe 15.13 for an introduction to UDP. An NTP message is constructed in the function named `sendNTPpacket` and you are unlikely to need to change the code in that function. The function takes the address of an NTP server; you can use the IP address in the preceding example or find a list of many more by using "NTP address" as a search term in Google. If you want more information about the purpose of the NTP fields, see the documentation at *http://www.ntp.org/*.

The reply from NTP is a message with a fixed format; the time information consists of four bytes starting at byte 40. These four bytes are a 32-bit value (an unsigned long integer), which is the number of seconds since January 1, 1900. This value (and the time converted into Unix time) is printed. If you want to convert the time from an NTP server to the friendlier format using hours, minutes, and seconds and days, months, and years, you can use the Arduino Time library (see Chapter 12). Here is a variation on the preceding code that prints the time as `14:32:56 Monday 18 Jan 2010`:

```
/*
 * Time_NTP sketch
 * Example showing time sync to NTP time source
 * This sketch uses the Time library
 * and the Arduino Ethernet library
 */

#include <Time.h>
#include <SPI.h>         // needed for Arduino versions later than 0018
#include <Ethernet.h>
#include <EthernetUDP.h>

byte mac[] = { 0xDE, 0xAD, 0xBE, 0xEF, 0xFE, 0xED };
byte ip[] = { 192, 168, 1, 44 }; // set this to a valid IP address (or use DHCP)

unsigned int localPort = 8888;      // local port to listen for UDP packets

IPAddress timeServer(192, 43, 244, 18); // time.nist.gov NTP server

const int NTP_PACKET_SIZE= 48; // NTP time stamp is in first 48 bytes of message
byte packetBuffer[ NTP_PACKET_SIZE]; // buffer to hold incoming/outgoing packets

time_t prevDisplay = 0;    // when the digital clock was displayed

// A UDP instance to let us send and receive packets over UDP
EthernetUDP Udp;

void setup()
{
  Serial.begin(9600);
  Ethernet.begin(mac,ip);
  Udp.begin(localPort);
  Serial.println("waiting for sync");
  setSyncProvider(getNtpTime);
  while(timeStatus()== timeNotSet)
     ; // wait until the time is set by the sync provider
}

void loop()
{
  if( now() != prevDisplay)   //update the display only if the time has changed
  {
    prevDisplay = now();
    digitalClockDisplay();
  }
}

void digitalClockDisplay(){
  // digital clock display of the time
  Serial.print(hour());
```

```
  printDigits(minute());
  printDigits(second());
  Serial.print(" ");
  Serial.print(dayStr(weekday()));
  Serial.print(" ");
  Serial.print(day());
  Serial.print(" ");
  Serial.print(monthShortStr(month()));
  Serial.print(" ");
  Serial.print(year());
  Serial.println();
}

void printDigits(int digits){
  // utility function for digital clock display: prints preceding
  // colon and leading 0
  Serial.print(":");
  if(digits < 10)
    Serial.print('0');
  Serial.print(digits);
}

/*-------- NTP code ----------*/

unsigned long getNtpTime()
{
  sendNTPpacket(timeServer); // send an NTP packet to a time server
  delay(1000);
  if ( Udp.parsePacket() ) {
     Udp.read(packetBuffer,NTP_PACKET_SIZE);  // read packet into buffer

     //the timestamp starts at byte 40, convert four bytes into a long integer
     unsigned long hi = word(packetBuffer[40], packetBuffer[41]);
     unsigned long low = word(packetBuffer[42], packetBuffer[43]);
     // this is NTP time (seconds since Jan 1 1900
     unsigned long secsSince1900 = hi << 16 | low;
     // Unix time starts on Jan 1 1970
     const unsigned long seventyYears = 2208988800UL;
     unsigned long epoch = secsSince1900 - seventyYears;  // subtract 70 years
     return epoch;
  }
  return 0; // return 0 if unable to get the time
}

// send an NTP request to the time server at the given address
unsigned long sendNTPpacket(IPAddress address)
{
  memset(packetBuffer, 0, NTP_PACKET_SIZE);  // set all bytes in the buffer to 0

  // Initialize values needed to form NTP request
  packetBuffer[0] = B11100011;   // LI, Version, Mode
  packetBuffer[1] = 0;     // Stratum
  packetBuffer[2] = 6;     // Max Interval between messages in seconds
  packetBuffer[3] = 0xEC;  // Clock Precision
  // bytes 4 - 11 are for Root Delay and Dispersion and were set to 0 by memset
```

```
    packetBuffer[12]  = 49;  // four-byte reference ID identifying
    packetBuffer[13]  = 0x4E;
    packetBuffer[14]  = 49;
    packetBuffer[15]  = 52;

    // send the packet requesting a timestamp:
    Udp.beginPacket(address, 123); //NTP requests are to port 123
    Udp.write(packetBuffer,NTP_PACKET_SIZE);
    Udp.endPacket();
  }
```

See Also

Chapter 12 provides more information on using the Arduino Time library.

Details on NTP are available at *http://www.ntp.org/*.

NTP code by Jesse Jaggars that inspired the sketch used in this recipe is available at *http://github.com/cynshard/arduino-ntp*.

If you are running an Arduino release prior to 1.0 you can download a UDP library from *https://bitbucket.org/bjoern/arduino_osc/src/tip/libraries/Ethernet/*.

15.15 Monitoring Pachube Feeds

Problem

You want Arduino to respond to information on a web service that offers security and data backup. Pachube is a web-based service that manages real-time data feeds; you want to activate a device or raise an alarm based on the value of data on a Pachube feed.

Solution

This sketch gets the first four data fields from feed number 504 and prints the results on the Serial Monitor:

```
/*
 * Monitor Pachube feed
 * Read feed using V2 API using CSV format
 */

#include <SPI.h>
#include <Ethernet.h>

const int  feedID          =    504; // this is the ID of the
                                      // remote Pachube feed that
                                      // you want to connect to
const int  streamCount     =      4; // Number of data streams to get
const long PACHUBE_REFRESH = 600000; // Update every 10 minutes
const long PACHUBE_RETRY   = 10000;  // if connection fails/resets
                                      // wait 10 seconds before trying
                                      // again - should not be less than 5
```

```
#define PACHUBE_API_KEY  "your key here . . ." // fill in your API key

// mac address, make sure this is unique on your network
byte mac[] = { 0xCC, 0xAC, 0xBE, 0xEF, 0xFE, 0x91 };
char serverName[] = "api.pachube.com";

int streamData[streamCount];    // change float to long if needed for your data

EthernetClient client;

void setup()
{
  Serial.begin(9600);
  if (Ethernet.begin(mac) == 0) {
    Serial.println(F("Failed to configure Ethernet using DHCP"));
    // no point in carrying on, so do nothing forevermore:
    for(;;)
      ;
  }
}

void loop()
{
   if( getFeed(feedID, streamCount) == true)
   {
      for(int id = 0; id < streamCount; id++){
        Serial.println( streamData[id]);
      }
      Serial.println("--");
      delay(PACHUBE_REFRESH);
   }
   else
   {
      Serial.println(F("Unable to get feed"));
      delay(PACHUBE_RETRY);
   }
}

// returns true if able to connect and get data for all requested streams
boolean getFeed(int feedId, int streamCount )
{
boolean result = false;
  if (client.connect(serverName, 80)>0) {
    client.print(F("GET /v2/feeds/"));
    client.print(feedId);
    client.print(F(".csv HTTP/1.1\r\nHost: api.pachube.com\r\nX-PachubeApiKey: "));
    client.print(PACHUBE_API_KEY);
    client.print("\r\nUser-Agent: Arduino 1.0");
    client.println("\r\n");
  }
  else {
    Serial.println("Connection failed");
  }
  if (client.connected()) {
```

```
    Serial.println("Connected");
    if(  client.find("HTTP/1.1") && client.find("200 OK") )
       result = processCSVFeed(streamCount);
    else
       Serial.println("Dropping connection - no 200 OK");
  }
  else {
    Serial.println("Disconnected");
  }
  client.stop();
  client.flush();
  return result;
}

int processCSVFeed(int streamCount)
{
  int processed = 0;
  client.find("\r\n\r\n"); // find the blank line indicating start of data
  for(int id = 0; id < streamCount; id++)
  {
    int id = client.parseInt(); // you can use this to select a specific id
    client.find(","); // skip past timestamp
    streamData[id] = client.parseInt();
    processed++;
  }
  return(processed == streamCount );  // return true if got all data
}
```

Discussion

To start using Pachube, you have to sign up for an account, and the Pachube Quickstart page explains how: *http://community.pachube.com/?q=node/4*. Once you're signed up, you will be emailed a username and API key. Add your key to the following line in the sketch:

```
#define PACHUBE_API_KEY  "your key here . . ." // fill in your API key
```

Every Pachube feed (data source) has an identifying ID; this example sketch uses feed 504 (environmental data from the Pachube office). In the sketch below, feeds are accessed using the `getFeed` method with the feed ID and the number of items of data to get passed as arguments. If this is successful, `getFeed` returns `true`, and you can process the data using the `processFeed` method. This returns the value for the data item you are interested in (each data item is called a *stream* in Pachube).

Pachube supports a number of data formats and the sketch above uses the simplest, CSV (comma-separated variables) (see: *http://api.pachube.com/v2/#data-formats* for more on Pachube data formats).

You can extract more information about a feed using the XML format. Here is an example of Pachube XML data for the stream used in this recipe:

```
<environment updated="2010-06-08T09:30:11Z" id="504"
    creator="http://www.pachube.com/users/hdr">
```

```
<title>Pachube Office environment</title>
<feed>http://api.pachube.com/v2/feeds/504.xml</feed>
<status>live</status>
<website>http://www.haque.co.uk/</website>
<tag>Tag1</tag>
<tag>Tag2</tag>
<location domain="physical" exposure="indoor" disposition="fixed">
  <name>office</name>
  <lat>51.5235375648154</lat>
  <lon>-0.0807666778564453</lon>
  <ele>23.0</ele>
</location>
<data id="0">
  <tag>humidity</tag>
  <min_value>0.0</min_value>
  <max_value>847.0</max_value>
  <current_value at="2010-06-08T09:30:11.000000Z">311</current_value>
</data>
</environment>
```

The title `Pachube Office environment` indicates the start of the data; each stream is indicated by the tag `data id=` followed by the numeric stream ID. The `processXMLFeed` function in the following sketch uses this information to find the desired feed ID and then extract readings for the min, max, and current value of the desired feed:

```
/*
 * Monitor Pachube feed
 * V2 API using XML format
 * controls a servo using value of a specified stream
 */

#include <SPI.h>
#include <Ethernet.h>

#include <Servo.h> // this sketch will control a servo

const int  feedID          =   504; // desired pachube feed
const int  streamToGet     =     0; // data id of the desired stream

const long PACHUBE_REFRESH = 600000; // Update every 10 minutes
const long PACHUBE_RETRY   = 10000;  // if connection fails/resets

#define PACHUBE_API_KEY  "your key here . . ." // fill in your API key

// mac address, make sure this is unique on your network
byte mac[] = { 0xCC, 0xAC, 0xBE, 0xEF, 0xFE, 0x91 };
char serverName[] = "api.pachube.com";

EthernetClient client;

// stream values returned from pachube will be stored here
int currentValue; // current reading for stream
int minValue;     // minimum value for stream
int maxValue;     // maximum value for stream
```

```
Servo myservo;  // create servo object to control a servo

void setup()
{
  Serial.begin(9600);
  myservo.attach(9);  // attaches the servo on pin 9 to the servo object

  if (Ethernet.begin(mac) == 0) {
    Serial.println("Failed to configure Ethernet using DHCP");
    // no point in carrying on, so do nothing forevermore:
    for(;;)
      ;
  }
}

void loop()
{
  if( getFeed(feedID, streamToGet) == true)
  {
    Serial.print(F("value="));
    Serial.println(currentValue);
    // position proportionaly within range of 0  to 90 degreees
    int servoPos = map(currentValue, minValue, maxValue, 0,90);
    myservo.write(servoPos);
    Serial.print(F("pos="));
    Serial.println(servoPos);
    delay(PACHUBE_REFRESH);
  }
  else
  {
    Serial.println(F("Unable to get feed"));
    delay(PACHUBE_RETRY);
  }
}

// returns true if able to connect and get data for requested stream
boolean getFeed(int feedId, int streamId )
{
  boolean result = false;
  if (client.connect(serverName, 80)>0) {
    Serial.print("Connecting feed ");
    Serial.print(feedId);
    Serial.print(" ... ");
    client.print("GET /v2/feeds/");
    client.print(feedId);
    client.print(".xml HTTP/1.1\r\nHost: api.pachube.com\r\nX-PachubeApiKey: ");
    client.print(PACHUBE_API_KEY);
    client.print("\r\nUser-Agent: Arduino 1.0");
    client.println("\r\n");
  }
  else {
    Serial.println("Connection failed");
  }
  if (client.connected()) {
    Serial.println("Connected");
```

```
    if(  client.find("HTTP/1.1") && client.find("200 OK") )
      result = processXMLFeed(streamId);
    else
      Serial.println("Dropping connection - no 200 OK");
  }
  else {
    Serial.println("Disconnected");
  }
  client.stop();
  client.flush();
  return result;
}

boolean processXMLFeed(int streamId)
{
  client.find("<environment updated=");
  for(int id = 0; id <= streamId; id++)
  {
    if( client.find( "<data id=" ) ){  // find next data field
      if(client.parseInt()== streamId){ // is this our stream?
        if(client.find("<min_value>")){
          minValue = client.parseInt();
          if(client.find("<max_value>")){
            maxValue = client.parseInt();
            if(client.find("<current_value ")){
              client.find(">"); // seek to the angle brackets
              currentValue = client.parseInt();
              return true; // found all the neeed data fields
            }
          }
        }
      }
    }
    else {
      Serial.print(F("unable to find data for ID "));
      Serial.println(id);
    }
  }
  return false;  // unable to parse the data
}
```

Arduino 1.0 Stream parsing is used to search for specific fields, see the Pachube API documentation for a list of all fields.

See Also

The Pachube API documentation is here: *http://api.pachube.com/v2/*.

An Arduino library to simplify Pachube access can be found here: *http://code.google.com/p/pachubelibrary/*.

15.16 Sending Information to Pachube

Problem

You want Arduino to update feeds on Pachube. For example, you want the values of sensors connected to Arduino to be published on a Pachube feed.

Solution

This sketch reads temperature sensors connected to the analog input ports (see Recipe 6.8) and sends the data to Pachube:

```
/*
 * Update Pachube feed
 * sends temperature read from (up to) six LM35 sensors
 * V2 API
 */

#include <SPI.h>
#include <Ethernet.h>

const int  feedID           =    2955; // this is the ID of this feed
const int  streamCount      =       6; // Number of data streams (sensors) to send
const long REFRESH_INTERVAL = 60000;   // Update every  minute
// if connection fails/resets wait 10 seconds before trying again
// should not be less than 5
const long RETRY_INTERVAL   = 10000;

#define PACHUBE_API_KEY  "Your key here . . . " // fill in your API key

// make sure this is unique on your network
byte mac[] = { 0xCC, 0xAC, 0xBE, 0xEF, 0xFE, 0x91 };
char serverName[] = "www.pachube.com";

EthernetClient client;

void setup()
{
  Serial.begin(9600);
  Serial.println("ready");
  if (Ethernet.begin(mac) == 0) { // start ethernet using mac & IP address
    Serial.println("Failed to configure Ethernet using DHCP");
    while(true)   // no point in carrying on, so stay in endless loop:
      ;
  }
}

void loop()
{
  String dataString = "";
  for (int id = 0; id < streamCount; id++)
  {
     int temperature = getTemperature(id);
```

```
    dataString += String(id);
    dataString  += ",";
    dataString += String(temperature);
    dataString  += "\n";
  }
  if ( putFeed(feedID, dataString, dataString.length()) == true)
  {
    Serial.println("Feed updated");
    delay(REFRESH_INTERVAL);
  }
  else
  {
     Serial.println("Unable to update feed");
     delay(RETRY_INTERVAL);
  }
}

// returns true if able to connect and send data
boolean putFeed(int feedId, String feedData, int length )
{
boolean result = false;
  if (client.connect(serverName, 80)>0) {
    Serial.print("Connecting feed "); Serial.println(feedId);
    client.print("PUT /v2/feeds/");
    client.print(feedId);
    client.print(".csv HTTP/1.1\r\nHost: api.pachube.com\r\nX-PachubeApiKey: ");
    client.print(PACHUBE_API_KEY);
    client.print("\r\nUser-Agent: Arduino 1.0");
    client.print("\r\nContent-Type: text/csv\r\nContent-Length: ");
    client.println(length+2, DEC); // allow for cr/lf
    client.println("Connection: close");
    client.println("\r\n");
    // now print the data:
    Serial.println(feedData); // optional echo to serial monitor
    client.print(feedData);
    client.println("\r\n");
  }
  else {
    Serial.println("Connection failed");
  }
  // response string
  if (client.connected()) {
    Serial.println("Connected");
    if(client.find("HTTP/1.1") && client.find("200 OK") ){
       result = true;
    }
    else
       Serial.println("Dropping connection - no 200 OK");
  }
  else {
    Serial.println("Disconnected");
  }
  client.stop();
  client.flush();
  return result;
```

```
}

// get the temperature rounded up to the nearest degree
int getTemperature(int pin)
{
   int value = analogRead(pin);
   int celsius = (value * 500L) / 1024; // 10mv per degree
   return celsius;
}
```

Discussion

This is similar to Recipe 15.15, but here you use the putFeed method to send your information to Pachube. This example sends information from temperature sensors; see the chapter covering the type of sensor you want to use to find code suitable for your application.

Pachube requires the number of characters in the data to be sent prior to the actual content. This is achieved using the string concatenation function in Recipe 2.5 to create a string containing all fields, and then using the String.length() method to get the number of characters.

The following sketch uses a different technique that does not require any RAM to store the string data. It uses a new capability introduced in Arduino 1.0 that returns the number of characters printed. The function outputCSV counts and returns the number of characters printed. It is first called to calculate the total character count by printing the output to serial; it's called again to output to the Ethernet client connected to Pachube:

```
/*
 * Update Pachube feed
 * sends floating point temperatures read from (up to) six LM35 sensors
 * V2 API
 */

#include <SPI.h>
#include <Ethernet.h>

const int  feedID           =    2955; // this is the ID of this feed
const int  streamCount      =       6; // Number of data streams (sensors) to send
const long REFRESH_INTERVAL = 60000;   // Update every  minute
// if connection fails/resets wait 10 seconds before trying again
// should not be less than 5
const long RETRY_INTERVAL   = 10000;

#define PACHUBE_API_KEY  "Your key here . . . " // fill in your API key

// make sure this is unique on your network
byte mac[] = { 0xCC, 0xAC, 0xBE, 0xEF, 0xFE, 0x91 };
char serverName[] = "www.pachube.com";

EthernetClient client;
```

```
void setup()
{
  Serial.begin(9600);
  Serial.println("ready");
  if(Ethernet.begin(mac) == 0) { // start ethernet using mac & IP address
    Serial.println("Failed to configure Ethernet using DHCP");
    while(true)   // no point in carrying on, so stay in endless loop:
      ;
  }
}

void loop()
{
  int contentLen = outputCSV(Serial); // get character count
  if( putFeed(feedID, contentLen) == true){
     Serial.println("Feed updated");
     delay(REFRESH_INTERVAL);
  }
  else {
      Serial.println("Unable to update feed");
      delay(RETRY_INTERVAL);
  }
}

// returns true if able to connect and send data
boolean putFeed(int feedId, int length)
{
boolean result = false;
  if (client.connect(serverName, 80)>0) {
    Serial.print("Connecting feed "); Serial.println(feedId);
    client.print("PUT /v2/feeds/");
    client.print(feedId);
    client.print(".csv HTTP/1.1\r\nHost: api.pachube.com\r\nX-PachubeApiKey: ");
    client.print(PACHUBE_API_KEY);
    client.print("\r\nUser-Agent: Arduino 1.0");
    client.print("\r\nContent-Type: text/csv\r\nContent-Length: ");
    client.println(length+2, DEC); // allow for cr/lf
    client.println("Connection: close");
    client.println("\r\n");
    outputCSV(client);
    client.println("\r\n");
  }
  else {
    Serial.println("Connection failed");
  }
  // response string
  if (client.connected()) {
    Serial.println("Connected");
    if(client.find("HTTP/1.1") && client.find("200 OK") ){
       result = true;
    }
    else
       Serial.println("Dropping connection - no 200 OK");
  }
  else {
```

```
    Serial.println("Disconnected");
  }
  client.stop();
  client.flush();
  return result;
}

int outputCSV(Stream &stream)
{
  int count = 0;
  for(int id = 0; id < streamCount; id++) {
    float temperature = getTemperature(id);
    count += stream.print(id,DEC);
    count += stream.print(',');
    count += stream.print(temperature,1); // one digit after decimal point
    count += stream.print("\n");
  }
  return count;
}

float getTemperature(int inPin)
{
   int value = analogRead(inPin);
   float millivolts = (value / 1024.0) * 5000;  //  see Recipe 6.8
   return millivolts / 10;                      // 10mV per degree C
}
```

CHAPTER 16

Using, Modifying, and Creating Libraries

16.0 Introduction

Libraries add functionality to the Arduino environment. They extend the commands available to provide capabilities not available in the core Arduino language. Libraries provide a way to add features that can be accessed from any of your sketches once you have installed the library.

The Arduino software distribution includes built-in libraries that cover common tasks. These libraries are discussed in Recipe 16.1.

Libraries are also a good way for people to share code that may be useful to others. Many third-party libraries provide specialized capabilities; these can be downloaded from the Arduino Playground and other sites. Libraries are often written to simplify the use of a particular piece of hardware. Many of the devices covered in earlier chapters use libraries to make it easier to connect to the devices.

Libraries can also provide a friendly wrapper around complex code to make it easier to use. An example is the Wire library distributed with Arduino, which hides much of the complexity of low-level hardware communications (see Chapter 13).

This chapter explains how to use and modify libraries. It also gives examples of how to create your own libraries.

16.1 Using the Built-in Libraries

Problem

You want to use the libraries provided with the Arduino distribution in your sketch.

Solution

This recipe shows you how to use Arduino library functionality in your sketch.

To see the list of available libraries from the IDE menu, click Sketch→Import Library. A list will drop down showing all the available libraries. The first dozen or so are the libraries distributed with Arduino. A horizontal line separates that list from the libraries that you download and install yourself.

Clicking on a library will add that library to the current sketch, by adding the following line to the top of the sketch:

```
#include <nameOfTheLibrarySelected.h>
```

This results in the functions within the library becoming available to use in your sketch.

The Arduino IDE updates its list of available libraries only when the IDE is first started on your computer. If you install a library after the IDE is running, you need to close the IDE and restart for that new library to be recognized.

The Arduino libraries are documented in the reference at *http://arduino.cc/en/Reference/Libraries* and each library includes example sketches demonstrating their use. Chapter 1 has details on how to navigate to the examples in the IDE.

The libraries that are included with Arduino as of version 1.0 are:

EEPROM
: Used to store and read information in memory that is preserved when power is removed; see Chapter 18

Ethernet
: Used to communicate with the Arduino Ethernet shield or for use with the Arduino Ethernet board; see Chapter 15

Firmata
: A protocol used to simplify serial communication and control of the board

LiquidCrystal
: For controlling compatible LCD displays; see Chapter 11

SD
: Supports reading and writing files to an SD card using external hardware

Servo
: Used to control servo motors; see Chapter 8

SoftwareSerial
: Enables additional serial ports

SPI
: Used for Ethernet and SPI hardware; see Chapter 13

Stepper
: For working with stepper motors; see Chapter 8

Wire
: Works with I2C devices attached to the Arduino; see Chapter 13

The following two libraries can be found in releases prior to Arduino 1.0 but are no longer included with the Arduino distribution:

Matrix
: Helps manage a matrix of LEDs; see Chapter 7

Sprite
: Enables the use of sprites with an LED matrix

Discussion

Libraries that work with specific hardware within the Arduino controller chip only work on predefined pins. The Wire and SPI libraries are examples of this kind of library. Libraries that allow user selection of pins usually have this specified in `setup`; Servo, LiquidCrystal, and Stepper are examples of that kind of library. See the library documentation for specific information on how to configure the library.

Including a library adds the library code to your sketch behind the scenes. This means the size of your sketch, as reported at the end of the compilation process, will increase, but the Arduino build process is smart enough to only include the code your sketch is actually using from the library, so you don't have to worry about the memory overhead for methods that are not being used. Therefore, you also don't have to worry about unused functions reducing the amount of code you can put into your sketch.

Libraries included with Arduino (and many contributed libraries) include example sketches that show how to use the library. They are accessed from the File→Examples menu.

See Also

The Arduino reference for libraries: *http://arduino.cc/en/Reference/Libraries*

16.2 Installing Third-Party Libraries

Problem

You want to use a library created for use with Arduino but not in the standard distribution.

Solution

Download the library. It will often be a *.zip* file. Unzip it and you will have a folder that has the same title as the name of the library. This folder needs to be put inside a folder called *libraries* inside your Arduino document folder. To find the Arduino document folder, open Preferences (Arduino→Preferences on Mac; File→Preferences on Windows) and note the sketchbook location. Navigate to that directory in a file system browser (such as Windows Explorer or Mac OS X Finder) or at the terminal. If no *libraries* folder exists, create one and put the folder you unzipped inside it.

If the Arduino IDE is still running, quit and restart it. The IDE scans this folder to find libraries only when it is launched. If you now go to the menu Sketch→Import Library, at the bottom, below the gray line and the word *Contributed*, you should see the library you have added.

If the libraries provide example sketches, you can view these from the IDE menu; click File→Examples, and the libraries examples will be under the libraries name in a section between the general examples and the Arduino distributed library example listing.

Discussion

A large number of libraries are provided by third parties. Many are very high quality, are actively maintained, and provide good documentation and example sketches. The Arduino Playground is a good place to look for libraries: *http://www.arduino.cc/playground/*.

Look for libraries that have clear documentation and examples. Check out the Arduino forums to see if there are any threads (discussion topics) that discuss the library. Libraries that were designed to be used with early Arduino releases may have problems when used with the latest Arduino version, so you may need to read through a lot of material (some threads for popular libraries contain hundreds of posts) to find information on using an older library with the latest Arduino release.

If the library examples do not appear in the Examples menu or you get a message saying "Library not found" when you try to use the library, check that the libraries folder is in the correct place with the name spelled correctly. A library folder named *<LibraryName>* (where *<LibraryName>* is the name for the library) must contain a file named *<LibraryName>.h* with the same spelling and capitalization. Check that additional files needed by the library are in the folder.

16.3 Modifying a Library

Problem

You want to change the behavior of an existing library, perhaps to extend its capability. For example, the TimeAlarms library in Chapter 12 only supports six alarms and you need more (see Recipe 12.5).

Solution

The Time and TimeAlarms libraries are described in Chapter 12, so refer to Recipe 12.5 to familiarize yourself with the standard functionality. The libraries can be downloaded from the website for this book (*http://shop.oreilly.com/product/0636920022244.do*), or from *http://www.arduino.cc/playground/uploads/Code/Time.zip* (this download includes both libraries).

Once you have the Time and TimeAlarms libraries installed, compile and upload the following sketch, which will attempt to create seven alarms—one more than the libraries support. Each `Alarm` task simply prints its task number:

```
/*
 multiple_alarms sketch
 has more timer repeats than the library supports out of the box -
 you will need to edit the header file to enable more than 6 alarms
 */

#include <Time.h>
#include <TimeAlarms.h>

int currentSeconds = 0;

void setup()
{
  Serial.begin(9600);

  // create 7 alarm tasks
  Alarm.timerRepeat(1, repeatTask1);
  Alarm.timerRepeat(2, repeatTask2);
  Alarm.timerRepeat(3, repeatTask3);
  Alarm.timerRepeat(4, repeatTask4);
  Alarm.timerRepeat(5, repeatTask5);
  Alarm.timerRepeat(6, repeatTask6);
  Alarm.timerRepeat(7, repeatTask7);  // 7th timer repeat
}


void repeatTask1()
{
  Serial.print("task 1  ");
}

void repeatTask2()
```

```
{
  Serial.print("task 2  ");
}
void repeatTask3()
{
  Serial.print("task 3  ");
}

void repeatTask4()
{
  Serial.print("task 4  ");
}

void repeatTask5()
{
  Serial.print("task 5  ");
}

void repeatTask6()
{
  Serial.print("task 6  ");
}

void repeatTask7()
{
  Serial.print("task 7  ");
}

void  loop()
{
  if(second() != currentSeconds)
  {
   // print the time for each new second
   // the task numbers will be printed when the alarm for that task is triggered
    Serial.println();
    Serial.print(second());
    Serial.print("->");
    currentSeconds = second();
    Alarm.delay(1); // Alarm.delay must be called to service the alarms
  }
}
```

Open the Serial Monitor and watch the output being printed. After nine seconds of output, you should see this:

```
1->task 1
2->task 1  task 2
3->task 1  task 3
4->task 1  task 2  task 4
5->task 1  task 5
6->task 1  task 2  task 3  task 6
7->task 1
8->task 1  task 2  task 4
9->task 1  task 3
```

The task scheduled for seven seconds did not trigger because the library only provides six timer "objects" that you can use.

You can increase this by modifying the library. Go to the libraries folder in your Arduino *Documents* folder.

You can locate the directory containing the sketchbook folder by clicking on the menu item File→Preferences (on Windows) or Arduino→Preferences (on a Mac) in the IDE. A dialog box will open, showing the sketchbook location.

If you have installed the Time and TimeAlarms libraries (both libraries are in the file you downloaded), navigate to the *Libraries\TimeAlarms* folder. Open the *TimeAlarms.h* header file (for more details about header files, see Recipe 16.4). You can edit the file with any text editor; for example, Notepad on Windows or TextEdit on a Mac.

You should see the following at the top of the *TimeAlarms.h* file:

```
#ifndef TimeAlarms_h
#define TimeAlarms_h

#include <inttypes.h>
#include "Time.h"
#define dtNBR_ALARMS 6
```

The maximum number of alarms is specified by the value defined for `dtNbr_ALARMS`.

Change:

```
#define dtNBR_ALARMS 6
```

to:

```
#define dtNMBR_ALARMS 7
```

and save the file.

Upload the sketch to your Arduino again, and this time the serial output should read:

```
1->task 1
2->task 1  task 2
3->task 1  task 3
4->task 1  task 2  task 4
5->task 1  task 5
6->task 1  task 2  task 3  task 6
7->task 1  task 7
8->task 1  task 2  task 4
9->task 1  task 3
```

You can see that task 7 now activates after seven seconds.

Discussion

Capabilities offered by a library are a trade-off between the resources used by the library and the resources available to the rest of your sketch, and it is often possible to change these capabilities if required. For example, you may need to decrease the amount of memory used for a serial library so that other code in the sketch has more RAM. Or you may need to increase the memory usage by a library for your application. The library writer generally creates the library to meet typical scenarios, but if your application needs capabilities not catered to by the library writer, you may be able to modify the library to accommodate them.

In this example, the TimeAlarms library allocates room (in RAM) for six alarms. Each of these consumes around a dozen bytes and the space is reserved even if only a few are used. The number of alarms is set in the library header file (the header is a file named *TimeAlarms.h* in the *TimeAlarms* folder). Here are the first few lines of *TimeAlarms.h*:

```
#ifndef TimeAlarms_h
#define TimeAlarms_h

#include <inttypes.h>

#include "Time.h"

#define dtNBR_ALARMS 6
```

In the TimeAlarms library, the maximum number of alarms is set using a `#define` statement. Because you changed it and saved the header file when you recompiled the sketch to upload it, it uses the new upper limit.

Sometimes constants are used to define characteristics such as the clock speed of the board, and when used with a board that runs at a different speed, you will get unexpected results. Editing this value in the header file to the correct one for the board you are using will fix this problem.

If you edit the header file and the library stops working, you can always download the library again and replace the whole library to return to the original state.

See Also

Recipe 16.4 has more details on how you can add functionality to libraries.

16.4 Creating Your Own Library

Problem

You want to create your own library. Libraries are a convenient way to reuse code across multiple sketches and are a good way to share with other users.

Solution

A library is a collection of methods and variables that are combined in a format that enables users to access functions and variables in a standardized way.

Most Arduino libraries are written as a class. If you are familiar with C++ or Java, you will be familiar with classes. However, you can create a library without using a class, and this recipe shows you how.

This recipe explains how you can transform the sketch from Recipe 7.1 to move the `BlinkLED` function into a library.

See Recipe 7.1 for the wiring diagram and an explanation of the circuit. The library will contain the `blinkLED` function from that recipe. Here is the sketch that will be used to test the library:

```
/*
 * blinkLibTest
 */

#include "blinkLED.h"

const int firstLedPin  = 3;         // choose the pin for each of the LEDs
const int secondLedPin = 5;
const int thirdLedPin  = 6;

void setup()
{
  pinMode(firstLedPin, OUTPUT);     // declare LED pins as output
  pinMode(secondLedPin, OUTPUT);    // declare LED pins as output
  pinMode(thirdLedPin, OUTPUT);     // declare LED pins as output
}

void loop()
{
  // flash each of the LEDs for 1000 milliseconds (1 second)
  blinkLED(firstLedPin, 1000);
  blinkLED(secondLedPin, 1000);
  blinkLED(thirdLedPin, 1000);
}
```

The `blinkLED` function from Recipe 7.1 should be removed from the sketch and moved into a separate file named *blinkLED.cpp* (see the Discussion for more details about *.cpp* files):

```
/* blinkLED.cpp
 * simple library to light an LED for a duration given in milliseconds
 */
#include "Arduino.h"    // use: Wprogram.h for Arduino versions prior to 1.0
#include "blinkLED.h"

// blink the LED on the given pin for the duration in milliseconds
void blinkLED(int pin, int duration)
{
```

```
    digitalWrite(pin, HIGH);      // turn LED on
    delay(duration);
    digitalWrite(pin, LOW);       // turn LED off
    delay(duration);
}
```

Most library authors are programmers who use their favorite programming editor, but you can use any plain text editor to create these files.

Create the *blinkLED.h* header file as follows:

```
/*
 * blinkLED.h
 * Library header file for BlinkLED library
 */
#include "Arduino.h"

void blinkLED(int pin, int duration);  // function prototype
```

Discussion

The library will be named "blinkLED" and will be located in the libraries folder (see Recipe 16.2); create a subdirectory named *blinkLED* in the libraries folder and move *blinkLED.h* and *blinkLED.cpp* into it.

The `blinkLED` function from Recipe 7.1 is moved out of the sketch and into a library file named *blinkLED.cpp* (the *.cpp* extension stands for "C plus plus" and contains the executable code).

The terms *functions* and *methods* are used in Arduino library documentation to refer to blocks of code such as `blinkLED`. The term *method* was introduced to refer to the functional blocks in a class. Both terms refer to the named functional blocks that are made accessible by a library.

The *blinkLED.cpp* file contains a `blinkLED` function that is identical to the code from Recipe 7.1 with the following two lines added at the top:

```
#include "Arduino.h"  // Arduino include
#include "blinkLED.h"
```

The `#include "Arduino.h"` line is needed by a library that uses any Arduino functions or constants. Without this, the compiler will report errors for all the Arduino functions used in your sketch.

Arduino.h was added in Release 1.0 and replaces *WProgram.h*. If you are compiling sketches using earlier releases, you can use the following conditional include to bring in the correct version:

```
#if ARDUINO >= 100
#include "Arduino.h  // for 1.0 and later
#else
#include "WProgram.h" // for earlier releases
#endif
```

The next line, `#include "blinkLED.h"`, contains the function definitions (also known as *prototypes*) for your library. The Arduino build process creates prototypes for all the functions within a sketch automatically when a sketch is compiled—but it does not create any prototypes for library code, so if you make a library, you must create a header with these prototypes. It is this header file that is added to a sketch when you import a library from the IDE (see Recipe 16.1).

Every library must have a file that declares the names of the functions to be exposed. This file is called a *header file* (also known as an *include* file) and has the form *<LibraryName>.h* (where *<LibraryName>* is the name for your library). In this example, the header file is named *blinkLED.h* and is in the same folder as *blinkLED.cpp*.

The header file for this library is simple. It declares the one function:

```
void blinkLED(int pin, int duration);  // function prototype
```

This looks similar to the function definition in the *blinkLED.cpp* file:

```
void blinkLED(int pin, int duration)
```

The difference is subtle but vital. The header file prototype contains a trailing semicolon. This tells the compiler that this is just a declaration of the form for the function but not the code. The source file, *blinkLED.cpp*, does not contain the trailing semicolon and this informs the compiler that this is the actual source code for the function.

Libraries can have more than one header file and more than one implementation file. But there must be at least one header and that must match the library name. It is this file that is included at the top of the sketch when you import a library.

A good book on C++ can provide more details on using header and *.cpp* files to create code modules. This recipe's See Also section lists some popular choices.

With the *blinkLED.cpp* and *blinkLED.h* files in the correct place within the libraries folder, close the IDE and reopen it.

The Arduino IDE updates its list of available libraries only when the IDE is first started on your computer. If you create a library after the IDE is running, you need to close the IDE and restart for that library to be recognized.

Upload the blinkLibTest sketch and you should see the three LEDs blinking.

It's easy to add additional functionality to the library. For example, you can add some constant values for common delays so that users of your libraries can use the descriptive constants instead of millisecond values.

Add the three lines with constant values, traditionally put just before the function prototype, to your header file as follows:

```
// constants for duration
const int BLINK_SHORT = 250;
const int BLINK_MEDIUM = 500;
const int BLINK_LONG = 1000;

void blinkLED(int pin, int duration);  // function prototype
```

Change the code in `loop` as follows and upload the sketch to see the different blink rates:

```
void loop()
{
  blinkLED(firstLedPin, BLINK_SHORT);
  blinkLED(secondLedPin, BLINK_MEDIUM);
  blinkLED(thirdLedPin, BLINK_LONG);
}
```

You need to close and restart the IDE when you first add the library to the libraries folder, but not after subsequent changes to the library. Libraries included in Arduino release 0017 and later are recompiled each time the sketch is compiled. Arduino releases earlier than 0017 required the deletion of the library object files to make the library recompile and for changes to be included.

New functions can be easily added. This example adds a function that continues blinking for the number of times given by the sketch. Here is the `loop` code:

```
void loop()
{
  blinkLED(firstLedPin,BLINK_SHORT, 5);      // blink 5 times
  blinkLED(secondLedPin,BLINK_MEDIUM, 3);    // blink 3 times
  blinkLED(thirdLedPin, BLINK_LONG);          // blink once
}
```

To add this functionality to the library, add the prototype to *blinkLED.h* as follows:

```
/*
 * blinkLED.h
 * Header file for BlinkLED library
```

```
 */
#include "Arduino.h"

// constants for duration
const int BLINK_SHORT = 250;
const int BLINK_MEDIUM = 500;
const int BLINK_LONG = 1000;

void blinkLED(int pin, int duration);

// new function for repeat count
void blinkLED(int pin, int duration, int repeats);
```

Add the function into *blinkLED.cpp*:

```
/*
 * blinkLED.cpp
 * simple library to light an LED for a duration given in milliseconds
 */
#include "Arduino.h"   // use: Wprogram.h for Arduino versions prior to 1.0
#include "blinkLED.h"

// blink the LED on the given pin for the duration in milliseconds
void blinkLED(int pin, int duration)
{
  digitalWrite(pin, HIGH);     // turn LED on
  delay(duration);
  digitalWrite(pin, LOW);      // turn LED off
  delay(duration);
}

/* function to repeat blinking */
void blinkLED(int pin, int duration, int repeats)
{
  while(repeats)
  {
     blinkLED(pin, duration);
     repeats = repeats -1;
  }
}
```

You can create a *keywords.txt* file if you want to add *syntax highlighting* (coloring the keywords used in your library when viewing a sketch in the IDE). This is a text file that contains the name of the keyword and the keyword type—each type uses a different color. The keyword and type must be separated by a tab (not a space). For example, save the following file as *keywords.txt* in the *blinkLED* folder:

```
#######################################
# Methods and Functions (KEYWORD2)
#######################################
blinkLED  KEYWORD2
#######################################
# Constants (LITERAL1)
#######################################
BLINK_SHORT    LITERAL1
```

```
BLINK_MEDIUM    LITERAL1
BLINK_LONG      LITERAL1
```

You need to quit and restart the IDE when you create a new library or when you add or modify a *keywords.txt* file. You do not need to restart after modifying library code (*.c* or *.cpp*) or header (*.h*) files.

See Also

See Recipe 16.5 for more examples of writing a library.

"Writing a Library for Arduino" reference document: *http://www.arduino.cc/en/Hacking/LibraryTutorial*

Also see the following books on C++:

- *Practical C++ Programming* by Steve Oualline (O'Reilly; search for it on oreilly.com)
- *C++ Primer Plus* by Stephen Prata (Sams)
- *C++ Primer* by Stanley B. Lippman, Josée Lajoie, and Barbara E. Moo (Addison-Wesley Professional)

16.5 Creating a Library That Uses Other Libraries

Problem

You want to create a library that uses functionality from one or more existing libraries. For example, to use the Wire library to get data from a Wii nunchuck game controller.

Solution

This recipe uses the functions described in Recipe 13.2 to communicate with a Wii Nunchuck using the Wire library.

Create a folder named *Nunchuck* in the libraries directory (see Recipe 16.4 for details on the file structure for a library). Create a file named *Nunchuck.h* with the following code:

```
/*
 * Nunchuck.h
 * Arduino library to interface with wii Nunchuck
 */

#ifndef Nunchuck_included
#define Nunchuck_included

// identities for each field provided by the wii nunchuck
enum nunchuckItems { wii_joyX, wii_joyY, wii_accelX, wii_accelY, wii_accelZ,
```

```
                        wii_btnC, wii_btnZ, wii_ItemCount };

// uses pins adjacent to I2C pins as power & ground for Nunchuck
void nunchuckSetPowerpins();

// initialize the I2C interface for the nunchuck
void nunchuckInit();

// Request data from the nunchuck
void nunchuckRequest();

// Receive data back from the nunchuck,
// returns true if read successful, else false
bool nunchuckRead();

// Encode data to format that most wiimote drivers except
char nunchuckDecode (uint8_t x);

// return the value for the given item
int nunchuckGetValue(int item);

#endif
```

Create a file named *Nunchuck.cpp* in the *Nunchuck* folder as follows:

```
/*
 * Nunchuck.cpp
 * Arduino library to interface with wii Nunchuck
 */

#include "Arduino.h"   // Arduino defines

#include "Wire.h"      // Wire (I2C) defines
#include "Nunchuck.h"  // Defines for this library

// defines for standard Arduino board (use 19 and 18 for mega)
const int vccPin = 17; // +v and gnd provided through these pins
const int gndPin = 16;


const int dataLength = 6;          // number of bytes to request
static byte rawData[dataLength];   // array to store nunchuck data

// uses pins adjacent to I2C pins as power & ground for Nunchuck
void nunchuckSetPowerpins()
{
  pinMode(gndPin, OUTPUT); // set power pins to the correct state
  pinMode(vccPin, OUTPUT);
  digitalWrite(gndPin, LOW);
  digitalWrite(vccPin, HIGH);
  delay(100); // wait for power to stabilize
}

// initialize the I2C interface for the nunchuck
void nunchuckInit()
{
```

```
  Wire.begin();                 // join i2c bus as master
  Wire.beginTransmission(0x52);// transmit to device 0x52
  Wire.write((byte)0x40);       // sends memory address
  Wire.write((byte)0x00);       // sends sent a zero.
  Wire.endTransmission();       // stop transmitting
}

// Request data from the nunchuck
void nunchuckRequest()
{
  Wire.beginTransmission(0x52);// transmit to device 0x52
  Wire.write((byte)0x00);// sends one byte
  Wire.endTransmission();// stop transmitting
}

// Receive data back from the nunchuck,
// returns true if read successful, else false
bool nunchuckRead()
{
  byte cnt=0;
  Wire.requestFrom (0x52, dataLength);// request data from nunchuck
  while (Wire.available ()) {
    byte x = Wire.read();
    rawData[cnt] = nunchuckDecode(x);
    cnt++;
  }
  nunchuckRequest();  // send request for next data payload
  if (cnt >= dataLength)
    return true;   // success if all 6 bytes received
  else
    return false; // failure
}

// Encode data to format that most wiimote drivers except
char nunchuckDecode (byte x)
{
   return (x ^ 0x17) + 0x17;
}

// return the value for the given item
int nunchuckGetValue(int item)
{
  if( item <= wii_accelZ)
    return (int)rawData[item];
  else if(item  == wii_btnZ)
    return bitRead(rawData[5], 0) ? 0: 1;
  else if(item  == wii_btnC)
    return bitRead(rawData[5], 1) ? 0: 1;
}
```

Connect the nunchuck as shown in Recipe 13.2 but use the following sketch to test the library (if Arduino was running while you created the previous two files, quit and restart it so it will see the new library):

```
/*
 * WiichuckSerial
 *
 *  Uses Nunchuck library to sends sensor values to serial port
 */

#include <Wire.h>
#include "Nunchuck.h"

void setup()
{
  Serial.begin(9600);
  nunchuckSetPowerpins();
  nunchuckInit(); // send the initialization handshake
  nunchuckRead(); // ignore the first time
  delay(50);
}

void loop()
{
  nunchuckRead();
  Serial.print("H,");        // header
  for(int i=0; i <  5; i++) // print values of accelerometers and buttons
  {
     Serial.print(nunchuckGetValue(wii_accelX+ i), DEC);
     Serial.write(',');
  }
  Serial.println();
  delay(20); // the time in milliseconds between sends
}
```

Discussion

To include another library, use its `include` statement in your code as you would in a sketch. It is sensible to include information about any additional libraries that your library needs in documentation if you make it available for others to use, especially if it requires a library that is not distributed with Arduino.

The major difference between the library code and the sketch from Recipe 13.2 is the addition of the *Nunchuck.h* header file that contains the function prototypes (Arduino sketch code silently creates prototypes for you, unlike Arduino libraries which require explicit prototypes).

Here is another example of creating a library; this one uses a C++ class to encapsulate the library functions. A class is a programming technique for grouping functions and variables together and is commonly used for most Arduino libraries.

This library can be used as a debugging aid by sending print output to a second Arduino board using the Wire library. This is particularly useful when the hardware serial port is not available and software serial solutions are not appropriate due to the timing delays they introduce. Here the core Arduino print functionality is used to create a new library that sends printed output I2C. The connections and code are covered in Recipe 13.9. The following description shows how that code can be converted into a library.

Create a folder named *i2cDebug* in the libraries directory (see Recipe 16.4 for details on the file structure for a library). Create a file named *i2cDebug.h* with the following code:

```
/*
 * i2cDebug.h
 */
#ifndef i2cDebug_included
#define i2cDebug_included

#include <Arduino.h>
#include <Print.h>   // the Arduino print class

class i2cDebugClass : public Print
{
  private:
    int i2cAddress;
    byte count;
    size_t write(byte c);
  public:
    i2cDebugClass();
    boolean begin(int id);
};

extern i2cDebugClass i2cDebug;   // the i2c debug object
#endif
```

Create a file named *i2cDebug.cpp* in the *i2cDebug* folder as follows:

```
/*
 * i2cDebug.cpp
 */

#include <i2cDebug.h>

#include <Wire.h>    // the Arduino I2C library


i2cDebugClass::i2cDebugClass()
{
}

boolean i2cDebugClass::begin(int id)
{
  i2cAddress = id;  // save the slave's address
  Wire.begin(); // join I2C bus (address optional for master)
  return true;
}

size_t i2cDebugClass::write(byte c)
{
  if( count == 0)
  {
```

```
    // here if the first char in the transmission
    Wire.beginTransmission(i2cAddress); // transmit to device
  }
  Wire.write(c);
  //  if the I2C buffer is full or an end of line is reached, send the data
  // BUFFER_LENGTH is defined in the Wire library
  if(++count >= BUFFER_LENGTH || c == '\n')
  {
    // send data if buffer full or newline character
    Wire.endTransmission();
    count = 0;
  }
  return 1; // one character written
}

i2cDebugClass i2cDebug;   // Create an I2C debug object
```

The `write` method returns `size_t`, a value that enables the print function to return the number of characters printed. This is new in Arduino 1.0—earlier versions did not return a value from `write` or `print`. If you have a library that is based on Stream or Print then you will need to change the return type to `size_t`.

Load this example sketch into the IDE:

```
/*
 * i2cDebug
 * example sketch for i2cDebug library
 */

#include <Wire.h>    // the Arduino I2C library
#include <i2cDebug.h>

const int address = 4;    // the address to be used by the communicating devices
const int sensorPin = 0;  // select the analog input pin for the sensor
int val;                  // variable to store the sensor value


void setup()
{
  Serial.begin(9600);
  i2cDebug.begin(address);
}

void loop()
{
  // read the voltage on the pot(val ranges from 0 to 1023)
  val = analogRead(sensorPin);
  Serial.println(val);
  i2cDebug.println(val);
}
```

Remember that you need to restart the IDE after creating the library folder. See Recipe 16.4 for more details on creating a library.

Upload the slave I2C sketch onto another Arduino board and wire up the boards as described in Recipe 13.9, and you should see the output from the Arduino board running your library displayed on the second board.

The following references provide an introduction to classes if C++ classes are new to you :

- *Programming Interactivity* by Joshua Noble (O'Reilly; search for it on oreilly.com)
- *C++ Primer* by Stanley B. Lippman, Josée Lajoie, and Barbara E. Moo (Addison-Wesley Professional)

16.6 Updating Third-Party Libraries for Arduino 1.0

Problem

You want to use a third-party library created for Arduino releases previous to 1.0.

Solution

Most libraries should only require the change of a few lines to work under Arduino 1.0. For example, any one or more of these header file includes:

```
#include "wiring.h"
#include "WProgram.h"
#include "WConstants.h"
#include "pins_arduino.h"
```

should be changed to a single include of:

```
#include "Arduino.h"
```

The file names may be enclosed in either angle brackets or quotes

Discussion

Older libraries that don't compile under Arduino 1.0 will usually generate one or more of these error messages:

```
source file: error: wiring.h: No such file or directory
source file: error: WProgram.h: No such file or directory
source file: error: WConstants.h: No such file or directory
source file: error: pins_arduino.h: No such file or directory
```

Source file is the full path the library file that needs to be updated. There will be a list of other errors following this due to the indicated file not being found in the 1.0 release, but these should disappear after you have replaced the old header names with `Arduino.h`. The definitions in these files are now included in `Arduino.h` and the solution is to replace includes for all of the above files with a single include for `Arduino.h`

If you want to run Arduino 1.0 alongside earlier compiles, you can use a conditional define (see Recipe 17.6):

```
#if ARDUINO >= 100
#include "Arduino.h"
#else
// These are the filenames that are used in the original version of library
#include "wiring.h"
#include "pins_arduino.h"
#endif
```

See Also

Third-party libraries that use Serial, Ethernet or other functionality that has changed syntax in Arduino 1.0 may require additional code changes. See Appendix H and specific chapters in this book covering the functionality for details.

CHAPTER 17

Advanced Coding and Memory Handling

17.0 Introduction

As you do more with your Arduino, your sketches need to become more efficient. The techniques in this chapter can help you improve the performance and reduce the code size of your sketches. If you need to make your sketch run faster or use less RAM, the recipes here can help. The recipes here are more technical than most of the other recipes in this book because they cover things that are usually concealed by the friendly Arduino wrapper.

The Arduino build process was designed to hide complex aspects of C and C++, as well as the tools used to convert a sketch into the bytes that are uploaded and run on an Arduino board. But if your project has performance and resource requirements beyond the capability of the standard Arduino environment, you should find the recipes here useful.

The Arduino board uses memory to store information. It has three kinds of memory: program memory, random access memory (RAM), and EEPROM. Each has different characteristics and uses. Many of the techniques in this chapter cover what to do if you do not have enough of one kind of memory.

Program memory (also known as *flash*) is where the executable sketch code is stored. The contents of program memory can only be changed by the *bootloader* in the upload process initiated by the Arduino software running on your computer. After the upload process is completed, the memory cannot be changed until the next upload. There is far more program memory on an Arduino board than RAM, so it can be beneficial to store values that don't change while the code runs (e.g., constants) in program memory. The bootloader takes up some space in program memory. If all other attempts to minimize the code to fit in program memory have failed, the bootloader can be removed to free up space, but an additional hardware programmer is then needed to get code onto the board.

If your code is larger than the program memory space available on the chip, the upload will not work and the IDE will warn you that the sketch is too big when you compile.

RAM is used by the code as it runs to store the values for the variables used by your sketch (including variables in the libraries used by your sketch). RAM is *volatile*, which means it can be changed by code in your sketch. It also means anything stored in this memory is lost when power is switched off. Arduino has much less RAM than program memory. If you run out of RAM while your sketch runs on the board (as variables are created and destroyed while the code runs) the board will misbehave (crash).

EEPROM (electrically erasable programmable read-only memory) is memory that code running on Arduino can read and write, but it is nonvolatile memory that retains values even when power is switched off. EEPROM access is significantly slower than for RAM, so EEPROM is usually used to store configuration or other data that is read at startup to restore information from the previous session.

To understand these issues, it is helpful to understand how the Arduino IDE prepares your code to go onto the chip and how you can inspect the results it produces.

Preprocessor

Some of the recipes here use the *preprocessor* to achieve the desired result. Preprocessing is a step in the first stage of the build process in which the source code (your sketch) is prepared for compiling. Various find and replace functions can be performed. Preprocessor commands are identified by lines that start with `#`. You have already seen them in sketches that use a library—`#include` tells the preprocessor to insert the code from the named library file. Sometimes the preprocessor is the only way to achieve what is needed, but its syntax is different from C and C++ code, and it can introduce bugs that are subtle and hard to track down, so use it with care.

See Also

AVRfreaks is a website for software engineers that is a good source for technical detail on the controller chips used by Arduino: *http://www.avrfreaks.net*.

Technical details on the C preprocessor are available at *http://gcc.gnu.org/onlinedocs/gcc-2.95.3/cpp_1.html*.

The memory specifications for all of the official boards can be found on the Arduino website (*http://www.arduino.cc/en/Main/hardware*).

17.1 Understanding the Arduino Build Process

Problem

You want to see what is happening under the covers when you compile and upload a sketch.

Solution

You can choose to display all the command-line activity that takes place when compiling or uploading a sketch through the Preferences dialog added in Arduino 1.0. Select File→Preferences to display the dialog box to check or uncheck the boxes to enable verbose output for compile or upload messages.

In releases earlier than 1.0, you can hold down the Shift key when you click on Compile or Upload. The console area at the bottom of the IDE will display details of the compile process.

In releases earlier than 1.0, you need to change a value in the Arduino *preferences.txt* file to make this detail always visible. This file should be in the following locations:

Mac
: */Users/<USERNAME>/Library/Arduino/preferences.txt*

Windows XP
: *C:\Documents and Settings\<USERNAME>\Application Data\Arduino\preferences.txt*

Windows Vista
: *c:\Users\<USERNAME>\AppData\Roaming\Arduino\ preferences.txt*

Linux
: *~/.arduino/preferences.txt*

Make sure the Arduino IDE is not running (changes made to *preferences.txt* will not be saved if the IDE is running). Open the file and find the line `build.verbose=false` (it is near the bottom of the file). Change `false` to `true` and save the file.

Discussion

When you click on Compile or Upload, a lot of activity happens that is not usually displayed on-screen. The command-line tools that the Arduino IDE was built to hide are used to compile, link, and upload your code to the board.

First your sketch file(s) are transformed into a file suitable for the compiler (*AVR-GCC*) to process. All source files in the sketch folder that have no file extension are joined together to make one file. All files that end in *.c* or *.cpp* are compiled separately. Header files (with an *.h* extension) are ignored unless they are explicitly included in the files that are being joined.

`#include "Arduino.h"` (`WProgram.h` in previous releases) is added at the top of the file to include the header file with all the Arduino-specific code definitions, such as `digitalWrite()` and `analogRead()`. If you want to examine its contents, you can find the file on Windows under the directory where Arduino was installed; from there, you can navigate to Hardware→Arduino→Cores→Arduino.

On the Mac, Ctrl+click the Arduino application icon and select Show Package Contents from the drop-down menu. A folder will open; from the folder, navigate to Contents→Resources→Java→Hardware→Arduino→Cores→Arduino.

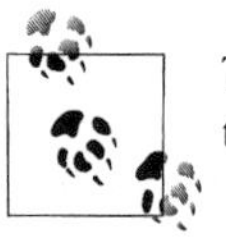

The Arduino directory structure may change in new releases, so check the documentation for the release you are using.

To make the code valid C++, the prototypes of any functions declared in your code are generated next and inserted.

Finally, the setting of the board menu is used to insert values (obtained from the *boards.txt* file) that define various constants used for the controller chips on the selected board.

This file is then compiled by AVR-GCC, which is included within the Arduino main download (it is in the *tools* folder).

The compiler produces a number of object files (files with an extension of *.o* that will be combined by the link tool). These files are stored in */tmp* on Mac and Linux. On Windows, they are in the applet directory (a folder below the Arduino install directory).

The object files are then linked together to make a hex file to upload to the board. Avrdude, a utility for transferring files to the Arduino controller, is used to upload to the board.

The tools used to implement the build process can be found in the *hardware\tools* directory.

Another useful tool for experienced programmers is `avr-objdump`, also in the *tools* folder. It lets you see how the compiler turns the sketch into code that the controller chip runs. This tool produces a disassembly listing of your sketch that shows the object code intermixed with the source code. It can also display a memory map of all the variables used in your sketch. To use the tool, compile the sketch and navigate to the folder containing the Arduino distribution. Then, navigate to the folder with all the intermediate files used in the build process (as explained earlier). The file used by `avr-objdump` is the one with the extension *.elf*. For example, if you compile the Blink sketch you could view the compiled output (the machine code) by executing the following on the command line:

```
..\hardware\tools\avr\bin\avr-objdump.exe -S blink.cpp.elf
```

It is convenient to direct the output to a file that can be read in a text editor. You can do this as follows:

```
..\hardware\tools\avr\bin\avr-objdump.exe -S blink.cpp.elf > blink.txt
```

This version adds a list of section headers (helpful for determining memory usage):

```
..\hardware\tools\avr\bin\avr-objdump.exe -S -h blink.cpp.elf > blink.txt
```

You can create a batch file to dump the listing into a file. Add the path of your Arduino installation to the following line and save it to a batch file:

```
hardware\tools\avr\bin\avr-objdump.exe -S -h -Tdata %1 > %1%.txt
```

See Also

For information on the Arduino build process, see *http://code.google.com/p/arduino/wiki/BuildProcess*.

The AVRfreaks website: *http://www.avrfreaks.net/wiki/index.php/Documentation:AVR_GCC*.

17.2 Determining the Amount of Free and Used RAM

Problem

You want to be sure you have not run out of RAM. A sketch will not run correctly if there is insufficient memory, and this can be difficult to detect.

Solution

This recipe shows you how you can determine the amount of free memory available to your sketch. This sketch contains a function called `memoryFree` that reports the amount of available RAM:

```
void setup()
{
  Serial.begin(9600);
}

void loop()
{
  Serial.print(memoryFree());  // print the free memory
  Serial.print(' ');           // print a space
  delay(1000);
}


// variables created by the build process when compiling the sketch
extern int __bss_end;
extern void *__brkval;

// function to return the amount of free RAM
int memoryFree()
```

```
{
  int freeValue;
  if((int)__brkval == 0)
     freeValue = ((int)&freeValue) - ((int)&__bss_end);
  else
    freeValue = ((int)&freeValue) - ((int)__brkval);

  return freeValue;
}
```

Discussion

The `memoryFree` function uses system variables to calculate the amount of RAM. System variables are not normally visible (they are created by the compiler to manage internal resources). It is not necessary to understand how the function works to use its output. The function returns the number of bytes of free memory.

The number of bytes your code uses changes as the code runs. The important thing is to ensure that you don't consume more memory than you have.

Here are the main ways RAM memory is consumed:

- When you initialize constants:

  ```
  #define ERROR_MESSAGE "an error has occurred"
  ```

- When you declare global variables:

  ```
  char myMessage[] = "Hello World";
  ```

- When you make a function call:

  ```
  void myFunction(int value)
  {
     int result;
     result = value * 2;
     return result;
  }
  ```

- When you dynamically allocate memory:

  ```
  String stringOne = "Arduino String";
  ```

The Arduino `String` class uses dynamic memory to allocate space for strings. You can see this by adding the following line to the very top of the code in the Solution:

```
String s = "\n";
```

and the following lines just before the `delay` in the `loop` code:

```
s = s + "Hello I am Arduino \n";
  Serial.println(s);            // print the string value
```

You will see the memory value reduce as the size of the string is increased each time through the `loop`. If you run the sketch long enough, the memory will run out—don't endlessly try to increase the size of a string in anything other than a test application.

Writing code like this that creates a constantly expanding value is a sure way to run out of memory. You should also be careful not to create code that dynamically creates different numbers of variables based on some parameter while the code runs, as it will be very difficult to be sure you will not exceed the memory capabilities of the board when the code runs.

Constants and global variables are often declared in libraries as well, so you may not be aware of them, but they still use up RAM. The Serial library, for example, has a 128-byte global array that it uses for incoming serial data. This alone consumes one-eighth of the total memory of an old Arduino 168 chip.

See Also

A technical overview of memory usage is available at *http://www.gnu.org/savannah-checkouts/non-gnu/avr-libc/user-manual/malloc.html*.

17.3 Storing and Retrieving Numeric Values in Program Memory

Problem

You have a lot of constant numeric data and don't want to allocate this to RAM.

Solution

Store numeric variables in program memory (the flash memory used to store Arduino programs).

This sketch adjusts a fading LED for the nonlinear sensitivity of human vision. It stores the values to use in a table of 256 values in program memory rather than RAM.

The sketch is based on Recipe 7.2; see Chapter 7 for a wiring diagram and discussion on driving LEDs. Running this sketch results in a smooth change in brightness with the LED on pin 5 compared to the LED on pin 3:

```
/*
 * ProgmemCurve sketch
 * uses table in program memory to convert linear to exponential output
 * See Recipe 7.2 and Figure 7-2
 */

#include <avr/pgmspace.h>  // needed for PROGMEM

// table of exponential values
// generated for values of i from 0 to 255 -> x=round( pow( 2.0, i/32.0) - 1);

const byte table[]PROGMEM = {
   0,   0,   0,   0,   0,   0,   0,   0,   0,   0,   0,   0,   0,   0,   0,   0,
   0,   0,   0,   1,   1,   1,   1,   1,   1,   1,   1,   1,   1,   1,   1,   1,
```

```
   1,   1,   1,   1,   1,   1,   1,   1,   1,   1,   1,   2,   2,   2,   2,   2,
   2,   2,   2,   2,   2,   2,   2,   2,   2,   2,   3,   3,   3,   3,   3,   3,
   3,   3,   3,   3,   3,   3,   4,   4,   4,   4,   4,   4,   4,   4,   4,   5,
   5,   5,   5,   5,   5,   5,   5,   6,   6,   6,   6,   6,   6,   6,   7,   7,
   7,   7,   7,   8,   8,   8,   8,   8,   9,   9,   9,   9,   9,  10,  10,  10,
  10,  11,  11,  11,  11,  12,  12,  12,  12,  13,  13,  13,  14,  14,  14,  15,
  15,  15,  16,  16,  16,  17,  17,  18,  18,  18,  19,  19,  20,  20,  21,  21,
  22,  22,  23,  23,  24,  24,  25,  25,  26,  26,  27,  28,  28,  29,  30,  30,
  31,  32,  32,  33,  34,  35,  35,  36,  37,  38,  39,  40,  40,  41,  42,  43,
  44,  45,  46,  47,  48,  49,  51,  52,  53,  54,  55,  56,  58,  59,  60,  62,
  63,  64,  66,  67,  69,  70,  72,  73,  75,  77,  78,  80,  82,  84,  86,  88,
  90,  91,  94,  96,  98, 100, 102, 104, 107, 109, 111, 114, 116, 119, 122, 124,
 127, 130, 133, 136, 139, 142, 145, 148, 151, 155, 158, 161, 165, 169, 172, 176,
 180, 184, 188, 192, 196, 201, 205, 210, 214, 219, 224, 229, 234, 239, 244, 250
};

const int rawLedPin  = 3;           // this LED is fed with raw values
const int adjustedLedPin = 5;       // this LED is driven from table

int brightness = 0;
int increment = 1;

void setup()
{
  // pins driven by analogWrite do not need to be declared as outputs
}

void loop()
{
  if (brightness > 254)
  {
     increment = -1; // count down after reaching 255
  }
  else if (brightness < 1)
  {
    increment =  1; // count up after dropping back down to 0
  }
  brightness = brightness + increment; // increment (or decrement sign is minus)

  // write the brightness value to the LEDs
  analogWrite(rawLedPin, brightness);  // this is the raw value
  int adjustedBrightness = pgm_read_byte(&table[brightness]);  // adjusted value
  analogWrite(adjustedLedPin, adjustedBrightness);

  delay(10); // 10ms for each step change means 2.55 secs to fade up or down
}
```

Discussion

When you need to use a complex expression to calculate a range of values that regularly repeat, it is often better to precalculate the values and include them in a table of values (usually as an array) in the code. This saves the time needed to calculate the values repeatedly when the code runs. The disadvantage concerns the memory needed to place these values in RAM. RAM is limited on Arduino and the much larger program memory space can be used to store constant values. This is particularly helpful for sketches that have large arrays of numbers.

At the top of the sketch, the table is defined with the following expression:

```
const byte table[]PROGMEM = {
  0, . . .
```

`PROGMEM` tells the compiler that the values are to be stored in program memory rather than RAM. The remainder of the expression is similar to defining a conventional array (see Chapter 2).

The low-level definitions needed to use `PROGMEM` are contained in a file named *pgmspace.h* and the sketch includes this as follows:

```
#include <avr/pgmspace.h>
```

To adjust the brightness to make the fade look uniform, this recipe adds the following lines to the LED output code used in Recipe 7.2:

```
  int adjustedBrightness = pgm_read_byte(&table[brightness]);
  analogWrite(adjustedLedPin, adjustedBrightness);
```

The variable `adjustedBrightness` is set from a value read from program memory. The expression `pgm_read_byte(&table[brightness]);` means to return the address of the entry in the `table` array at the index position given by `brightness`. Each entry in the table is one byte, so another way to write this expression is:

```
pgm_read_byte(table + brightness);
```

If it is not clear why `&table[brightness]` is equivalent to `table + brightness`, don't worry; use whichever expression makes more sense to you.

Another example is from Recipe 6.5, which used a table for converting an infrared sensor reading into distance. Here is the sketch from that recipe converted to use a table in program memory instead of RAM:

```
/* ir-distance_Progmem sketch
 * prints distance & changes LED flash rate depending on distance from IR sensor
 * uses progmem for table
 */

#include <avr/pgmspace.h> // needed when using Progmem

// table entries are distances in steps of 250 millivolts
const int TABLE_ENTRIES = 12;
const int firstElement = 250; // first entry is 250 mV
const int interval  = 250; // millivolts between each element
// the following is the definition of the table in Program Memory
const int distanceP[TABLE_ENTRIES] PROGMEM = { 150,140,130,100,60,50,
40,35,30,25,20,15 };

// This function reads from Program Memory at the given index
int getTableEntry(int index)
{
```

```
  int value =  pgm_read_word(&distanceP[index]);
  return value;
}
```

The remaining code is similar to Recipe 6.5, except that the `getTableEntry` function is used to get the value from program memory instead of accessing a table in RAM. Here is the revised `getDistance` function from that recipe:

```
int getDistance(int mV)
{
   if( mV >  interval * TABLE_ENTRIES )
      return getTableEntry(TABLE_ENTRIES-1); // the minimum distance
   else
   {
      int index = mV / interval;
      float frac = (mV % 250) / (float)interval;
      return getTableEntry(index) - ((getTableEntry(index) -
getTableEntry(index+1)) * frac);
   }
}
```

See Also

See Recipe 17.4 for the technique introduced in Arduino 1.0 to store strings in flash memory.

17.4 Storing and Retrieving Strings in Program Memory

Problem

You have lots of strings and they are consuming too much RAM. You want to move string constants, such as menu prompts or debugging statements, out of RAM and into program memory.

Solution

This sketch creates a string in program memory and prints its value to the Serial Monitor using the `F("text")` expression introduced in Arduino 1.0. The technique for printing the amount of free RAM is described in Recipe 17.2:

```
/*
 * Write strings using Program memory (Flash)
 */

void setup()
{
  Serial.begin(9600);
}

void loop()
{
  Serial.print(memoryFree());  // print the free memory
```

```
  Serial.print(' ');           // print a space

  Serial.print(F("arduino duemilanove ")); // print the string
  delay(1000);
}

// variables created by the build process when compiling the sketch
extern int __bss_end;
extern void *__brkval;

// function to return the amount of free RAM
int memoryFree()
{
  int freeValue;

  if ((int)__brkval == 0)
     freeValue = ((int)&freeValue) - ((int)&__bss_end);
  else
    freeValue = ((int)&freeValue) - ((int)__brkval);

  return freeValue;
}
```

Discussion

Strings are particularly hungry when it comes to RAM. Each character uses a byte, so it is easy to consume large chunks of RAM if you have lots of words in strings in your sketch. Inserting your text in the `F("text")` expression stores the text in the much larger flash memory instead of RAM.

If you are using an earlier Arduino release you can still store text in program memory, but you need to add a little more code to your sketch. Here is the same functionality implemented for releases earlier than 1.0:

```
#include <avr/pgmspace.h> // for progmem

//create a string of 20 characters in progmem
const prog_uchar myText[] PROGMEM = "arduino duemilanove ";

void setup()
{
  Serial.begin(9600);
}

void loop()
{
  Serial.print(memoryFree()); // print the free memory
  Serial.print(' ');          // print a space

  printP(myText);             // print the string
  delay(1000);
}

// function to print a PROGMEM string
```

```
void printP(const prog_uchar *str)
{
char c;

   while((c = pgm_read_byte(str++)))
     Serial.write(c);
}

// variables created by the build process when compiling the sketch
extern int __bss_end;
extern void *__brkval;

// function to return the amount of free RAM
int memoryFree(){
  int freeValue;

  if((int)__brkval == 0) freeValue = ((int)&freeValue) - ((int)&__bss_end);
  else freeValue = ((int)&freeValue) - ((int)__brkval);
  return freeValue;
}
```

See Also

See Recipe 15.11 for an example of flash memory used to store web page strings.

17.5 Using #define and const Instead of Integers

Problem

You want to minimize RAM usage by telling the compiler that the value is constant and can be optimized.

Solution

Use `const` to declare values that are constant throughout the sketch.

For example, instead of:

```
int ledPin=13;
```

use:

```
const int ledPin=13;
```

Discussion

We often want to use a constant value in different areas of code. Just writing the number is a really bad idea. If you later want to change the value used, it's difficult to work out which numbers scattered throughout the code also need to be changed. It is best to use named references.

Here are three different ways to define a value that is a constant:

```
int ledPin =  13;           // a variable, but this wastes RAM
const int ledPin =  13;     // a const does not use RAM
#define ledPin 13           // using a #define
                            // the preprocessor replaces ledPin with 13

pinMode(ledPin, OUTPUT);
```

Although the first two expressions look similar, the term `const` tells the compiler not to treat `ledPin` as an ordinary variable. Unlike the ordinary `int`, no RAM is reserved to hold the value for the `const`, as it is guaranteed not to change. The compiler will produce exactly the same code as if you had written:

```
pinMode(13, OUTPUT);
```

You will sometimes see `#define` used to define constants in older Arduino code, but `const` is a better choice than `#define`. This is because a `const` variable has a *type*, which enables the compiler to verify and report if the variable is being used in ways not appropriate for that type. The compiler will also respect C rules for the scope of a `const` variable. A `#define` value will affect all the code in the sketch, which may be more than you intended. Another benefit of `const` is that it uses familiar syntax—`#define` does not use the equals sign, and no semicolon is used at the end.

See Also

See this chapter's introduction section for more on the preprocessor.

17.6 Using Conditional Compilations

Problem

You want to have different versions of your code that can be selectively compiled. For example, you may need code to work differently when debugging or when running with different boards.

Solution

You can use the conditional statements aimed at the preprocessor to control how your sketch is built.

This example from sketches in Chapter 15 includes the *SPI.h* library file that is only available for and needed with Arduino versions released after 0018:

```
#if ARDUINO > 18
#include <SPI.h>          // needed for Arduino versions later than 0018
#endif
```

This example, using the sketch from Recipe 5.6, displays some debug statements only if `DEBUG` is defined:

```
/*
 Pot_Debug sketch
 blink an LED at a rate set by the position of a potentiometer
 Uses Serial port for debug if DEBUG is defined
 */

const int potPin = 0;    // select the input pin for the potentiometer
const int ledPin = 13;   // select the pin for the LED
int val = 0;             // variable to store the value coming from the sensor

#define DEBUG

void setup()
{
  Serial.begin(9600);
  pinMode(ledPin, OUTPUT);      // declare the ledPin as an OUTPUT
}

void loop() {
  val = analogRead(potPin);     // read the voltage on the pot
  digitalWrite(ledPin, HIGH);  // turn the ledPin on
  delay(val);                   // blink rate set by pot value
  digitalWrite(ledPin, LOW);   // turn the ledPin off
  delay(val);                   // turn LED off for same period as it was turned on
#if defined DEBUG
  Serial.println(val);
#endif
}
```

Discussion

This recipe uses the preprocessor used at the beginning of the compile process to change what code is compiled. The first example tests if the value of the constant `ARDUINO` is greater than 18, and if so, the file *SPI.h* is included. The value of the `ARDUINO` constant is defined in the build process and corresponds to the Arduino release version. The syntax for this expression is not the same as that used for writing a sketch. Expressions that begin with the # symbol are processed before the code is compiled—see this chapter's introduction section for more on the preprocessor.

You have already come across `#include`:

```
#include <library.h>
```

The `< >` brackets tell the compiler to look for the file in the location for standard libraries:

```
#include "header.h"
```

The compiler will also look in the sketch folder.

You can have a conditional compile based on the controller chip selected in the IDE. For example, the following code will produce different code when compiled for a Mega board that reads the additional analog pins that it has:

```
/*
 * ConditionalCompile sketch
 * This sketch recognizes the controller chip using conditional defines
 */


int numberOfSensors;
int val = 0;                  // variable to store the value coming from the sensor

void setup()
{
   Serial.begin(9600);

#if defined(__AVR_ATmega1280__)   // defined when selecting Mega in the IDE
  numberOfSensors = 16;           // the number of analog inputs on the Mega
#else                             // if not Mega then assume a standard board
  numberOfSensors = 6;            // analog inputs on a standard Arduino board
#endif

 Serial.print("The number of sensors is ");
 Serial.println(numberOfSensors);
}

void loop() {
  for(int sensor = 0; sensor < numberOfSensors; sensor++)
  {
     val = analogRead(sensor);    // read the sensor value
     Serial.println(val);         // display the value
  }
  Serial.println();
  delay(1000);         // delay a second between readings
}
```

See Also

Technical details on the C preprocessor are available at *http://gcc.gnu.org/onlinedocs/gcc-2.95.3/cpp_1.html*.

See the Discussion section of Recipe 16.4 for an example of conditional compilation used to handle differences between Arduino 1.0 and previous releases.

CHAPTER 18

Using the Controller Chip Hardware

18.0 Introduction

The Arduino platform simplifies programming by providing easy-to-use function calls to hide complex, low-level hardware functions. But some applications need to bypass the friendly access functions to get directly at hardware, either because that's the only way to get the needed functionality or because higher performance is required. This chapter shows how to access and use hardware functions that are not fully exposed through the documented Arduino language.

Changing register values can change the behavior of some Arduino functions (e.g., `millis`). The low-level capabilities described in this chapter require care, attention, and testing if you want your code to function correctly.

Registers

Registers are variables that refer to hardware memory locations. They are used by the chip to configure hardware functions or for storing the results of hardware operations. The contents of registers can be read and written by your sketch. Changing register values will change the way the hardware operates, or the state of something (such as the output of a pin). Some registers represent a numerical value (the number a timer will count to). Registers can control or report on hardware status; for example, the state of a pin or if an interrupt has occurred. Registers are referenced in code using their names—these are documented in the data sheet for the microcontrollers. Setting a register to a wrong value usually results in a sketch functioning incorrectly, so carefully check the documentation to ensure that you are using registers correctly.

Interrupts

Interrupts are signals that enable the controller chip to stop the normal flow of a sketch and handle a task that requires immediate attention before continuing with what it was doing. Arduino core software uses interrupts to handle incoming data from the serial port, to maintain the time for the `delay` and `millis` functions, and to trigger the `attachInterrupt` function. Libraries, such as Wire and Servo, use interrupts when an event occurs, so the code doesn't have to constantly check to see if the event has happened. This constant checking, called *polling*, can complicate the logic of your sketch. Interrupts can be a reliable way to detect signals of very short duration. Recipe 18.2 explains how to use interrupts to determine if a digital pin has changed state.

Two or more interrupts may occur before the handling of the first interrupt is completed; for example, if two switches are pressed at the same time and each generates a different interrupt. The interrupt handler for the first switch must be completed before the second interrupt can get started. Interrupts should be brief, because an interrupt routine that takes too much time can cause other interrupt handlers to be delayed or to miss events.

Arduino services one interrupt at a time. It suspends pending interrupts while it deals with an interrupt that has happened. Code to handle interrupts (called the *interrupt handler*, or *interrupt service routine*) should be brief to prevent undue delays to pending interrupts. An interrupt routine that takes too much time can cause other interrupt handlers to miss events. Activities that take a relatively long time, such as blinking an LED or even serial printing, should be avoided in an interrupt handler.

Timers

A standard Arduino board has three hardware timers for managing time-based tasks (the Mega has six). The timers are used in a number of Arduino functions:

Timer0
: Used for `millis` and `delay`; also `analogWrite` on pins 5 and 6

Timer1
: `analogWrite` functions on pins 9 and 10; also driving servos using the Servo library

Timer2
: `analogWrite` functions on pins 3 and 11

The Servo library uses the same timer as `analogWrite` on pins 9 and 10, so you can't use `analogWrite` with these pins when using the Servo library.

The Mega has three additional 16-bit timers and uses different pin numbers with `analogWrite`:

Timer0
: `analogWrite` functions on pins 4 and 13

Timer1
: `analogWrite` functions on pins 11 and 12

Timer2
: `analogWrite` functions on pins 9 and 10

Timer3
: `analogWrite` functions on pins 2, 3, and 5

Timer4
: `analogWrite` functions on pins 6, 7, and 8

Timer5
: `analogWrite` functions on pins 45 and 46

Timers are counters that count pulses from a time source, called a *timebase*. The timer hardware consists of 8-bit or 16-bit digital counters that can be programmed to determine the mode the timer uses to count. The most common mode is to count pulses from the timebase on the Arduino board, usually 16 MHz derived from a crystal; 16 MHz pulses repeat every 62.5 nanoseconds, and this is too fast for many timing applications, so the timebase rate is reduced by a divider called a *prescaler*. Dividing the timebase by 8, for example, increases the duration of each count to half a microsecond. For applications in which this is still too fast, other prescale values can be used (see Table 18-1).

Timer operation is controlled by values held in registers that can be read and written by Arduino code. The values in these registers set the timer frequency (the number of system timebase pulses between each count) and the method of counting (up, down, up and down, or using an external signal).

Here is an overview of the timer registers (*n* is the timer number):

Timer Counter Control Register A (TCCRnA)
: Determines the operating mode

Timer Counter Control Register B (TCCRnB)
: Determines the prescale value

Timer Counter Register (TCNTn)
: Contains the timer count

Output Compare Register A (OCRnA)
: Interrupt can be triggered on this count

Output Compare Register B (OCRnB)
: Interrupt can be triggered on this count

Timer/Counter Interrupt Mask Register (TIMSKn)
    Sets the conditions for triggering an interrupt

Timer/Counter 0 Interrupt Flag Register (TIFRn)
    Indicates if the trigger condition has occurred

Table 18-1 is an overview of the bit values used to set the timer precision. Details of the functions of the registers are explained in the recipes where they are used.

Table 18-1. Timer prescale values (16 MHz clock)

Prescale factor	CSx2, CSx1, CSx0	Precision	Time to overflow	
			8-bit timer	16-bit timer
1	B001	62.5 ns	16 µs	4.096 ms
8	B010	500 ns	128 µs	32.768 ms
64	B011	4 µs	1,024 µs	262.144 ms
256	B100	16 µs	4,096 µs	1048.576 ms
1,024	B101	64 µs	16,384 µs	4194.304 ms
	B110	External clock, falling edge		
	B111	External clock, rising edge		

All timers are initialized for a prescale of 64.

Precision in nanoseconds is equal to the CPU period (time for one CPU cycle) multiplied by the prescale.

Analog and Digital Pins

Chapter 5 described the standard Arduino functions to read and write (to/from) digital and analog pins. This chapter explains how you can control pins faster than using the Arduino read and write functions and make changes to analog methods to improve performance.

Some of the code in this chapter is more difficult to understand than the other recipes in this book, as it is moving beyond Arduino syntax and closer to the underlying hardware. These recipes work directly with the tersely named registers in the chip and use bit shifting and masking to manipulate bits in them. The benefit from this complexity is enhanced performance and functionality.

See Also

Overview of hardware resources: *http://code.google.com/p/arduino/wiki/HardwareResourceMap*

Timer1 (and Timer3) library: *http://www.arduino.cc/playground/Code/Timer1*

Tutorial on timers and PWM: *http://arduino.cc/en/Tutorial/SecretsOfArduinoPWM*

The Atmel ATmega 168/328 data sheets: *http://www.atmel.com/dyn/resources/prod_documents/doc8271.pdf*

Atmel application note on how to set up and use timers: *http://www.atmel.com/dyn/resources/prod_documents/DOC2505.PDF*

A thorough summary of information covering 8-bit timers: *http://www.cs.mun.ca/~rod/Winter2007/4723/notes/timer0/timer0.html*

Diagrams showing register settings for timer modes: *http://web.alfredstate.edu/weimandn/miscellaneous/atmega168_subsystem/atmega168_subsystem_index.html*

Wikipedia article on interrupts: *http://en.wikipedia.org/wiki/Interrupts*

18.1 Storing Data in Permanent EEPROM Memory

Problem

You want to store values that will be retained even when power is switched off.

Solution

Use the EEPROM library to read and write values in EEPROM memory. This sketch blinks an LED using values read from EEPROM and allows the values to be changed using the Serial Monitor:

```
/*
  based on Blink without Delay
  uses EEPROM to store blink values
 */

#include <EEPROM.h>

// these values are saved in EEPROM
const byte EEPROM_ID = 0x99;   // used to identify if valid data in EEPROM
byte ledPin =  13;             // the number of the LED pin
int interval = 1000;           // interval at which to blink (milliseconds)

// variables that do not need to be saved
int ledState = LOW;             // ledState used to set the LED
long previousMillis = 0;        // will store last time LED was updated

//constants used to identify EEPROM addresses
const int ID_ADDR = 0;       // the EEPROM address used to store the ID
const int PIN_ADDR = 1;      // the EEPROM address used to store the pin
const int INTERVAL_ADDR = 2; // the EEPROM address used to store the interval

void setup()
{
  Serial.begin(9600);
  byte id = EEPROM.read(ID_ADDR); // read the first byte from the EEPROM
  if( id == EEPROM_ID)
```

```
    {
      // here if the id value read matches the value saved when writing eeprom
      Serial.println("Using data from EEPROM");
      ledPin = EEPROM.read(PIN_ADDR);
      byte hiByte =  EEPROM.read(INTERVAL_ADDR);
      byte lowByte =  EEPROM.read(INTERVAL_ADDR+1);
      interval =  word(hiByte, lowByte); // see word function in Recipe 3.15
    }
    else
    {
      // here if the ID is not found, so write the default data
      Serial.println("Writing default data to EEPROM");
      EEPROM.write(ID_ADDR,EEPROM_ID); // write the ID to indicate valid data
      EEPROM.write(PIN_ADDR, ledPin); // save the pin in eeprom
      byte hiByte = highByte(interval);
      byte loByte = lowByte(interval);
      EEPROM.write(INTERVAL_ADDR, hiByte);
      EEPROM.write(INTERVAL_ADDR+1, loByte);

    }
    Serial.print("Setting pin to ");
    Serial.println(ledPin,DEC);
    Serial.print("Setting interval to ");
    Serial.println(interval);

    pinMode(ledPin, OUTPUT);
  }

  void loop()
  {
    // this is the same code as the BlinkWithoutDelay example sketch
    if (millis() - previousMillis > interval)
    {
      previousMillis = millis();     // save the last time you blinked the LED
      // if the LED is off turn it on and vice versa:
      if (ledState == LOW)
        ledState = HIGH;
      else
        ledState = LOW;
      digitalWrite(ledPin, ledState);   // set LED using value of ledState
    }
    processSerial();
  }

  // function to get duration or pin values from Serial Monitor
  // value followed by  i is interval, p is pin number
  int value = 0;

  void processSerial()
  {
     if( Serial.available())
    {
      char ch = Serial.read();
      if(ch >= '0' && ch <= '9') // is this an ascii digit between 0 and 9?
      {
```

```
        value = (value * 10) + (ch - '0'); // yes, accumulate the value
      }
      else if (ch == 'i')  // is this the interval
      {
        interval = value;
        Serial.print("Setting interval to ");
        Serial.println(interval);
        byte hiByte = highByte(interval);
        byte loByte = lowByte(interval);
        EEPROM.write(INTERVAL_ADDR, hiByte);
        EEPROM.write(INTERVAL_ADDR+1, loByte);
        value = 0; // reset to 0 ready for the next sequence of digits
      }
      else if (ch == 'p')  // is this the pin number
      {
        ledPin = value;
        Serial.print("Setting pin to ");
        Serial.println(ledPin,DEC);
        pinMode(ledPin, OUTPUT);
        EEPROM.write(PIN_ADDR, ledPin); // save the pin in eeprom
        value = 0; // reset to 0 ready for the next sequence of digits
      }
    }
  }
```

Open the Serial Monitor. As the sketch starts, it tells you whether it is using values previously saved to EEPROM or defaults, if this is the first time the sketch is started.

You can change values by typing a number followed by a letter to indicate the action. A number followed by the letter *i* changes the blink interval; a number followed by a *p* changes the pin number for the LED.

Discussion

Arduino contains EEPROM memory that will store values even when power is switched off. There are 512 bytes of EEPROM in a standard Arduino board, 4K bytes in a Mega.

The sketch uses the EEPROM library to read and write values in EEPROM memory.

Once the library is included in the sketch, an EEPROM object is available that accesses the memory. The library provides methods to `read`, `write`, and `clear`. `EEPROM.clear()` is not used in this sketch because it erases all the EEPROM memory.

The EEPROM library requires you to specify the address in memory that you want to read or write. This means you need to keep track of where each value is written so that when you access the value it is from the correct address.

To write a value, you use `EEPROM.write(address, value)`. The address is from 0 to 511 (on a standard Arduino board), and the value is a single byte.

To read, you use `EEPROM.read(address)`. The byte content of that memory address is returned.

The sketch stores three values in EEPROM. The first value stored is an `ID` value that is used only in `setup` to identify if the EEPROM has been previously written with valid data. If the value stored matches the expected value, the other variables are read from EEPROM and used in the sketch. If it doesn't match, this sketch has not been run on this board (otherwise, the `ID` would have been written), so the default values are written, including the `ID` value.

The sketch monitors the serial port, and new values received are written to EEPROM.

The sketch stores the `ID` value in EEPROM address 0, the pin number in address 1, and the two bytes for the interval start in address 2. The following line writes the pin number to EEPROM. The variable `ledPin` is a byte, so it fits into a single EEPROM address:

```
EEPROM.write(PIN_ADDR, ledPin); // save the pin in eeprom
```

Because interval is an `int`, it requires two bytes of memory to store the value:

```
byte hiByte = highByte(interval);
byte loByte = lowByte(interval);
EEPROM.write(INTERVAL_ADDR, hiByte);
EEPROM.write(INTERVAL_ADDR+1, loByte);
```

The preceding code splits the value into two bytes that are stored in two consecutive addresses. Any additional variables to be added to EEPROM would need to be placed in addresses that follow these two bytes.

Here is the code used to rebuild the `int` variable from EEPROM:

```
ledPin = EEPROM.read(PIN_ADDR);
byte hiByte =  EEPROM.read(INTERVAL_ADDR);
byte lowByte =  EEPROM.read(INTERVAL_ADDR+1);
interval =  word(hiByte, lowByte);
```

See Chapter 3 for more on using the `word` expression to create an integer from two bytes.

For more complicated use of EEPROM, it is advisable to draw out a map of what is being saved where, to ensure that no address is used by more than one value, and that multibyte values don't overwrite other information.

See Also

Recipe 3.14 provides more information on converting 16- and 32-bit values into bytes.

18.2 Using Hardware Interrupts

Problem

You want to perform some action when a digital pin changes value and you don't want to have to constantly check the pin state.

Solution

This sketch monitors pulses on pin 2 and stores the duration in an array. When the array has been filled (64 pulses have been received), the duration of each pulse is displayed on the Serial Monitor:

```
/*
  Interrupts sketch
  see Recipe 10.1 for connection diagram
 */

const int irReceiverPin = 2;         // pin the receiver is connected to
const int numberOfEntries = 64;      // set this number to any convenient value

volatile unsigned long microseconds;
volatile byte index = 0;
volatile unsigned long results[numberOfEntries];

void setup()
{
  pinMode(irReceiverPin, INPUT);
  Serial.begin(9600);
  attachInterrupt(0, analyze, CHANGE);  // encoder pin on interrupt 0 (pin 2);
  results[0]=0;
}

void loop()
{
  if(index >= numberOfEntries)
  {
    Serial.println("Durations in Microseconds are:") ;
    for( byte i=0; i < numberOfEntries; i++)
    {
      Serial.println(results[i]);
    }
    index = 0; // start analyzing again
  }
  delay(1000);
}

void analyze()
{
  if(index < numberOfEntries  )
  {
    if(index > 0)
    {
      results[index] = micros() - microseconds;
    }
    index = index + 1;
  }
  microseconds = micros();
}
```

If you have an infrared receiver module, you can use the wiring in Recipe 10.1 to measure the pulse width from an infrared remote control. You could also use the wiring in

Recipe 6.12 to measure pulses from a rotary encoder or connect a switch to pin 2 (see Recipe 5.1) to test with a push button.

Discussion

In `setup`, the `attachInterrupt(0, analyze, CHANGE);` call enables the sketch to handle interrupts. The first number in the call specifies which interrupt to initialize. On a standard Arduino board, two interrupts are available: number 0, which uses pin 2, and number 1 on pin 3. The Mega has four more: number 2, which uses pin 21, number 3 on pin 20, number 4 on pin 19, and number 5 on pin 18.

The next parameter specifies what function to call (sometimes called an *interrupt handler*) when the interrupt event happens; `analyze` in this sketch.

The final parameter specifies what should trigger the interrupt. `CHANGE` means whenever the pin level changes (goes from low to high, or high to low). The other options are:

`LOW`
: When the pin is low

`RISING`
: When the pin goes from low to high

`FALLING`
: When the pin goes from high to low

When reading code that uses interrupts, bear in mind that it may not be obvious when values in the sketch change because the sketch does not directly call the interrupt handler; it's called when the interrupt conditions occur.

In this sketch, the main `loop` checks the `index` variable to see if all the entries have been set by the interrupt handler. Nothing in `loop` changes the value of `index`. `index` is changed inside the `analyze` function when the interrupt condition occurs (pin 2 changing state). The `index` value is used to store the time since the last state change into the next slot in the `results` array. The time is calculated by subtracting the last time the state changed from the current time in microseconds. The current time is then saved as the last time a change happened. (Chapter 12 describes this method for obtaining elapsed time using the `millis` function; here `micros` is used to get elapsed microseconds instead of milliseconds.)

The variables that are changed in an interrupt function are declared as `volatile`; this lets the compiler know that the values could change at any time (by an interrupt handler). Without using the `volatile` keyword, the compiler would may store the values in registers that can be accidentally overwritten by an interrupt handler. To prevent this, the `volatile` keyword tells the compiler to store the values in RAM rather than registers.

Each time an interrupt is triggered, `index` is incremented and the current time is saved. The time difference is calculated and saved in the array (except for the first time the interrupt is triggered, when `index` is `0`). When the maximum number of entries has occurred, the inner block in `loop` runs, and it prints out all the values to the serial port. The code stays in the `while` loop at the end of the inner block, so you need to reset the board when you want to do another run.

See Also

Recipe 6.12 has an example of external interrupts used to detect movement in a rotary encoder.

18.3 Setting Timer Duration

Problem

You want to do something at periodic intervals, and you don't want to have your code constantly checking if the interval has elapsed. You would like to have a simple interface for setting the period.

Solution

The easiest way to use a timer is through a library. The following sketch uses the MsTimer2 library (*http://www.arduino.cc/playground/Main/MsTimer2*) to generate a pulse with a period that can be set using the Serial Monitor. This sketch flashes pin 13 at a rate that can be set using the Serial Monitor:

```
/*
  pulseTimer2
  pulse a pin at a rate set from serial input
 */

#include <MsTimer2.h>

const int pulsePin = 13;
const int NEWLINE = 10; // ASCII value for newline

int period = 100; // 10 milliseconds
boolean output = HIGH; // the state of the pulse pin

void setup()
{
  pinMode(pulsePin, OUTPUT);
  Serial.begin(9600);

  MsTimer2::set(period/2, flash);
  MsTimer2::start();

  period= 0; // reset to zero, ready for serial input
}

void loop()
```

```
{
   if( Serial.available())
  {
    char ch = Serial.read();
    if( isDigit(ch) ) // is this an ascii digit between 0 and 9?
    {
       period = (period * 10) + (ch - '0'); // yes, accumulate the value
    }
    else if (ch == NEWLINE)  // is the character the newline character
    {
       Serial.println(period);
       MsTimer2::set(period/2, flash);
       MsTimer2::start();
       period = 0; // reset to 0, ready for the next sequence of digits
    }
  }
}

void flash()
{
  digitalWrite(pulsePin, output);
  output = !output;  // invert the output
}
```

Run this with the Serial Monitor drop-down for appending a newline character at the end of every send (see "Discussion" on page 15).

Discussion

Enter digits for the desired period in milliseconds using the Serial Monitor. The sketch accumulates the digits and divides the received value by 2 to calculate the duration of the on and off states (the period is the sum of the on time and off time, so the smallest value you can use is 2). Bear in mind that an LED flashing very quickly may not appear to be flashing to the human eye.

This library uses Timer2, so it will prevent operation of `analogWrite` on pins 3 and 11.

This library enables you to use Timer2 by providing the timing interval and the name of the function to call when the interval has elapsed:

```
MsTimer2::set(period/2, flash);
```

This sets up the timer. The first parameter is the time for the timer to run in milliseconds. The second parameter is the function to call when the timer gets to the end of that time (the function is named `flash` in this recipe):

```
MsTimer2::start();
```

As the name implies, `start` starts the timer running. Another method, named `stop`, stops the timer.

As in Recipe 18.2, the sketch code does not directly call the function to perform the action. The LED is turned on and off in the `flash` function that is called by MsTimer2 each time it gets to the end of its time setting. The code in `loop` deals with any serial messages and changes the timer settings based on it.

Using a library to control timers is much easier than accessing the registers directly. Here is an overview of the inner workings of this library: Timers work by constantly counting to a value, signaling that they have reached the value, then starting again. Each timer has a *prescaler* that determines the counting frequency. The prescaler divides the system timebase by a factor such as 1, 8, 64, 256, or 1,024. The lower the prescale factor, the higher the counting frequency and the quicker the timebase reaches its maximum count. The combination of how fast to count, and what value to count to, gives the time for the timer. Timer2 is an 8-bit timer; this means it can count up to 255 before starting again from 0. (Timer1 and Timers 3, 4, and 5 on the Mega use 16 bits and can count up to 65,535.)

The MsTimer2 library uses a prescale factor of 64. On a 16 MHz Arduino board, each CPU cycle is 62.5 nanoseconds long, and when this is divided by the prescale factor of 64, each count of the timer will be 4,000 nanoseconds (62.5 * 64 = 4,000, which is four microseconds).

Remember that when you directly use a timer in your sketch, built-in functions that use that timer, such as `analogWrite`, may no longer work correctly.

See Also

An easy-to-use library for interfacing with Timer2: *http://www.arduino.cc/playground/Main/MsTimer2*

A collection of routines for interfacing with Timer1 (also Timer3 on the Mega): *http://www.arduino.cc/playground/Code/Timer1*

18.4 Setting Timer Pulse Width and Duration

Problem

You want Arduino to generate pulses with a duration and width that you specify.

Solution

This sketch generates pulses within the frequency range of 1 MHz to 1 Hz using Timer1 PWM on pin 9:

```
#include <TimerOne.h>

#define pwmRegister OCR1A    // the logical pin, can be set to OCR1B
const int   outPin =  9;     // the physical pin

long period = 10000;       // the period in microseconds
long pulseWidth = 1000;    // width of a pulse in microseconds

int prescale[] = {0,1,8,64,256,1024}; // the range of prescale values

void setup()
{
  Serial.begin(9600);
  pinMode(outPin, OUTPUT);
  Timer1.initialize(period);         // initialize timer1, 1000 microseconds
  setPulseWidth(pulseWidth);
}


void loop()
{
}

bool setPulseWidth(long microseconds)
{
  bool ret = false;

  int prescaleValue = prescale[Timer1.clockSelectBits];
  // calculate time per counter tick in nanoseconds
  long  precision = (F_CPU / 128000)  * prescaleValue  ;
  period = precision * ICR1 / 1000; // period in microseconds
  if( microseconds < period)
  {
    int duty = map(microseconds, 0,period, 0,1024);
    if( duty < 1)
      duty = 1;
    if(microseconds > 0 && duty < RESOLUTION)
    {
       Timer1.pwm(outPin, duty);
       ret = true;
    }
  }
  return ret;
}
```

Discussion

You set the pulse period to a value from 1 to 1 million microseconds by setting the value of the period at the top of the sketch. You can set the pulse width to any value in microseconds that is less than the period by setting the value of `pulseWidth`.

The sketch uses the Timer1 library from *http://www.arduino.cc/playground/Code/Timer1*.

Timer1 is a 16-bit timer (it counts from 0 to 65,535). It's the same timer used by `analogWrite` to control pins 9 and 10 (so you can't use this library and `analogWrite` on those pins at the same time). The sketch generates a pulse on pin 9 with a period and pulse width given by the values of the variables named `period` and `pulseWidth`. If you want to use pin 10 instead of pin 9, you can make the following change:

```
#define pwmRegister OCR1B    // the logical pin
const int   outPin =  10;    // the physical pin - OCRIB is pin 10
```

`OCR1A` and `OCR1B` are constants that are defined in the code included by the Arduino core software (OCR stands for Output Compare Register). Many different hardware registers in the Arduino hardware are not usually needed by a sketch (the friendly Arduino commands hide the actual register names). But when you need to access the hardware directly to get at functionality not provided by Arduino commands, these registers need to be accessed. Full details on the registers are in the Atmel data sheet for the chip.

The sketch in this recipe's Solution uses the following registers:

`ICR1` (Input Compare Register for Timer1) determines the period of the pulse. This register contains a 16-bit value that is used as the maximum count for the timer. When the timer count reaches this value it will be reset and start counting again from 0. In the sketch in this recipe's Solution, if each count takes 1 microsecond and the `ICR1` value is set to `1000`, the duration of each count cycle is 1,000 microseconds.

`OCR1A` (or `OCR1B` depending on which pin you want to use) is the Output Compare Register for Timer1. When the timer count reaches this value (and the timer is in PWM mode as it is here), the output pin will be set low—this determines the pulse width. For example, if each count takes one microsecond and the `ICR1` value is set to `1000` and `OCR1A` is set to `100`, the output pin will be `HIGH` for 100 microseconds and `LOW` for 900 microseconds (the total period is 1,000 microseconds).

The duration of each count is determined by the Arduino controller timebase frequency (typically 16 MHz) and the prescale value. The prescale is the value that the timebase is divided by. For example, with a prescale of 64, the timebase will be four microseconds.

The Timer1 library has many useful capabilities—see the Playground article (*http://www.arduino.cc/playground/Code/Timer1*) for details—but it does not provide for the setting of a specific pulse width. This functionality is added by the function named `setPulseWidth`.

This function uses a value of ICR1 to determine the period:

```
int prescaleValue = prescale[Timer1.clockSelectBits];
```

The prescale value is set by a variable in the library named clockSelectBits. This variable contains a value between 1 and 7—this is used as an index into the prescale array to get the current prescale factor.

The duration for each count (precision) is calculated by multiplying the prescale value by the duration of a timebase cycle:

```
// time per counter tick in ns
long  precision = (F_CPU / 128000)  * prescaleValue  ;
```

The period is the precision times the value of the ICR1 register; it's divided by 1,000 to give the duration in microseconds:

```
period = precision * ICR1 / 1000; // period in microseconds
```

The Timer1 library has a function named pwm that expects the duty cycle to be entered as a ratio expressed by a value from 1 to 1,023 (where 1 is the shortest pulse and 1,023 is the longest). This value is calculated using the Arduino map function to scale the microseconds given for the period into a proportional value of the period that ranges from 1 to 1,023:

```
int duty = map(microseconds, 0,period, 1,1023);
```

See Also

See "See Also" on page 602 for links to data sheets and other references for timers.

18.5 Creating a Pulse Generator

Problem

You want to generate pulses from Arduino and control the characteristics from the Serial Monitor.

Solution

This is an enhanced version of Recipe 18.4 that enables the frequency, period, pulse width, and duty cycle to be set from the serial port:

```
#include <TimerOne.h>

const char SET_PERIOD_HEADER      = 'p';
const char SET_FREQUENCY_HEADER   = 'f';
const char SET_PULSE_WIDTH_HEADER = 'w';
const char SET_DUTY_CYCLE_HEADER  = 'c';


#define pwmRegister OCR1A    // the logical pin, can be set to OCR1B
```

```
const int   outPin =  9;     // the physical pin

long period = 1000;      // the period in microseconds
int duty = 512;          // duty as a range from 0 to 1024, 512 is 50% duty cycle

int prescale[] = {0,1,8,64,256,1024}; // the range of prescale values

void setup()
{
  Serial.begin(9600);
  pinMode(outPin, OUTPUT);
  Timer1.initialize(period);        // initialize timer1, 1000 microseconds
  Timer1.pwm(9, duty);              // setup pwm on pin 9, 50% duty cycle
}


void loop()
{
   processSerial();
}

void processSerial()
{
 static long val = 0;

  if ( Serial.available())
  {
    char ch = Serial.read();

    if(ch >= '0' && ch <= '9')               // is ch a number?
    {
       val = val * 10 + ch - '0';            // yes, accumulate the value
    }
    else if(ch == SET_PERIOD_HEADER)
    {
      period = val;
      Serial.print("Setting period to ");
      Serial.println(period);
      Timer1.setPeriod(period);
      Timer1.setPwmDuty(outPin, duty);  // don't change the duty cycle
      show();
      val = 0;
    }
    else if(ch == SET_FREQUENCY_HEADER)
    {
      if(val > 0)
      {
        Serial.print("Setting frequency to ");
        Serial.println(val);
        period = 1000000 / val;
        Timer1.setPeriod(period);
        Timer1.setPwmDuty(outPin, duty);  // don't change the duty cycle
      }
      show();
      val = 0;
```

```
    }
    else if(ch ==  SET_PULSE_WIDTH_HEADER)
    {
      if( setPulseWidth(val) ) {
         Serial.print("Setting Pulse width to ");
         Serial.println(val);
      }
      else
         Serial.println("Pulse width too long for current period");
      show();
      val = 0;
    }
    else if(ch == SET_DUTY_CYCLE_HEADER)
    {
     if( val >0 && val < 100)
     {
        Serial.print("Setting Duty Cycle to ");
        Serial.println(val);
        duty = map(val,1,99, 1, ICR1);
        pwmRegister = duty;
        show();
     }
     val = 0;
    }
  }
}

bool setPulseWidth(long microseconds)
{
  bool ret = false;

  int prescaleValue = prescale[Timer1.clockSelectBits];
  // calculate time per tick in ns
  long  precision = (F_CPU / 128000)  * prescaleValue  ;
  period = precision * ICR1 / 1000; // period in microseconds
  if( microseconds < period)
  {
    duty = map(microseconds, 0,period, 0,1024);
    if( duty < 1)
      duty = 1;
    if(microseconds > 0 && duty < RESOLUTION)
    {
       Timer1.pwm(outPin, duty);
       ret = true;
    }
  }
  return ret;
}



void show()
{
   Serial.print("The period is ");
   Serial.println(period);
```

```
    Serial.print("Duty cycle is ");
    // pwmRegister is ICR1A or ICR1B
    Serial.print( map( pwmRegister, 0,ICR1, 1,99));
    Serial.println("%");
    Serial.println();
}
```

Discussion

This sketch is based on Recipe 18.4, with the addition of serial code to interpret commands to receive and set the frequency, period, pulse width, and duty cycle percent. Chapter 4 explains the technique used to accumulate the variable `val` that is then used for the desired parameter, based on the command letter.

You can add this function if you want to print instructions to the serial port:

```
void instructions()
{
    Serial.println("Send values followed by one of the following tags:");
    Serial.println(" p - sets period in microseconds");
    Serial.println(" f - sets frequency in Hz");
    Serial.println(" w - sets pulse width in microseconds");
    Serial.println(" c - sets duty cycle in %");
    Serial.println("\n(duty cycle can have one decimal place)\n");
}
```

See Also

Recipe 18.4

See "See Also" on page 602 for links to data sheets and other references for timers.

18.6 Changing a Timer's PWM Frequency

Problem

You need to increase or decrease the Pulse Width Modulation (PWM) frequency used with `analogWrite` (see Chapter 7). For example, you are using `analogWrite` to control a motor speed and there is an audible hum because the PWM frequency is too high, or you are multiplexing LEDs and the light is uneven because PWM frequency is too low.

Solution

You can adjust the PWM frequency by changing a register value. The register values and associated frequencies are shown in Table 18-2.

Table 18-2. Adjustment values for PWM

Timer0 (pins 5 and 6)		
TCCR0B value	Prescale factor (divisor)	Frequency
32 (1)	1	62500
33 (2)	8	7812.5
34	**64**	**976.5625**
35	256	244.140625
36	1,024	61.03515625
Timer1 (pins 9 and 10)		
TCCR1B prescale value	**Prescale factor (divisor)**	**Frequency**
1	1	312500
2	8	3906.25
3	**64**	**488.28125**
4	256	122.0703125
5	1,024	30.517578125
Timer2 (pins 11 and 3)		
TCCR2B value	**Prescale factor (divisor)**	**Frequency**
1	1	312500
2	8	3906.25
3	**64**	**488.28125**
4	256	122.0703125
5	1,024	30.517578125

All frequencies are in hertz and assume a 16 MHz system timebase. The default prescale factor of 64 is shown in bold.

This sketch enables you to select a timer frequency from the Serial Monitor. Enter a digit from 1 to 7 using the value in the lefthand column of Table 18-2 and follow this with character *a* for Timer0, *b* for Timer1, and *c* for Timer2:

```
const byte mask = B11111000; // mask bits that are not prescale values
int prescale = 0;

void setup()
{
  Serial.begin(9600);
  analogWrite(3,128);
  analogWrite(5,128);
  analogWrite(6,128);
```

```
  analogWrite(9,128);
  analogWrite(10,128);
  analogWrite(11,128);
}

void loop()
{
  if ( Serial.available())
  {
    char ch = Serial.read();
    if(ch >= '0' && ch <= '9')                // is ch a number?
    {
      prescale = ch - '0';
    }
    else if(ch == 'a')  // timer 0;
    {
      TCCR0B = (TCCR0B & mask)  | prescale;
    }
    else if(ch == 'b')  // timer 1;
    {
      TCCR1B = (TCCR1B & mask)  | prescale;
    }
    else if(ch == 'c')  // timer 2;
    {
      TCCR2B = (TCCR2B & mask)  | prescale;
    }
  }
}
```

Avoid changing the frequency of Timer0 (used for `analogWrite` pins 5 and 6) because it will result in incorrect timing using `delay` and `millis`.

Discussion

If you just have LEDs connected to the analog pins in this sketch, you will not see any noticeable change to the brightness as you change the PWM speed. You are changing the speed as they are turning on and off, not the ratio of the on/off time. If this is unclear, see the introduction to Chapter 7 for more on PWM.

You change the PWM frequency of a timer by setting the `TCCRnB` register, where *n* is the register number. On a Mega board you also have `TCCR3B`, `TCCR4B`, and `TCCR5B` for timers 3 through 5.

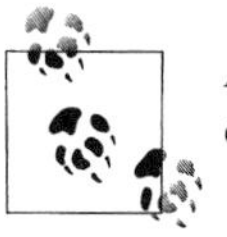

All analog output (PWM) pins on a timer use the same frequency, so changing timer frequency will affect all output pins for that timer.

See Also

See "See Also" on page 602 for links to data sheets and other references for timers.

18.7 Counting Pulses

Problem

You want to count the number of pulses occurring on a pin. You want this count to be done completely in hardware without any software processing time being consumed.

Solution

Use the pulse counter built into the Timer1 hardware:

```
/*
 * HardwareCounting sketch
 *
 * uses pin 5 on 168/328
 */

const int hardwareCounterPin = 5;   // input pin fixed to internal Timer
const int ledPin             = 13;

const int samplePeriod = 1000;   // the sample period in milliseconds
unsigned int count;


void setup()
{
  Serial.begin(9600);
  pinMode(ledPin,OUTPUT);
  // hardware counter setup (see ATmega data sheet for details)
  TCCR1A=0;         // reset timer/counter control register A
}

void loop()
{
  digitalWrite(ledPin, LOW);
  delay(samplePeriod);
  digitalWrite(ledPin, HIGH);
  // start the counting
  bitSet(TCCR1B ,CS12);  // Counter Clock source is external pin
  bitSet(TCCR1B ,CS11);  // Clock on rising edge
  delay(samplePeriod);
  // stop the counting
  TCCR1B = 0;
  count = TCNT1;
  TCNT1 = 0;  // reset the hardware counter
  if(count > 0)
     Serial.println(count);
}
```

Discussion

You can test this sketch by connecting the serial receive pin (pin 0) to the input pin (pin 5 on a standard Arduino board). Each character sent should show an increase in the count—the specific increase depends on the number of pulses needed to represent the ASCII value of the characters (bear in mind that serial characters are sandwiched between start and stop pulses). Some interesting character patterns are:

```
'u' = 01010101
'3' = 00110011
'~' = 01111110
'@' = 01000000
```

If you have two Arduino boards, you can run one of the pulse generator sketches from previous recipes in this chapter and connect the pulse output (pin 9) to the input. The pulse generator also uses Timer1 (the only 16 bit timer on a standard Arduino board), so you can combine the functionality using a single board.

Hardware pulse counting uses a pin that is internally wired within the hardware and cannot be changed. Use pin 5 on a standard Arduino board. The Mega uses Timer5 that is on pin 47; change `TCCR1A` to `TCCR5A` and `TCCR1B` to `TCCR5B`,

The Timer's `TCCR1B` register controls the counting behavior, setting it so 0 stops counting. The values used in the `loop` code enable count in the rising edge of pulses on the input pin. `TCNT1` is the Timer1 register declared in the Arduino core code that accumulates the count value.

In `loop`, the current count is printed once per second. If no pulses are detected on pin 5, the values will be `0`.

See Also

The FrequencyCounter library using the method discussed in this recipe: *http://interface.khm.de/index.php/lab/experiments/arduino-frequency-counter-library/*

See "See Also" on page 602 for links to data sheets and other references for timers.

18.8 Measuring Pulses More Accurately

Problem

You want to measure the period between pulses or the duration of the on or off time of a pulse. You need this as accurate as possible, so you don't want any delay due to calling an interrupt handler (as in Recipe 18.2), as this will affect the measurements.

Solution

Use the hardware pulse measuring capability built in to the Timer1 hardware:

```
/*
 * InputCapture
 * uses timer hardware to measure pulses on pin 8 on 168/328
 */

/* some interesting ASCII bit patterns:
 u 01010101
 3 00110011
 ~ 01111110
 @ 01000000
 */

const int inputCapturePin = 8;      // input pin fixed to internal Timer
const int ledPin          = 13;

const int prescale = 8;             // prescale factor (each tick 0.5 us @16MHz)
const byte prescaleBits = B010;     // see Table 18-1 or data sheet
// calculate time per counter tick in ns
const long  precision = (1000000/(F_CPU/1000)) * prescale  ;

const int numberOfEntries  = 64;    // the max number of pulses to measure
const int gateSamplePeriod = 1000;   // the sample period in milliseconds

volatile byte index = 0; // index to the stored readings
volatile byte gate  = 0; // 0 disables capture, 1 enables
volatile unsigned int results[numberOfEntries]; // note this is 16 bit value


/* ICR interrupt vector */
ISR(TIMER1_CAPT_vect)
{
  TCNT1 = 0;                            // reset the counter
  if(gate)
  {
    if( index != 0 || bitRead(TCCR1B ,ICES1) == true)   // wait for rising edge
    {                                     // falling edge was detected
      if(index < numberOfEntries)
      {
        results[index] = ICR1;              // save the input capture value
        index++;
      }
    }
  }
  TCCR1B ^= _BV(ICES1);            // toggle bit to trigger on the other edge
}

void setup() {
  Serial.begin(9600);
  pinMode(ledPin, OUTPUT);
  pinMode(inputCapturePin, INPUT); // ICP pin (digital pin 8 on Arduino) as input
```

```
  TCCR1A = 0 ;                      // Normal counting mode
  TCCR1B = prescaleBits ;           // set prescale bits
  TCCR1B |= _BV(ICES1);             // enable input capture

  bitSet(TIMSK1,ICIE1);             // enable input capture interrupt for timer 1

  Serial.println("pulses are sampled while LED is lit");
  Serial.print( precision);     // report duration of each tick in microseconds
  Serial.println(" microseconds per tick");

}

// this loop prints the number of pulses in the last second, showing min
// and max pulse widths
void loop()
{
  digitalWrite(ledPin, LOW);
  delay(gateSamplePeriod);
  digitalWrite(ledPin, HIGH);
  gate = 1; // enable sampling
  delay(gateSamplePeriod);
  gate = 0;  // disable sampling
  if(index > 0)
  {
    Serial.println("Durations in Microseconds are:") ;
    for( byte i=0; i < numberOfEntries; i++)
    {
      long duration;
      duration = results[i] * precision; // pulse duration in nanoseconds
      if(duration >0)
        Serial.println(duration / 1000);    // duration in microseconds
    }
    index = 0;
  }
}
```

Discussion

This sketch uses a timer facility called Input Capture to measure the duration of a pulse. Only 16-bit timers support this capability and this only works with pin 8 on a standard Arduino board.

Input Capture uses a pin that is internally wired within the hardware and cannot be changed. Use pin 8 on a standard Arduino board and pin 48 on a Mega (using Timer5 instead of Timer1).

Because Input Capture is implemented entirely in the controller chip hardware, no time is wasted in interrupt handling, so this technique is more accurate for very short pulses (less than tens of microseconds).

The sketch uses a gate variable that enables measurements (when nonzero) every other second. The LED is illuminated to indicate that measurement is active. The input capture interrupt handler stores the pulse durations for up to 64 pulse transitions.

The edge that triggers the timer measurement is determined by the ICES1 bit of the TCCR1B timer register. The line:

```
TCCR1B ^= _BV(ICES1);
```

toggles the edge that triggers the handler so that the duration of both high and low pulses is measured.

If the count goes higher than the maximum value for the timer, you can monitor overflow to increment a variable to extend the counting range. The following code increments a variable named overflow each time the counter overflows:

```
volatile int overflows = 0;

/* Overflow interrupt vector */
ISR(TIMER1_OVF_vect)                 // here if no input pulse detected
{
  overflows++;                       // increment overflow count
}
```

Change the code in setup as follows:

```
TIMSK1 =  _BV(ICIE1);           // enable input capture interrupt for timer 1
TIMSK1 |= _BV(TOIE1);           // Add this line to enable overflow interrupt
```

See Also

See "See Also" on page 602 for links to data sheets and other references for timers.

18.9 Measuring Analog Values Quickly

Problem

You want to read an analog value as quickly as possible without decreasing the accuracy.

Solution

You can increase the analogRead sampling rate by changing register values that determine the sampling frequency:

```
const int sensorPin = 0;             // pin the receiver is connected to
const int numberOfEntries = 100;

unsigned long microseconds;
unsigned long duration;

int results[numberOfEntries];
```

```
void setup()
{
  Serial.begin(9600);

  // standard analogRead performance (prescale =  128)
  microseconds = micros();
  for(int i = 0; i < numberOfEntries; i++)
  {
      results[i] = analogRead(sensorPin);
  }
  duration = micros() - microseconds;
  Serial.print(numberOfEntries);
  Serial.print(" readings took ");
  Serial.println(duration);

 // running with high speed clock (set prescale to 16)
  bitClear(ADCSRA,ADPS0) ;
  bitClear(ADCSRA,ADPS1) ;
  bitSet(ADCSRA,ADPS2) ;
  microseconds = micros();
  for(int i = 0; i < numberOfEntries; i++)
  {
      results[i] = analogRead(sensorPin);
  }
  duration = micros() - microseconds;
  Serial.print(numberOfEntries);
  Serial.print(" readings took ");
  Serial.println(duration);
}

void loop()
{
}
```

Running the sketch on a 16 MHz Arduino will produce output similar to the following:

```
100 readings took 11308
100 readings took 1704
```

Discussion

`analogRead` takes around 110 microseconds to complete a reading. This may not be fast enough for rapidly changing values, such as capturing the higher range of audio frequencies. The sketch measures the time in microseconds for the standard `analogRead` and then adjusts the timebase used by the analog-to-digital converter (ADC) to perform the conversion faster. With a 16 MHz board, the timebase rate is increased from 125 kHz to 1 MHz. The actual performance improvement is slightly less than eight times because there is some overhead in the Arduino `analogRead` function that is not improved by the timebase change. The reduction of time from 113 microseconds to 17 microseconds is a significant improvement.

The ADCSRA register is used to configure the ADC, and the bits set in the sketch (ADPS0, ADPS1, and ADPS2) set the ADC clock divisor to 16.

See Also

Atmel has an application note that provides a detailed explanation of performance aspects of the ADC: *http://www.atmel.com/dyn/resources/prod_documents/DOC2559.PDF*.

18.10 Reducing Battery Drain

Problem

You want to reduce the power used by your application by shutting down Arduino until a period of time has elapsed or until an external event takes place.

Solution

This Solution uses a library by Arduino guru Peter Knight. You can download the library from *http://code.google.com/p/narcoleptic/*:

```
#include <Narcoleptic.h>

void setup() {
  pinMode(2,INPUT);
  digitalWrite(2,HIGH);
  pinMode(13,OUTPUT);
  digitalWrite(13,LOW);
}

void loop() {
  int a;

  // Merlin the cat is snoozing... Connect digital pin 2 to ground to wake him up.
  Narcoleptic.delay(500); // During this time power consumption is minimized

  while (digitalRead(2) == LOW) {
    // Wake up CPU. Unfortunately, Merlin does not like waking up.

    // Swipe claws left
    digitalWrite(13,HIGH);
    delay(50);

    // Swipe claws right
    digitalWrite(13,LOW);
    delay(50);
  }

  // Merlin the cat goes to sleep...
}
```

Discussion

A standard Arduino board would run down a 9-volt alkaline battery in a few weeks (the Duemilanove typically draws more than 25 milliamperes [mA], excluding any external devices that may be connected). You can reduce this consumption by half if you use a board that does not have a built-in USB interface chip, such as the Arduino Mini, LilyPad, Fio, or one of the Modern Device Bare Bones Boards that require the use of an external USB interface for uploading sketches. Significantly greater power savings can be achieved if your application can suspend operation for a period of time—Arduino hardware can be put to sleep for a preset period of time or until a pin changes state, and this reduces the power consumption of the chip to less than one one-hundredth of 1 percent (from around 15 mA to around 0.001 mA) during sleep.

The library used in this recipe provides easy access to the hardware sleep function. The sleep time can range from 16 to 8,000 milliseconds (eight seconds). To sleep for longer periods, you can repeat the delay intervals until you get the period you want:

```
void longDelay(long milliseconds)
{
   while(milliseconds > 0)
   {
      if(milliseconds > 8000)
      {
         milliseconds -= 8000;
         Narcoleptic.delay(8000);
      }
      else
      {
        Narcoleptic.delay(milliseconds);
        break;
      }

   }
}
```

Sleep mode can reduce the power consumption of the controller chip, but if you are looking to run for as long as possible on a battery, you should minimize current drain through external components such as inefficient voltage regulators, pull-up or pull-down resistors, LEDs, and other components that draw current when the chip is in sleep mode.

See Also

See the Arduino hardware page for links to information on the LilyPad and Fio boards: *http://www.arduino.cc/en/Main/Hardware*.

For an example of very low power operation, see *http://interface.khm.de/index.php/lab/experiments/sleep_watchdog_battery/*.

18.11 Setting Digital Pins Quickly

Problem

You need to set or clear digital pins much faster than enabled by the Arduino `digital Write` command.

Solution

Arduino `digitalWrite` provides a safe and easy-to-use method of setting and clearing pins, but it is more than 30 times slower than directly accessing the controller hardware. You can set and clear pins by directly setting bits on the hardware registers that are controlling digital pins.

This sketch uses direct hardware I/O to send Morse code (the word *arduino*) to an AM radio tuned to approximately 1 MHz. The technique used here is 30 times faster than Arduino `digitalWrite`:

```
/*
 * Morse sketch
 *
 * Direct port I/O used to send AM radio carrier at 1MHz
 */

const int sendPin = 2;

const byte WPM = 12;                    // sending speed in words per minute
const long repeatCount = 1200000 / WPM; // count determines dot/dash duration
const byte dot = 1;
const byte dash = 3;
const byte gap = 3;
const byte wordGap = 7;
byte letter = 0; // the letter to send

char *arduino = ".- .-. -.. ..- .. -. ---";


void setup()
{
  pinMode(sendPin, OUTPUT);
  Serial.begin(9600);
}


void loop()
{
  sendMorse(arduino);
  delay(2000);
}

void sendMorse(char * string)
{
```

```
  letter = 0 ;
  while(string[letter]!= 0)
  {
    if(string[letter] == '.')
    {
      sendDot();
    }
    else if(string[letter] == '-')
    {
      sendDash();
    }
    else if(string[letter] == ' ')
    {
      sendGap();
    }
    else if(string[letter] == 0)
    {
      sendWordGap();
    }
    letter = letter+1;
  }
}

void  sendDot()
{
  transmitCarrier( dot *  repeatCount);
  sendGap();
}

void  sendDash()
{
  transmitCarrier( dash *  repeatCount);
  sendGap();
}

void  sendGap()
{
  transmitNoCarrier( gap *  repeatCount);
}

void  sendWordGap()
{
  transmitNoCarrier( wordGap *  repeatCount);
}

void transmitCarrier(long count)
{
  while(count--)
  {
    bitSet(PORTD, sendPin);
    bitSet(PORTD, sendPin);
    bitSet(PORTD, sendPin);
    bitSet(PORTD, sendPin);
    bitClear(PORTD, sendPin);
  }
```

```
}

void transmitNoCarrier(long count)
{
  while(count--)
  {
    bitClear(PORTD, sendPin);
    bitClear(PORTD, sendPin);
    bitClear(PORTD, sendPin);
    bitClear(PORTD, sendPin);
    bitClear(PORTD, sendPin);
  }
}
```

Connect one end of a piece of wire to pin 2 and place the other end near the antenna of a medium wave AM radio tuned to 1 MHz (1,000 kHz).

Discussion

The sketch generates a 1 MHz signal to produce dot and dash sounds that can be received by an AM radio tuned to this frequency. The frequency is determined by the duration of the `bitSet` and `bitClear` commands that set the pin `HIGH` and `LOW` to generate the radio signal. `bitSet` and `bitClear` are not functions, they are *macros*. Macros substitute an expression for executable code—in this case, code that changes a single bit in register `PORTD` given by the value of `sendPin`.

Digital pins 0 through 7 are controlled by the register named `PORTD`. Each bit in `PORTD` corresponds to a digital pin. Pins 8 through 13 are on register `PORTB`, and pins 14 through 19 are on `PORTA`. The sketch uses the `bitSet` and `bitClear` commands to set and clear bits on the port (see Recipe 3.12). Each register supports up to eight bits (although not all bits correspond to Arduino pins). If you want to use Arduino pin 13 instead of pin 2, you need to set and clear `PORTB` as follows:

```
const int sendPin = 13;

bitSet(PORTB, sendPin - 8);
bitClear(PORTB, sendPin - 8);
```

You subtract 8 from the value of the pin because bit 0 of the `PORTB` register is pin 8, bit 1 is pin 9, and so on, to bit 5 controlling pin 13.

Setting and clearing bits using `bitSet` is done in a single instruction of the Arduino controller. On a 16 MHz Arduino, that is 62.5 nanoseconds. This is around 30 times faster than using `digitalWrite`.

The transmit functions in the sketch actually need more time updating and checking the `count` variable than it takes to set and clear the register bits, which is why the `transmitCarrier` function has four `bitSet` commands and only one `bitClear` command—the additional `bitClear` commands are not needed because of the time it takes to update and check the `count` variable.

18.12 Uploading Sketches Using a Programmer

Problem

You want to upload sketches using a programmer instead of the bootloader. Perhaps you want the shortest upload time, or you don't have a serial connection to your computer suitable for bootloading, or you want to use the space normally reserved for the bootloader to increase the program memory available to your sketch.

Solution

Connect an external in-system programmer (ISP) to the Arduino programming ICSP (In-Circuit Serial Programming) connector. Programmers intended for use with Arduino have a 6-pin cable that attaches to the 6-pin ICSP connector as shown in Figure 18-1.

Ensure that pin 1 from the programmer (usually marked with different color than the other wires) is connected to pin 1 on the ICSP connector. The programmer may have a switch or jumper to enable it to power the Arduino board; read the instructions for your programmer to ensure that the Arduino is powered correctly.

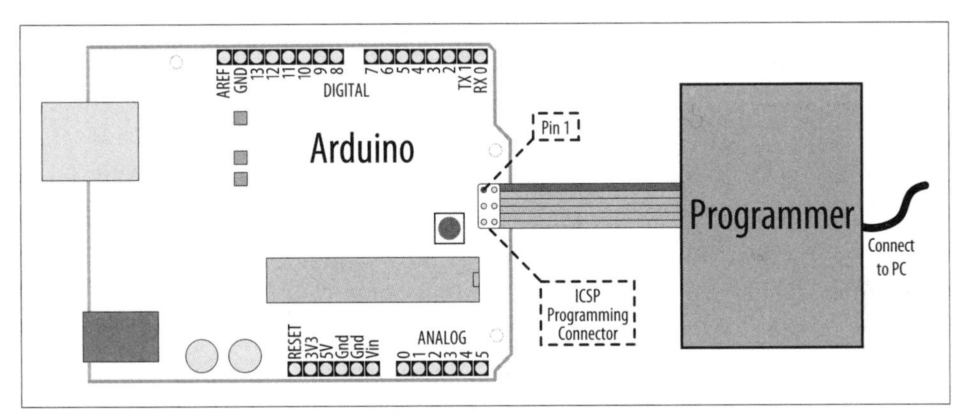

Figure 18-1. Connecting a programmer to Arduino

Select your programmer from the Tools menu. (AVRISP, AVRISPII, USBtinyISP, Parallel programmer, or Arduino as ISP) and double check that you have the correct Arduino board selected. From the File menu, select Upload Using Programmer to perform the upload.

Discussion

There are a number of different programmers available, from expensive devices aimed at professional developers that offer various debugging options, to low-cost self-build kits, or programming an additional Arduino to perform this function. The programmer may be a native USB device, or appear as a serial port. Check the documentation for your device to see what kind it is, and whether you need to install drivers for it.

The serial Rx and Tx LEDs on the Arduino will not flicker during upload because the programmer is not using the hardware serial port.

Uploading using a programmer removes the bootloader code from the chip. This frees up the space the bootloader occupies and gives a little more room for your sketch code.

See Also

Code to convert an Arduino into an ISP programmer can be found in the sketch example named ArduinoISP. The comments in the sketch describe the connections to use.

See Recipe 18.13.

Suitable hardware programmers include:

- USBtinyISP (*http://www.ladyada.net/make/usbtinyisp/*)
- Atmel avrisp2 (*http://parts.digikey.com/1/parts/408608-programmer-avr-system-atavrisp2.html*)
- CrispAVR_USB STK500 (*http://shop.chip45.com/epages/es10644620.sf/en_US/?ObjectPath=/Shops/es10644620/Products/CrispAVR-USB*)

18.13 Replacing the Arduino Bootloader

Problem

You want to replace the bootloader. Perhaps you can't get the board to upload programs and suspect the bootloader is not working. Or you want to replace an old bootloader with one with higher performance or different features.

Solution

Connect a programmer and select it as discussed in Recipe 18.12. Double check you have the correct board selected and click "Burn Bootloader" from the Tools menu.

A message will appear in the IDE saying "Burning bootloader to I/O board (this may take a minute)..." Programmers with status lights should indicate that the bootloader is being written to the board. You should see the LED connected to pin 13 flash as the board is programmed (pin 13 happens to be connected to one of the ICSP signal pins). If all goes well, you should get a message saying "Done Loading Bootloader."

Disconnect the programmer and try uploading code through the IDE to verify it is working.

Discussion

The bootloader is a small program that runs on the chip and briefly checks each time the chip powers up to see if the IDE is trying upload code to the board. If so, the bootloader takes over and replaces the code on the chip with new code being uploaded through the serial port. If the bootloader does not detect a request to upload, it relinquishes control to the sketch code already on the board.

If you have used a serial programmer, you will need to switch the serial port back to the correct one for your Arduino board as described in Recipe 1.4.

See Also

Optiloader (*https://github.com/WestfW/OptiLoader*), maintained by Bill Westfield, is another way to update or install the bootloader. It uses an Arduino connected as an ISP programmer, but all the bootloaders are included in the Arduino sketch code. This means an Arduino with Optiloader can program another chip automatically when power is applied—no external computer needed. The code identifies the chip and loads the correct bootloader onto it.

18.14 Reprogram the Uno to Emulate a Native USB device

Problem

You want your Arduino Uno to appear like a native USB device instead of as a serial port, for example as a MIDI USB device to communicate directly with music programs on your computer.

Solution

Replace the code running on the Uno USB controller (ATmega8U2) chip so that it communicates with the computer as a native USB device rather than a serial port.

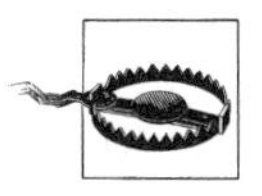

If the reprogramming is not done carefully, or a different firmware is used that does not include the DFU firmware, you can get the board into a state where you will need an external programmer to fix it using a command-line utility named `avrdude`. If you are not familiar with running command-line tools, you should think carefully before trying out this recipe.

Start by programing the Uno board with the sketch that will be talking to the 8U2, as once you have reprogrammed the 8U2 it will be more difficult to change the sketch. Darran Hunt has written suitable code for this that you can download from: *http://hunt*

.net.nz/users/darran/weblog/52882/attachments/1baa3/midi_usb_demo.pde (at the time of writing, this sketch used the old *.pde* extension but it is compatible with Arduino 1.0). Upload this to the Uno from the IDE in the usual way. This sketch will send commands to the 8U2 that will tell it what MIDI messages to send back to the computer.

Download the code to reprogram the 8U2 chip from *http://hunt.net.nz/users/darran/weblog/52882/attachments/e780e/Arduino-usbmidi-0.2.hex*.

You will also need programming software that can talk to the 8U2 chip:

On Windows
: Install the Atmel Flip program: *http://www.atmel.com/dyn/products/tools_card.asp?tool_id=3886*.

Mac
: Install the command line tool `dfu-programmer`. A handy install script for installing is here: *http://www.uriahbaalke.com/?p=106*.

Linux
: From terminal, type: `sudo apt-get install dfu-programme` or `sudo aptitude install dfu-programmer` depending on your distribution.

Set the 8U2 into its reprogram mode: if your Uno has the 6-pin connector by the 8U2 chip populated with pins, then you just need to short the lefthand pair of pins (closest to the USB connector) together to put the chip in DFU mode.

The first Uno boards (revision 1) did not have a resistor needed to reset the 8U2. If you are unable to reset your board, follow the instructions at *http://arduino.cc/en/Hacking/DFUProgramming8U2*. Halfway down the page it describes what to do if your board needs to have an external resistor added to enable resetting the 8U2 chip.

On Windows
: When the board is put into DFU mode for the first time, the Found New Hardware Wizard will appear. If the board installs without error then carry on. If the hardware installation fails (in the same way the Uno installation does) then you need to go into Device Manager and highlight the entry for Arduino DFU (it will have a yellow warning triangle next to it), right-click, and select update drivers. Navigate to the Flip 3.4.3 folder in *Program Files/Atmel* and select the USB folder. The drivers should now successfully install.

: Launch the Flip program.

: Select the device type AT90USB82 from the drop-down menu (it is the only active option when you first run the program). Click on the icon of a lead and select USB. If you get the error message `AtLibUsbDfu.dll not found`, the drivers have not installed. Follow the instructions above.

Click on the Select EEPROM button at the bottom of the window and open `Arduino-usbmidi-0.2.hex`. Select Run to the left of this button, and the program should go through the cycle listed above the button: Erase, Program, Verify. Unplug the board and plug it back in and it will show up as a MIDI device on your computer.

Mac and Linux
In terminal, `cd` into the folder with the hex file.

Clear the chip by typing **`sudo dfu-programmer at90usb82 erase`**.

When this has finished, type
`sudo dfu-ptogrammer at90usb82 flash Arduino-usbmidi-0.2.hex`.

Unplug the board and plug it back in to get the new firmware to run in the 8U2.

The operating system should now recognize the device as a MIDI device. Hook it up to a music program and you should hear a string of notes.

Discussion

Once the 8U2 is reprogrammed, the messages that are sent to the computer are still controlled by the sketch running on the main chip, but your computer sees the Arduino board as a MIDI device instead of a serial port. The sketch running on the main chip determines what gets sent to your computer, allowing Arduino to respond to switches and sensors to control what is played.

The IDE will not see the standard bootloader when the 8U2 has been reprogrammed as described in this Recipe, so to change the sketch you use an external programmer (see Recipe 18.12).

If you want to return your 8U2 to its original state, you can obtain the required `.HEX` file at *https://github.com/arduino/Arduino/tree/master/hardware/arduino/firmwares*. Put this on the 8U2 using the procedure described above, but use this hex file instead of the MIDI one.

If you have used other firmware that does not include the DFU loader (not all firmware found on the internet include it), or something has gone wrong and the board will not go into DFU mode, then you need to use an external programmer to replace the firmware.

This needs to be done from the command line using the upload utility named AVR-dude (it cannot be done using the Arduino IDE).

In order for the following command to work, you need to supply the full path to `avrdude`, not just the name. `avrdude` is located inside your Arduino program folder: *Arduino.app/Contents/Resources/Java/hardware/tools/avr/bin* on a Mac; *hardware/tools/avr/bin* inside the Arduino folder on Windows. (Or you can add this location to your PATH environment; Google "set path environment" for your operating system for details.)

At the command line from the folder where the hex file is located, execute the following command:

For the Uno

```
avrdude -p at90usb82 -F -P usb -c avrispmkii -U flash:w:UNO-dfu_and_usbse
rial_combined.hex -U lfuse:w:0xFF:m -U hfuse:w:0xD9:m -U efuse:w:0xF4:m -U
lock:w:0x0F:m
```

For the Mega 2560

```
avrdude -p at90usb82 -F -P usb -c avrispmkii -U flash:w:MEGA-dfu_and_usbse
rial_combined.hex -U lfuse:w:0xFF:m -U hfuse:w:0xD9:m -U efuse:w:0xF4:m -U
lock:w:0x0F:m
```

If your programming device is a serial device rather than USB you will need to change `-P usb` to specify which serial port (e.g., `-P \\.\COM19` on Windows; `-P /dev/tty.usbserial-XXXXXX` on Mac (check the Arduino serial port menu for the name it appears as, and what values `XXXXXX` are). Set the `-c avrispmkii` based on the type of programmer you have. For more details on this, see Recipe 18.12.

See Also

See Recipe 18.12.

Darran Hunt's ATmega8U2 blog: *http://hunt.net.nz/users/darran/*

Updating the Atmega8U2 on an Uno or Mega2560 using DFU: *http://arduino.cc/en/Hacking/DFUProgramming8U2*

The Teensy and Teensy++ boards can emulate USB HID devices: *http://www.pjrc.com/teensy/*.

The Arduino Leonardo board supports emulation of USB HID devices. Leonardo had not been released when this book was printed; check the Arduino hardware page to see if it is available: *http://www.arduinocc/en/Main/hardware*.

See Recipe 9.6 for the conventional way to control MIDI from Arduino.

A tutorial covering the low-level `avrdude` programming tool: *http://www.ladyada.net/make/usbtinyisp/avrdude.html*

APPENDIX A
Electronic Components

If you are just starting out with electronic components, you may want to purchase a beginner's starter kit that contains the basic components needed for many of the recipes in this book. These usually include the most common resistors, capacitors, transistors, diodes, LEDs, and switches.

Here are some popular choices:

Getting Started with Arduino kit
: *http://www.makershed.com/ProductDetails.asp?ProductCode=MSGSA*

Starter Kit for Arduino-Flex (SKU: DEV-10174)
: *http://www.sparkfun.com/products/10174*

Adafruit Starter Pack for Arduino-1.0 (product ID #68)
: *http://www.adafruit.com/index.php?main_page=product_info&products_id=68*

Oomlout Starter Kit for Arduino (ARDX)
: *http://oomlout.co.uk/arduino-experimentation-kit-ardx-p-183.html*

Arduino Sidekick Basic Kit
: *http://www.seeedstudio.com/depot/arduino-sidekick-basic-kit-p-775.html*

You can also purchase the individual components for your project, as shown in Figure A-1. The following sections provide an overview of common electronic components—part numbers can be found on this book's website (*http://shop.oreilly .com/product/0636920022244.do*).

Capacitor

Capacitors store an electrical charge for a short time and are used in digital circuits to filter (smooth out) dips and spikes in electrical signals. The most commonly used capacitor is the nonpolarized ceramic capacitor; for example, a 100 nF disc capacitor used for decoupling (reducing noise spikes). Electrolytic capacitors can generally store more charge than ceramic caps and are used for higher current circuits, such as power

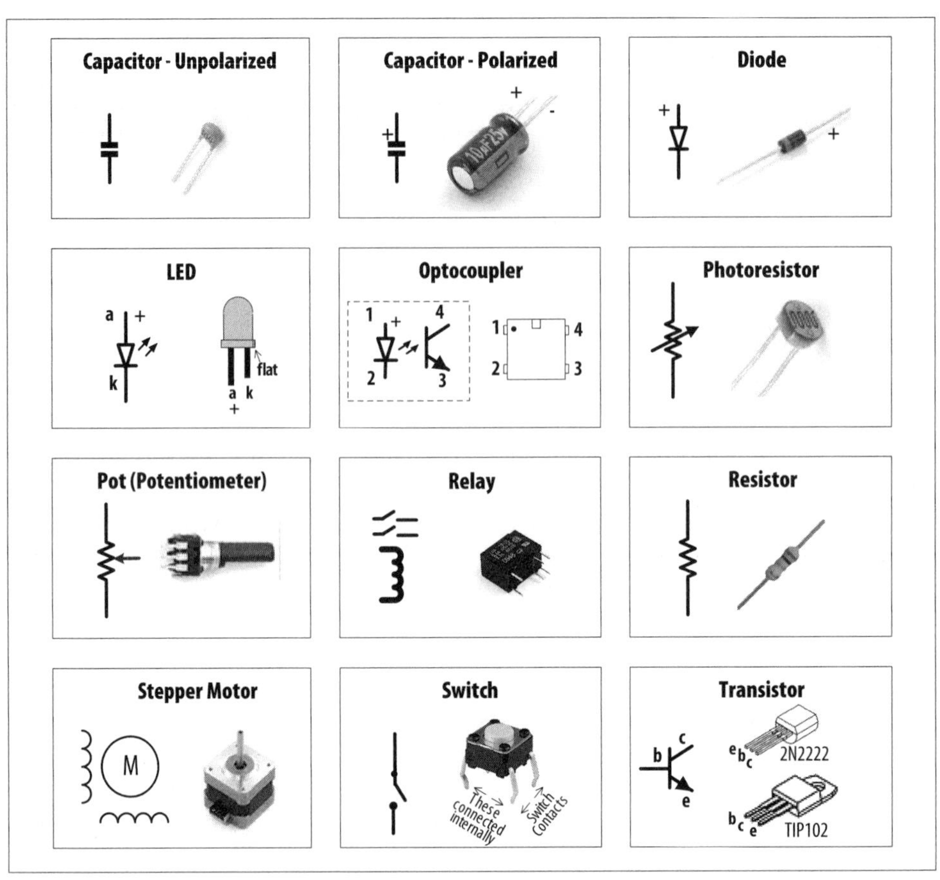

Figure A-1. Schematic representation of common components

supplies and motor circuits. Electrolytic capacitors are usually polarized, and the negative leg (marked with a minus sign) must be connected to ground (or to a point with lower voltage than the positive leg). Chapter 8 contains examples showing how capacitors are used in motor circuits.

Diode

Diodes permit current to flow in one direction and block it in the other direction. Most diodes have a band (see Figure A-1) to indicate the cathode (negative) end.

Diodes such as the 1N4148 can be used for low-current applications such as the levels used on Arduino digital pins. The 1N4001 diode is a good choice for higher currents (up to 1 amp).

Integrated Circuit

Integrated circuits contain electronic components packaged together in a convenient chip. These can be complex, like the Arduino controller chip that contains thousands of transistors, or as simple as the optical isolator component used in Chapter 10 that contains just two semiconductors. Some integrated circuits (such as the Arduino chip) are sensitive to static electricity and should be handled with care.

Keypad

A keypad is a matrix of switches used to provide input for numeric digits. See Chapter 5.

LED

An LED (light-emitting diode) is a diode that emits light when current flows through the device. As they are diodes, LEDs only conduct electricity in one direction. See Chapter 7.

Motor (DC)

Motors convert electrical energy into physical movement. Most small direct current (DC) motors have a speed proportional to the voltage, and you can reverse the direction they move by reversing the polarity of the voltage across the motor. Most motors need more current than the Arduino pins provide, and a component such as a transistor is required to drive the motor. See Chapter 8.

Optocoupler

Optocouplers (also called optoisolators) provide electrical separation between devices. This isolation allows devices that operate with different voltage levels to work safely together. See Chapter 10.

Photocell (Photoresistor)

Photocells are variable resistors whose resistance changes with light. See Chapter 6.

Piezo

A small ceramic transducer that produces sound when pulsed, a Piezo is polarized and may have a red wire indicating the positive end and a black wire indicating the side to be connected to ground. See Chapter 9.

Pot (Potentiometer)

A potentiometer (pot for short) is a variable resistor. The two outside terminals act as a fixed resistor. A movable contact called a wiper (or slider) moves across the resistor, producing a variable resistance between the center terminal and the two sides. See Chapter 5.

Relay

A relay is an electronic switch—circuits are opened or closed in response to a voltage on the relay coil, which is electrically isolated from the switch. Most relay coils require more current than Arduino pins provide, so they need a transistor to drive them. See Chapter 8.

Resistor

Resistors resist the flow of electrical current. A voltage flowing through a resistor will limit the current proportional to the value of the resistor (see Ohm's law). The bands on a resistor indicate the resistor's value. Chapter 7 contains information on selecting a resistor for use with LEDs.

Solenoid

A solenoid produces linear movement when powered. Solenoids have a metallic core that is moved by a magnetic field created when passing current through a coil. See Chapter 8.

Speaker

A speaker produces sound by moving a diaphragm (the speaker cone) to create sound waves. The diaphragm is driven by sending an audio frequency electrical signal to a coil of wire attached to the diaphragm. See Chapter 9.

Stepper Motor

A stepper motor rotates a specific number of degrees in response to control pulses. See Chapter 8.

Switch

A switch makes and breaks an electrical circuit. Many of the recipes in this book use a type of push button switch known as a *tactile switch*. Tactile switches have two pairs of contacts that are connected together when the button is pushed. The pairs are wired together, so you can use either one of the pair. Switches that make contact when pressed are called Normally Open (NO) switches. See Chapter 5.

Transistor

Transistors are used to switch on high currents or high voltages in digital circuits. In analog circuits, transistors are used to amplify signals. A small current through the transistor base results in a larger current flowing through the collector and emitter.

For currents up to .5 amperes (500 mA) or so, the 2N2222 transistor is a widely available choice. For currents up to 5 amperes, you can use the TIP120 transistor.

See Chapters 7 and 8 for examples of transistors used with LEDs and motors.

See Also

For more comprehensive coverage of basic electronics, see the following:

- *Make: Electronics* by Charles Platt (O'Reilly; search for it on oreilly.com)
- *Getting Started in Electronics* by Forrest Mims (Master Publishing)
- *Physical Computing* by Tom Igoe (Cengage)
- *Practical Electronics for Inventors* by Paul Scherz (McGraw-Hill)

APPENDIX B

Using Schematic Diagrams and Data Sheets

A *schematic diagram*, also called a *circuit diagram*, is the standard way of describing the components and connections in an electronic circuit. It uses iconic symbols to represent components, with lines representing the connections between the components.

A circuit diagram represents the connections of a circuit, but it is not a drawing of the actual physical layout. Although you may initially find that drawings and photos of the physical wiring can be easier to understand than a schematic, in a complicated circuit it can be difficult to clearly see where each wire gets connected.

Circuit diagrams are like maps. They have conventions that help you to orient yourself once you become familiar with their style and symbols. For example, inputs are usually to the left, outputs to the right; 0V or ground connections are usually shown at the bottom of simple circuits, the power at the top.

Figure A-1 in Appendix A shows some of the most common components, and the symbols used for them in circuit diagrams. Figure B-1 is a schematic diagram from Recipe 8.8 that illustrates the symbols used in a typical diagram.

Components such as the resistor and capacitor used here are not polarized—they can be connected either way around. Transistors, diodes, and integrated circuits are polarized, so it is important that you identify each lead and connect it according to the diagram.

Figure B-2 shows how the wiring could look when connected using a breadboard. This drawing was produced using a tool called Fritzing that enables the drawing of electronic circuits. See *http://fritzing.org/*.

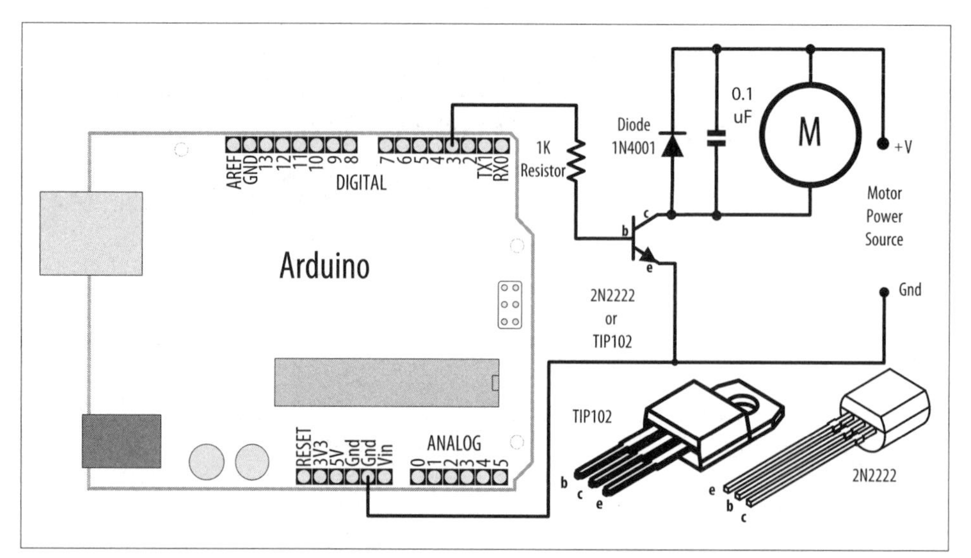

Figure B-1. Typical schematic diagram

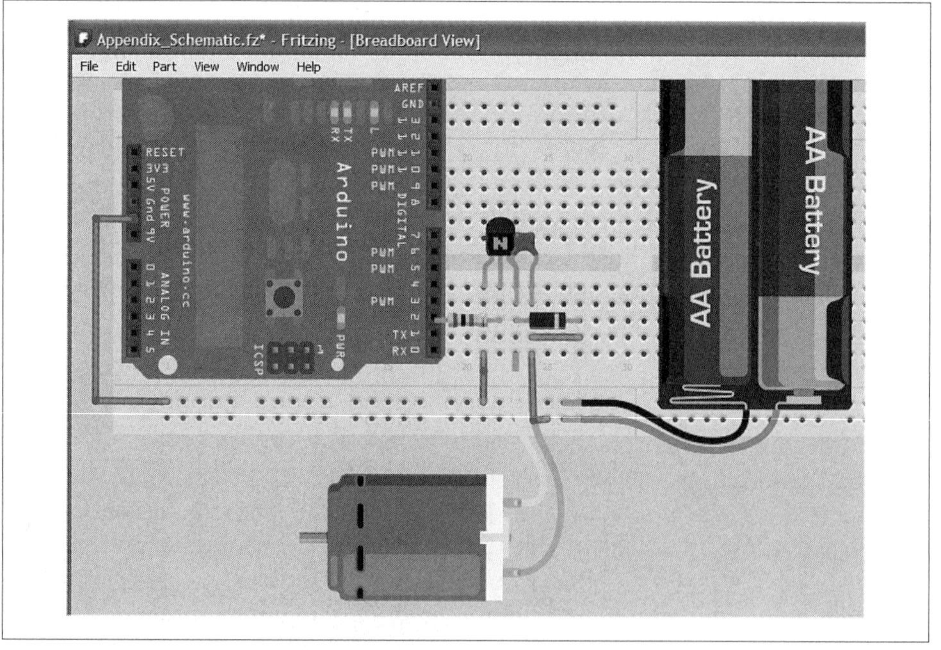

Figure B-2. Physical layout of the circuit shown in Figure B-1

Wiring a working breadboard from a circuit diagram is easy if you break the task into individual steps. Figures B-3 and B-4 show how each step of breadboard construction is related to the circuit diagram. The circuit shown is from Recipe 1.6.

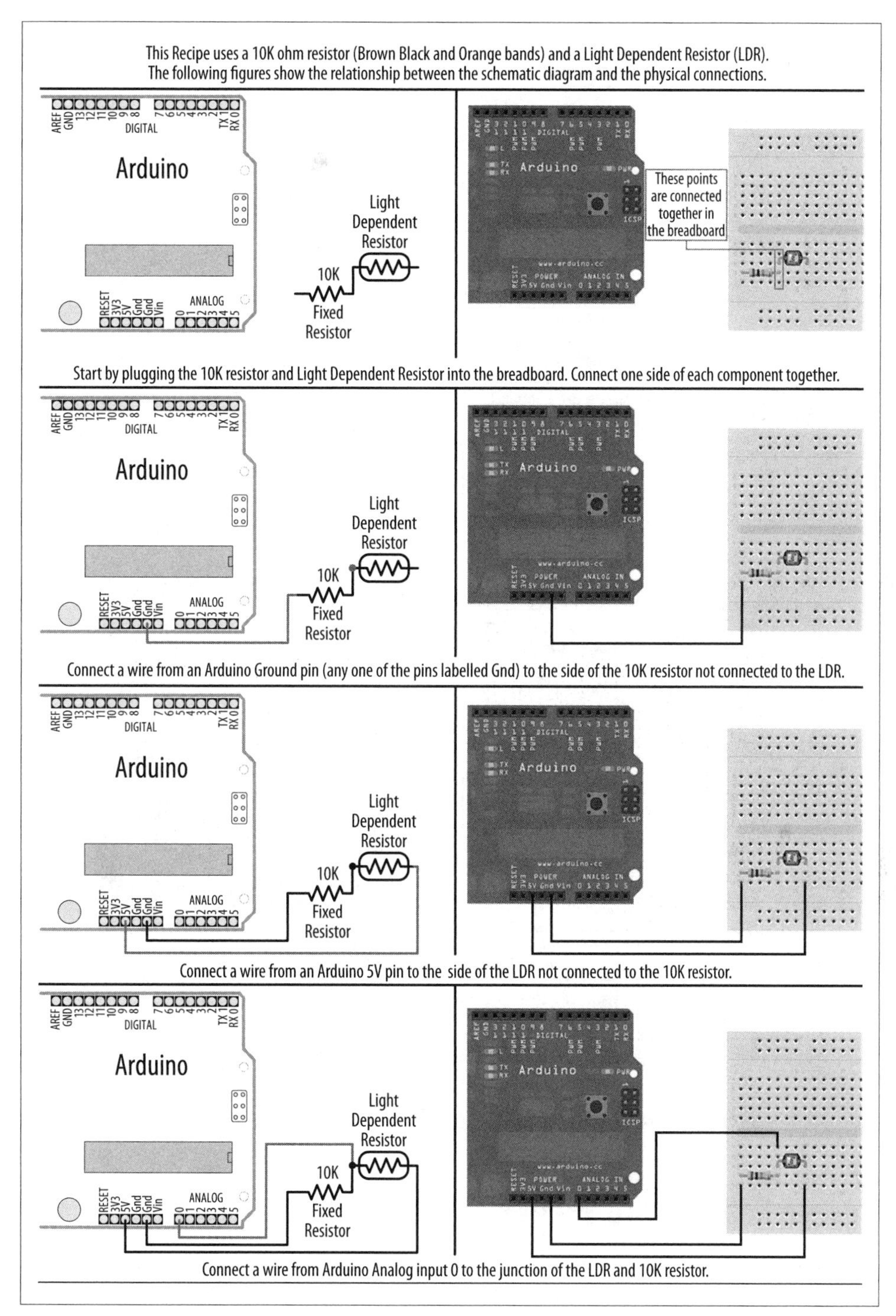

This Recipe uses a 10K ohm resistor (Brown Black and Orange bands) and a Light Dependent Resistor (LDR).
The following figures show the relationship between the schematic diagram and the physical connections.
Arduino
DIGITAL
ANALOG
Light Dependent Resistor
10K
Fixed Resistor
These points are connected together in the breadboard
Start by plugging the 10K resistor and Light Dependent Resistor into the breadboard. Connect one side of each component together.
Connect a wire from an Arduino Ground pin (any one of the pins labelled Gnd) to the side of the 10K resistor not connected to the LDR.
Connect a wire from an Arduino 5V pin to the side of the LDR not connected to the 10K resistor.
Connect a wire from Arduino Analog input 0 to the junction of the LDR and 10K resistor.

Figure B-3. Transferring a schematic diagram to a breadboard

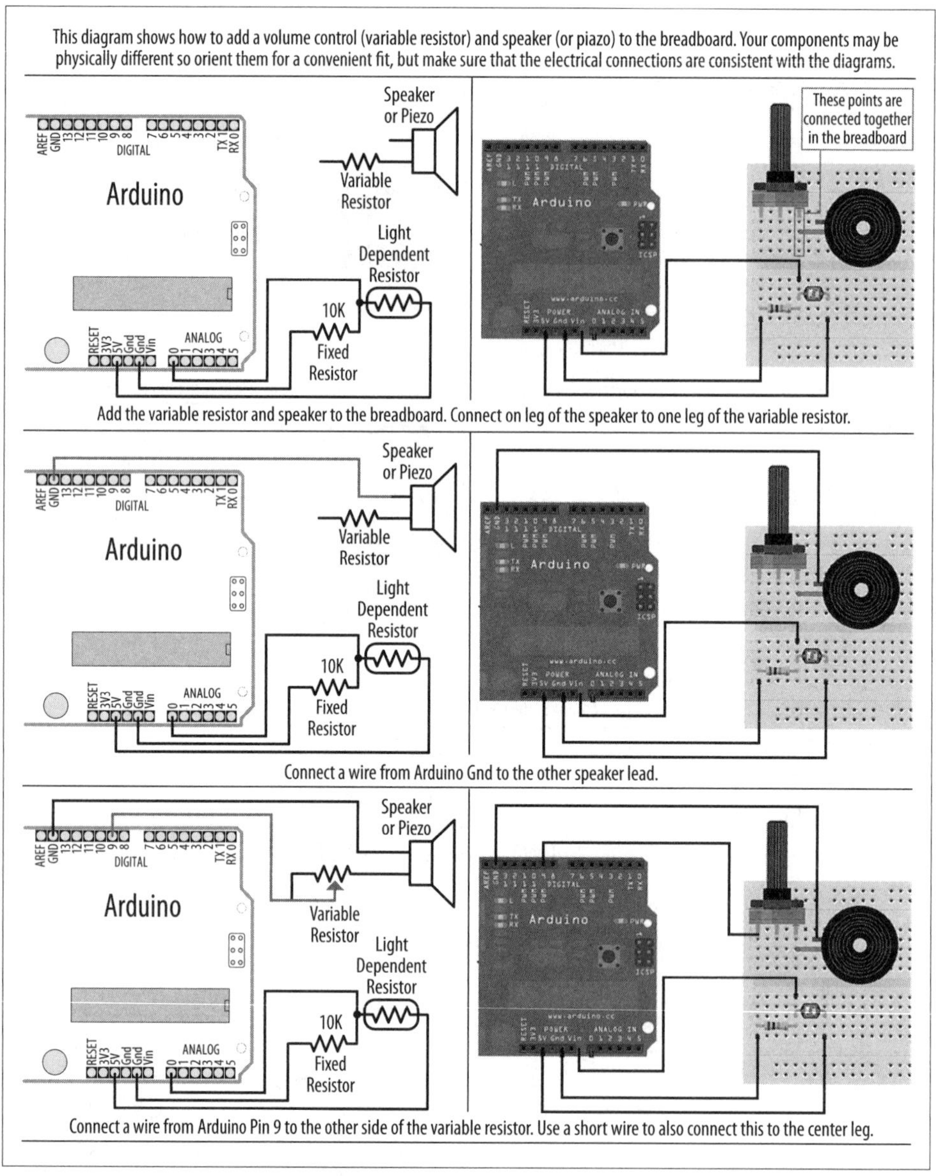
This diagram shows how to add a volume control (variable resistor) and speaker (or piazo) to the breadboard. Your components may be physically different so orient them for a convenient fit, but make sure that the electrical connections are consistent with the diagrams.
Speaker or Piezo
Variable Resistor
Light Dependent Resistor
10K
Fixed Resistor
Arduino
DIGITAL
ANALOG
These points are connected together in the breadboard
Add the variable resistor and speaker to the breadboard. Connect on leg of the speaker to one leg of the variable resistor.
Connect a wire from Arduino Gnd to the other speaker lead.
Connect a wire from Arduino Pin 9 to the other side of the variable resistor. Use a short wire to also connect this to the center leg.

Figure B-4. Adding to the breadboard

How to Read a Data Sheet

Data sheets are produced by the manufacturers of components to summarize the technical characteristics of a device. Data sheets contain definitive information about the performance and usage of the device; for example, the minimum voltage needed for the device to function and the maximum voltage that it can reliably tolerate. Data sheets contain information on the function of each pin and advice on how to use the device.

For more complicated devices, such as LCDs, the data sheet covers how to initialize and interact with the device. Very complex devices, such as the Arduino controller chip, require hundreds of pages to explain all the capabilities of the device.

Data sheets are written for design engineers, and they usually contain much more information than you need to get most devices working in an Arduino project. Don't be intimidated by the volume of technical information; you will typically find the important information in the first couple of pages. There will usually be a circuit diagram symbol labeled to show how the connections on the device correspond to the symbols. This page will typically have a general description of the device (or family of devices) and the kinds of uses they are suitable for.

After this, there is usually a table of the electrical characteristics of the device.

Look for information about the maximum voltage and the current the device is designed to handle to check that it is in the range you need. For components to connect directly to a standard Arduino board, devices need to operate at +5 volts. To be powered directly from the pin of the Arduino, they need to be able to operate with a current of 40 mA or less.

Some components are designed to operate on 3.3 volts and can be damaged if connected to a 5V Arduino board. Use these devices with a board designed to run from a 3.3V supply (e.g., the LilyPad, Fio, or 3.3V Mini Pro), or use a logic-level converter such as the SparkFun BOB-08745. More information on logic-level conversion is available at *http://ics.nxp.com/support/documents/interface/pdf/an97055.pdf*.

Choosing and Using Transistors for Switching

The Arduino output pins are designed to handle currents up to 40 mA (milliamperes), which is only 1/25 of an amp. You can use a transistor to switch larger currents. This section provides guidance on transistor selection and use.

The most commonly used transistors with Arduino projects are bipolar transistors. These can be of two types (named NPN and PNP) that determine the direction of current flow. NPN is more common for Arduino projects and is the type that is illustrated in the recipes in this book. For currents up to .5 amperes (500 mA) or so, the 2N2222 transistor is a widely available choice; the TIP120 transistor is a popular choice for currents up to 5 amperes.

Figure B-1 shows an example of a transistor connected to an Arduino pin used to drive a motor.

Transistor data sheets are usually packed with information for the design engineer, and most of this is not relevant for choosing transistors for Arduino applications. Table B-1 shows the most important parameters you should look for (the values shown are for a typical general-purpose transistor). Manufacturing tolerances result in varying performance from different batches of the same part, so data sheets usually indicate the minimum, typical, and maximum values for parameters that can vary from part to part.

Here's what to look for:

Collector-emitter voltage
: Make sure the transistor is rated to operate at a voltage higher than the voltage of the power supply for the circuit the transistor is controlling. Choosing a transistor with a higher rating won't cause any problems.

Collector current
: This is the absolute maximum current the transistor is designed to handle. It is a good practice to choose a transistor that is rated at least 25 percent higher than what you need.

DC current gain
: This determines the amount of current needed to flow through the base of the transistor to switch the output current. Dividing the output current (the maximum current that will flow through the load the transistor is switching) by the gain gives the amount of current that needs to flow through the base. Use Ohms's law (Resistance = Voltage / Current) to calculate the value of the resistor connecting the Arduino pin to the transistor base. For example, if the desired collector current is 1 amp and the gain is 100, you need at least 0.01 amps (10 mA) through the transistor base. For a 5 volt Arduino: 5 / .01 = 500 ohms (500 ohms is not a standard resistor value so 470 ohms would be a good choice).

Collector-emitter saturation voltage
: This is the voltage level on the collector when the transistor is fully conducting. Although this is usually less than 1 volt, it can be significant when calculating a series resistor for LEDs or for driving high-current devices.

Table B-1. Example of key transistor data sheet specifications

Absolute maximum ratings				
Parameter	**Symbol**	**Rating**	**Units**	**Comment**
Collector-emitter voltage	Vceo	40	Volts	The maximum voltage between the collector and emitter
Collector current	Ic	600	mA or A	The maximum current that the transistor is designed to handle
Electrical characteristics				
DC current gain	Ic	90 @ 10 mA		Gain with 10 mA current flowing
	Ic	50 @ 500 mA		Gain with 500 mA current flowing
Collector-emitter saturation voltage	Vce (sat)	0.3 @ 100 mA 1.0 @ 500 mA	Volts Volts	Voltage drop across collector and emitter at various currents

APPENDIX C

Building and Connecting the Circuit

Using a Breadboard

A breadboard enables you to prototype circuits quickly, without having to solder the connections. Figure C-1 shows an example of a breadboard.

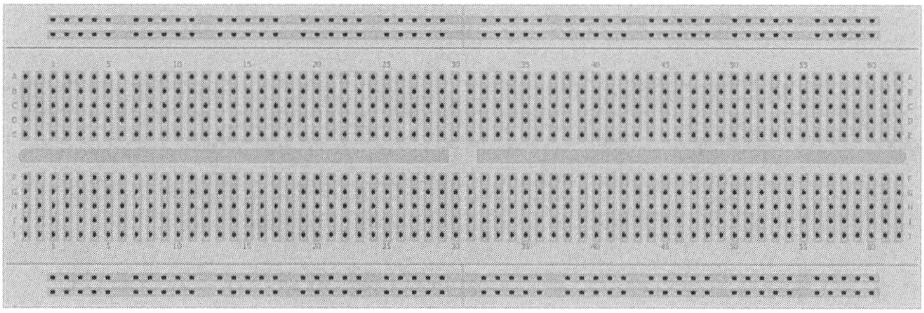

Figure C-1. Breadboard for prototyping circuits

Breadboards come in various sizes and configurations. The simplest kind is just a grid of holes in a plastic block. Inside are strips of metal that provide electrical connections between holes in the shorter rows. Pushing the legs of two different components into the same row joins them together electrically. A deep channel running down the middle indicates that there is a break in connections there, meaning you can push a chip in with the legs at either side of the channel without connecting them together.

Some breadboards have two strips of holes running along the long edges of the board that are separated from the main grid. These have strips running down the length of the board inside, and provide a way to connect a common voltage. They are usually in pairs for +5 volts and ground. These strips are referred to as *rails* and they enable you to connect power to many components or points in the board.

While breadboards are great for prototyping, they have some limitations. Because the connections are push-fit and temporary, they are not as reliable as soldered connections. If you are having intermittent problems with a circuit, it could be due to a poor connection on a breadboard.

Connecting and Using External Power Supplies and Batteries

The Arduino can be powered from an external power source rather than through the USB lead. You may need more current than the USB connection can provide (the maximum USB current is 500 mA; some USB hubs only supply 100 mA), or you may want to run the board without connection to the computer after the sketch is uploaded.

The standard Arduino board has a socket for connecting external power. This can be an AC-powered power supply or a battery pack.

> These details relate to the Uno, Duemilanove, and Mega boards. Other Arduino and compatible boards may not protect the board from reverse connections, or they may automatically switch to use external power and may not accept higher voltages. If you are using a different board, check before you connect power or you may damage the board.

If you are using an AC power supply, you need one that produces a DC voltage between 7 and 12 volts. Choose a power supply that provides at least as much current as you need (there is no problem in using a power supply with a higher current than you need). Wall wart–style power supplies come in two broad types: regulated and unregulated. A regulated power supply has a circuit that maintains the specified voltage, and this is a good choice to use with Arduino. An unregulated power supply will produce a higher voltage when run at a lower current and can sometimes produce twice the rated voltage when driving low-current devices such as Arduino. Voltages higher than 12 volts can overheat the regulator on the Arduino, and this can cause intermittent operation or even damage the board.

Battery voltage should also be in the range of 7 to 12 volts. Battery current is rated in mAh (the amount of milliamperes the battery can supply in one hour). A battery with a rating of 500 mAh (a typical alkaline 9V battery) should last around 20 hours with an Arduino board drawing 25 mAh. If your project draws 50 mA, the battery life will be halved, to around 10 hours. How much current your board uses depends mostly on the devices (such as LEDs and other external components) that you use. Bear in mind that the Uno and Duemilanove boards are designed to be easy to use and robust, but they are not optimized for low power use with a battery. See Recipe 18.10 for advice on reducing battery drain.

The positive (+) connection from the power supply should be connected to the center pin of the Arduino power plug. If you connect it the wrong way around on an Uno, Duemilanove, or Mega, the board will not break, but it will not work until the connection is reversed. These boards automatically detect that an external power supply is connected and use that to power the board. You can still have the USB lead plugged in, so serial communication and code uploading will still work.

Using Capacitors for Decoupling

Digital circuits switch signals on and off quickly, and this can cause fluctuations in the power supply voltage that can disrupt proper operation of the circuit. Properly designed digital circuits use decoupling capacitors to filter these fluctuations. Decoupling capacitors should be connected across the power pins of each IC in your circuit with the capacitor leads kept as short as possible. A ceramic capacitor of 0.1 uF is a good choice for decoupling—that value is not critical (20 percent tolerance is OK).

Using Snubber Diodes with Inductive Loads

Inductive loads are devices that have a coil of wire inside. This includes motors, solenoids, and relays. The interruption of current flow in a coil of wire generates a spike of electricity. This voltage can be higher than +5 volts and can damage sensitive electronic circuits such as Arduino pins. Snubber diodes are used to prevent that by conducting the voltage spikes to ground. Figure A-1 in Appendix A shows an example of a snubber diode used to suppress voltage spikes when driving a motor.

Working with AC Line Voltages

When working with an AC line voltage from a wall socket, the first thing you should consider is whether you can avoid working with it. Electricity at this voltage is dangerous enough to kill *you*, not just your circuit, if it is used incorrectly. It is also dangerous for people using whatever you have made if the AC line voltage is not isolated properly.

Hacking controllers for devices that are manufactured to work with mains voltage, or using devices designed to be used with microcontrollers to control AC line voltages, is safer (and often easier) than working with mains voltage itself. See Chapter 10 for recipes on controlling external devices for examples of how to do this.

APPENDIX D

Tips on Troubleshooting Software Problems

As you write and modify code, you will get code that doesn't work for some reason (this reason is usually referred to as a *bug*). There are two broad areas of software problems: code that won't compile and code that compiles and uploads to the board but doesn't behave as you want.

Code That Won't Compile

Your code might fail to compile when you click on the Verify (checkbox) button or the Upload button (see Chapter 1). This is indicated by red error messages in the black console area at the bottom of the Arduino software window and a yellow highlight in the code if there is a specific point where the compilation failed. Often the problem in the code is in the line immediately before the highlighted line. The error messages in the console window are generated by the command-line programs used to compile and link the code (see Recipe 17.1 for details on the build process). This message may be difficult to understand when you first start.

One of the most common errors made by people new to Arduino programming is omission of the semicolon at the end of a line. This can produce various different error messages, depending on the next line. For example, this code fragment:

```
void loop()
{
  digitalWrite(ledPin, HIGH)    // <- BUG: missing semicolon
  delay(1000);
}
```

produces the following error message:

```
In function 'void loop()':
error: expected ';' before 'delay
```

A less obvious error message is:

```
expected unqualified-id before numeric constant
```

Although the cause is similar, a missing semicolon after a constant results in the preceding error message, as in this fragment:

```
const int ledPin =  13    // <- BUG: missing semicolon after constant
```

The combination of the error message and the line highlighting provides a good starting point for closer examination of the area where the error has occurred.

Another common error is misspelled words, resulting in the words not being recognized. This includes incorrect capitalization—`LedPin` is different from `ledPin`. This fragment:

```
const int ledPin =  13;

digitalWrite(LedPin, HIGH);   // <- BUG: the capitalization is different
```

results in the following error message:

```
In function 'void loop()':
error: 'LedPin' was not declared in this scope
```

The fix is to use exactly the same spelling and capitalization as the variable declaration.

You must use the correct number and type of parameters for function calls (see Recipe 2.10). The following fragment:

```
digitalWrite(ledPin);   // <- BUG: this is missing the second parameter
```

generates this error message:

```
error: too few arguments to function 'void digitalWrite(uint8_t, uint8_t)'
error: at this point in file
```

The cursor in the IDE will point to the line in the sketch that contains the error.

Functions in sketches that are missing the return type will generate an error. This fragment:

```
loop()               // <- BUG: loop is missing the return type
{
}
```

produces this error:

```
error: ISO C++ forbids declaration of 'loop' with no type
```

The error is fixed by adding the missing return type:

```
void loop()               // <- return type precedes function name
{
}
```

Incorrectly formed comments, such as this fragment that is missing the second "/":

```
digitalWrite(ledPin, HIGH);   / set the LED on (BUG: missing //)
```

result in this error:

```
error: expected primary-expression before '/' token
```

It is good to work on a small area of code, and regularly verify/compile to check the code. You don't need to upload to check that the sketch compiles (just click Verify button in the IDE). The earlier you become aware of a problem, the easier it is to fix it, and the less impact it will have on other code. It is much easier to fix code that has one problem than it is to fix a large section of code that has multiple errors in it.

Code That Compiles but Does Not Work as Expected

There is always a feeling of accomplishment when you get your sketch to compile without errors, but correct syntax does not mean the code will do what you expect.

This is usually a subtler problem to isolate. You are now in a world where software and hardware are interacting. It is important to try to separate problems in hardware from those in software. Carefully check the hardware (see Appendix E) to make sure it is working correctly.

Troubleshooting Interrelated Hardware/Software Problems

Some problems are not due strictly to software or hardware errors, but to the interplay between them.

The most common of these is connecting the circuit to one pin and in software reading or writing a different pin. Hardware and software are both correct in isolation—but together they don't work. You can change either the hardware or the software to fix this: change the pin in software or move the connection to the pin number declared in your sketch.

If you are sure the hardware is wired and working correctly, the first step in debugging your sketch is to carefully read through your code to review the logic you used. Pausing to think carefully about what you have written is usually a faster and more productive way to fix problems than diving in and adding debugging code. It can be difficult to see faulty reasoning in code you have just written. Walking away from the computer not only helps prevent repetitive strain injury, but it also refreshes your troubleshooting abilities. On your return, you will be looking at the code afresh, and it is very common for the cause of the error to jump out at you where you could not see it before.

If this does not work, move on to the next technique: use the Serial Monitor to watch how the values in your sketch are changed when the program runs and whether conditional sections of code run. Chapter 4 explains how to use Arduino serial print statements to display values on your computer.

To troubleshoot, you need to find out what is actually happening when the code runs. `Serial.print()` lines in your sketch can display what part of the code is running and the values of your variables. These statements are temporary, so you should remove them once you have fixed your problem. The following sketch reads an analog value and is based on the Solution from Recipe 5.6. The sketch should change the blink rate based on the setting of a variable resistor (see the Discussion for Recipe 5.6 for more details on how this works). If the sketch does not function as expected, you can see if the software is working correctly by using a `serial.print()` statement to display the value read from the analog pin:

```
const int potPin = 0;
const int ledPin = 13;
int val = 0;

void setup()
{
  Serial.begin(9600);        // <- add this to initialize Serial
  pinMode(ledPin, OUTPUT);
}

void loop() {
  val = analogRead(potPin);    // read the voltage on the pot
  Serial.println(val);         // <- add this to display the reading
  digitalWrite(ledPin, HIGH);
  delay(val);
  digitalWrite(ledPin, LOW);
  delay(val);
}
```

If the value displayed on the Serial Monitor does not vary from 0 to 1023 when the pot (variable resistor) is changed, you probably have a hardware problem—the pot may be faulty or not wired correctly. If the value does change but the LED does not blink, the LED may not be wired correctly.

APPENDIX E

Tips on Troubleshooting Hardware Problems

Hardware problems can have more immediate serious ramifications than software problems because incorrect wiring can damage components. The most important tip is *always disconnect power when making or changing connections, and double-check your work before connecting power.*

Unplug Arduino from power while building and modifying circuits.

Applying power is the last thing you do to test a circuit, not the first.

For a complicated circuit, build it a bit at a time. Often a complicated circuit consists of a number of separate circuit elements, each connected to a pin on the Arduino. If this is the case, build one bit and test it, then the other bits, one at a time. If you can, test each element using a known working sketch such as one of the example sketches supplied with Arduino or on the Arduino Playground. It usually takes much less time getting a complex project working if you test each element separately.

For some of the techniques in this appendix, you will need a multimeter (any inexpensive digital meter that can read volts, current, and resistance should be suitable).

The most effective test is to carefully inspect the wiring and check that it matches the circuit you are trying to build. Take particular care that power connections are the correct way around and there are no short circuits, +5 volts accidentally connected to 0 volts, or legs of components touching where they should not. If you are unsure how much current a device connected to an Arduino pin will draw, test it with a multimeter before connecting it to a pin. If the circuit draws more than 40 mA, the pin on the Arduino can get damaged.

You can find a video tutorial and PDF explaining how to use a multimeter at *http://blog.makezine.com/archive/2007/01/multimeter_tutorial_make_1.html.*

You may be able to test output circuits (LEDs or motors) by connecting to the positive power supply instead of the Arduino pin. If the device does not function, it may be faulty or not wired correctly.

If the device tests OK, but when you connect to the pin and run the code you don't get the expected behavior, the pin might be damaged or the problem is in software.

To test a digital output pin, hook up an LED with a resistor (see Chapter 7) or connect a multimeter to read the voltage and run the Blink sketch on that pin. If the LED does not flash, or doesn't jump between 0 volts and 5 volts on the multimeter, the output pin is probably damaged.

Take care that your wiring does not accidentally connect the power line to ground. If this happens on a board that is powered from USB, all the lights will go out and the board will become unresponsive. The board has a component, called a *polyfuse*, which protects the computer from excessive current being drawn from the USB port. If you draw too much current, it will "trip" and switch off power to the board. You can reset it by unplugging the board from the USB hub (you may also need to restart your computer). Before reconnecting the power, check your circuits to find and fix the faulty wiring; otherwise, the polyfuse will trip again when you plug it back in.

Still Stuck?

After trying everything you can think of, you still may not be able to get your project to work. If you know someone who is using Arduino or similar boards, you could ask him for help. But if you don't, use the Internet—particularly the Arduino forum site at *http://www.arduino.cc/*. This is a place where people of all experience levels can ask questions and share knowledge. Use the forum search box (it's in the top-right corner) to try to find information relating to your project. A related site is the Arduino Playground, a wiki for user-contributed information about Arduino.

If a search doesn't yield the information you need, you can post a question to the Arduino forum. The forum is very active, and if you ask your question clearly, you are likely to get a quick answer.

To ask your question well, identify which forum section the question should go in and choose a title for your thread that reflects the specific problem you want to solve. Post in only one place—most people who are likely to answer will check all the sections that have new posts, and multiple posts will irritate people and make it less likely that you will get help.

Explain your problem, and the steps you have taken to try to fix it. It's better to describe what happens than to explain why you think it is happening. Include all relevant code, but try to produce a concise test sketch that does not contain code that you know is not related to the problem. If your problem relates to a device or component that is external to the Arduino board, post a link to the data sheet. If the wiring is complex, post a diagram or photo showing how you have connected things up.

APPENDIX F

Digital and Analog Pins

Tables F-1 and F-2 show the digital and analog pins for a standard Arduino board and the Mega board.

The "Port" column lists the physical port used for the pin—see Recipe 18.11 for information on how to set a pin by writing directly to a port. The introduction to Chapter 18 contains more details on timer usage. The table shows:

- USART RX is hardware serial receive
- USART RX is hardware serial transmit
- Ext Int is external interrupt (followed by the interrupt number)
- PWM TnA/B is the Pulse Width Modulation (`analogWrite`) output on timer n
- MISO, MOSI, SCK and SS are SPI control signals
- SDA and SCL are I2C control signals

Table F-1. Analog and digital pin assignments common to popular Arduino boards

	Arduino 168/328			Arduino Mega (pins 0–19)		
Digital pin	**Port**	**Analog pin**	**Usage**	**Port**	**Analog pin**	**Usage**
0	PD 0		USART RX	PE 0		USART0 RX, Pin Int 8
1	PD 1		USART TX	PE 1		USART0 TX
2	PD 2		Ext Int 0	PE 4		**PWM** T3B, INT4
3	PD 3		**PWM** T2B, Ext Int 1	PE 5		**PWM** T3C, INT5
4	PD 4			PG 5		**PWM** T0B
5	PD 5		**PWM** T0B	PE 3		**PWM** T3A
6	PD 6		**PWM** T0A	PH 3		**PWM** T4A
7	PD 7			PH 4		**PWM** T4B
8	PB 0		Input capture	PH 5		**PWM** T4C
9	PB 1		**PWM** T1A	PH 6		**PWM** T2B

	Arduino 168/328			Arduino Mega (pins 0–19)		
Digital pin	Port	Analog pin	Usage	Port	Analog pin	Usage
10	PB 2		**PWM** T1B, SS	PB 4		**PWM** T2A, Pin Int 4
11	PB 3		**PWM** T2A, MOSI	PB 5		**PWM** T1A, Pin Int 5
12	PB 4		SPI MISO	PB 6		**PWM** T1B, Pin Int 6
13	PB 5		SPI SCK	PB 7		**PWM** T0A, Pin Int 7
14	PC 0	0		PJ 1		USART3 TX, Pin Int 10
15	PC 1	1		PJ 0		USART3 RX, Pin Int 9
16	PC 2	2		PH 1		USART2 TX
17	PC 3	3		PH 0		USART2 RX
18	PC 4	4	I2C SDA	PD 3		USART1 TX, Ext Int 3
19	PC 5	5	I2C SCL	PD 2		USART1 RX, Ext Int 2

Table F-2. Assignments for additional Mega pins

Arduino Mega (pins 20–44)			Arduino Mega (pins 45–69)			
Digital pin	Port	Usage	Digital pin	Port	Analog pin	Usage
20	PD 1	I2C SDA, Ext Int 1	45	PL 4		**PWM** 5B
21	PD 0	I2C SCL, Ext Int 0	46	PL 3		**PWM** 5A
22	PA 0	Ext Memory addr bit 0	47	PL 2		T5 external counter
23	PA 1	Ext Memory bit 1	48	PL 1		ICP T5
24	PA 2	Ext Memory bit 2	49	PL 0		ICP T4
25	PA 3	Ext Memory bit 3	50	PB 3		SPI MISO
26	PA 4	Ext Memory bit 4	51	PB 2		SPI MOSI
27	PA 5	Ext Memory bit 5	52	PB 1		SPI SCK
28	PA 6	Ext Memory bit 6	53	PB 0		SPI SS
29	PA 7	Ext Memory bit 7	54	PF 0	0	
30	PC 7	Ext Memory bit 15	55	PF 1	1	
31	PC 6	Ext Memory bit 14	56	PF 2	2	
32	PC 5	Ext Memory bit 13	57	PF 3	3	
33	PC 4	Ext Memory bit 12	58	PF 4	4	
34	PC 3	Ext Memory bit 11	59	PF 5	5	
35	PC 2	Ext Memory bit 10	60	PF 6	6	
36	PC 1	Ext Memory bit 9	61	PF 7	7	
37	PC 0	Ext Memory bit 8	62	PK 0	8	Pin Int 16
38	PD 7		63	PK 1	9	Pin int 17

Arduino Mega (pins 20–44)			Arduino Mega (pins 45–69)			
Digital pin	Port	Usage	Digital pin	Port	Analog pin	Usage
39	PG 2	ALE Ext Mem	64	PK 2	10	Pin Int 18
40	PG 1	RD Ext Mem	65	PK 3	11	Pin Int 19
41	PG 0	Wr Ext Mem	66	PK 4	12	Pin Int 20
42	PL 7		67	PK 5	13	Pin Int 21
43	PL 6		68	PK 6	14	Pin Int 22
44	PL 5	PWM 5C	69	PK 7	15	Pin Int 23

Table F-3 is a summary of timer modes showing the pins used with popular Arduino chips.

Table F-3. Timer modes

Timer	Arduino 168/328	Mega
Timer 0 mode (8-bit)	Fast PWM	Fast PWM
Timer0A `analogWrite` pin	Pin 6	Pin 13
Timer0B `analogWrite` pin	Pin 5	Pin 4
Timer 1 (16-bit)	Phase correct PWM	Phase correct PWM
Timer1A `analogWrite` pin	Pin 9	Pin 11
Timer1B `analogWrite` pin	Pin 10	Pin 12
Timer 2 (8-bit)	Phase correct PWM	Phase correct PWM
Timer2A `analogWrite` pin	Pin 11	Pin 10
Timer2B `analogWrite` pin	Pin 3	Pin 9
Timer 3 (16-bit)	N/A	Phase correct PWM
Timer3A `analogWrite` pin		Pin 5
Timer3B `analogWrite` pin		Pin 2
Timer3C `analogWrite` pin		Pin 3
Timer 4 (16-bit)	N/A	Phase correct PWM
Timer4A `analogWrite` pin		Pin 6
Timer4B `analogWrite` pin		Pin 7
Timer4C `analogWrite` pin		Pin 8
Timer 5 (16-bit)	N/A	Phase correct PWM
Timer5A `analogWrite` pin		Pin 46
Timer5B `analogWrite` pin		Pin 45
Timer5C `analogWrite` pin		Pin 44

Note that the Arduino column is for the ATmega 168/328, and the Mega column is for the ATmega 1280/2560.

Full details of the Arduino controller chips can be found in the data sheets:

- The data sheet for standard boards (Atmega168/328) can be downloaded from *http://www.atmel.com/dyn/resources/prod_documents/doc8271.pdf*.
- The mega (ATmega1280/2560) data sheet can be downloaded from *http://www.atmel.com/dyn/resources/prod_documents/doc2549.pdf*.

APPENDIX G

ASCII and Extended Character Sets

ASCII stands for American Standard Code for Information Interchange. It is the most common way of representing letters and numbers on a computer. Each character is represented by a number—for example, the letter *A* has the numeric value 65, and the letter *a* has the numeric value 97 (lowercase letters have a value that is 32 greater than their uppercase versions).

Values below 32 are called *control codes*—they were defined as nonprinting characters to control early computer terminal devices. The most common control codes for Arduino applications are listed in Table G-1.

Table G-1. Common ASCII control codes

Decimal	Hex	Escape code	Description
0	0x0	`'\o '`	Null character (used to terminate a C string)
9	0x9	`'\t '`	Tab
10	0xA	`'\n'`	New line
13	0xD	`'\r '`	Carriage return
27	0x1B		Escape

Table G-2 shows the decimal and hexadecimal values of the printable ASCII characters.

Table G-2. ASCII table

	Dec	Hex		Dec	Hex		Dec	Hex
Space	32	20	@	64	40	`	96	60
!	33	21	A	65	41	a	97	61
"	34	22	B	66	42	b	98	62
#	35	23	C	67	43	c	99	63
$	36	24	D	68	44	d	100	64
%	37	25	E	69	45	e	101	65

	Dec	Hex		Dec	Hex		Dec	Hex
&	38	26	F	70	46	f	102	66
'	39	27	G	71	47	g	103	67
(	40	28	H	72	48	h	104	68
)	41	29	I	73	49	i	105	69
*	42	2A	J	74	4A	j	106	6A
+	43	2B	K	75	4B	k	107	6B
,	44	2C	L	76	4C	l	108	6C
-	45	2D	M	77	4D	m	109	6D
.	46	2E	N	78	4E	n	110	6E
/	47	2F	O	79	4F	o	111	6F
0	48	30	P	80	50	p	112	70
1	49	31	Q	81	51	q	113	71
2	50	32	R	82	52	r	114	72
3	51	33	S	83	53	s	115	73
4	52	34	T	84	54	t	116	74
5	53	35	U	85	55	u	117	75
6	54	36	V	86	56	v	118	76
7	55	37	W	87	57	w	119	77
8	56	38	X	88	58	x	120	78
9	57	39	Y	89	59	y	121	79
:	58	3A	Z	90	5A	z	122	7A
;	59	3B	[	91	5B	{	123	7B
<	60	3C	\	92	5C	|	124	7C
=	61	3D	]	93	5D	}	125	7D
>	62	3E	^	94	5E	~	126	7E
?	63	3F	_	95	5F			

Characters above 128 are non-English characters or special symbols and are displayed in the Serial Monitor using the UTF-8 standard (*http://en.wikipedia.org/wiki/UTF-8*). Table G-3 lists the UTF-8 extended character set.

Table G-3. UTF-8 extended characters

	Dec	Hex		Dec	Hex		Dec	Hex
Space	160	A0	À	192	C0	à	224	E0
¡	161	A1	Á	193	C1	á	225	E1
¢	162	A2	Â	194	C2	â	226	E2
£	163	A3	Ã	195	C3	ã	227	E3
¤	164	A4	Ä	196	C4	ä	228	E4
¥	165	A5	Å	197	C5	å	229	E5
¦	166	A6	Æ	198	C6	æ	230	E6
§	167	A7	Ç	199	C7	ç	231	E7
¨	168	A8	È	200	C8	è	232	E8
©	169	A9	É	201	C9	é	233	E9
ª	170	AA	Ê	202	CA	ê	234	EA
«	171	AB	Ë	203	CB	ë	235	EB
¬	172	AC	Ì	204	CC	ì	236	EC
	173	AD	Í	205	CD	í	237	ED
®	174	AE	Î	206	CE	î	238	EE
¯	175	AF	Ï	207	CF	ï	239	EF
°	176	B0	Ð	208	D0	ð	240	F0
±	177	B1	Ñ	209	D1	ñ	241	F1
²	178	B2	Ò	210	D2	ò	242	F2
³	179	B3	Ó	211	D3	ó	243	F3
´	180	B4	Ô	212	D4	ô	244	F4
µ	181	B5	Õ	213	D5	õ	245	F5
¶	182	B6	Ö	214	D6	ö	246	F6
·	183	B7	×	215	D7	÷	247	F7
¸	184	B8	Ø	216	D8	ø	248	F8
¹	185	B9	Ù	217	D9	ù	249	F9
º	186	BA	Ú	218	DA	ú	250	FA
»	187	BB	Û	219	DB	û	251	FB
¼	188	BC	Ü	220	DC	ü	252	FC
½	189	BD	Ý	221	DD	ý	253	FD
¾	190	BE	Þ	222	DE	þ	254	FE
¿	191	BF	ß	223	DF	ÿ	255	FF

You can view the entire character set in the Serial Monitor using this sketch:

```
/*
 * display characters from 1 to 255
 */

void setup()
{
  Serial.begin(9600);
  for(int i=1; i < 256; i++)
  {
     Serial.write(i);
     Serial.print(", dec: ");
     Serial.print(i,DEC);
     Serial.print(", hex: ");
     Serial.println(i, HEX);
  }
}

void loop()
{
}
```

Note that some devices, such as LCD displays (see Chapter 11), may use different symbols for the characters above 128, so check the data sheet for your device to see the actual character supported.

APPENDIX H

Migrating to Arduino 1.0

Although it should not be difficult to get sketches written for previous Arduino versions working with Arduino 1.0, that release has important changes you need to be aware of. The first thing you will notice when launching the software is the look of the IDE. Some icons are different from previous versions of the software and there are changes and additions in the menus. The error messages when dealing with selecting boards have been improved and the new ADK, Ethernet, and Leonardo boards have been added.

More significant are changes in the underlying core software and libraries. The stated purpose of 1.0 is to introduce disruptive changes that will smooth the way for future enhancements but break some code written for older software. New header files mean that older contributed libraries will need updating. Methods in Ethernet and Wire have been changed and there are subtle differences in the print functionality.

New functionality has been added to Streams (the underlying class for anything that uses `.print()` statements), Ethernet, Wire (I2C), and low-level input/output.

Improvements have been made to the way libraries handle dependencies and to simplify the support for new boards. Because of these changes, third-party libraries will need updating, although many popular ones may already have been updated.

The file extension used for sketches has been changed from *.pde* to *.ino* to differentiate Processing files from Arduino and to remove the inconvenience of accidental opening of a file in the wrong IDE.

Sketches opened in the 1.0 IDE will be renamed from *.pde* to *.ino* when the file is saved. Once renamed, you will not be able to open them in older versions without changing the extension back. There is an option in the File→Preferences dialog to disable this behavior if you don't want the files renamed

The following is a summary of the changes you need to make for 1.0 to compile sketches written for earlier releases. You will find examples of these in the chapters covering Serial, Wire, Ethernet, and Libraries.

Migrating Print Statements

There are a few changes in how `print()` (or `println`) is handled:

Working with byte datatypes

`print(byte)` now prints the integer value of the byte as ASCII characters; previous releases sent the actual character. This affects Serial, Ethernet, Wire or any other library that has a class derived from the Print class.

Change:

```
Serial.print(byteVal)
```

to:

```
Serial.write(val); // send as char
```

The BYTE keyword

The `BYTE` keyword is no longer supported.

Change:

```
Serial.print(val, BYTE)
```

to:

```
Serial.write(val); // sends as char
```

Return values from write() methods

Classes derived from Print must implement a `write` method to write data to the device that the class supports. The signature of the `write` method has changed from `void` to `size_t` to return the number of characters written. If you have a class derived from Print you need to modify the write method as follows and return the number of characters written (typically 1). See the discussion on the i2cDebug library in Recipe 16.5 for an example of a 1.0 write method.

Change:

```
void write
```

to:

```
size_t write
```

Migrating Wire (I2C) Statements

You'll need to make some changes when working with the Wire library.

Method name changes

Wire method names have been changed to make them consistent with other services based on Streams.

Change:

```
Wire.send()
```

to:

```
Wire.write()
```

Change:

```
Wire.receive()
```

to:

```
Wire.read()
```

The write method requires types for constant arguments
: You now need to specify the type for literal constant arguments to write.

Change:

```
write(0x10)
```

to:

```
write((byte)0x10)
```

Migrating Ethernet Statements

Arduino 1.0 changes a number of things in the Ethernet library.

Client class
: The client Ethernet classes and methods have been renamed.

Change:

```
client client(server, 80)
```

to:

```
EthernetClient client;
```

Change:

```
if(client.connect())
```

to:

```
if(client.connect(serverName, 80)>0)
```

`client.connect` should test for values >0 to ensure that errors returned as negative values are detected.

Server class
: Change:

```
Server server(80)
```

to:

```
EthernetServer server(80)
```

Change:

```
UDP
```

to:

```
EthernetUDP
```

Migrating Libraries

If your sketch includes any libraries that have not been designed for 1.0 then you will need to change the library if it uses any of the old header files that have been replaced with the new *Arduino.h* file.

If you include any of these libraries, change:

```
#include "wiring.h"
#include "WProgram.h"
#include "WConstants.h"
#include "pins_arduino.h"
```

To:

```
#include "Arduino.h"
```

You can use a conditional include (see Recipe 17.6) to enable libraries to also compile in earlier versions. For example, you could replace `#include "WProgram.h"` with the following:

```
#if ARDUINO >= 100
  #include "Arduino.h"
#else
  #include "WProgram.h"
#endif
```

New Stream Parsing Functions

Arduino 1.0 introduced a simple parsing capability to enable finding and extracting strings and numbers from any of the objects derived from Stream, such as: Serial, Wire, and Ethernet.

These functions include:

```
find(char *target);
findUntil(char *target,char *term)
readBytesUntil(term,buffer,length);
parseInt();
parseFloat();
```

See the discussion of Recipe 4.5 for an example of Stream parsing with Serial. Many recipes in Chapter 15 demonstrate the use of Stream parsing; see Recipe 15.4 and Recipe 15.7 for examples.

Index

Symbols

A

B

C

D

E

F

G

H

I

J

K

L

M

N

R

S

T

U

V

W

X

Z

About the Author

Michael Margolis is a technologist in the field of real-time computing with expertise in developing and delivering hardware and software for interacting with the environment. He has over 30 years of experience in a wide range of relevant technologies, working with Sony, Microsoft, Lucent/Bell Labs, and most recently as Chief Technical Officer with Avaya.

Colophon

The animal on the cover of *Arduino Cookbook* is a toy rabbit. Mechanical toys like this one move by means of springs, gears, pulleys, levers, or other simple machines, powered by mechanical energy. Such toys have a long history, with ancient examples known from Greece, China, and the Arab world.

Mechanical toy making flourished in early modern Europe. In the late 1400s, German inventor Karel Grod demonstrated flying wind-up toys. Prominent scientists of the day, including Leonardo da Vinci, Descartes, and Galileo Galilei, were noted for their work on mechanical toys. Da Vinci's famed mechanical lion, built in 1509 for Louis XII, walked up to the king and tore open its chest to reveal a fleur-de-lis.

The art of mechanical toy making is considered to have reached its pinnacle in the late eighteenth century, with the famed "automata" of the Swiss watchmaker Pierre Jaquet-Droz and his son Henri-Louis. Their human figures performed such lifelike actions as dipping a pen in an inkwell, writing full sentences, drawing sketches, and blowing eraser dust from the page. In the nineteenth century, European and American companies turned out popular clockwork toys that remain collectible today.

Because these original toys, which had complicated works and elaborate decorations, were costly and time-consuming to make, they were reserved for the amusement of royalty or the entertainment of adults. Only since the late nineteenth century, with the appearance of mass production and cheap materials (tin, and later, plastic), have mechanical toys been considered playthings for children. These inexpensive moving novelties were popular for about a century, until battery-operated toys superseded them.

The cover image is from the Dover Pictorial Archive. The cover font is Adobe ITC Garamond. The text font is Linotype Birka; the heading font is Adobe Myriad Condensed; and the code font is LucasFont's TheSansMonoCondensed.